Neural and Intelligent Systems Integration

Sixth-Generation Computer Technology Series

Branko Souček, Editor
University of Zagreb

Neural and Massively Parallel Computers: The Sixth Generation
Branko Souček and Marina Souček

Neural and Concurrent Real-Time Systems: The Sixth Generation
Branko Souček

Neural and Intelligent Systems Integration: Fifth and Sixth Generation Integrated Reasoning Information Systems
Branko Souček and the IRIS Group

Dynamic, Genetic, and Chaotic Programming: The Sixth Generation (in preparation)
Branko Souček and the IRIS Group

Fuzzy, Holographic, Invariant, and Parallel Intelligence: The Sixth-Generation Breakthrough (in preparation)
Branko Souček and the IRIS Group

Neural and Intelligent Systems Integration

Fifth and Sixth Generation Integrated Reasoning Information Systems

BRANKO SOUČEK
and
The IRIS GROUP

A Wiley-Interscience Publication
JOHN WILEY & SONS, INC.
New York-Chichester-Brisbane-Toronto-Singapore

In recognition of the importance of preserving what has been
written, it is a policy of John Wiley & Sons, Inc., to have books
of enduring value published in the United States printed on
acid-free paper, and we exert our best efforts to that end.

Copyright © 1991 by John Wiley & Sons, Inc.

All rights reserved. Published simultaneously in Canada.

Reproduction or translation of any part of this work
beyond that permitted by Section 107 or 108 of the
1976 United States Copyright Act without the permission
of the copyright owner is unlawful. Requests for
permission or further information should be addressed to
the Permissions Department, John Wiley & Sons, Inc.

Library of Congress Cataloging in Publication Data:

Souček, Branko.
   Neural and intelligent systems integration: fifth and sixth
generation integrated reasoning information systems / Branko Soucek
and the IRIS Group.
     p. cm.—(Sixth-generation computer technology series)
  "A Wiley-Interscience publication."
  Includes bibliographical references and index.
   1. Neural networks (Computer science) 2. Expert systems(Computer
science) I. IRIS Group. II. Title. III. Series.
QA76.87.S64 1991
006.3—dc20   91-26616
   ISBN 0-471-53676-8   CIP

Printed in the United States of America

10 9 8 7 6 5 4 3 2 1

To Sneška
and
our daughter, Amalia

CONTRIBUTORS

The IRIS Group presents a forum for international cooperation in research, development, and applications of intelligent systems. The IRIS Group is involved in projects, design, measurements and experiments as well as in teaching courses and workshops and consulting. The IRIS Group invites inquiries and operates under the auspices of the Zagrebačka Banka D.D., IRIS, Vijenac 3, 41000 Zagreb, Croatia, Yugoslavia. The group's coordinator is Professor Branko Souček.

PAUL BASEHORE
*Micro Devices
Division of Chip Supply
Orlando, Florida*

VAHE BEDIAN
*Department of Human Genetics
University of Pennsylvania
Philadelphia, Pennsylvania*

MARK CHIGNELL
*Intelligence Ware Inc.
Los Angeles, California*

HUGO DE GARIS
*CADEPS Artificial Intelligence
and Artificial Life Research Unit
Université Libre de Bruxelles
Brussels, Belgium*

MARTIN J. DUDZIAK
*NCAP, INMOS Technology
 Center
Columbia, Maryland*

PETER FANDEL
*Flavors Technology Inc.
Amherst, New Hampshire*

LAWRENCE O. HALL
*Department of Computer
 Science and Engineering
University of South Florida
Tampa, Florida*

YUEH-MIN HUNG
*Department of Electrical and
 Computer Engineering,
University of Arizona,
Tucson, Arizona*

TAKAO ICHIKO
Department of Information
 Engineering
Yamagata National University
Yonezawa City, Yamagata 992
Japan

RYOTARO KAMIMURA
Information Science Laboratory
Tokai University
Hiratsuka Kanagawa 259-12
Japan

SETRAG KHOSHAFIAN
Intelligence Ware Inc.
Los Angeles, California

TAKASHI KUROZUMI
ICOT, Institute for New
 Generation Computer
 Technology
Mita Kokusai Building
Tokyo 108, Japan

JAMES S. J. LEE
Neopath Inc.
Bellevue, Washington

JAMES F. LYNCH
Department of Mathematics and
 Computer Science
Clarkson University
Potsdam, New York

SEUNG R. MAENG
Department of Computer
 Science
Center for Artificial Intelligence
 Research
Korea Advanced Institute of
 Science and Technology
Seoul, Korea

JONG H. NANG
Department of Computer
 Science
Center for Artificial Intelligence
 Research
Korea Advanced Institute of
 Science and Technology
Seoul, Korea

DZIEM D. NGUYEN
Boeing High Technology Center
Seattle, Washington

KAMRAN PARSAYE
Intelligence Ware Inc.
Los Angeles, California

MARKUS F. PESCHL
Department of Epistemology
 and Cognitive Science
University of Vienna
Vienna, Austria

GERALD REED
Micro Devices
Division of Chip Supply
Orlando, Florida

ALI REZGUI
Department of Electrical and
 Computer Engineering
Florida Institute of Technology
Melbourne, Florida

MICHAEL H. ROBERTS
Department of Biology
Clarkson University
Potsdam, New York

STEVE G. ROMANIUK
Department of Computer
 Science and Engineering
University of South Florida
Tampa, Florida

RAMONA ROSARIO
Department of Electrical and
 Computer Engineering
Florida Institute of Technology
Melbourne, Florida

JERZY W. ROZENBLIT
Department of Electrical and
 Computer Engineering
University of Arizona
Tucson, Arizona

ROBERT SCALERO
Grumman Melbourne Systems
Melbourne, Florida

PHILIPPE G. SCHYNS
Department of Cognitive and Linguistic Sciences
Brown University
Providence, Rhode Island

BRANKO SOUČEK
Department of Mathematics
University of Zagreb
Zagreb, Yugoslavia

RON SUN
Honeywell SSDC
Minneapolis, Minnesota

NAZIF TEPEDELENLIOGLU
Department of Electrical and Computer Engineering
Florida Institute of Technology
Melbourne, Florida

MIOMIR VUKOBRATOVIĆ
Robotics and Flexible Automation Laboratory
Mihajlo Pupin Institute
Belgrade, Yugoslavia

DAVID WALTZ
Thinking Machines Corporation and Brandeis University,
Waltham, Massachusetts

STEPHEN S. WILSON
Applied Intelligent Systems, Inc.
Ann Arbor, Michigan

HARRY WONG
Intelligence Ware Inc.
Los Angeles, California,

JOSEPH YESTREBSKY
Micro Devices
Division of Chip Supply
Orlando, Florida

HYUNSOO YOON
Department of Computer Science and Center for Artificial Intelligence
Korea Advanced Institute of Science and Technology
Seoul, Korea

FENGMAN ZHANG
Department of Mathematics and Computer Science
Clarkson University
Potsdam, New York

MICHAEL ZIEMACKI
Micro Devices
Division of Chip Supply
Orlando, Florida

CONTENTS

Preface   xv

PART I NEURAL, GENETIC, AND INTELLIGENT ALGORITHMS AND COMPUTING ELEMENTS

1. **From Modules to Application-Oriented Integrated Systems**   1

   Branko Souček

2. **A Biologically Realistic and Efficient Neural Network Simulator**   37

   Vahe Bedian, James F. Lynch, Fengman Zhang, and Michael H. Roberts

3. **Neural Network Models of Concept Learning**   71

   Philippe G. Schyns

4. **Fast Algorithms for Training Multilayer Perceptrons**   107

   Nazif Tepedelenlioglu, Ali Rezgui, Robert Scalero, and Ramona Rosario

5. **Teaching Network Connections for Real-Time Object Recognition** — 135

   Stephen S. Wilson

6. **The Discrete Neuronal Model and the Probabilistic Discrete Neuronal Model** — 161

   Ron Sun

7. **Experimental Analysis of the Temporal Supervised Learning Algorithm with Minkowski-r Power Metrics** — 179

   Ryotaro Kamimura

8. **Genetic Programming: Building Artificial Nervous Systems with Genetically Programmed Neural Network Modules** — 207

   Hugo De Garis

9. **Neural Networks on Parallel Computers** — 235

   Hyunsoo Yoon, Jong H. Nang, and Seung R. Maeng

10. **Neurocomputing on Transputers** — 281

    Martin J. Dudziak

11. **Neural Bit-Slice Computing Element** — 311

    Joseph Yestrebsky, Paul Basehore, and Gerald Reed

PART II INTEGRATED NEURAL–KNOWLEDGE–FUZZY HYBRIDS

12. **Data Processing for AI Systems: A Maximum Information Viewpoint** — 323

    Joseph Yestrebsky and Michael Ziemacki

13. **Fuzzy Data Comparator with Neural Network Postprocessor: A Hardware Implementation** — 333

    Paul Basehore, Joseph Yestrebsky, and Gerald Reed

14. **A Neurally Inspired Massively Parallel Model of Rule-Based Reasoning** — 341

    Ron Sun and David Waltz

15. **Injecting Symbol Processing into a Connectionist Model**   383

    Steve G. Romaniuk and Lawrence O. Hall

16. **Adaptive Algorithms and Multilevel Processing Control in Intelligent Systems**   407

    James S.J. Lee and Dziem D. Nguyen

PART III INTEGRATED REASONING, INFORMING, AND SERVING SYSTEMS

17. **Architectures for Distributed Knowledge Processing**   437

    Yueh-Min Huang and Jerzy Rozenblit

18. **Current Results in Japanese Fifth-Generation Computer Systems (FGCS)**   457

    Takao Ichiko and Takashi Kurozumi

19. **An Advanced Software Paradigm for Intelligent Systems Integration**   503

    Takao Ichiko

20. **Integrated Complex Automation and Control: Parallel Inference Machine and Paracell Language**   529

    Peter Fandel

21. **Robotics and Flexible Automation Simulator/Integrator**   563

    Miomir Vukobratovic

22. **Intelligent Data Base and Automatic Discovery**   615

    Kamran Parsaye, Mark Chignell, Setrag Khoshafian, and Harry Wong

23. **Artificial Intelligence: Between Symbol Manipulation, Commonsense Knowledge, and Cognitive Processes**   629

    Markus F. Peschl

Index   663

PREFACE

This book presents a new and rapidly growing discipline: Integration of Reasoning Information Systems. In other words, it describes Integration of Reasoning, Informing, and Serving (IRIS) functions and modules into hybrid systems. These functions and modules include:

Reasoning: generalization; knowledge; heuristic engines; expert systems; learning and adaptation; neural networks; mapping; transformations; holographic networks; genetic selection; intelligent, fuzzy and chaotic algorithms; self-organization; artificial life.

Informing: Local, global and distributed memories; data bases; knowledge bases; input/output; sensors; image; speech.

Serving: Data processing; computing; control; communication; robotics; data delivery; decision making; real-time services.

These modules are available now, on the market and in laboratories, and present fifth- and sixth-generation building blocks for users and system designers. The user works with these modules dealing with high-level constructs, as just one application-specific object. As a result, the designer's emphasis shifts from computer hardware and software to applications, where users play their own productive role in creating their intelligent information systems. Users seek integrated, application-oriented solutions, based on arrays of tightly focused, customer-oriented products and modules.

Recently, remarkable results have been achieved through integration of modules into hybrid systems. This text discusses professional and everyday life applications which include: business, management and stock control; process control and automation; surveillance; robotics; flexible manufacturing; data delivery; and information services.

To be efficient, integration must be automated, supported with proper tools, and based on newly discovered paradigms. These include: *automation* of software development based on expert systems, simulators and new languages; *adaptation* based on learning in neural networks; module *selection* based on genetic programming; *self-organization* based on artificial life ideas; and *automated discovery* based on intelligent data bases. IRIS and related techniques described in this book, present the key for future better business and highly efficient and clean technology and services.

The book unifies material that is otherwise scattered over many publications and research reports. Previously unpublished methods and results based on the research of international IRIS Group are presented. The IRIS Group brings together the results from leading American, European, Japanese, and Korean laboratories and projects. In particular, the results of the 10-year long Japanese Fifth-Generation Project are presented and compared with American solutions.

IRIS paradigms present the base for new information systems which are able to think, reason, and judge like human beings. They deal with fuzziness, uncertainty, and incompleteness and operate in a massively concurrent distributed mode. The Japanese Ministry of International Trade and Industry (MITI) is ready to launch a new project in this direction. America and Europe are driving toward the same goal. Problems of intelligence integration and their first results and concrete applications are also identified in this book.

The book is divided into three parts: PART I: Neural, Genetic and Intelligent Algorithms and Computing Elements, deals with the basic modules. It starts with the description of a software package for biological neural networks simulation. Neural network modeling of human concept is described. Concept learning is divided into subtasks and solved by independent modules.

Fast algorithms have been developed which perform considerably better than classical back propagation. New algorithms use a momentum term, conjugate gradient, and adapt slopes of the sigmoid functions and the Kalman filter.

An intelligent method is described for the automatic training of objects to be recognized by a machine vision system. Objects to be trained by this method include integrated circuit wafer patterns, marks on printed circuit boards, objects to be located by robots, and alphanumeric characters to be recognized or verified. Learning in discrete and recurrent neural network models is described.

Genetic programming is described as it applies to the genetic algorithm which finds the signs and weights of fully connected neural network modules (called Gen Nets) so that a desired output over time is obtained. Several functional Gen Nets connected in an ensemble present a new higher-order module.

Neural network simulations on parallel computers are presented. Various implementation methods such as ones based on coprocessors, systolic arrays, SIMD, and MIMD are studied. Transputer-based systems supporting concurrent neural and intelligent modules are presented as are neural bit-slice building blocks for the construction of neural networks and of parallel processing units. Slice architecture and neural software modules allow devices and programs to be interconnected efficiently, allowing many different neural networks to be implemented.

PART II: Integrated Neural-Knowledge-Fuzzy Hybrids, deals with the module mix. Data transformation preprocessors and artificial intelligence units combined with neural networks are discussed, from a maximum information viewpoint. Fuzzy-set comparators (FSC) for adaptive ranking of fuzzy data in groups are described. FSC are intended to simplify the implementation of systems where decisions must be made rapidly from inaccurate, noisy, or real-time variable data. Hybrid connectionist networks for constructing fuzzy expert systems are described. In all cases, hybrid learning mechanism requires only one pass through the examples, which makes it significantly faster than classic connectionist learning algorithms.

Integration of rapid LMS neural algorithms and multilevel processing control leads to new effective solutions. Examples of automatic target recognition are shown in detail, using the data obtained from real target tracking systems, based on infrared images as inputs.

PART III: Integrated Reasoning, Informing and Serving Systems presents complex, parallel, and distributed systems, composed of knowledge, data base, control, and robot modules.

Distributed knowledge processing and Japanese Fifth-Generation Computer Systems (FGCS) are described. FGCS targets are easy-to-use computers, support of intellectual activities, and increase software productivity. FGCS performance increases several hundred times more than the value of conventional inference mechanisms, thereby realizing a feasible environment for many pragmatic uses of knowledge processing.

A new software design approach is described that uses an expert system shell for effective human interface during the design and verification processes.

Massively parallel real-time automation and process control are described which is based on the Parallel Inference Machine. The Paracell language offers an interactive interface that assists control engineers in breaking up large applications into increasingly smaller parts.

User-oriented software modules for simulation and intergration of robots and flexible manufacturing cells (FMC) are presented. In this way one can test whether the selected robot and its controller are capable of satisfying all requirements for specific FMC.

Intelligent data bases and automatic discovery is described. Relations between knowledge processing in humans, neural networks, symbolic and hybrid systems are discussed, pointing to future research avenues.

The book has been written as a textbook for students, as well as a reference for practicing engineers and scientists. The treatment is kept as straightforward as possible, with emphasis on functions, systems, and applications. The background for this book is presented in:

B. Souček and M. Souček, *Neural and Massively Parallel Computers: The Sixth Generation,* Wiley, New York, 1988.

B. Souček, *Neural and Concurrent Real-Time Systems: The Sixth Generation,* Wiley, New York, 1989.

These three books are independent, mutually supporting volumes.

BRANKO SOUČEK

Zagreb, Croatia
August 1991

Neural and Intelligent Systems Integration

PART I
Neural, Genetic, and Intelligent Algorithms and Computing Elements

From Modules to Hybrid Systems
Neural Network Simulator
Neural Network Models of Concept Learning
Fast Algorithms
Real-Time Object Recognition
The Discrete Neuronal Model
Temporal Supervised Learning Algorithms
Genetic Programming
Neural Networks on Parallel Computers
Neurocomputing on Transputers
Neural Bit-Slice Computing Elements

CHAPTER 1

From Modules to Application-Oriented Integrated Systems

BRANKO SOUČEK

1.1 INTRODUCTION

Living systems are composed of three distinct groups of functions: reasoning, informing, and serving (see Figure 1.1). The same functions are desirable in manmade systems (see Figure 1.2). Remarkable results have been recently achieved through integration of reasoning, informing, and serving modules. The modules include advanced neural networks and neural computing elements; fuzzy data comparators and correlators; expert systems; intelligent algorithms, programs and data bases; off-the-shelf software building blocks; local, global, and distributed memories; parallel processing blocks; data acquisition, control, and robot units.

Integration involves software, hardware, and system levels. To be efficient, integration must be automated, supported with proper tools, and based on newly discovered methods. These include knowledge-system-based software developers; simulators; special languages; development kits; intelligent data bases; genetic programming and artificial life ideas.

This chapter introduces a new discipline: Integration of Reasoning, Informing and Serving (IRIS). Integration paradigms are identified, including automation; adaptation; selection; self-organization; and automatic discovery. These paradigms present the road from modules to hybrid, application-

---

Neural and Intelligent Systems Integration, By Branko Souček and the IRIS Group.
ISBN 0-471-53676-8 ©1991 John Wiley & Sons, Inc.

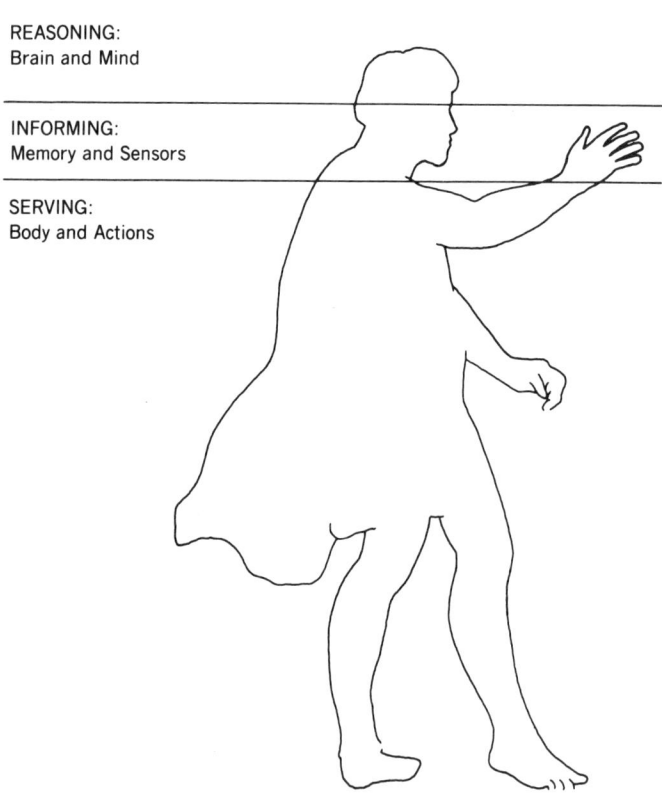

Figure 1.1 Integration of reasoning, informing, and serving modules and functions in a living organism. Silhouette adapted from P.P. Rubens.

oriented, integrated intelligent systems. Similar paradigms have been identified in living systems.

Concrete results and examples of applications are described. They include business and stock market; process control, automation and surveillance; robotics and flexible manufacturing; data delivery and information services. Detailed descriptions appear in Chapters 2 to 23.

Integration of Reasoning, Informing and Serving (IRIS) is here to stay. It presents the key for better business and more efficient technology and services, in the near future and in the third millenium. For these reasons, integration of reasoning, informing, and serving penetrates into education as well as in technology and business.

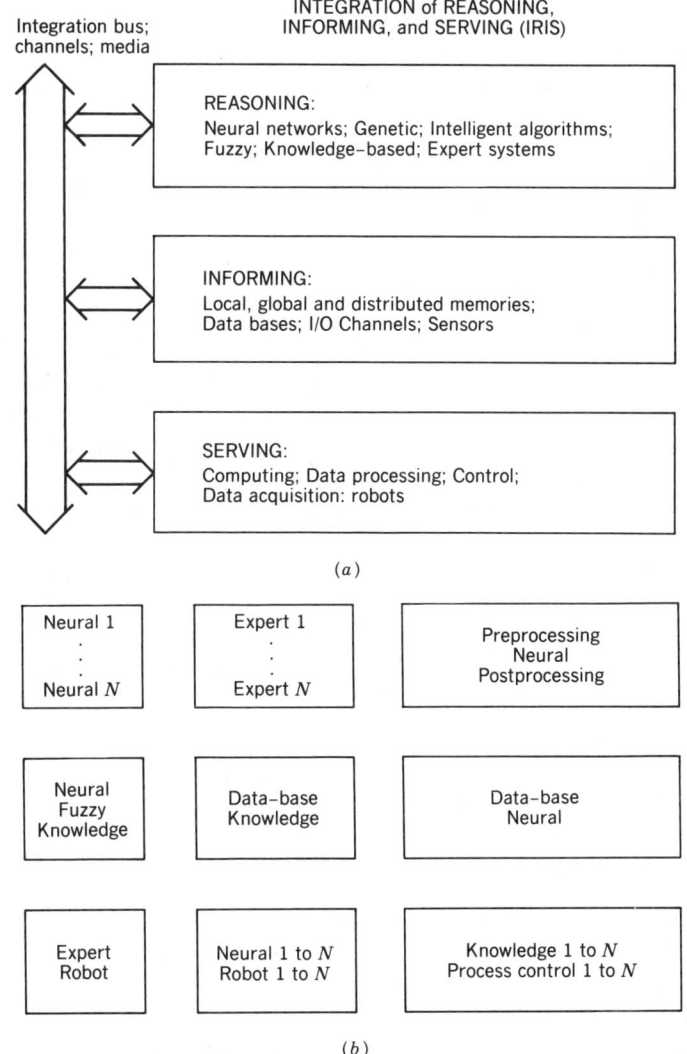

Figure 1.2 *(a) Integration of reasoning, informing, and serving modules and functions in a manmade system. (b) Examples of integrated hybrid systems.*

1.2 LEARNING FROM BIOLOGY AND COGNITIVE SCIENCE

The impressive calculating speeds of electronic computers make it possible to complete, in a few seconds, computations that would have otherwise taken many years. Many computers and their programs are exhibiting some kind of intelligence, involving the solution of problems, prediction, and ability to learn. To make more intelligent computers, we need to know more about the brain.

We know very little about the brain's detailed microcircuitry on the neural network, cellular, and molecular levels. In some cases, each cell can be considered to be an oscillator, and the entire network a system of coupled oscillators. (See the brain window theory [1].)

Current connectionist models of neural networks are oversimplified in terms of the internal mechanisms of individual neurons and communication between them. The most frequently seen neural network model is the weighted-sum, or thresholding, model. Its biological validity and computational sufficiency are questionable. Many alternative neural models may provide better computational and biological properties. Souček and Souček [1] examine information processing models of the brain that could directly influence the development of neural networks and of other sixth-generation systems. Biological coding layers and brain theories are summarized in Table 1.1.

It seems that natural neural networks are not rigid hard-wired circuits. They are flexible circuits that can be modified by the actions of chemical, hormonal, and electrical modulators. Circuits can adapt not only by changing synaptic strength but by altering virtually every physiological parameter available to it. Bullock [2] composed a comprehensive list of physiological parameters with numerous variables. From this list, we take only two examples:

1. *Synaptic Transmission:*
    chemical or electrical,
    Excitatory or inhibitory,
    Polarized or unpolarized, and
    High gain or low gain.
2. *Postsynaptic Response:*
    Potential change or conductance change,
    Monophasic or biphasic,
    Slow or fast,
    Only passive or partly active,
    With increased or decreased conductance, and
    Facilitating, antifacilitating, or neither.

Only a few of the preceding parameters have been considered so far in designing artificial neural networks and intelligent systems.

According to Souček [3], different brain theories could be used to design brainlike computers and intelligent systems, as shown in Table 1.2.

In the human brain the number of neurons is at least 10^{11}, and the number of synaptic connections is about 10^{14}. The average connectivity in the brain

TABLE 1.1 Brain Theories and Coding Layers

Genetic code	Corresponds to computer ROM
Microtubule code	Corresponds to programmable read/write memory/processor layer. It could present the basis to develop biochip and molecular intelligence systems
Neuron	Presents the basic unit of the brain structure
Synaptic connections	Present the basis for a "connectionist" theory of memory and intelligence. Artificial neural networks have been designed on this principle.
Feature Extractors	For a particular sensory experience, the pattern of activity over the brain and neural system consists of only a few activated areas, whereas the rest of the elements in the field are silent.
Brain windows	Adaptive, fuzzy receive and send windows are interleaved in the sequence. Each receive window is related to specific group of stimuli. Each send window is related to specific behavior/action.
Cooperative action	Thoughts and perceptions are encoded in the changing patterns of an electromagnetic field rather than in the impulses of individual neurons.
Goal-directed hierarchy	Describes the brain and behavior in the form of a hierarchical pyramid. The pyramid, sensors, and actuators are cross-coupled through multivariable feedback systems

is fairly small, but the brain is able to solve complicated tasks. According to Banzhaf et al. [4], the natural conclusion is that the brain follows a very clever strategy in order to perform well. Braitenberg [5] suggests that the brain is based on modular and hierarchically structured architectures, which enable a filtering and recycling of relevant information.

According to Gierer [6], given the mere number of synapses, there cannot be enough genetic information to contain detailed instructions for all the synapses to form. Rather, after birth a portion of the neurons were randomly connected. The part of the brain's organization is therefore provided by the environment to which it is exposed.

Neurons are interconnected in elaborate networks in the nervous system. Such networks can generate positive and negative feedback loops, produce oscillations, and allow the animal to integrate inputs and choose between alternative behaviors under the influence of diverse stimuli. Experimental analysis of single neurons and neural networks is a difficult task that can be aided by the availability of a realistic, predictive computer model. To approach this problem, Bedian, Lynch, Zhang, and Roberts have constructed a biophysically realistic neuron simulator. The simulator, described in Chapter 2, can account for many observed properties of neurons. A network editor program, which allows convenient creation and modification of neural networks, is used to construct and test neural networks. Using this program, one could get closer to biologically realistic neurons and neural networks.

TABLE 1.2 Taxonomy Tree Based on Intelligence Granularity

Field computers
  ultrafine grain,
  continuous
	Optical
	Electrical
	Molecular

Neural computers very fine grain	Mapping networks Projection networks Probabilistic logic neurons

High-order neural units
  fine grain
	Sigma-Pi units; Holographic neural
	  technology
	Brain-window logic/language
	High-order conjective connections

Feature extractors medium grain	Genetic selection systems Rule-based expert systems Boolean logic neurons
Schema, frame, FCM Coarse grain	Schemas, frames Truth maintenance systems, worlds Fuzzy cognitive maps (FCM)

Hybrid systems
  fine, medium, and
  coarse grain
	Field, neural, genetic, fuzzy,
	  holographic;
	Brain window, rule-based,
	  artificial life, parallel
	  processing

According to Pellionisz [7,8], brain theory or the mathematical theory of biological neural networks is only in its infancy. The belief is becoming increasingly widespread that the mathematics underlying brain function is not algebra nor even calculus, but geometry, particularly fractal geometry in the neuron structure. Analysis and computer simulation of biological brains is the only way to develop the necessary mathematical theory and to open the road toward more powerful artificial neural networks.

The mechanisms by which humans separate distinct objects into classes are based on concept learning. A concept is the mental representation of a category of objects in the world. Categorization is the operation by which an object is identified as the member of a particular category.

In Chapter 3, Schyns presents the major neural network architectures for concept learning. Schyns opens a direction for future research in the domain, which is based on modularity: an unsupervised learning module that learns the categories and a supervised learning module that learns to tag arbitrary symbols—category names—to the output of the unsupervised module.

According to Schyns, the current neural networks are not modular enough. Modular, bottom-up strategy should be used for concept learning. The overall idea is to construct hierarchies of modules that extract gradually more and more high-level features from the raw input. The outputs of low-level modules should be used by different modules at higher levels.

Biology has solved profound and difficult problems, such as visual and auditory pattern recognition, task planning, and navigation, by the computational abilities of synapses and neurons and by evolutionary mechanisms. Cognitive sciences have long studied the relation between the real world and its knowledge representation. Computer technology can profit from an analysis of problems solved by biology and the cognitive sciences. Chapters 2 and 3 present examples of the tools needed to build the bridges between computer science, biology, and cognitive science.

1.3 ADVANCED NEURAL NETWORKS

Layered neural networks based on a back-propagation learning algorithm present the basis for many practical applications. The back-propagation algorithm has been developed independently by Werbos [9,10] in 1974, Parker [11] in 1982, and Rumelhart, Hinton, and Williams [12] in 1986. The back-propagation algorithm is the most important algorithm today. The main thrust of the whole field of neurocomputing is back propagation.

Back propagation is a slow and unstable algorithm. Numerous stability and speed-up techniques, such as momentum, batching, conjugate gradients, Kalman filtering, and delta-bar-delta have been applied with varying success.

Jurik [13] has developed a modified algorithm called *back percolation*. Whereas back propagation adjusts weights in proportion to a cell's error gradient, back percolation assigns each cell's true output error, by which the cell's weights then adjust to reduce it.

A typical training cycle of the Jurik algorithm would be as follows:

1. A signal is fed forward through the network, producing outputs and errors.
2. Gradients back-propagate through the network, thereby assigning each cell an output gradient.
3. Simple errors back percolate through the network, thereby assigning each cell an output error.
4. Weights are modified by a function designed to reduce cell output error.

Jurik applied the back-percolation algorithm to two tasks: the 8-2-8 encoder/decoder problem and the six-cell symmetry detection problem. His experiments show that in both tasks back percolation performed much better than back propagation.

Tependelenlioglu, Rezgui, Scalero and Rosario developed several algorithms that perform better than classical back propagation. Their fast algorithms are described in Chapter 4. They include generalized momentum, adapting slopes of sigmoids, conjugate gradients, and Kalman's algorithm. These new algorithms have been compared with back propagation on two problems: circle in a square, and parity. In the first example, back propagation converged in 1850 iterations, and a new algorithm in 320 iterations. In the second example, the number of iterations is 20,455 for back propagation and 4800 for the Kalman algorithm.

Wilson has developed a new method for object recognition by a machine vision system. Wilson's method is described in Chapter 5. In this method, neural training is used to find an optimum connectivity pattern for a fixed number of inputs that have fixed weights, rather than employing the usual technique of finding the optimum weight for a fixed connectivity. In industrial applications on a massively parallel architecture, objects can be trained by an unskilled user in less than 1 minute, and, after training, parts can be located in about 10 ms.

Sun is studying neuronal models of discrete states. In Chapter 6, he describes discrete models, their biological plausibility, utility, and learning algorithms.

Kamimura has developed a new method to accelerate the temporal supervised learning algorithm. Experimental results are discussed concerning the generation of natural language. Chapter 7 shows that the recurrent neural network can recognize the hierarchical structure found in natural language.

New methods and algorithms described in Chapters 4 to 7 point the way toward practical and industrial applications of advanced neural networks.

Most recently the research concentrates on the mathematical theory of generalization (MTG) and on holographic neural technology (HNET).

MTG has been developed by Wolper [14]. It explains the ability to generalize using the learning set in neural networks, memory-based reasoning systems, genetic systems, and so on. In MTG, any generalizer is defined as a heuristic binary engine, HERBIE. Wolpert reports that for the task of reading aloud, HERBIE performs substantially better than back propagation, used in the well-known NET talk system.

HNET has been developed by Sutherland [15]. The stimulus field is expanded to the level of higher order statistics ($\Sigma\Pi$). Learning is accomplished by a straightforward mathematical transformation related to that used for enfolding the associations in memory, avoiding multiple layers. The system is designed to run on transputers. Sutherland reports accuracy and storage density well beyond other known neural network paradigms.

Backpropagation, backpercolation, HERBIE, and HNET present different and competing paradigms for learning and generalization. Which way will this go? Intensive research is now in progress. Realistic comparison of results will be presented in Souček et al. [16].

1.4 GENETIC ALGORITHMS AND GENETIC MODULES

The genetic algorithm is a form of artificial evolution. Originally it was introduced by Holland [17] in 1975. Genetic algorithms in search, optimization, and machine learning have been described by Goldberg [18].

The genetic algorithm implies the existence of a set of different network configurations (population); a method of configuration quality estimation (fitness); a way of concise configuration representation (genotype); and the operator for genotype modification (such as crossover and mutation).

Following De Jong [19], we present the basic idea of the genetic algorithm in Table 1.3. The key steps are the parent selection and the formation of new genotypes. Parents are selected based on their fitness. The better ones have a greater chance of being selected. New genotypes are formed by genetic operators, the most important of which are crossover and mutation. The crossover operator builds a new genotype of bits from each parent. In this way an offspring is formed that has characteristics of both, already well-adapted, parents. The mutation operator changes genotype bits in a random manner with a low probability, sustaining in this way population diversity.

Steels [20] has identified a few examples where the genetic algorithm and related selectionist processes could be an alternative to both connectionism and symbolic structures.

1. *Recognition:* To recognize a word, we almost need to know which word we are looking for. Elements of recognition ignite the process of domination. This process, after a while, will feed back to the elements that caused it to happen in the first place.
2. *Context:* Context is needed to understand the text, but the text provides information on what the context is. Hence, we start with an initial population of many possible contexts. Specific words in the text enforce the domination of a particular context, and, conversely, the context enforces the words in the text.
3. *Goal Selection:* All potentially executable goals form the initial population. Initial success in execution of the goal will further enforce the goal; the more resources and supports from the environment will go toward it.
4. *Rule Learning:* The process starts with a diverse set of rules. When rules are successful, they gain in strength; otherwise they become weaker.

In Chapter 8, De Garis describes the way to build artificial neural systems using genetically programmed neural network modules, called *GenNets*. De Garis introduces the technique of genetic programming, which enables one to build hierarchical structures, using basic GenNet modules. Chapter 8

TABLE 1.3 The Basic Idea of Genetic Algorithms

Create initial population of genotypes and structures to be optimized;
Evaluate structures in the population;
REPEAT
    Select genotypes, forming the new population;
    Form new genotypes, recombining the old ones;
    Evaluate structures created according to new genotypes;
UNTIL   terminating condition is satisfied.

presents three concrete examples of application: pointer (robot arm); walker (stick legs); and LIZZY (lizardlike creature, artificial life). De Garis predicts a "grand synthesis" between traditional artificial intelligence, genetic algorithms, neural networks, and artificial life.

1.5 NEURAL NETWORK HARDWARE

Long computation times required in most real-word applications have been a critical problem in neural network research. The problem can be resolved through neural network hardware. Hardware design follows two lines: special-purpose systems and general-purpose systems.

The special-purpose neurocomputer is a specialized neural network implementation dedicated to a specific model and, therefore, potentially has a very high performance. The general-purpose neurocomputer could emulate a range of neural network models. Its performance is moderate, but it is attractive because of its flexibility, especially in developing and testing new neural network models and new applications.

Chapter 9, by Yoon, Nang, and Maeng, gives an overview of basic neural networks and their implementations. Coprocessor, SIMD, MIMD, and DMM neurocomputers are analyzed and compared. The authors distributed the back-propagation algorithm on a network of transputers. According to them, the architecture for a neural network should have a very large address space and a few specific instructions that are dominant in any neural network simulation.

In Chapter 10, Dudziak concentrates on a specific class of problems: neurocomputing on transputers. The key points of his arguments rest upon three foundations:

1. The dual special features of the transputer (full on-chip computation facilities and multiple point-to-point communications);
2. The dual nature of what is parallel processing (algorithmic performance gain and multiple-source, nondeterministic input processing); and

3. The fundamental architecture of transputer modules (TRAMs), which allow for integration of specific-purpose coprocessor silicon with transputers.

Chapter 11, by Yestrebsky, Basehore, and Reed, describes the neural bit slice (NBS) kit. The NBS presents a hardward and integrated circuit (IC) solution for the construction of neural networks. The kit consists of

- A completely functioning evaluation circuit contained on an IBM XT, AT, or compatible printed circuit card,
- An evaluation program software, and
- A "broomstick" balancing fixture (for experiments).

NBS provides performance equivalent to that of a 55 million instructions per second (MIPS) processor.

Chapters 9 to 11 answer the questions of when, why, and how to implement neural networks, based on the latest advances in hardware technology.

1.6 NEURAL-KNOWLEDGE-FUZZY HYBRIDS

In order to develop high-quality practical intelligent systems, we need a methodology that integrates all available information and processing technology. In practice, many data sources, such as those from optical, speech, telemetry, and judgmental sources, tend to be inaccurate, noisy, or fuzzy. Such data are therefore difficult to use directly. Defuzzification, preprocessing, signal-to-symbol converting, and so on, must take place to make the data useful in any decision-making process. This leads to solutions in the form of multilevel neural-knowledge-fuzzy hybrids.

Fuzzy logic was first discussed in the 1920s and more recently in the 1960s and 1980s. Fuzzy logic is explicitly programmable and yet can be used for tasks for which digital algorithms are hard to develop. The Japanese Laboratory for International Fuzzy Engineering (LIFE) plans to spend $70 million over the next five years for research and development, with an eye toward commercialization of fuzzy systems. Already in use, the Sendai subway relies on a fuzzy system to control the acceleration, cruising, and breaking of cars. It is expected that fuzzy systems will find applications in investment, decision making, process control, and robotics.

The problem in robotics is that the environmental complexity in which the robot has to act cannot be fully represented in the model. Much uncertainty and fuzziness are employed in such robotic systems because of inadequate robot receptors and effectors or because it is impossible to represent objects, to locate objects, or to perform actions on objects with sufficient accuracy. Hence, it is not possible to construct a precise functional mapping between the state-space of the model and the state-space of the external world. To solve this problem without a deterministic approach, we need fuzzy logic.

Yestrebsky and Ziemacki, in Chapter 12, present a hybrid system composed of a preprocessing transformation and a neural network classifier. Several data transforms commonly used in communication and information theory were considered as the preprocessing operators. In experiments with image recognition, systems that included preprocessing performed substantially better than systems without preprocessing.

Basehore, Yestrebsky, and Reed, in Chapter 13, describe the fuzzy data comparator with neural network postprocessor. The system is fully hardware implemented. The described monolithic IC is used to rank the fuzzy data sets, allowing the device to operate in high-speed applications.

In Chapter 14, Sun and Waltz present the topic of combining conceptual level reasoning and subconceptual level reasoning to solve the brittleness problem in typical rule-based systems. This is a neural-symbolic hybrid.

Romaniuk and Hall, in Chapter 15, show how to use connectionist neural networks for constructing rule-based expert systems. The resulting learning system is significantly faster than classical back-propagation learning.

In Chapter 16, Lee and Nguyen present several systems that combine adaptive algorithms with multilevel processing control. Incremental learning characteristics of neural networks are combined with rule-based reasoning. Performance is bench-marked for target detection, and it meets real-time requirements of tracking applications.

Chapters 12 to 16 give concrete examples of high-performance, practical, hybrid intelligent systems.

1.7 KNOWLEDGE SYSTEMS INTEGRATION

The research, development and application of knowledge systems is in full swing. An increasing number of companies are beginning to achieve substantial returns from applications using artificial intelligence in areas such as expert systems to aid in manufacturing operations, scheduling, equipment repair, design, and financial planning, as well as in rapid prototyping of complex software systems. Solutions to complex problems require a combination of

- New architectures,
- Powerful software environment, and
- Performance beyond the capabilities of traditional computers.

The first major step in high-speed architecture was the development of Texas Instruments' LISP microprocessor, one of the most complex integrated circuits ever produced. The LISP microprocessor is designed to support a superset of the common LISP language. Its pipelined architecture executes microinstructions and many of the more complex LISP macroinstructions in a single clock cycle. Fabricated with 1-micron CMOS (complementary metal-oxide semiconductor) technology, the chip packs 553,687 transistors in a 1-cm^2 area. Six on-chip random-access memories (RAMs) totaling 114 K bits provide the memory bandwidth needed for the high instruction execution rate.

The second major step in high-speed architecture is distributed knowledge processing. In Chapter 17, Hung and Rozenblit review and analyze architectures for distributed knowledge processing.

Powerful software environment increases productivity by allowing the developer to focus on the conceptual level of the problem, without being preoccupied with lower-level details. The software environment should include sophisticated networking and communications capabilities, permitting easy integration of knowledge-based applications with other systems and applications.

In Japan, 10 years were allotted for the fifth-generation computer system project. The system will have three major functions: problem solving and inference, knowledge-based management, and intelligent interface.

In Chapter 18, Ichiko and Kurozumi present current research and development results in Japanese fifth-generation computer systems. A performance of several hundred times more than the value of conventional interference mechanisms is expected, realizing a feasible environment for many pragmatic uses in knowledge information processing. The Institute for New Generation Computer Technology (ICOT) believes that parallel architecture based on logic programming will be a computer generation oriented toward advanced information processing in the 1990s. The main focus is to develop a symbol-crunching parallel supercomputer and knowledge information processing within this framework.

1.8 SOFTWARE INTEGRATION

Joining the worlds of information, robotics, and control is achieved by interfacing diverse systems. *Interfacing* means the design of in-between software or hardware modules. This is expensive and time-consuming. A better solution is integration. Ideally, *integration* is a process of directly attaching modules in a system. Modules must have both integrated and stand-alone capabilities. The system should be expandable to meet the growing integration of information, control, and robotics technologies.

In this section we analyze the problem of integration of software modules, with two goals:

1. To inform intelligent system integration specialists of how the integration has been done in more conventional fields;

2. to inform business, control, robotics, and data delivery specialists of what they could achieve by integrating newly developed neural and knowledge intelligent techniques into their systems.

Software integration has long been a problem. A hardware designer can confidently rely on many standard building blocks, to minimize design time and cost and to improve product's reliability. Yet embedded software is still done the old-fashioned, hard way—by custom crafting.

Program development, modification, and start-up are considerably simplified by breaking programs into small, clearly defined blocks. Program development costs can be reduced by using standard function blocks for recurrent complex functions. Standard function blocks are off-the-shelf software modules that can be merged into programs written by the user. They can be called many times during program execution and supplied with the required parameters. Good examples are the standard function block libraries for programmable controllers, including the Allen-Bradley Pyramid Integrator™ family and the Siemens Simatic S5™ family, among others. Siemens claims over 100,000 Simatic installations worldwide, running the same software modules.

The ultimate modules are software components delivered as absolute, binary code on electrically programmable read-only memories (EPROMs), disks, or tapes. This eliminates source code manipulations, recompilations, system generation (SYSGEN), and other hazards and labor required to integrate with more traditional, off-the shelf software. Each absolute software component has a private configuration table, supplied by the user and containing application- and hardware-related parameters relevant to its operation. Installation involves placing the component (pure code) and its configuration table somewhere in RAM or ROM and plugging the address of its configuration table into the software component interconnect bus. This kind of approach is supported, for example, by the Software Components Group Inc.

The major step forward is the introduction of artificial intelligence and expert system techniques in software development. Although the system cannot synthesize programs automatically, it can help programmers in various stages of program production, such as design, coding, debugging, and testing. As a rule, the system is composed of several major units, each an expert system for a subdomain of the program development process.

Ichiko, in Chapter 19, presents an advanced paradigm for software integration. The aim is to construct an interactive design aid environment with a highly semantic description facility and a logically effective verification facility for increased software production oriented toward reutilization. This can be achieved by semantically representing an entity model and performing predicate calculus on its entity characteristics, using rule-based inference from the data base for conceptual software description. Concrete examples are presented, including the complex stock control software for the Information Processing Society of Japan.

Fandel, in Chapter 20, describes the Paracell language developed for control software integration and the parallel inference machine (PIM). PIM is a highly parallel, plantwide, process control system organized as an inference engine. It follows the concepts of artificial intelligence systems. PIM can act on English language rules, as in expert systems. It can fire more than 1,000,000 rules per second, which is 2000 times faster than the best graphics workstations. It is interesting to compare the American concepts of PIM and those of the Japanese fifth-generation computer systems project, described in Chapter 18.

1.9 SOFTWARE PROTECTION AND MORI SUPERDISTRIBUTION*

For further progress in software integration, we must protect software against piracy. Mori and Kawahara [21] define protection levels 0 to 4, as shown in Table 1.4. The most advanced level is the ABYSS architecture developed by White, Commerford, and Weingart at the IBM Thomas J. Watson Research Center in Yorktown Heights, New York [22,27]. ABYSS is based on the notion of a use-once authorization mechanism called a "token" that provides the "right to execute" a software product. All or part of the software product is executed within a protected processor and is distributed in encrypted form. Physical security for an ABYSS processor is provided by a dense cocoon of wires with resistance constantly monitored by the processor. A change in resistance indicates a likely attempt to penetrate the system.

In 1983 Mori [28,29] invented an approach called *superdistribution*. According to Mori and Kawahara [21], the chief difference between superdistribution and the ABYSS scheme is that superdistribution does not require the physical distribution of tokens or anything else to users of a software product. In other words, ABYSS requires that software be paid for in advance, whereas superdistribution does not. Since 1987, work on superdistribution has been carried out by a committee of the Japan Electronics Industry Development Association (JEIDA), a nonprofit industrywide organization. That committee is now known as the Superdistribution Technology Research Committee.

Superdistribution of software has the following novel combination of desirable properties:

1. Software products are freely distributed without restriction. The user of a software product pays for using that product, not for possessing it.
2. The vendor of a software product can set the terms and conditions of its use and the schedule of fees, if any, to be charged for its use.

---

* Section 1.9 is based on R. Mori and M. Kawahara, "Superdistribution: The Concept and the Architecture"; *The Transactions of the IEICE*, **Vol. E73;** No. 7; July 1990; pp. 1133–1145.

TABLE 1.4 Levels of Software Protection Technology, Following Mori and Kawahara [21]

Level	Key Concept	Remarks
4	Superdistribution	Sda (Superdistribution architecture) 1983, R. Mori (University of Tsukuba)
3	Execution privileges	"Right-to-Execute" ABYSS (a basic Yorktown security system) 1987 S. R. White (IBM) [22]
2	Customizing software with a computer ID	Customizing deciphering key (software is common) 1986 A. Herzberg PPS (public protection of software) [23] 1984 D. J. Albert (enciphering and key management) [24] 1982 G. B. Purdy SPS (software protection scheme [25] Customizing each copy of the software
1	Hardware protection	Controling execution: hardware key (e.g., ADAPSO Key-ring [26]) Inhibiting duplication: copy protection, noncompatible ROM
0	No physical protection	Law & ethics

3. Software products can be executed by any user having the proper equipment, provided only that the user adheres to the conditions of use set by the vendor and pays the fees charged by the vendor.
4. The proper operation of the superdistribution system, including the enforcement of the conditions set by the vendors, is ensured by tamper-resistant electronic devices, such as digitally protected modules.

From a different viewpoint, the needs of users and the needs of vendors have until now been in irreconcilable conflict because the protective measures needed by vendors have been viewed by users as an intolerable burden. The superdistribution architecture resolves that conflict to the interests of both parties. Wide distribution benefits vendors because it increases usage of their products at little added cost and thus brings them more income. It benefits users because it makes more software available, and the lower unit costs lead to lower prices. It also creates the possibilities of additional value-added services to be provided by the software industry. Moreover, users themselves become distributors of programs that they like, since with superdistribution there is absolutely nothing wrong with giving a copy of a program to a friend or colleague.

It might seem at first that publicly distributed software such as freeware and shareware already solves the problem addressed by superdistribution. But the likelihood of the authors being paid is too small for public domain software to play a leading role in the software industry. Superdistribution software is much like public domain software for which physical measures are used to ensure that the software producer is fairly compensated and that the software is protected against modification. Although public domain software might achieve the aims of superdistribution in an idealized world where all users paid for software voluntarily and none of them abused it, Mori and Kawahara see little hope that such an idealized world will ever come to exist.

The superdistribution architecture Mori and Kawahara have developed provides three principal functions:

1. Administrative arrangements for collecting accounting information on software usage and fees for software usage;
2. An accounting process that records and accumulates usage charges, payments, and the allocation of usage charges among different software vendors;
3. A defense mechanism, utilizing digitally protected modules, that protects the system against interference with its proper operation.

In order to participate in superdistribution, a computer must be equipped with a device known as an *S-box* (superdistribution box). An S-box is a protected module containing microprocessors, RAM, ROM, and a real-time clock. It preserves secret information, such as a deciphering key, and manages the proprietary aspects of a superdistribution system. An S-box can be installed on nearly any computer, although it must be specialized to the computer's central processor unit (CPU) type. It is also possible to integrate the S-box directly into the design of a computer. We call a computer equipped with an S-box an *S-computer*.

Programs designed for use with superdistribution are known as *S-programs*. They can be distributed freely since they are maintained in encrypted form.

In order to make it acceptable to users, software vendors, and hardware manufacturers, the superdistribution architecture has been designed to satisfy the following requirements:

- The presence of the S-box must not prevent the host computer from executing software not designed for the S-box. The presence of the S-box must be invisible while such software is active.
- The modifications needed to install an S-box in an existing computer must be simple and inexpensive.
- The initial investment required to make S-programs generally available must be small.
- The execution speed of an S-program must not suffer a noticeable performance penalty in comparison with an ordinary program.
- The S-box and its supporting protocols must be unobtrusive to users and to programmers.

- The S-box must be compatible with multiprogramming environments, since we anticipate that such environments will become very common in personal computers.

In current design a program can be written without considering the S-box, but the program must be explicitly encrypted before it is distributed if it is to gain the protection of superdistribution. This encryption can be done by a programmer, using the S-box itself.

Although Mori directed the work toward superdistribution of computer software, the architecture can be applied to other forms of digital information as well, for example music as it is recorded on digital audio tapes.

Containers for tangible items reflect the value and importance of their contents. No similar containers exist for digital information despite the progress of microelectronics and communications networks. The lack of such containers has invited problems, such as computer viruses and software piracy.

Digitally protected modules promise to provide such containers and thus to become one of the most fundamental applications of integrated circuits in the information society. Superdistribution should provide unrestricted distribution and integration of commercial software.

1.10 INTEGRATION IN BUSINESS AND IN DATA DELIVERY

Business professionals daily face problem solving and decision making based on extensive but incomplete, uncertain, and even contradictory data and knowledge. Introduction and integration of neural, expert, and other intelligent modules helps businesses reorganize into more efficient and effective organizations. It does this by helping individuals solve problems more quickly and efficiently than they did before. It is expected that middle managers of corporations will be equipped with a "staff" of 15 to 20 automated experts and neural networks that will always be available to answer questions and give advice about problems those managers face.

A good example is a buying and selling timing prediction for a stock market. Kimoto et al. [30] developed a number of learning algorithms and prediction methods for the TOPIX (Tokyo Stock Exchange Price Index) prediction system. The system achieved accurate predictions, and the simulation on stock trading showed an excellent profit. The input to TOPIX consists of several technical and economical indexes. Several modular neural networks learned the relationships between the past technical and economic indexes and the timing for when to buy and sell. Stock price fluctuation factors could be extracted by analyzing the networks.

The basic architecture of the prediction system and its performance are presented in Figures 1.3 and 1.4, respectively.

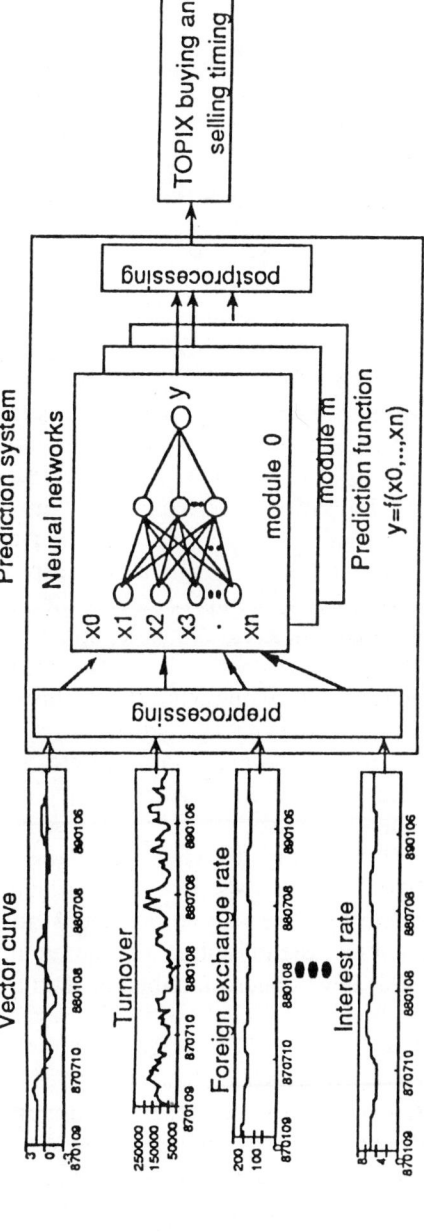

Figure 1.3 *Basic architecture of prediction system. Reprinted by permission from Kimoto, Asakawa, Yoda, and Takeoka [30].*

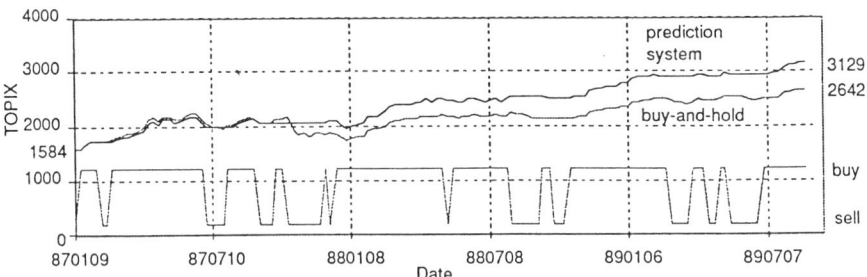

Figure 1.4 *Performance of the prediction system. Reprinted by permission from Kimoto, Asakawa, Yoda, and Takeoka [30].*

Stock prices are determined by time-space patterns of economic indexes, such as foreign exchange rates, New York Dow-Jones average, and interest rates and technical indexes, such as vector curves and turnover. The prediction system uses a moving average of weekly average data of each index for minimizing influence due to random walk. The time-space patterns of the indexes were converted into space patterns. The converted indexes are analog values in the 0–1 range. The timing for when to buy and sell is indicated as an analog value in the 0–1 range in one output unit. The system learns data for the past M months, then predicts for the next L months. The system advances while repeating this. Four independent modular networks learn for four types of different learning data. The average of prediction outputs from these networks becomes the prediction output from the system.

To verify the effectiveness of the prediction system, Kimoto et al. simulated the buying and selling of stock. Buying and selling according to the prediction system made a greater profit than buying and holding. Buying and selling was simulated by the one-point buying and selling strategy, so performance could be clearly evaluated. One-point buying and selling means all available money is used to buy stocks and all stocks held are sold at one time. In the prediction system, an output of 0.5 of more indicates buy, and an output less than 0.5 indicates sell. Signals are intensified as they get close to 0 or 1.

The buying and selling simulation considered "buy" to be an output above some threshold and "sell" to be below some threshold. Figure 1.4 shows an example of the simulation results. In the upper diagram, the buy-and-hold performance (that is, the actual TOPIX) is shown as dotted lines, and the prediction system's performance is shown as solid lines. The TOPIX index of January 1987 was considered as 1.00; it was 1.67 by buy-and-hold at the end of September 1989. It was 1.98 by the buying and selling operation according to the prediction system. Use of the system showed an excellent profit.

The prediction system is a good example of the integration of intelligent modules.

In stock tradings, so-called triangle patterns indicate an important clue to the trend of future change in stock prices, but the patterns are not clearly defined by rule-based approaches. Kamijo and Tanigawa [31] use a recurrent neural network to solve the problem of triangle recognition. Subsequently, test triangles were appropriately recognized.

Bigus and Goolsbey [32] report a hybrid system including IBM AS/400 data base facilities, AS/400 Knowledge Tool (a rule-based expert system), and self-organizing neural networks. This hybrid system is used in a commercial office business environment. The problem studied is that of dispatching delivery trucks under weight and volume constraints to minimize the number of trucks required and the total miles traveled.

The IBM AS/400 has over 125,000 systems installed worldwide and is used heavily in commercial office environments. Thus, it serves as a good test bed for evaluating how neural networks perform in this environment. The AS/400 integrated data base services were used to store customer information such as name, location, order information related to the day of delivery, and size and weight of the shipments.

AS/400 Knowledge Tool is a rule-based forward-chaining inference engine with development and run-time environments on AS/400 systems. It is an OPS/5-like tool with procedural extensions that allow it to be integrated with COBOL and RPG III applications. Both Knowledge Tool interfacing and neural networks routines are callable from RPG III or COBOL programs.

Bigus and Goolsbey's [32] solution proceeds in four steps. First, the minimum number of trucks that will be needed is determined. Second, an initial assignment of deliveries to each truck is made. This assignment is valid, but not necessarily good. This step also determines if more trucks are needed. Next, the assignments are improved by swapping deliveries between trucks to reduce the distance the truck must travel. Finally, the traveling salesman problem is solved for each truck to determine the actual driving route. It is based on a self-organizing neural map, with output nodes competing for each delivery location. From this, the shortest delivery route was selected.

This is a very good example of a hybrid intelligent system for commercial office business environment. Hybrid intelligent systems have been used in several transaction processing applications, such as credit scoring and risk analysis. Moreover, the data base facilities of most data processing systems can be used to provide training data for the neural network.

The complex stock control problem of the IPSJ (Information Processing Society of Japan) has been studied over a long period. Due to the regularity of processing, an expert system-like solution is possible. Ichiko, in Chapter 19, describes intelligent software integration to solve this and similar problems.

Note that a lot of activity in intelligent business systems come from Japan. America and Europe follow the same trend.

Data delivery is rapidly growing into an important information service for business, science, technology, and everyday life.

According to Martinez and Harston [33], tying the information obtained from data bases to a neural network can create a new system with expertise in any field. The system can provide many answers to a particular input in a well-organized manner. As the number of associations increases by hundreds, many of these associations become similar, underscoring the need for a very accurate system. The system can find a single matching solution to a particular input if there is only one solution, but it can also respond with alternative solutions.

An important characteristic of the system is that its knowledge is growing as the number of associations increases without being saturated. The system can become an expert adviser using information from the data base. The other characteristics are that memory size is relative to the data base attributes and the number of entries, it can perform in real time, the input can be weighted differently, and it works with inconsistent data.

Parsaye, Chignell, Khoshafian, and Wong, in Chapter 22, describe in detail the integration of intelligent reasoning and informing modules into commercial data delivery systems. The results are intelligent data bases and automated discovery. Such new tools open the road to direct and fast integration of knowledge.

Chapters 19 and 22 present concrete examples of integrated reasoning, informing, and serving (IRIS) systems, with applications in business and data delivery.

1.11 INTEGRATION IN PROCESS CONTROL AND IN ROBOTICS

Neural, concurrent, and intelligent modules bring new blood to process control and robotics. Expert systems have proven to be effective tools, when the expert can explicitly state the necessary rules. Neural networks are fully capable of discovering hidden information by observing samples of a system's behavior during a problem-solving task. For the background on neural and concurrent process control and robotics systems, see Souček [3].

Integration of neural and intelligent modules into *process control* results in hybrid systems. Integration aims toward the following goals:

- Proper tools for approaching complex control problems;
- Improving programmer productivity;
- Establishing a common meeting ground for the process engineer and the control engineer;
- Library creation for reuse of past software;

- Environment for debugging and troubleshooting;
- Graphical presentation for easy use;
- The need to adopt a format with some familiarity to the large pool of existing control designers;
- Real-time, input/output orientation that excels at handling large numbers of I/O, be they discrete points, analog signals, or numerical values.

Fandel, in Chapter 20, analyzes complex automation and control systems. For small scale, distribution computer systems (DCS) are quite adequate. When the size of DCS grow to include more than 10 or so nodes, inherent communication and memory requirements of these systems render them impractical. Fandel presents the solution as a parallel inference machine and Paracell language. Such a solution brings together parallel processors with a software organization of inference engines, acting on expert systemlike rules.

A special problem is integration of intelligent functions into *robot systems*. No matter how much mechanical and electrical hardware is built up around a conventional robot, it is still only a sophisticated machine tool. The major step forward is achieved now by the introduction of sophisticated real-time software and neural networks. Standardized real-time software with the following capabilities are now available: fast response to real-world events, priority-driven multitasking, reentrant programming languages, extremely fast response to interrupts, and support for special I/O devices and for special I/O programming.

Real-time operating systems provide important functions needed in reprogrammable systems. Real-time kernels provide minimal functions needed in embedded applications.

Standardization is a major issue. Development of proprietary software requires large investments and produces systems that are incompatible with the rest of the world. Designers are therefore using well-established software tools and systems, such as DEC's VAXELN operating system, Intel's DEM modular systems operations, and Lynx Unix-compatible real-time operating system.

Neural networks open the road toward learning robots. The robot learns from what is happening rather than from what it has been told to expect. A new class of neural robots requires self-adapting sensing feedback systems and hand-eye coordination. The first neural robots consisted of a stereo-camera, a machine-vision system, and an industrial robot arm. The operation of a neural robot is similar to neural process control, pole balancing, and graded learning networks. Conventional robots cannot correct for process variations or for part preparation errors. To compensate for such variations or error, robots must include real-time sensor feedback and the features of adaptation and learning. These features might result in a widespread use of neural robots in automated manufacturing laboratories and services. Hybrid neural, expert, and concurrent robot systems promise the best characteristics.

Nagata, Sekiguchi, and Asakawa [34] developed a mobile robot controlled by a neural network. The robot moves freely in confined areas and senses environmental changes. This ability requires a control circuit for movement and sensing. Figure 1.5 shows a structured hierarchical neural network that can be thought of as a converter having many inputs and outputs. The structure combines two types of layered networks: a reason network and an instinct network. The input signals from sensors and the output signals to motors are one bit each (0 or 1). They control motors simply by turning them on or off. Sensor signals are fed directly to the input layer of the reason network. The reason network determines the correspondence between sensory input and behavior pattern and outputs the behavior pattern. An example of such a behavior pattern is "move forward when the infrared sensor on the head detects light." The instinct network determines the correspondence between sensory input and a series of behavior patterns the robot should take over a certain period of time. An example is "repeat a cycle of right and left turns until the infrared sensor on the head receives light." Robots must often perform such motions in sequence. Therefore, the function of the instinct network is essential when applying a neural network to a robot.

The instinct network has three short-term memory units, two of which have a mutually inhibitory link. A short-term memory unit holds an excited and inhibited state for a certain period of time. As illustrated in Figure 1.5, some of the output units of the reason network are connected to the instinct network by using excitatory and inhibitory signals. The inhibitory signal from the reason network to the instinct network is reset when there are no input signals from the sensors or when only the tactile sensor is active. In these cases, the excitatory signal is transmitted from the reason network via short-term memory unit 2 to the instinct network. The instinct network becomes active and takes control. The robot then begins to avoid obstacles or search for ultrasonic waves and infrared rays from other robots.

When a signal from a tactile sensor is received, the robot must take action to avoid the obstacle. This action is a behavior pattern series consisting of several motions in sequence, such as moving backward for a certain distance, then moving forward and circling the obstacle. For this action, the reason network transmits an excitatory signal to short-term memory unit 2, which holds the excitatory signal for a certain period of time. The instinct network continues to transmit a motor control signal to turn the robot and move it in reverse for as long as the short-term memory unit is excited, thus allowing the robot to avoid the obstacle.

When a torsion limit sensor turns on, the instinct network is also activated. When the robot turns to the left or right and reaches the limit of its rotation, a sensor signal corresponding to the limit position is transmitted to the proper input unit of the instinct network by using short-term memory 1,

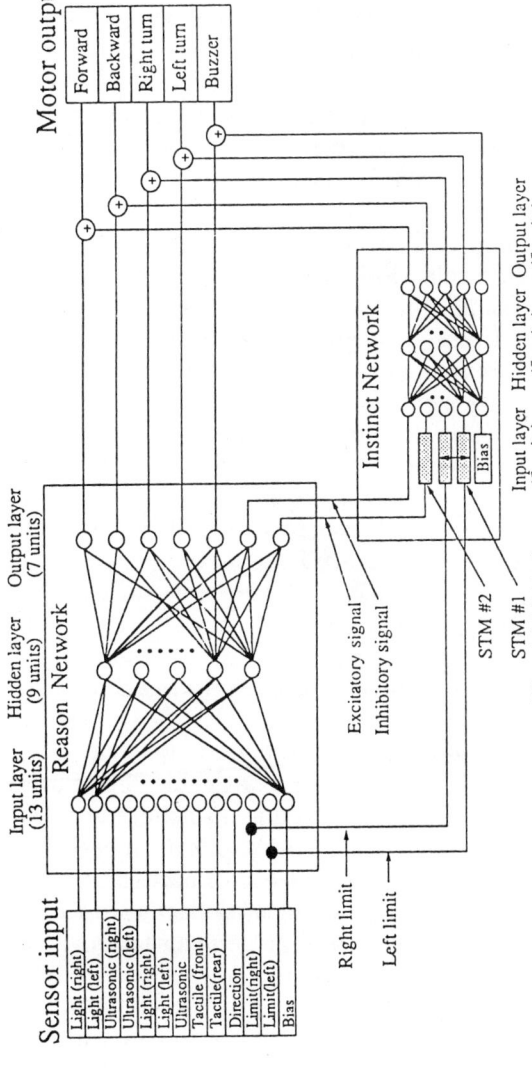

Figure 1.5 *Structured hierarchical network model. Adapted from Naqata, Sekiguchi, and Asakawa [34]. Copyright © 1990 IEEE. Reprinted with permission.*

which is also equipped with a mutually inhibitory link. The instinct network generates a motor control signal for right or left rotation in response to the left or right limit signal to rotate the robot away from its limit.

The robot behaves according to the reason network when it is given sensory information on which its action should be based, and behaves according to the instinct network when it is not receiving such sensory information, and when it must take a series of actions for a certain period of time. Nagata et al.'s [34] robot control is an example of integration of intelligent modules into hierarchical structures.

The hierarchical network of Nagata et al. [34] is an example of integration of intelligent modules *within* one robot control system. The next level is integration of many robots into a complex automation system and the split of tasks *among* several robots.

Vukobratović in Chapter 21 discusses the integration of robots into complex flexible automation cells (FMC). The number of robots applied within various FMC is increasing. One of the major problems arising in these applications is the fast and efficient selection of the adequate robots for each FMC. Vukobratović presents a set of software tools to simulate the performance of the FMC using the dynamic model of the robots. In this way one can test whether the selected robots, controllers, and neural networks are capable of satisfying all requirements in specific FMC or in an intelligent system.

Chapters 21 and 22 present concrete examples of system integration, with applications to process control and robotics.

1.12 INTEGRATION OF SCIENTIFIC DISCIPLINES

Intelligent systems integration is here to stay. Many devices, modules, systems, applications, and services are now available. For further advance, strong interaction between several scientific disciplines is necessary.

Peschl, in Chapter 23, calls for an interdisciplinary cooperation. To represent and process commonsense knowledge, it is necessary to integrate epistemological, biological, psychological, linguistic, and computer scientific aspects. Hence, a new, highly automated and efficient technology is needed: *integration* of scientific *disciplines*. Peschl's suggestions for science integration include well-defined terminology, restriction of the scientific subdomains, use of icons, private user interface, learning by using self-organization techniques, artificial organism constructing its own reality and acting in the environment, and others.

The ultimate goal is to integrate scientific disciplines within a particular subdomain, such as intelligent systems. Fifth- and sixth-generation computer technologies offer the tools needed for automated integration of scientific disciplines. Some of the already existing tools are presented by Parsaye, Chignell, Khoshafian, and Wong in Chapter 22.

1.13 INTEGRATION OF REASONING, INFORMING, AND SERVING (IRIS)—A NEW DISCIPLINE

Users' attention is shifting from components to system development. As a result, integration of reasoning, informing, and serving is receiving increased attention in engineering practice as well as in research and education. Integration of reasoning, informing, and serving is based on mixed intelligence granularity and on specific new methods and sets of tools. Table 1.5 defines

TABLE 1.5 Integration of Reasoning, Informing and Serving, IRIS

Discipline Foundations	Examples	Described in Chapters
1. Mixed intelligence granularity	Neural networks; expert systems; fuzzy systems; genetic modules; pre/postprocessors	1–23
2. New integration paradighms	Automation based on expert systems Adaptation based on learning; Selection based on genetic programming Self-organization based on artificial life	1–23
3. Standard software modules	Open library blocks; closed, binary object blocks	1
4. Special languages	Ladder; Paracel; Grafcet	20
5. Software development tools	Expert-system-based software development environment; neural network simulators; control/robotics systems simulators	19–21
6. Automated discovery and interdiscipline	Intelligent, interactive data bases and user interfaces	22, 23
7. Standard control, robot and automation modules	The Pyramid Integrator; Simatic S5; PIM	1, 20
8. Case studies and working systems	Applications in banking and stock control; process control and automation; surveillance; robotics and manufacturing systems; information services	2–22
9. Concurrency	Transputer modules, megaframe modules; OCCAM language; Helios operating systems	9, 10, 17–21
10. Signal-to-symbol transformation; pattern to category mapping	Instrument to expert system; communication signal to language; neural network to postprocessor; frame grabber to preprocessor.	1–16

integration of reasoning, informing, and serving as a new discipline, based on 10 foundations, which are described in this book.

From the implementation point of view, foundation 9, concurrency, is of special interest. Concurrency is so much a feature of the universe that we are not normally concerned with it at all. To adequately model the concurrency of the real world, we need many processors working simultaneously on the same program. Concurrent clusters and superclusters provide supercomputer power for one-tenth the price. This is accomplished by organizing a system in a number of local computing nodes. Adding more nodes means adding channels, and therefore communication bandwidth, so the overall balance of computing power and communication bandwidth is always maintained. For introduction to concurrent systems, see Souček [1]. Advanced concurrent systems and details of design, programming, and applications are given in Souček [3].

Foundation 10 deals with the signal-to-symbol transformation. The underlying principle of this technique is the transformation of row numerical data to high-level descriptions that are essential for further symbolic manipulations.

Signal-to-symbol transformation assigns qualitative values to the subsets of a large data set based on low-level primitives, such as signals from instruments or effectors, numbers from neural networks, or pixels from frame grabbers. The criterion for detecting specific subsets is usually highly domain-dependent. There is no general solution to the problem. Udupa and Murthy [35] give the solution for ECG analysis. Souček and Carlson [36,37] describe the transformation of firefly communication signals into coded language. Filipič and Konvalinka [38] present transformation of power spectrum data into symbols needed for expert systems analysis.

Input pattern to output category mapping in neural networks is related to the signal-to-symbol transformation. Transformation and mapping are essential processes, needed to integrate intelligent systems and, particularly in mixing modules of different granularity.

Foundations 1 to 10 integrate subdomains of several disciplines. In this way a new discipline, Integration of Reasoning, Informing, and Serving (IRIS), has been formed. IRIS presents the key for better world in the near future and in the third millenium.

1.14 PRESENT AND FUTURE LEVELS OF INTEGRATION: 6 C

According to Winograd [39], the 1990s will see an acceleration of the progress from connectivity, to compatibility, to coordination. With widespread networking and universal low-level protocols, soon the basic *connectivity* of computers will be taken for granted in the same way, as the connectivity of roads and of telephones. Next comes *compatibility*: the task developed on one computer can work with the task developed on another computer. This becomes possible through the creation of comprehensive standards for languages, databases, communication protocols, and interaction architectures.

According to Winograd [39], the third level, *coordination* is a central issue. For users, the purpose of integration is not to share data or connect processors but to integrate the work: customers, orders, products, designs and the myriad of specialized things that appear in different lines of endeavor. Over the next few years, the development of conversation/action systems will create people-centered systems that bring new clarity to the coordination of human action.

In the field of intelligent systems we define three more levels of integration, aiming at cognition, conception, and conscience.

Cognition means integration of knowledge and of intelligence that come from many sources. The results is a system with higher intelligence quotient (IQ). Cognitive integrated systems should have increased ability to categorize objects into classes of equivalence; to recognize associations; to learn new concepts; to generalize and reason.

Conception means integration that automatically and adaptively creates new observables and new devices: artificial life. Artificial life (AL) is a new scientific and engineering discipline (see Langton [40]).

Biological organisms evolve new sensors, allowing them to make more distinction on the world, thereby expanding their set of available observables in an apparently open-ended way.

According to Cariani [41], in terms of their adaptability, the major difference between biological organisms and contemporary computers and robots is that the former have the capacity to construct new observables, thereby constructing their relationship to the world around them, but the latter are prisoners of their fixed, prespecified observable domain. If we wish to construct devices that are to act creatively, to solve problems that even we cannot yet adequately characterize, then our device must be able to search for possible solutions outside of the realm of what we already know and can specify. Our device must be capable of some degree of autonomous action relative to our specifications; it must have emergent properties relative to our model—our expectations—of its behavior. Such devices would be useful in ill-defined problem situations, where we lack even a plausible rudimentary definition of the problem at hand, where creativity on a fundamental level is needed.

Cariani [41] defines four types of devices: formal-computational, formal-robotic, adaptive, and evolutionary (see Table 1.6). Evolutionary devices would be very difficult to construct. In the field of neural networks, Cariani suggests looking at the following mechanisms: change blocking properties of presynaptic branches rather than changing synaptic strengths; construct neural networks as assemblages of tunable oscillators; use variably coupled oscillators via tunable filters and pulse-coded communication channels.

Conscience is the ultimate integration level. Intelligent systems should be created, having in mind noble applications. The system should be able to

TABLE 1.6 Device Taxonomy, Following Cariani [41]

Formal computational devices	Can only deal with problems already completely encoded into symbols. They are syntactically static.
Formal robotics devices	Are similar to formal-computational ones except that they have fixed sensors and or effectors that connect them directly to the world. They are nonadaptive and cannot alter their internal structure to optimize their performance.
Adaptive devices	Alter their syntactic computational parts based on their past experience and performance. Because they learn, they are flexible and could adapt to unforeseen situations. However, the behavior of the device is semantically bounded: It must take place within the confines of what its sensors and effectors can do.
Evolutionary device	Would adaptively construct new sensors and effectors. New sensors and effectors mean new primitive features and actions. The device would adaptively construct its own signal types, thereby deciding what aspects of external world it considered relevant. They would lead to new concept formation.

make distinction between good and evil. Its most important feature should be global thinking. Souček[3] has defined name for this ultimate intelligence: *Noble intelligence*.

The six integration levels are presented in Figure 1.6. Integrated intelligent systems are beginning to reach the level of cognition.

The levels of cognition and conception ask for the development of a new class of integration media. This includes: intelligent dialogues; definition of intelligent message quanta; the use of context; intelligence response and transfer functions; fuzzy receiving and sending windows; cognition/concep-

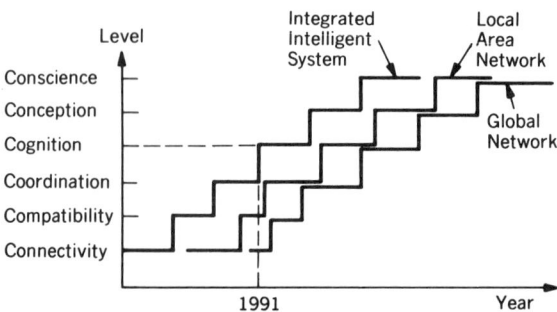

Figure 1.6 *Six levels of reasoning, informing, and serving integration (6C). Integrated systems are beginning to reach the level of cognition. Systems connected through local area networks and global networks follow.*

TABLE 1.7 Brain-Window-Based Artificial Life

Adaptive sensors connected to a set of fuzzy receive windows.
Adaptive effectors connected to a set of fuzzy send windows.
Signal transformation to standard domain
Signals-to-symbols-to-language conversion
Neural networks, remembering the past history of stimulation
Natural selection system controlled by the neural networks and by a punishment/reward logic
Context formation and context switching
Genetic crossover/mutation system
Random start leads nowhere. The proposed system has the structural features adapted to the problem to begin with, including hierarchical organization, brain window fuzzy logic, and brain window language. Starting from this structure, evolution, adaptation, selection, and self-organization would construct new receive windows sensing the environment and new send windows controlling effectors. This would lead to modified or new life concept formation.

---

tion channels, links and pools; intelligent bases; broadcasting of intelligence; intelligence hidden in chaotic attractors and in fractal geometry.

A natural effect that could be used as the base to build cognition/conception devices is brain windows. Brain windows have been observed by Souček and Carlson [36,37] in computer analysis of firefly communication and behavior. Brain window theory is described in Souček in Souček [1]. Based on brain window theory, we suggest here the basic ingredients necessary to build a class of artificial life systems (see Table 1.7).

The basic set of receive/send windows would be genetically inherited from the parents. Each receive window is responsible for a particular class of input stimuli. Each send window controls a particular behavior. During the life of the system, some old windows would be widened, other would be narrowed or closed. Also, some new windows would be opened. This process depends on interaction with the environment and on punishment/reward logic. In this way, combining genetic processes and natural selection under the pressure of the environment, a new species (new artificial life device) would be automatically formed.

In short, in brain windows a transition from continuous signaling to discrete coding is achieved, and a communication language is formed. Brain windows are seen as a natural transition from fuzzy information processing to symbolic reasoning. The response of the system is a continuous (analog) function of the stimulus and of the content of the memory. The response is at the same time a discrete (discontinuous) function of the context stored in the memory. In this way the response can be switched from one window (and corresponding behavior) to another, although the windows are far away.

Brain windows are discrete but flexible. The windows change shape, depending on behavior and external stimulation. If the behavior is "courtship," the windows get narrowed and the system becomes highly selective.

Only the partner with very good characteristics is acceptable. In this way, natural selection is in action already in the process of choosing a mate. This selection differs from that based on genetic algorithms.

If the behavior is "feeding," the windows get widened. In this way the system increases its chance of catching the prey and surviving.

The windows are "plastic" at the beginning of the stimulation sequence, and they readily adjust. However, when the windows are pushed far away from their normal width, they resist further change.

For a detailed description of how nature does this, see Souček and Souček [1]. The realization of practical devices is seen in the integration of the following technologies:

Intelligence processing: symbolic, fuzzy, neural, genetic; holographic;

Parallel processing: associative, cellular, farms, clusters;

Direct biology inspired processing: brain windows, event trains, tunable oscillators.

IRIS paradigms present the base for new information systems which can think, reason, and judge like humans; deal with fuzziness, uncertainty and incompleteness; and operate in a massively concurrent distributed mode. The Japanese Ministry of International Trade and Industry (MITI) is ready to launch a new project in this direction [42]. America and Europe drive towards the same goal.

This book identifies the problems of intelligence integration and presents the first results and concrete applications.

For further development see Souček and the IRIS Group [16, 43].

REFERENCES

[1] B. Souček and M. Souček, *Neural and Massively Parallel Computers: The Sixth Generation*, Wiley, New York, 1988.

[2] T. H. Bullock, "In Search of Principles in Neural Integration," in Fentress, Ed., *Simpler Networks and Behavior*, Sinauer Associates, Sunderland, Mass., 1976.

[3] B. Souček, *Neural and Concurrent Real-Time Systems: The Sixth Generation*, Wiley, New York, 1989.

[4] W. Banzhaf, T. Ishii, S. Nara, and T. Nakayama, A sparcely connected asymmetric neural network and its possible application to the processing of transient spatio-temporal signals, *Proc. Internat. Neural Network Conference*, Paris, 1990.

[5] V. J. Braitenberg, *J. Theor. Biology*, 46, 421 (1974).

[6] A. Gierer, *Biological Cybernetics*, 59, 13 (1988).

[7] A. J. Pellionisz, About the geometry intrinsic to neural nets, *Proc. Internat. Joint Conference on Neural Networks*, 15, (1990).

REFERENCES

[8] A. J. Pellionisz, *The Geometry of Brain Function: Tenzor Network Theory,* Cambridge University Press, Cambridge, 1990.

[9] P. J. Werbos, "Beyond Regression: New Tools for Prediction and Analysis in Behavioral Sciences," Unpublished doctoral dissertation, Harvard University, Cambridge, Mass., 1974.

[10] P. J. Werbos, Generalization of backpropagation with application to a recurrent gas market model, *Neural Networks,* 1, 339–356 (1988).

[11] D. Parker, "Learning logic," Invention report, 581–64, File 1, Office of Technology Licensing, Stanford University, Stanford, Calif., 1982.

[12] D. E. Rumelhart, G. E. Hinton, and R. J. Williams, "Learning internal representations by error propagation" in D. E. Rumelhart and J. L. McClelland, Eds., *Parallel Distributed Processing I,* MIT Press, Cambridge, Mass., 1986, pp. 318–362.

[13] M. Jurik, "A Summary Report on Backpercolation," International Workshop, Parallel Problem Solving from Nature, University of Dortmund, Germany, October 1990.

[14] D. Wolpert, A mathematical theory of generalization. *Complex Systems,* (in press).

[15] J. Sutherland, A holographic model of memory, learning and expression. *International Journal of Neural Systems,* Vol 1, 3, (199) 259–267.

[16] B. Souček and IRIS Group, *Fuzzy, Holographic, Invariant and Parallel Intelligence, The Sixth Generation,* Wiley, New York (in press).

[17] J. H. Holland, *Adaptation in Natural and Artificial Systems,* University of Michigan Press, Ann Arbor, Mich. 1975.

[18] D. E. Goldberg, *Genetic Algorithms in Search, Optimization and Machine Learning,* Addison-Wesley, Reading, Mass., 1989.

[19] K. A. De Jong, "Genetic Algorithms: A 10 year perspective," Proc. International Conference on Genetic Algorithms and Their Applications, Pittsburgh, Pa., 1985, pp. 169–177.

[20] L. Steels, Self-organization through selection, *Proc. IEEE First. Internat. Conf. on Neural Networks,* June 21–24, Vol 2, 1987, pp. 55–62.

[21] R. Mori and M. Kawahara, Superdistribution, The concept and the architecture, *IEICE Trans.,* E 73(7), 1133–1146 (1990).

[22] S. R. White, ABYSS: A trusted architecture for software protection, *IEEE Symp. on Security and Privacy Proc.,* 38–51 (April 1987).

[23] A. Herzberg and G. Karmi, "Public protection of software," in *Advances in Cryptology; Proc. Crypto 85,* H. C. Williams, Ed., 158 (1986).

[24] D. J. Albert and S. P. Morse, Combating software piracy by encryption and key management, *IEEE Computer,* 68–73 (April 1984).

[25] G. B. Purdy, G. J. Simmons and J. A. Studier, A software protection scheme, *IEEE Symp. on Security and Privacy Proc.* 99–103 (April 1982).

[26] "Proposal for Software Authorization System Standards," Ver. 0.30, ADAPSO, Association of Data Processing Service Organization, (1985).

[27] S. H. Weingart, Physical security for the ABYSS system, *IEEE Symp. on Security and Privacy, Proc.* 52–58 (April 1987).

[28] R. Mori and S. Tashiro, The concept of Software Service System (SSS), *IEICE Trans.*, J70-D (1) 70–81 (January 1987).

[29] S. Tashiro and R. Mori, The implementation of a small scale prototype for Software Service System (SSS), *IEICE Trans.*, J70-D(2) 335–345 (February 1987).

[30] T. Kimoto, K. Asakawa, M. Yoda, and M. Takeoka, Stock market prediction system with modular neural networks, *Internat. Joint Conference on Neural Networks Proc.*, I-1 San Diego, (June 1990).

[31] K. Kamijo and T. Tanigawa, Stock price pattern recognition: A recurrent neural network approach, *Internat. Joint Conference on Neural Networks Proc.*, I-215 San Diego, (June 1990).

[32] J. P. Bigus and K. Goolsbey, Integrating neural networks and knowledge-based systems in a commercial environment, *Proc. Internat Joint Conference on Neural Networks*, II-463, Washington, (January 1990).

[33] O. E. Martinez and C. Harston, Interfacing data base to find the best and alternative solutions to problems by obtaining the knowledge from the data base, *Internat. Joint Conference on Neural Networks Proc.*, II-663, Washington (January 1990).

[34] S. Nagata, M. Sekiguchi, and K. Asakawa, Mobil robot control by a structured hierarchical neural network, *IEEE Control Systems Magazine*, 69–76 (April 1990).

[35] J. K. Udupa and I. S. N. Murthy, Syntactic approach to ECG rhythm analysis, *IEEE Trans. Biomed. Eng.*, BME-27, 370–375 (July 1980).

[36] B. Souček and A. D. Carlson, Brain windows in firefly communication, *J. Theor. Biol.*, 119, 47–65 (1986).

[37] B. Souček and A. D. Carlson, Brain windows language of fireflies, *J. Theor. Biol.*, 125, 93–103 (1987).

[38] B. Filipič and I. Konvalinka, Signal-to-symbol transformation as a prerequisite for knowledge-based spectrum estimation, *Proc. Internat. Workshop, Parallel Problem Solving from Nature*, University of Dortmund, 1990.

[39] T. Winograd, "What lies ahead?" *Byte*, January 1989, p. 350.

[40] C. G. Langton, Ed., *Artificial Life*, Addison-Wesley, Reading, Mass., 1989.

[41] P. Cariani, "On the design of devices with emergent semantic functions," Doctoral dissertation, State University of New York at Binghamton, 1989.

[42] T. J. Schwartz, Japan looks at new information processing, *IEEE Expert*, 67–69, (Dec. 1990).

[43] B. Souček and IRIS Group, Dynamic, Genetic and Chaotic Programming, The Sixth Generation. Wiley, New York (in press).

CHAPTER 2

A Biologically Realistic and Efficient Neural Network Simulator

VAHE BEDIAN
JAMES F. LYNCH
FENGMAN ZHANG
AND
MICHAEL H. ROBERTS

2.1 INTRODUCTION

The ability of higher animals to produce sophisticated behaviors, such as appropriate responses to a variety of environmental stimuli, strengthening or weakening of responses due to multiple stimulation (sensitization and habituation), associative memory, and pattern recognition, have held a strong fascination for biologists and psychologists alike. In addition, the ability to manipulate symbolic and natural languages, and demonstration of intelligence and free will, at least in humans, has posed conceptual challenges for linguists, computer scientists, and philosophers. All of these behaviors are derived from functions of the nervous system, and the traditional view has held that they are mostly outcomes of the complex architecture of the nervous system, which is presumed to be made of binary, threshold elements. The large number of neurons and their interconnections are thought to be responsible for the ability to process complex inputs, associate and analyze different types of information, and lead to pattern recognition, learning, and creative thinking. A predominant role for network features has been shown

---

Neural and Intelligent Systems Integration, By Branko Souček and the IRIS Group.
ISBN 0-471-53676-8 ©1991 John Wiley & Sons, Inc.

in pattern recognition and the generation of rhythmic motor activity that control movement, respiration, and digestion [1–7].

Recent evidence from invertebrate and vertebrate experimental systems has shifted the spotlight from neural networks to single neurons. The complex structure of many neurons, including dendritic arborizations [8], the soma [9], and branched axons terminating in multiple synapses, leads to temporal and spatial integration of information converging on different regions of the neuron, and delivery of the neuron's response to a large number of physically separated targets. The prototypic neuronal response is a train of action potentials, the frequency of spikes increasing with excitatory stimulation. However, single neurons are also capable of producing alternative responses, such as bursting (a short train of action potentials when the stimulus is turned on) and self-sustained oscillations [10]. The large variety of physiological responses exhibited by different neurons appears to be related to the diversity of channel types and their distribution on the neuron. Many factors are now known to produce subtle modulations in neuronal properties, leading to changes in neuronal excitability that can have profound effects on behavior [11,12]. Interactions between ligand-receptor activated second messenger systems and ion channels are responsible for repetitive firing [13], several forms of learning [14], and synaptic plasticity [15]. One of the surprising conclusions of recent experiments on invertebrate and vertebrate systems is that the processes that lead to learning and memory appear to be based on modulation of electrophysiological properties of single neurons or synaptic connections. Yet these critical sites of neuromodulation reside in the framework of complex neural networks that define the basic behavior of the system. Thus, models for complex animal and human behavioral features should account for electrophysiological details and complexity of the single neuron as well as the connectivity and architecture of neural networks (see Selverston [16] for a comprehensive discussion of essential properties). With this goal in mind, we have constructed a biophysically realistic neuronal model that can account for many basic properties of single neurons, and, with the aid of a network editor [17], we have used such neurons to construct biologically relevant neural networks. This chapter describes the neuronal properties represented in the model, their mathematical formulation, and incorporation into networks, followed by simulation results of single neurons and neural networks. Since the single-neuron model does not account for modifiable neuronal properties necessary for learning, the applications emphasize rhythmic activity and complex behavior, but features of the model relevant to learning and memory are also presented.

2.2 BASIC NEURONAL PROPERTIES

The neuron, like all other cell types, has a semipermeable membrane that contains specific ion channels and transport systems and allows the cell to have an internal ionic environment distinct from that of the surrounding extracellular medium. The ubiquitous sodium-potassium pump uses biochemical energy to transport sodium ions out of the cell and potassium ions into the cell. This pump, along with a passive potassium leakage channel and a variety of other channels, accounts for the *resting voltage* of the neuron, where the chemical potential due to ion concentration differences across the membrane is balanced by the electrical potential difference between the outside and inside of the neuron. Generally, a neuron will have lower intracellular sodium, calcium, and chloride ion concentrations compared with the outside, but the intracellular potassium concentration will be higher. The resting voltage of a neuron is typically 80 to 100 mV lower than the outside, and is arbitrarily set to zero and used as a reference point in our model.

The lipid bilayer constituting the plasma membrane of the neuron has resistive and capacitive properties, since it allows selective flow of ions and can also store charges. These features, along with the corresponding properties of the neuronal cytoplasm, give the neuron a characteristic *RC time constant*. This time constant, which is modified by changes in the voltage and permeability of the neuron, determines the shape of the exponential approach to a new voltage due to ion currents flowing across the membrane.

The presence of voltage-activated ion channels in the membrane of a neuron make it an excitable cell. When the cell is depolarized (i.e., the potential difference across the membrane is decreased, corresponding to a positive internal potential in the model), voltage-gated channels open and allow the flow of ions across the membrane. When the neuron is depolarized beyond a critical voltage, known as its *threshold,* opening of voltage-gated sodium (and sometimes calcium) channels causes further depolarization of the cell, which causes more channels to open. This positive feedback loop results in a sudden increase in the neuron's voltage. Slower, voltage-gated potassium channels allow outward flow of potassium, quickly repolarizing the cell. These changes in the neuron's voltage are known as *spikes* or *action potentials*.

After being activated and opened, voltage-gated sodium channels go through an inactive state before returning to the active and closed state. As a result, the neuron has a *refractory period*. Immediately after the rising phase of the action potential, most of the voltage-gated sodium channels are inactivated, and the neuron is in its absolute refractory period during which no amount of depolarizing stimulation can generate an action potential. A relative refractory period follows, during which larger than normal stimulation is required to produce an action potential, corresponding to an increase in the threshold of the neuron. The threshold progressively returns to its normal value as a function of time from the action potential. During repeated firings of a neuron, a decrease in the spike height may also be observed. This decrease occurs because only a proportion of voltage-gated channels has returned to the active state and, possibly, because of metabolic depletion of the cell. This phenomenon is known as *accommodation*.

The flow of ions across the membrane depends not only on ion permeability and concentration gradient but also on the potential difference across the membrane. Each ion has a characteristic equilibrium potential, given by the Nernst equation, at which the chemical gradient is balanced by the electrical potential difference. For instance, the flow of sodium ions across the membrane is inward at the resting potential of the neuron, since the equilibrium potential of sodium is around 140 mV above rest. The sodium influx decreases as the voltage of the cell increases. At the equilibrium potential of sodium, the voltage difference counteracts the chemical gradient, and no flow of sodium ions will take place even when sodium channels are opened. This voltage is known as the *reversal potential* for sodium. If the voltage of the cell is above the reversal potential, an outward flow of sodium ions will be observed when the channels are opened. Thus, ion flows and voltage changes due to opening of gated channels depend on the present voltage of the neuron.

Neurons communicate with each other through *synapses,* which can be *chemical* or *electrotonic* in nature. Chemical synapses are mediated by the release of neurotransmitters, which control ligand-gated channels on the postsynaptic neuron. The effect of a neurotransmitter can be *excitatory* (causes depolarization of the neuron and brings it closer to its threshold) or *inhibitory* (causes hyperpolarization of the postsynaptic neuron). Electrotonic synapses are mediated by gap junctions that provide direct communication channels between two cells. They generally act bidirectionally (*nonrectifying*), with the current flow proportional to the potential difference between the two cells, although some are *rectifying* and allow current flow in one direction only.

The model presented here accounts for most of these basic neuronal properties, with some limitations. In some cases, the processes are represented by empirical or plausible functions rather than complete, mechanistic detail. Second, only sodium and potassium conductances are represented in the model, and other important ions, such as calcium and chloride, would need to be added for a more general model of a neuron. Last, there is no representation of modifiable channels or synapses in this version of the model. The mathematical formulation of the model and its applications are presented in the next section.

2.3 SINGLE-NEURON MODEL

The neural network model incorporates a single-neuron model that accounts for many of the observed biophysical properties of biological neurons. The voltage of a neuron is a time-dependent variable updated at each iteration. Other variables, such as the conductance and time constant of a neuron, depend on the voltage and are also updated at each iteration. The firing history of a neuron is used to determine the absolute refractory period, increase in threshold during the relative refractory period, and effects on firing rate and peak spike height due to accommodation. The model consists mostly of iterative integration of a set of differential equations representing the temporal dependence of these variables, using a 1-ms iteration step size. In addition, a discrete decision is made to fire a neuron when the voltage is above the present threshold. The variables and parameters used in the model and their mathematical relationships are described here.

The electrical state of a neuron is represented by its voltage, and the resting voltage is used as a reference point and arbitrarily set to zero. At every iteration, the voltage of a neuron decays passively toward rest and can be modified by external tonic inputs or inputs from other elements. The voltage of neuron i at time $t + \Delta t$ is given by

$$V_i(t + \Delta t) = V_i(t)e^{-g_i \Delta t/\tau_i} + E_i(1 - e^{-g_i \Delta t/\tau_i})/g_i \tag{2.1}$$

where $V_i(t)$ = deviation of the neuron's voltage from resting potential at time t (mv);
Δt = iteration step size (ms), set at 1 in this version of the model;
τ_i = resting RC time constant of the neuron (ms);
g_i = dimensionless factor representing the voltage-dependent conductance changes of the neuron;
E_i = total effect of external tonic inputs and connections from other neurons.

The first term represents passive decay toward rest. The passive decay is calculated in exponential form, to avoid approximation errors when the exponent $g_i \Delta t/\tau_i$ is not small. The term τ_i/g_i represents the time constant of the neuron at a given voltage. The model accounts for the two major voltage-dependent conductances, sodium and potassium. Since both of these conductances increase sigmoidally with voltage, have a 50 percent transition point at 45 mV above rest, and reach saturating values around 90 mV above rest, a single sigmoidal curve, representing the combined effect of both conductances, was used. Given the present voltage of the neuron, the voltage-dependent factor g_i, which has a minimum value of 1.0 at rest and a maximum value of 40.0, is looked up from a table.

The combined effect of external and neuronal inputs (E_i) is derived in the next section, in the context of the network model. It is multiplied by the factor $1/g_i$ to account for the fact that all of these inputs amount to current injections into the neuron, and their effect on voltage is inversely proportional to g_i. The effect of the time constant is represented by the factor $1 - e^{-g_i \Delta t/\tau_i}$, accounting for exponential approach of the neuron voltage toward the final value. When all changes to the voltage of a neuron are made, the voltage is compared with the neuron's threshold.

The threshold U_i of neuron i is given by

$$U_i = U_i^0 + H_i/\rho_i \tag{2.2}$$

where U_i^0 = basal threshold of the neuron;
H_i = factor representing the cumulative firing history of the neuron, and accounting for the increase in threshold during the relative refractory period of the neuron;
ρ_i = parameter that controls the maximum firing rate of the neuron.

The cumulative firing history factor H_i is set to zero at the beginning of a simulation, and successive values are calculated from the expression

$$H_i(t + \Delta t) = H_i(t)(1 - \beta_i \Delta t) + K_i \tag{2.3}$$

where K_i = zero, except if the neuron has fired on the previous iteration, in which case it is set to a positive value representing the contribution to accommodation due to a single action potential;
β_i = rate constant for recovery from accommodation.

Thus, the value of H_i is increased by K_i after a neuron has fired, and it decays back toward zero if no additional firings occur. Here, the linear approximation for exponential decay is used, since $\beta_i \Delta t$ is generally much smaller than 1.

If the voltage of a neuron is above threshold, the firing condition is satisfied and the neuron is set to a peak voltage P_i, given by

$$P_i = P_i^0 (\sigma_i)^{H_i} \tag{2.4}$$

where P_i^0 = peak spike height of an unaccommodated neuron, generally set to 100 mV in our simulations;
σ_i = dimensionless parameter (<1) for spike height lowering due to accommodation.

Each neuron is also characterized by a threshold W_i for opening calcium channels. This parameter, generally set at 10 mV above rest, is used in the neural network model (see the next section) to determine the duration of the action potential above the calcium channel threshold and, thus, the efficacy of chemical synapses made by the neuron. A listing of the FORTRAN program used for simulating the single-neuron model is presented in Appendix B.

2.4 NEURAL NETWORK MODEL

The model neurons described here can receive tonic external inputs and/or be connected in a neural network, constructed with the aid of the network editor described in Appendix A. A chemical synapse from neuron j to neuron i is characterized by a connection strength T_{ij} and can be excitatory (positive

T_{ij}) or inhibitory (negative T_{ij}). Electrical synapses are characterized by a connection strength L_{ij} and can be nonrectifying ($L_{ij} = L_{ji}$) or rectifying ($L_{ji} = 0$ if L_{ij} is nonzero). Both types of synapses are associated with a transmission delay D_{ij}, accounting for transmission along axons and/or polysynaptic connections. The total change in the voltage of neuron i due to these inputs is

$$E_i = I_i + \sum_j (A_{ij} + B_{ij} + C_{ij}) \tag{2.5}$$

where I_i = strength of tonic external input, positive or negative, to the neuron;
A_{ij}, B_{ij}, C_{ij} = voltage changes in neuron i due to chemical excitatory, chemical inhibitory, and electrotonic inputs, respectively, from neuron j.

The voltage change in neuron i due to an excitatory chemical synapse from neuron j is

$$A_{ij} = T_{ij} (1 - 0.01 V_i(t))(V_j(t - D_{ij}) - W_j)e^{-\Delta t/\tau_j} \tag{2.6}$$

where T_{ij} = dimensionless parameter representing the connection strength from neuron j to neuron i;
D_{ij} = transmission delay (ms) from neuron j to neuron i;
W_j = threshold for opening calcium channels in neuron j (mV).

The term $1 - 0.01 V_i(t)$ accounts for the reversal potential of neuron i. Assuming that excitatory chemical synapses are mediated through sodium channels, and the reversal potential for Na^+ is 100 mV above rest, this term gives linear decrease in efficacy of an excitatory synapse with voltage of the postsynaptic neuron. The effect of such a synapse would be zero when the neuron is at 100 mV, and negative if the neuron is above 100 mV. The remaining two factors in Eq. (2.6) account for the effect of spike width on synaptic efficacy. The duration of an action potential above the calcium threshold of the synaptic region would increase with spike height above the calcium threshold ($V_j(t - D_{ij}) - W_j$), as well as with the time constant of the presynaptic neuron ($e^{-\Delta t/\tau_j}$).

A similar expression is used to describe the effect of inhibitory chemical synapses, except the assumption here is that potassium channels of the postsynaptic neuron are involved, and the K^+ reversal potential is 10 mV below rest:

$$B_{ij} = T_{ij} (1 + 0.1 V_i(t))(V_j(t - D_{ij}) - W_j)e^{-\Delta t/\tau_j} \tag{2.7}$$

Here, the connection strength T_{ij} is negative. The voltage change of the postsynaptic cell increases as the neuron is depolarized, decreases linearly

44 A BIOLOGICALLY REALISTIC AND EFFICIENT NEURAL NETWORK SIMULATOR

toward zero as it is hyperpolarized from rest to -10 mV, and is positive when it is below -10 mV.

The effect of electrotonic synapses is assumed to be proportional to the difference in voltage between the presynaptic and postsynaptic cells and is

$$C_{ij} = L_{ij}(V_j(t - D_{ij}) - V_i(t)) \qquad (2.8)$$

where L_{ij} = electrotonic connection strength (dimensionless) from neuron j to neuron i.

2.5 SINGLE-NEURON SIMULATION RESULTS

The model neuron described produces the expected behaviors, such as hyperpolarization due to negative tonic input, subthreshold depolarization due

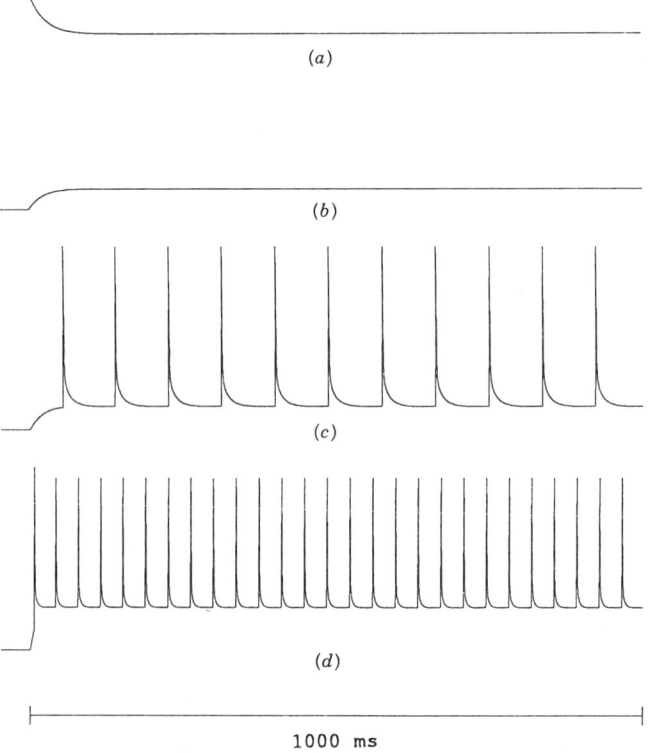

Figure 2.1 Effect of tonic input on neuronal response. (a) Inhibitory tonic input hyperpolarizes the neuron. (b) Weak excitatory input depolarizes the neuron to subthreshold levels and does not result in action potentials. (c) Stronger excitatory input produces a train of action potentials at around 10 Hz. (d) Maximal stimulation of the neuron increases the responses to around 30 Hz, and accommodation of the spike height is observed.

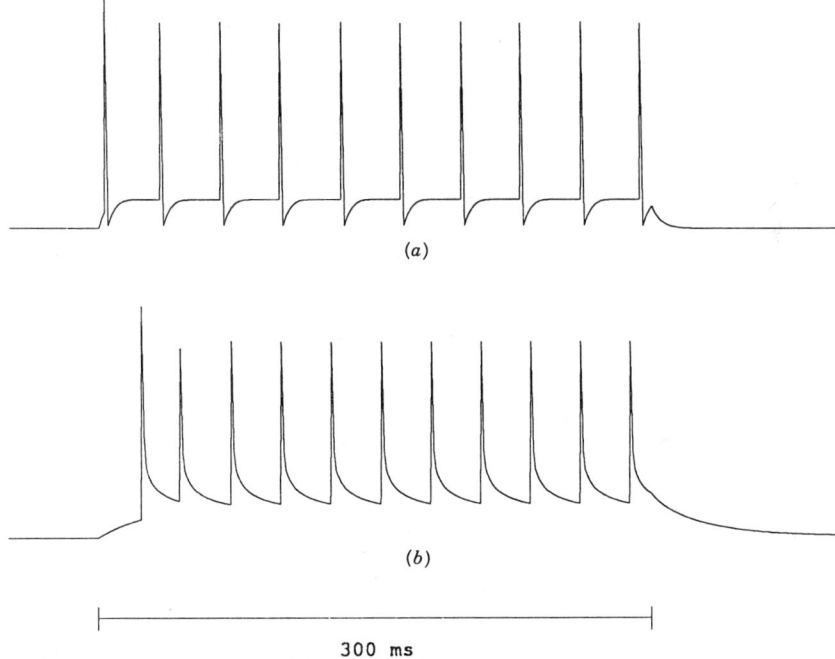

Figure 2.2 Effect of time constant on neuronal response. (a) A model neuron with a 5-ms time constant produces sharp action potentials. (b) A neuron with 30-ms time constant has broader action potentials and a slight increase in the frequency of action potentials, due to persistent depolarization, which make it more excitable.

to small positive inputs, and tonic firing due to stronger inputs (Fig. 2.1). As seen in the figure, firing rate increases with increasing input, and increasing effects of accommodation can be observed. Spike broadening due to increasing time constant of the neuron is demonstrated in Figure 2.2. Under particular choices of parameters, such as low accommodation rates, the model also produces bursting behavior and oscillation due to tonic stimulation (Fig. 2.3). The bursting behavior appears when recovery from accommodation is very slow. Under these conditions the neuron is capable of producing a short train of action potentials, the increased threshold interrupts the firing, and a relatively long time is required for the neuron to return to its resting state. Oscillatory behavior is observed when the rate of recovery from accommodation is moderate, enabling the neuron to produce successive bursts of action potential. Since biological oscillatory neurons generally utilize calcium spikes in dendritic regions and calcium channels are not explicitly represented in our model, the significance of this oscillatory behavior is not clear. However, we have used our model neurons to represent spatially distinct regions of a complex neuron, and obtained oscillations between the dendritic tree and soma (see the next section).

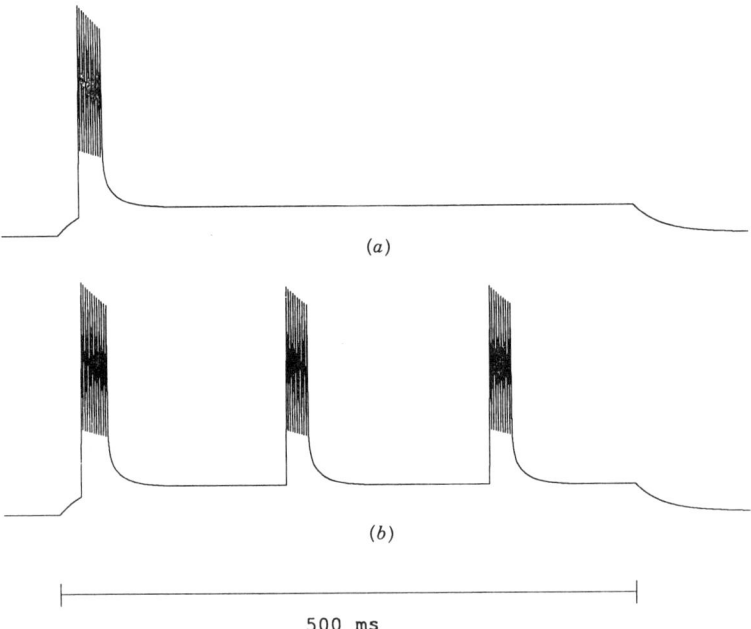

Figure 2.3 Alternative responses observed in a model neuron with low accommodation rate. (a) Bursting ("on" response) is observed when the recovery from accommodation is also slow. (b) Oscillatory response is observed when the rate of recovery from accommodation is moderate.

2.6 SIMULATION OF AN ANATOMICALLY COMPLEX NEURON

The neuronal model presented here is a "point" model; that is, the neuron is assumed to be a spatially homogeneous, localized unit, with the interneuron transmission delay representing physical distance between the soma and synaptic terminals. However, most neurons are anatomically complex, with many distinct regions, such as dendrites, soma, axon, and synapses. Such neurons can easily be modeled within our formulation by simply reinterpreting our basic "neuron" as a node representing a spatially homogeneous region of a complex neuron. An appropriate network of such nodes, electrotonically connected, can then represent a complex neuron. An example of such a model is shown in Figure 2.4. Nodes 1 to 3, representing terminal branches of the dendritic tree, as well as node 6, representing the soma and axon, are excitable, whereas nodes 4 and 5 represent intermediate dendritic branches with high threshold (not easily excitable, passive transmission). Such a neuron can integrate inputs in different regions (Fig. 2.5a) and be controlled by synaptic connection along the axon (Fig. 2.5b). In addition, this complex neuron can exhibit oscillatory behavior, with the terminal dendritic branches and the soma out of phase with each other (Fig. 2.6).

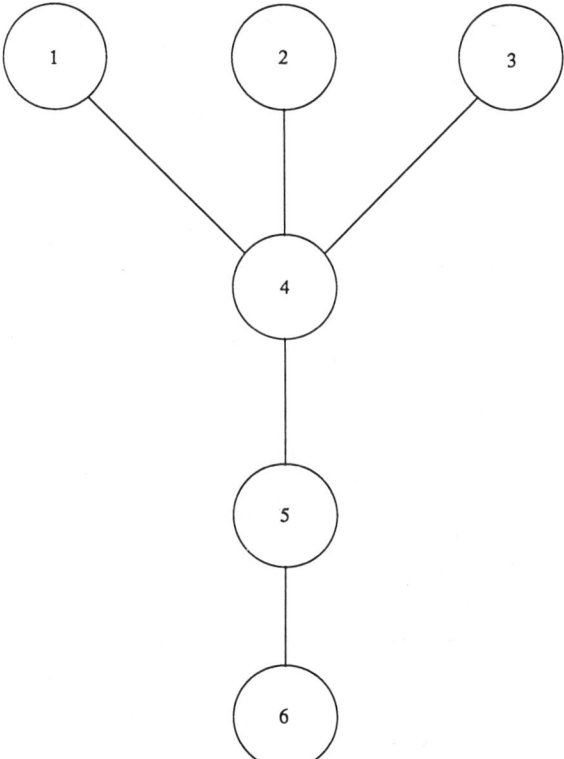

Figure 2.4 Model of an anatomically complex neuron. Nodes 1, 2, and 3 represent distal branches of the dendritic tree and are excitable; nodes 4 and 5 represent high threshold, nonexcitable intermediate dendritic regions; and node 6 represents the soma. The lines connecting the nodes are nonrectifying electrotonic connections.

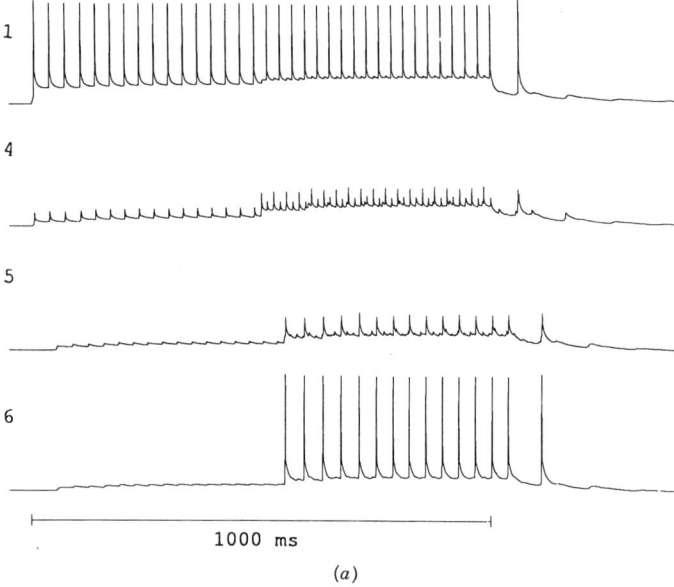

Figure 2.5a *Responses of the complex neuron model. During the first 500 ms only node 1 is stimulated, producing action potentials which are transmitted passively through 4 and 5, but are not capable of raising the soma above its threshold. During the second half of the trace shown, all three terminal nodes (1, 2, and 3) are stimulated, and the integrated response produces action potentials in the soma.*

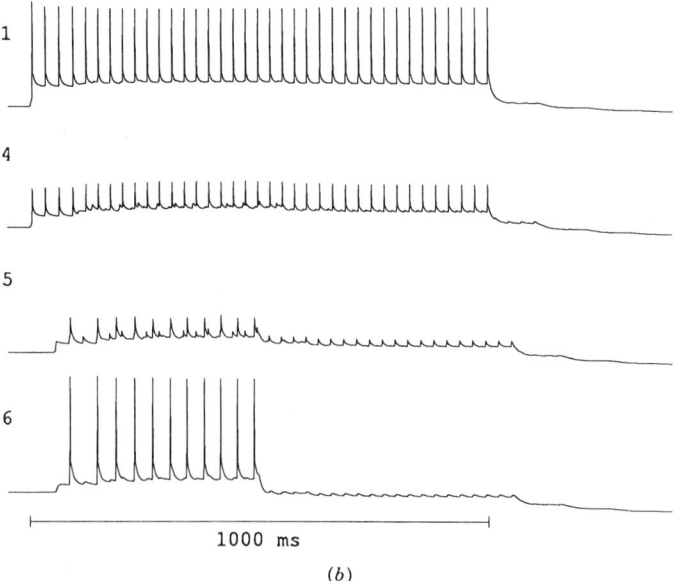

Figure 2.5b *Responses of the anatomically complex neuron model. In the first half of the simulation, inputs at the terminal nodes (1, 2, and 3) produce action potentials in the soma (6), but an inhibitory signal arriving at the soma during the second half of the experiment overrides the excitatory inputs at the dendrites.*

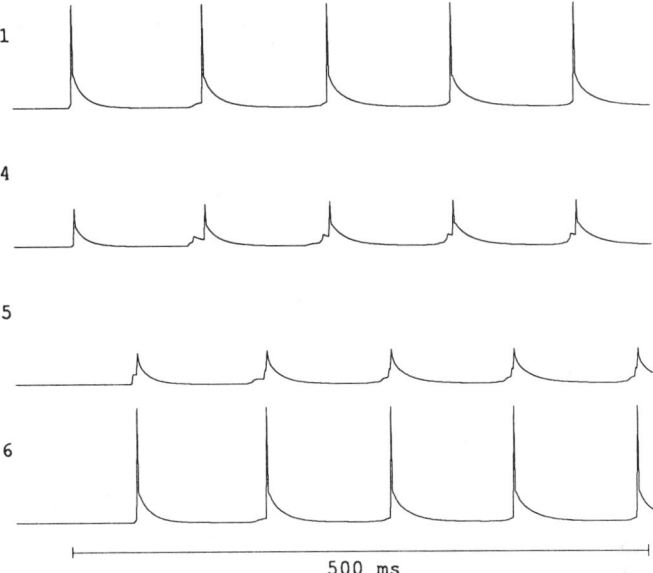

Figure 2.6 *Oscillatory response in a complex neuron: A short (5-ms) stimulus to the dendritic branches initiates a self-sustaining wave of excitation, which travels from the terminal dendritic branches to the soma and back.*

2.7 NEURAL NETWORK SIMULATIONS

The effect of reversal potentials is best seen in two neuron models, where the first element provides excitatory or inhibitory inputs, and the second neuron is given a high threshold (to observe postsynaptic potentials (PSPs) without the complication of firing) and maintained at a voltage away from rest by an appropriate tonic input. Figure 2.7 shows the dependence of PSPs in neuron 2 due to excitatory inputs from neuron 1 as the voltage of neuron 2 is increased. As expected, the PSPs approach zero as the voltage of neuron 2 approaches 100 mV (Na^+ reversal potential). The corresponding results for an inhibitory synapse are seen in Figure 2.8.

Another example of a two-neuron network is given by the negative feedback delay oscillator (Fig. 2.9a). The first neuron has a positive tonic input and stimulatory connections with long delay to the second, which in turn inhibits the first neuron. This gives rise to oscillations with a period twice the total delay in the loop (Fig. 2.9b). A bistable flip-flop circuit can also be constructed with two neurons (Fig. 2.10a). This is useful as a model of neural circuits that can generate two mutually exclusive behaviors. Figure 2.10b shows that the ratio of tonic inputs to the two neurons determines which neuron will fire, and the network switches sharply from one stable state to the other. Both the feedback oscillator and the flip-flop are part of

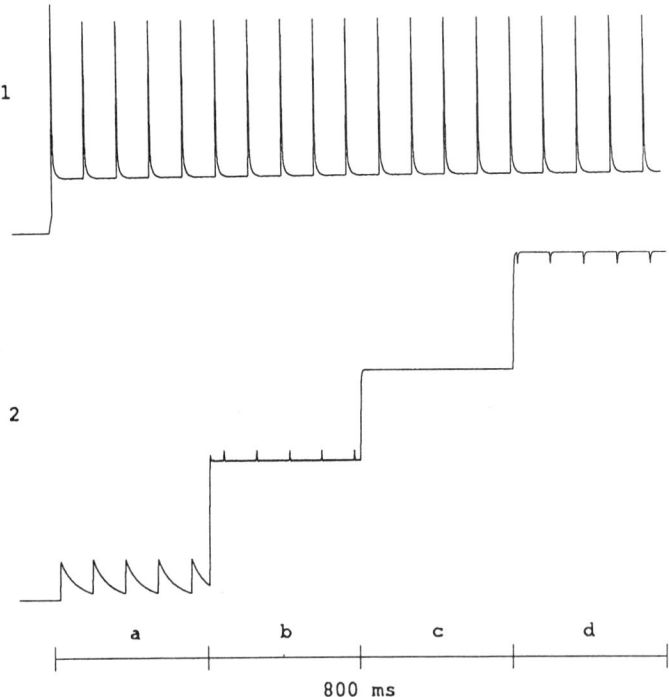

Figure 2.7 Dependence of excitatory postsynaptic potentials (EPSPs) on the voltage of neuron 2, which is receiving action potentials from neuron 1 and has been set at a high threshold to eliminate complications from action potentials. Strong, positive EPSPs are observed when the neuron is clamped at its resting potential (a); but the PSPs decrease in size when the neuron is depolarized (b), since it is closer to the reversal potential of sodium. No PSPs are produced when the neuron is at the sodium reversal potential (c), and downward PSPs are produced when it is above the reversal potential (d).

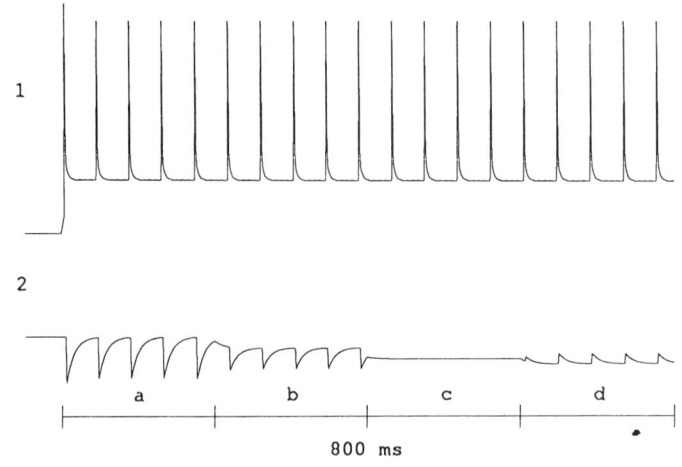

Figure 2.8 Dependence of inhibitory PSPs on the voltage of the postsynaptic neuron (2). The IPSPs decrease in size, become zero, and reverse polarity as the voltage of neuron 2 approaches the potassium reversal potential (b and c) and goes below it (d).

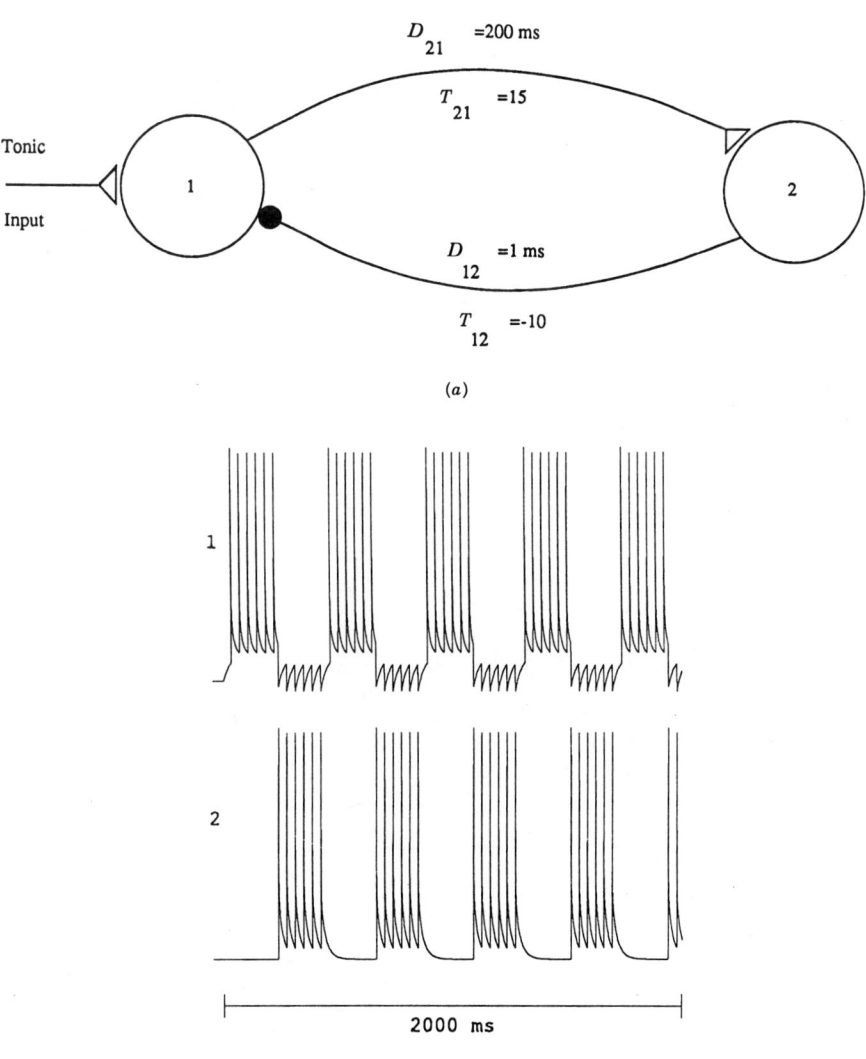

Figure 2.9 (a) Two-neuron, negative feedback with delay oscillator model. Open triangle represents an excitatory synapse, while closed circle represents an inhibitory synapse. (b) Responses of the two neurons when a tonic input is applied to 1.

52 A BIOLOGICALLY REALISTIC AND EFFICIENT NEURAL NETWORK SIMULATOR

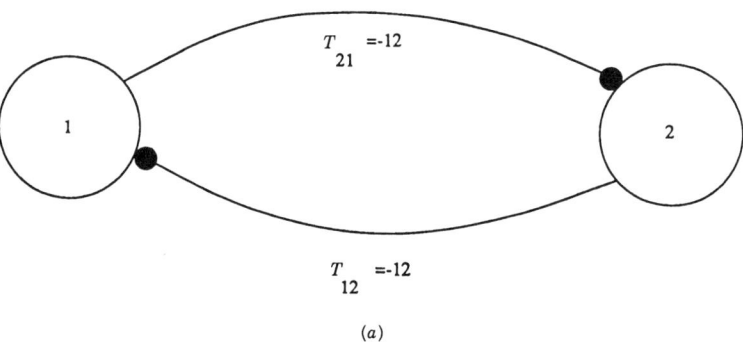

(a)

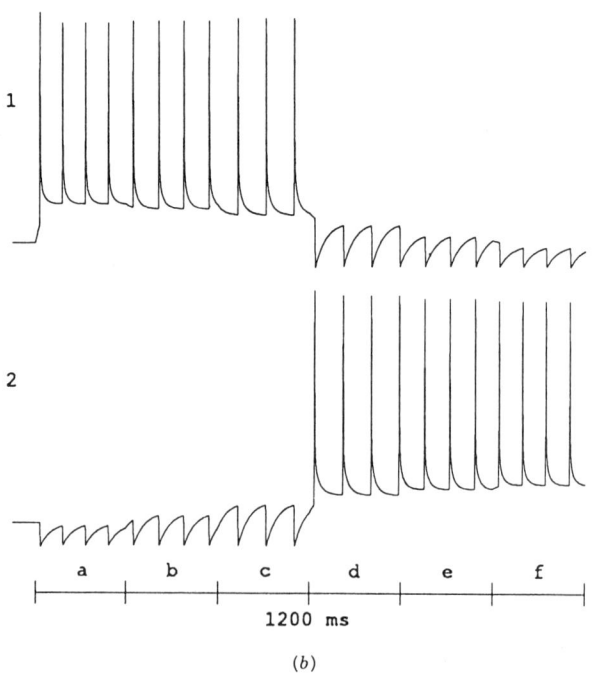

(b)

Figure 2.10 (a) Mutually inhibitory bistable flip-flop network. (b) Output of neurons 1 and 2 when the ratio of their inputs is varied. Ratio of inputs (neuron 1:neuron 2) is (a) 10 : 0; (b) 8 : 2; (c) 6 : 4; (d) 4 : 6; (e) 2 : 8; (f) 0 : 10.

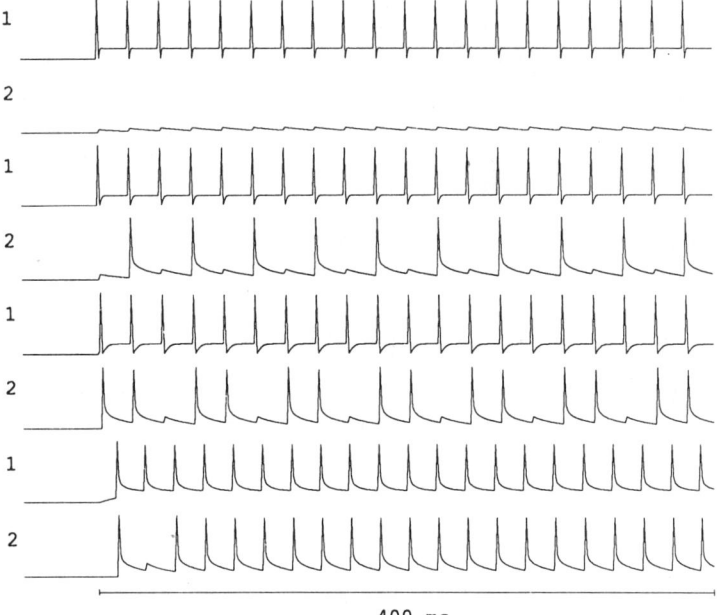

Figure 2.11 *Effect of action potential broadening on efficacy of an excitatory synapse from neuron 1 to 2: Four pairs of traces are shown, with the time constant and spike width of neuron 1 increasing from top to bottom. In the top trace, the EPSPs in neuron 2 are subthreshold, but as the time constant of neuron 1 is increased, progressively higher frequencies of action potentials are produced in neuron 2.*

the more complex circuit that controls the feeding behavior of the marine gastropod *Pleurobranchaea*, presented in Section 2.8.

Another important aspect of the model can be demonstrated with a two-neuron network. We have incorporated the length of time the spike in the presynaptic neuron spends above the calcium threshold as a factor determining the efficacy of synapses from this neuron. This factor is the basis of learning and memory in *Aplysia*, where modification of potassium conductances result in presynaptic spike broadening and an increase in the firing rate of the postsynaptic neuron. Figure 2.11 shows that, as the time constant of neuron 1 is increased (while keeping its firing rate constant), the firing rate of neuron 2 increases due to the spike broadening. Another form of learning is observed in *Hermissenda*, where prolonged afterpotentials due to modification of potassium channels make the postsynaptic neuron more excitable and account for the associative memory of the animal. This situation is simulated by increasing the time constant of the postsynaptic neuron, producing higher firing rates in response to constant input from the presynaptic neuron (Fig. 2.12). Increase in efficacy of synapses, which would be represented by the T_{ij} parameters in our model, appears to account for another

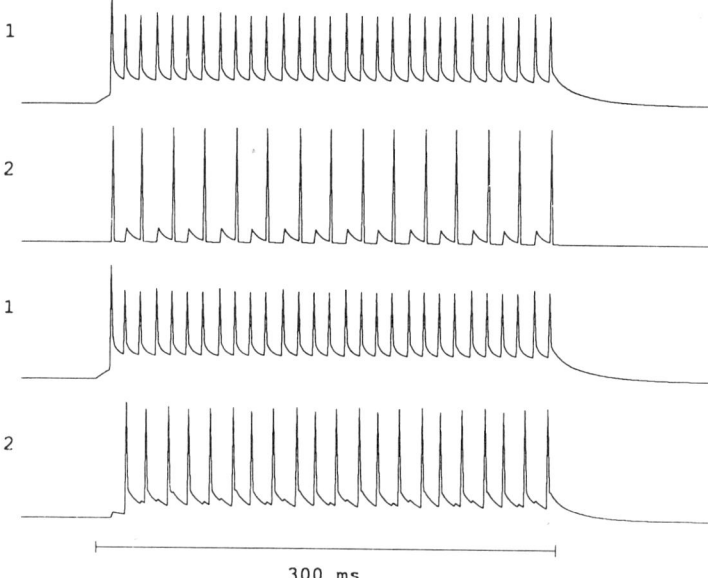

Figure 2.12 Effect of time constant of postsynaptic neuron (2) on the efficacy of an excitatory synapse from neuron 1. In the top pair of traces, neuron 2 has a short time constant and drops down to its resting potential soon after an action potential. In the bottom trace, the time constant of neuron 2 is longer, resulting in a long-lasting depolarization that effectively makes the neuron more excitable.

form of learning, potentiation in the vertebrate hippocampus. This could be represented by modifying the values of the T_{ij} parameters as a function of the firing history of the neurons. Thus, although we have not incorporated the detailed mechanisms responsible for different forms of learning in our model, the basic processes responsible for learning and memory can be accounted for by the model.

2.8 APPLICATION TO ANIMAL BEHAVIOR

To test our model in the context of a neural network involved in the control of more complex behavior, we chose to simulate the rhythmic feeding activity observed in *Pleurobranchaea*. This carnivorous mollusk will initiate a rhythmic protraction and retraction of its proboscis when presented with food stimuli, culminating in bites and swallowing of the food. Apart from the fact that this feeding behavior obviously requires an oscillator, there are several additional features of the behavior that are of interest and present a challenge to the model.

2.8 APPLICATION TO ANIMAL BEHAVIOR   55

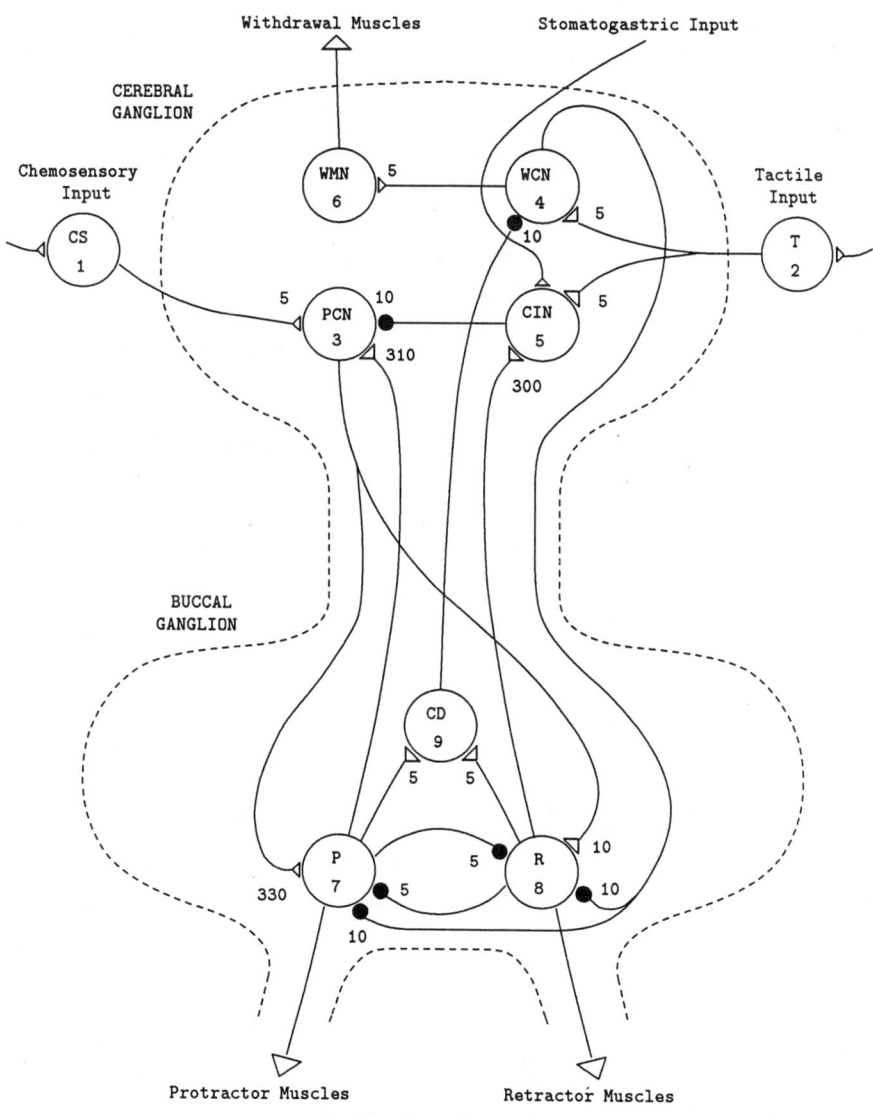

Figure 2.13 *Simplified neural network for simulation of* Pleurobranchaea's *behavior. Elements in the network do not correspond to single neurons but to groups of functionally equivalent neurons distributed in the cerebral and buccal ganglia of this carnivorous mollusk. Many of the connections have long associated delays (shown adjacent to each connection), representing long axonal pathways and/or polysynaptic connections. See text for a description of the network.*

1. The rhythmic behavior, once initiated, can persist even if the food stimulus is taken away.
2. The feeding behavior is harder to initiate or is suppressed completely when the animal is satiated.
3. The animal exhibits choice behavior between feeding and withdrawal due to a noxious stimulus. When both types of stimuli are present, it chooses one or the other, depending on the stimulus strengths.

The neural circuitry responsible for these behaviors has been analyzed in some detail [18], and the essential features are shown in the network of Figure 2.13. The neural elements responsible for the animal's behavior are distributed between the cerebral and buccal ganglia. Individual neurons in the model represent groups of functionally equivalent neurons in the animal, and many of the connections in the network have long delays, indicating that they are polysynaptic. Neurons 1 and 2 represent chemosensory (CS) and

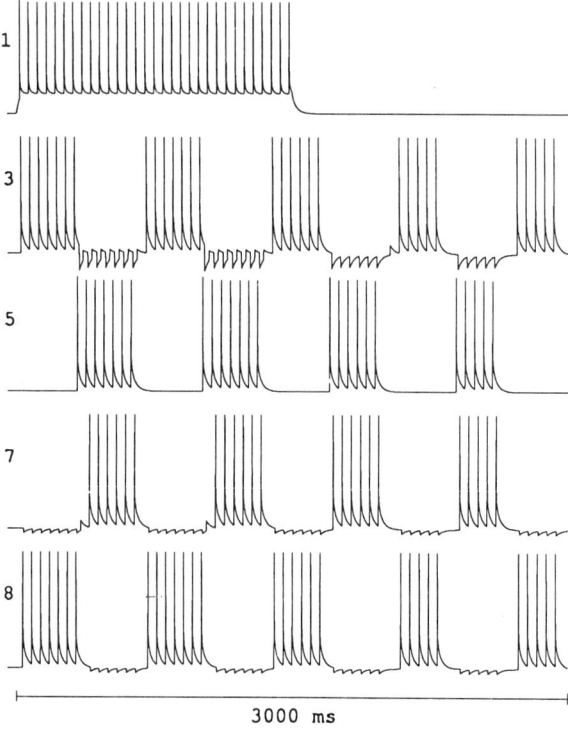

Figure 2.14 *Rhythmic feeding behavior (output patterns of elements 1, 3, 5, 7, and 8 are shown) produced when food stimulus is presented to neuron 1. The protractor and retractor neurons fire in alternating trains of action potentials, and the feeding activity, once initiated, persists even when food stimulus is removed (after the first 1500 ms in the simulation), corresponding to behavior observed in the animal.*

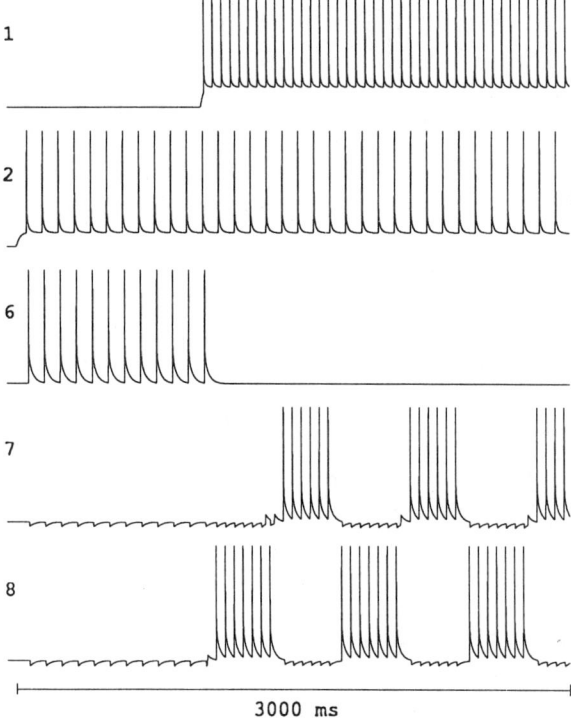

Figure 2.15 Choice behavior by Pleurobranchaea. Withdrawal response due to mild tactile stimulation is suppressed by food stimulation. A constant tactile stimulus is presented throughout the duration of the simulation, but an additional food stimulus is presented after 1000 ms have elapsed. Withdrawal response is apparent by the activity of element 6 in the first 1000 ms of the stimulation, but this activity is suppressed and feeding rhythm of elements 7 and 8 is observed in the remainder.

tactile (T) receptors, respectively. Neuron 3 represents the group of paracerebral command neurons (PCN) that control the rhythmic feeding activity. Neuron 5 represents the central inhibitory network (CIN), which receives inputs from the tactile sensory neuron 2 and the stomatogastric pathways that signal satiation. Tactile inputs are also transmitted to the withdrawal command neurons (neuron 4, WCN), which transmit their outputs to the withdrawal motor neurons (neuron 6, WMN). The PCN (neuron 3) provides excitatory inputs to the protractor and retractor motor neurons (P and R, neurons 7 and 8, respectively) in the buccal ganglion. Neurons P and R are mutually inhibitory and act as a flip-flop circuit, thus providing the feeding motor output. The corollary discharge neuron (neuron 9, CD) in the buccal ganglion integrates the overall feeding activity and provides inhibitory inputs to WCN in the cerebral ganglion. The loop formed by PCN, R, and CIN constitutes a negative feedback oscillator with delay and is primarily responsible for rhythmic motor output.

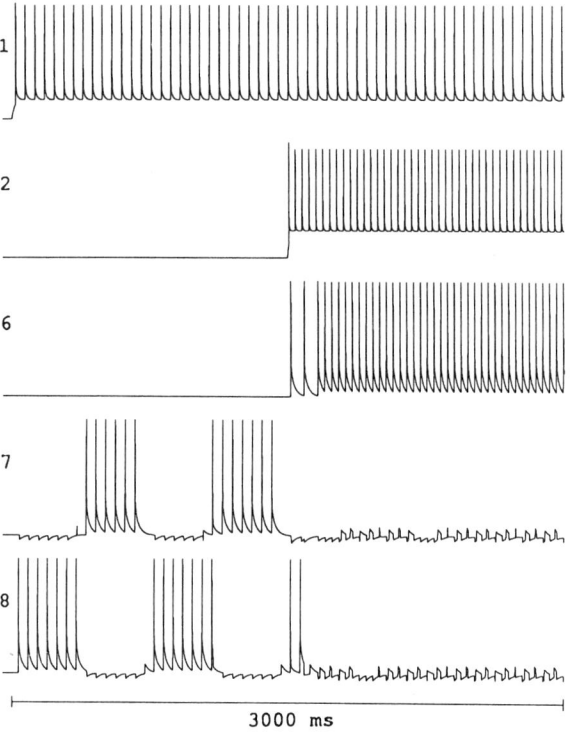

Figure 2.16 Choice behavior by Pleurobranchaea. Feeding activity is suppressed by strong tactile stimulation. Food stimulus is presented during all of the simulation, and a strong tactile stimulus, which suppresses the feeding rhythm of 7 and 8, and produces withdrawal response in 6, is presented during the second half.

With chemosensory (food) stimulation, the network produces rhythmic firing in PCN (neuron 3), as well as alternate firing of protractor and retractor motor neurons (7 and 8), corresponding to the normal feeding behavior of the animal. The rhythmic firing persists even when the food stimulus is taken away, as observed in the animal. These results are demonstrated in Figure 2.14. If a mild tactile stimulus, which normally produces withdrawal response, is presented along with food, feeding behavior continues, demonstrating the fact that *Pleurobranchaea* chooses to continue feeding in spite of the tactile stimulus (Fig. 2.15). However, stronger tactile stimulation does suppress the feeding behavior and results in a withdrawal response (Fig. 2.16). Furthermore, a satiated animal receiving stomatogastric input requires stronger than normal chemosensory stimulation to initiate the rhythmic feeding behavior (results not shown).

These results demonstrate that the neural network model, wired according to experimentally determined pathways in the animal's nervous system, is capable of producing multiple behavioral patterns, including self-sustaining rhythmic firing of the feeding motor network, modulation of this behavior due to satiation, and choice between feeding and escape behaviors. In contrast to the phenomena of learning and associative memory in mollusks, which depend more strongly on modulation of single-neuron properties, *Pleurobranchaea's* behavioral complexity appears to be derived mostly from the connectivity of the nervous system. Furthermore, a single neural network is capable of producing multiple behavioral patterns, integration of diverse stimuli, and choice of appropriate behavior when antagonistic stimuli are presented. The fact that a single network and a single choice of neuronal parameters can account for the animal's behavioral complexity supports the validity and usefulness of our model.

2.9 DISCUSSION

The model presented here was motivated by two considerations: first, recent experimental evidence indicates that the diversity of neuronal characteristics and their modifiable nature contributes to many interesting features of nervous systems; second, our ability to process, integrate, and respond to multiple stimuli is based on the complex connectivity of neural networks in the nervous system. Since the experimental analysis of the neural basis of human and animal behavior is making impressive progress, the need for realistic models that can account for relevant properties of neurons and networks is becoming more acute. Such models can serve two important functions: They can help identify relevant variables and the hierarchy of organization in systems of enormous complexity; and they can have predictive value in suggesting experimentally verifiable relationships and results.

From the pioneering works of Hodgkin and Huxley [19] and McCulloch and Pitts [20], both the fundamental electrophysiology of the neuron and networks of binary threshold elements have been modeled extensively (see MacGregor [21] for a review). We have chosen to take the middle ground between detailed biophysical description of single neurons and large-scale connectivity of neural networks. With semiempirical formulations, the single neuron model accounts for important neuronal features such as absolute and relative refractory periods and accommodation, which modulate the neuron's response on the basis of its firing history, and conductance changes and reversal potentials, which describe the dependence of neuronal responses on the present electrical state. Our model neuron produces the prototypic response of a train of action potentials, as well as the less common responses of bursting and oscillation observed in some neurons. Furthermore, synaptic efficacy is not a fixed parameter but is influenced by the width of presynaptic action potential and postsynaptic time constant, features that appear to be the mechanisms of learning and memory. Although we have generally used our neuromime as a point model of a neuron, anatomically complex neurons can easily be modeled by using a set of electrotonically connected nodes. While the neuromime we have constructed has been successful in many respects, it is limited by the fact that only sodium and potassium conductances are represented, and explicit rules of neuronal modifiability have not been incorporated in the model. A more realistic and versatile model, which we hope to develop from the current one, should provide for a variety of ion conductances as well as second messengers. A specific neuronal type would then be defined by prescribing the particular composition of channels on the membrane and the signal transduction mechanisms that regulate and modulate its responses.

The neuromime presented is simple enough to allow construction of biologically relevant networks. As we have demonstrated with the simplified network of *Pleurobranchaea*'s nervous system, such a network is capable of reproducing individual behavioral features, integration of multiple stimuli, and choice between alternative behaviors. Simulation of networks of 100 or more elements for tens of seconds of real time can be achieved on a workstation or personal computer. These features should make the model a useful tool in analyzing the nervous system and behavior of invertebrates and vertebrates, and lead to realistic models of learning and memory in mollusks and the vertebrate hippocampus.

APPENDIX 2A: SYSTEM OVERVIEW

The neural net editor and simulator consists of three major levels. The top, or outermost, level is the editor. The user interacts with this part of the system. Essentially, it executes commands that create, modify, and save data files that describe neural networks and experiments performed on them.

The middle level of the system is the network simulator. It may be regarded as a language interpreter, where the network files are programs written in that language. When invoked, the simulator will read a specified network file, perform the simulation described by the file, and generate various kinds of output as requested by the user. Built into the network simulator is the core of the system—a biophysically realistic neuromime capable of representing major electrical neuronal properties, such as threshold, time constant, absolute and relative refractory periods, and accommodation. The simulator was used to model the rhythmic feeding and choice behavior observed in *Pleurobranchaea*. Preliminary results were reported in Bedian et al. [22]. This nine-element network required 2 min of CPU time per 1000 iterations (1 s of real time) on a SUN 3/50 workstation. The source code of the simulator is listed in Appendix 2B.

The version of the system used in the initial investigation of *Pleurobranchaea* did not have the network editor. A text editor was used to make modifications to the network. Nevertheless, even with a table-driven simulator and an interactive text editor, the procedure was tedious. Also, there was no systematic way to do sensitivity analysis by performing a search through

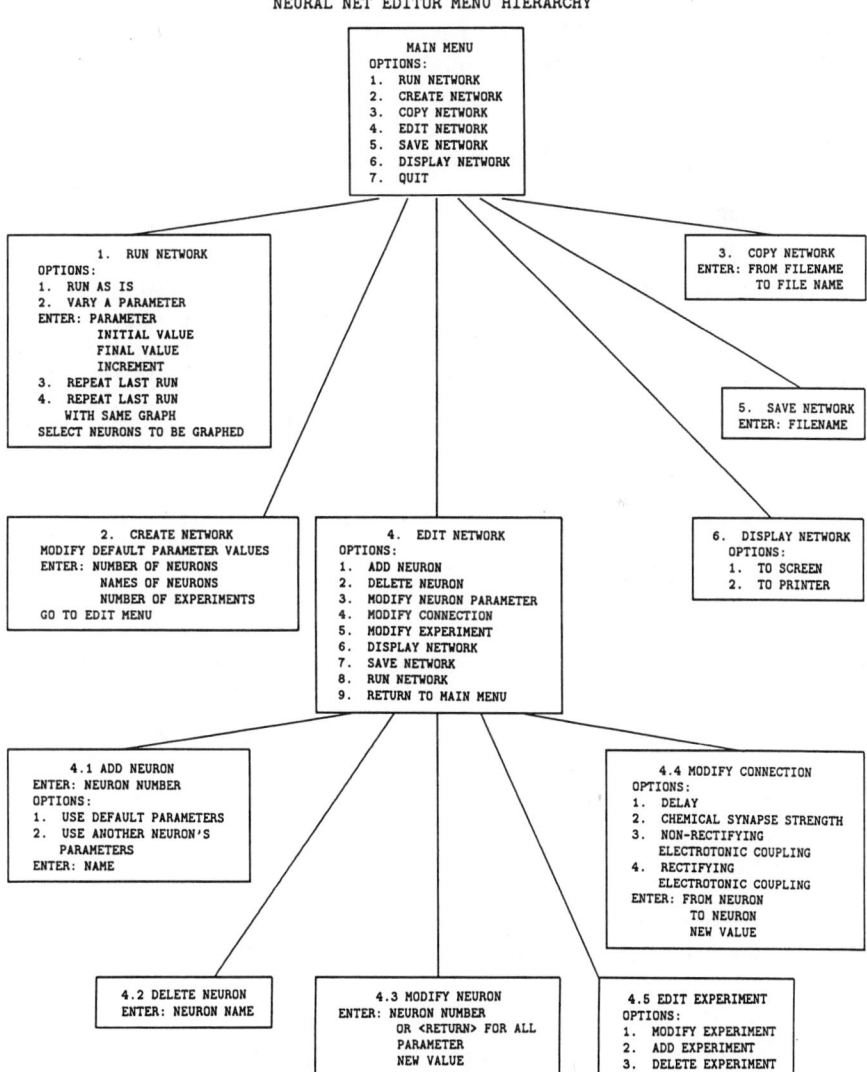

Figure 2.17 Flow chart/block diagram of network editor and its menu screens.

the parameter space. Thus, considerable effort was devoted to the design and implementation of a special-purpose editor that allows rapid modification and testing of networks.

The editor is a hierarchical, menu-driven system (see Fig. 2.17). At every step, the user responds to a prompt from the editor. All the responses are brief: single keystrokes followed by a RETURN for menu options and yes/no queries, or short character strings for neuron numbers and names, experiment labels, and file names. Most of the prompts are self-explanatory, but there is an extensive HELP function available. In a number of situations, the user will want to issue a command repeatedly, for instance, the ADD NEURON command. The editor uses the convention that such a command is repeated until the user enters a null line, that is, a line with RETURN as the first character. Then the editor returns to the menu that invoked the command.

Each user response activates a new menu or performs some operation on a network file. The logical structure of a network file is similar to the structure of the files manipulated by combinatorial graph editing systems such as CABRI [23] and TYGES [24]. The two main categories of data in those systems are node parameters and edge parameters. In our system, the analogous categories are neuron parameters and connection parameters. The neuron parameters are

Threshold voltage,
Peak spike voltage,
Recovery from accommodation,
Time constant,
Accommodation rate,
Threshold for opening calcium channels,
Tonic input.

The connection parameters are

Synaptic delay,
Chemical synapses strength,
Nonrectifying electrotonic coupling,
Rectifying electrotonic coupling.

The network file also contains a description of the experiments to be run. Each experiment applies input stimuli to selected neurons for a specified length of time.

All neuron parameters have default values, which are saved in a file separate from the network files. All connection parameters have default value 0. When the CREATE NETWORK command is invoked, the user has the option of overriding any of the neuron parameter default values. The changes may then be saved by writing them to the default file, or they may take effect only for the duration of the command. The user then specifies the number of neurons in the network, their names, and the number of experiments. The new network consists of homogeneous collection of neurons, all with the current default values and no connections. The EDIT menu is then entered to allow the user to change neuron parameters and add connections via the MODIFY NEURON and MODIFY CONNECTION commands. Alternatively, the user may create a network with zero neurons, and then repeatedly use the ADD NEURON command to generate the neurons. ADD NEURON also allows the user to copy parameters from the default file or from some existing neuron.

When the user is satisfied with the network, the simulator may be invoked by the RUN SIMULATOR command. Various options are allowed. A sequence of experiments may be run, where the input stimuli to the neurons are varied. Systematic search of parameter space is possible by specifying a parameter to be varied, its initial and final values, and its increment value. With this option, the simulator runs repeatedly, varying the selected parameter after each run. Output choices are also given. Different neurons can be selected for plotting their voltage levels as a function of time.

The effort expended on development of the editor has amply repaid itself. The simulations described in this chapter were all done with the aid of the editor. In many cases, a very large number of parameter variations had to be tested before the final results were found. The use of the editor speeded this up greatly. Further use of the system is anticipated. Modeling networks of other reasonably well-understood organisms is an obvious possibility, as is modeling learning in such organisms. This would require relatively minor additions to the editor and simulator.

APPENDIX 2B: SIMULATOR SOURCE CODE

```
C
C       Variable Definitions:
C       ------------------------
C       ca(i)       = Threshold for opening calcium channels.
C                     Controls effectiveness of a synapse.
C       counter(i)  = Counter for total number of firings of neuron i.
C       d(i,j)      = Delay in synaptic transmission from neuron j to neuron i,
C                     in milliseconds.
C       deltat      = Iteration step size, one millisecond.
C       delv(i)     = Change in voltage of neuron i due to synapses and outside
C                     inputs.
C       ffrv(i)     = Fractional approach of neuron i towards a new voltage
C                     due to ion current fluxes, corrected for conductance:
C                     ffrv(i) = (1-frv(i))/gv(index)
C       fire(i)     = Counter for firing history of neuron i; increment up by
C                     s(i,1) and decay by (1-rec(i)*deltat):
C                     fire(i) = (fire(i) + s(i,1))*(1-rec(i)*deltat)
C       fr(i)       = Fractional retention of voltage of neuron i:
C                     fr(i) = exp(-deltat/tau(i))
C       frv(i)      = Fractional retention of neuron i corrected for change
C                     in conductance: frv(i) = fr(i)**(gv(index/1.6)
C       ftijn(i)    = Factor controlling effectiveness of negative chemical
C                     synapses.
```

```
C       ftijp(i)    = Factor controlling effectiveness of positive chemical
C                     synapses.
C       ftion(i)    = Factor controlling effectiveness of negative tonic input.
C       ftiop(i)    = Factor controlling effectiveness of positive tonic input.
C       gv(index)   = Conductance as a function of voltage.
C       index       = Nearest integer rounding of v(i,1)
C       peak(i)     = Factor controlling drop in spike height of neuron i, due
C                     to accomodation; between 0 and 1:
C                     v(i) = vmax(i)*(peak(i)**fire(i))
C       rate(i)     = Factor used in ufact(i) to control maximum firing rate of
C                     neuron i. High rate(i) ---> high firing rate.
C       rec(i)      = Recovery rate of neuron i from accomodation; between 0 and 1;
C                     contributes to firing rate and spike peak.
C       s(i,j)      = State of neuron i (j-1) iterations ago. Present state =
C                     s(i,1); s(i,1) = 0 if not fired; s(i,1) = 1 if fired.
C       t(i,j)      = Chemical synapse strength from neuron j to neuron i,
C                     for j = 1 to n;
C                     external injected current strength into neuron i,
C                     for j = n+1.
C                     Can be positive (excitatory) or negative(inhibitory).
C                     Positive t(i,j) assumed due to sodium influx,
C                     multiplied by reversal potential expression (1-0.01v(i,1)).
C                     Negative t(i,j) assumed due to potassium efflux,
C                     multiplied by reversal potential expression (1+0.1v(i,1)).
C       tau(i)      = RC time constant of neuron i, in milliseconds.
C       te(i,j)     = Non-rectifying electrotonic coupling from neuron j to neuron
C                     i, te(i,j) = te(j,1)
C       tr(i,j)     = Rectifying electrotonic synapse from neuron j to neuron i,
C                     tr(j,i) = 0 when t(i,j) is non-zero.
C       u(i)        = Basal threshold voltage of neuron i. Modified by ufact(i)
C                     due to accommodation: u(i)+ufact(i)
C       ufact(i)    = Factor added to basal threshold of neuron i during
C                     relative refractory period.
C                     ufact(i) = fire(i)/rate(i)
C       v(i,j)      = Voltage of neuron i (j-1) iterations ago. Present
C                     voltage = v(i,1); resting voltage = 0; in millivolts.
C       vjd         = Effectiveness of synapse from neuron j to i:
C                     vjd = (v(j,d(i,j) + 1) - ca(j)) * fr(j)
C                     First factor is spike height above calcium threshold;
C                     proportional to spike width (symmetrical part).
C                     Multiplication by fr(j) corrects for additional
C                     spike width due to time constant.
C       vmax(i)     = Peak spike voltage produced by neuron i when not accommodated
C
        subroutine simulator
C ***********************************
C *     Declarations                *
C ***********************************
        integer i, j, m, s
        integer size,nstates,iter,ios, nin
        integer outunit,test
        parameter (size=15,nstates=610,nin=1)
        logical valid
        integer sinp(size),d(size,size),s(size,nstates),ite(size)
```

```
      integer n,s2,no,eleno,invary
      integer counter(40),index
      real ufact(size),t(size,size+nin),u(size),fr(size)
      real ftijp(size),ftijn(size),ftiop(size),ftion(size),vmax(size)
      real fteijp(size),ftrijp(size),tin(size,size)
      real te(size,size), tr(size,size),vjd, delv(size), ffrv(size)
      real v(size,5000), fire(size), acc(size), rec(size),ca(size),p
      integer mfr(size),mt(size,size),l,int,test,s
      integer sl,il
      real frv(100),gv(100), rate(size), tau(size), peak(size)
      character*10 labela(size)
      integer neu(200)
      common /block2/v,u,fr,d,mfr,s,t,mt,te,tr,h
      common /general1/valid,ios
      common /block3/sinp
      common /general2/n
      common /general3/iter,test,p
      common /block1/inunit,outunit,int
      common /block4/labela
      common /block8/deltat,fire,rec,peak,ftijp,ftijn
      common /block9/ftiop,ftion,fteijp,ftrijp
      common /block10/tau,acc,rate,ca,vmax
      common /block11/ite
      common /block5/l,nexpt,neuno
      common /block6/counter
      common /block17/gv, frv
      common /block18/neu
      common /block21/tin
      common /block25/il
      common /block26/eleno
      common /block30/it
      common /block32/s2
      common /block34/no
      common /block36/invary
c
      write(*,6)
6     format('------------------------------')
C ***************************************
C *    Initialize    parameters         *
C ***************************************
      do 12 i=1,n
      do 12 j=1,n+1
      d(i,j) = d(i,j)/deltat
12    continue
      do 14 i=1,n
      fire(i)=0
14    continue
      do 303 i=i,n
 do 404 j=1,it
 v(i,j)=0.0
404       continue
303    continue
C ***********************************
C *       Calculate      fr         *
```

```
C ****************************************
      do 331 i=1,n
fr(i) = exp(-deltat/tau(i))
331   continue
C ****************************************
C *        Begin experiments              *
C ****************************************
      do 2222 l = 1, nexpt
      do 17 i=1,n
      counter(i) = 0
17    continue
      iter = 0
      maxiter = ite(l)/deltat
      if(l .gt. 1)il = il+ite(l-1)
      do 3333 i = 1,n
  t(i,n+1) = tin(l,i)
3333  continue
      if(invary .eq. 1)then
 if(no .eq. 1)t(eleno,n+1) = p
      end if
           m = 1
C ****************************************
C *   Iterate network maxiter times       *
C ****************************************
      do 304 s=1,maxiter
  do 345 i = 1,n
    index = nint(v(i,1))
    if (index .lt. 1)index = 1
    if (index .gt. 100) index =100
    frv(i) = fr(i)**(gv(index)/1.6)
    ffrv(i) = (1.0 - frv(i))/gv(index)
345      continue
C ****************************************
C *   Calculate change in voltage         *
C ****************************************
           do 505 i = 1, n
    delv(i)=0.0
              do 606 j = 1, n
    vjd=0.0
               if(s(j,d(i,j)+1).eq.1)vjd=(v(j,d(i,j)+1)-ca(j))*fr(j)
    if(vjd.lt.0.0)vjd=0.0
                if (t(i,j).ne. 0.0)then
                  if((t(i,j).lt.0.0).and.vjd.gt.0.0)delv(i)=delv(i)+ftijn(i)
     1*vjd*ffrv(i)*t(i,j)*(1+0.1*v(i,1))
                  if((t(i,j).gt.0.0).and.vjd.gt.0.0)delv(i)=delv(i)+ftijp(i)
     1*vjd*ffrv(i)*t(i,j)*(1-0.01*v(i,1))
    endif
                if (tr(i,j) .ne. 0.0.and.v(j,1) .gt. v(i,1))delv(i)=
     1(v(j,1)-v(i,1))*ffrv(i)*tr(i,j)*ftrijp(i)+ delv(i)
                if (te(i,j) .ne. 0.0)delv(i)=(v(j,d(i,j)+1)-v(i,1))*ffrv(i)
     1*te(i,j)*fteijp(i)+ delv(i)
606            continue
                if (t(i,n+1) .lt. 0.0 )delv(i)=
     1delv(i) + ftion(i)*ffrv(i)*t(i,n+1)
```

```
              if (t(i,n+1) .gt. 0.0 )delv(i)=
     ldelv(i)+ftiop(i)*ffrv(i)*t(i,n+1)
C ****************************************
C *    Increase in threshold             *
C ****************************************
              s(i,1) = 0
              if(s(i,2) .eq. 0)then
     ufact(i) = fire(i)/rate(i)
              end if
505   continue
         do 123 i = 1,n
     v(i,1) = v(i,1)*frv(i) + delv(i)
              if ((v(i,1).lt.(ufact(i)+u(i))).or.(s(i,2).eq.1))then
                 s(i,1) = 0
              else
                 s(i,1) = 1
                 counter(i) = counter(i) + 1
                 v(i,1)=vmax(i)*peak(i)**(fire(i))
              endif
     fire(i) = (fire(i) + acc(i)*s(i,1))*(1-rec(i)*deltat)
123   continue
      if (iter .le. maxiter) iter = iter + 1
      if (maxiter .le. m*nstates .and. iter .eq. maxiter)then
      do 323 sl=1,neuno-1
      do 77 i = (m-1)*nstates+1, maxiter
      write(neu(sl)+10*ios,*)i+il,v(neu(sl),maxiter-i+1)
     1+(sl-1)*120.0+(ios-1)*(neuno-1)*120.0
77    continue
323     continue
        end if
        if (iter .eq. m*nstates)then
        do 343 sl=1,neuno-1
        do 234 j =m*nstates, (m-1)*nstates+1, -1
        write(neu(sl)+10*ios,*)il+(2*m-1)*nstates-j+1,
     lv(neu(sl),j-(m-1)*nstates)+(sl-1)*120.0+(ios-1)*(neuno-1)*120.0
234     continue
343     continue
        m = m + 1
        end if
           do 808 i = 1, n
              do 909 j = nstates, 2, -1
                 s(i, j) = s(i, j-1)
                 v(i, j) = v(i, j-1)
909           continue
808        continue
304     continue
        write(*,318)labela(1),s2
318     format(''the firing rate in experiment '',a6,'' of run '',i2,
     +'' is:'')
        write(*,*)(counter(i),i=1,n)
2222    continue
        return
        end
```

REFERENCES

[1] F. Delcomyn, Neural basis of rhythmic behavior in animals, *Science*, 210, 492–498 (1980).

[2] M. I. Cohn and J. L. Feldman, Discharge properties of dorsal medullary inspiratory neurons: Relation to pulmonary afferent and phrenic efferent discharge, *J. Neurophysiol.*, 51, 753–776 (1984).

[3] G. A. Carpenter and S. Grossberg, A neural theory of circadian rhythms: The gated pacemaker, *Biological Cybernetics*, 48, 35–59 (1983).

[4] R. Lara and M. A. Arbib, A model for the neural mechanisms responsible for pattern recognition and stimulus specific habituation in toads, *Biological Cybernetics*, 51, 223–237 (1985).

[5] S. Grossberg and E. Mingolla, Neural dynamics of perceptual grouping: textures, boundaries, and emergent segmentation, *Percept. Psychophys*, 38, 141–171 (1985).

[6] D. Marr, *Vision. A Computational Investigation into the Human Representation and Processing of Visual Information*, W. H. Freeman, San Francisco, 1983.

[7] W. D. Knowles, R. D. Traub, R. K. S. Wong, and R. Miles, Properties of neural networks: Experimentation and modeling of the epileptic hippocampal slice, *Trends Neurosci.*, 8, 61–67 (1985).

[8] D. H. Perkel and D. J. Perkel, Dendritic spines: Role of active membrane in modulating synaptic efficacy, *Brain Res.*, 325, 331–335 (1985).

[9] D. Noble, Conductance mechanisms in excitable cells, *Biomembranes*, 3, 427–447 (1973).

[10] R. R. Llinas, The intrinsic electrophysiological properties of mammalian neurons: Insights into central nervous system functions, *Science*, 242, 1654–1664 (1988).

[11] L. K. Kaczmarek and I. B. Levitan, "What is neuromodulation?" in L. K. Kaczmarek and I. B. Levitan, Eds., *Neuromodulation*, Oxford Univ. Press, New York, 1987, pp. 3–17.

[12] I. B. Levitan and L. K. Kaczmarek, "Ion currents and ion channels: Substrates for neuromodulation," in L. K. Kaczmarek and I. B. Levitan, Eds., *Neuromodulation*, Oxford Univ. Press, New York, 1987, pp. 18–38.

[13] J. A. Benson and W. B. Adams, "The control of rhythmic neuronal firing," in L. K. Kaczmarek and I. B. Levitan, Eds., *Neuromodulation*, Oxford Univ. Press, New York, 1987, 100–118.

[14] E. R. Kandel and J. H. Schwartz, Molecular biology of learning: Modulation of transmitter release, *Science*, 218, 433–443 (1982).

[15] G. Lynch and M. Baudry, The biochemistry of memory: A new and specific hypothesis, *Science*, 224, 1057–1063 (1984).

[16] A. I. Selverston, A consideration of invertebrate central pattern generators as computational databases, *Neural Networks*, 1, 109–117 (1988).

[17] V. Bedian, J. F. Lynch, and F. Zhang, "A neural net editor with biological applications," in H. Caudill, Ed., *Proc Internat. Joint Conference on Neural Networks*, Lawrence Erlbaum Associates, Inc. Hillsdale, N.J. 35–38 (1989).

[18] W. J. Davis, "Neural mechanisms of behavioral plasticity in an invertebrate model system," in A. I. Selverston, Ed., *Model Neural Networks and Behavior,* Plenum Press, New York, 1987, pp. 263–282.

[19] A. L. Hodgkin and A. F. Huxley, A quantitative description of membrane current and its application to conduction and excitation in nerve, *J. Physiol.,* 117, 500–544 (1952).

[20] W. S. McCulloch and W. Pitts, A logical calculus of the ideas immanent in nervous activity, *Bull. Math. Biophys.,* 9, 127–147 (1943).

[21] R. J. MacGregor, *Neural and Brain Modeling,* Academic Press, San Diego, 1987.

[22] V. Bedian, F. Zhang, J. F. Lynch, and M. H. Roberts, A neural network model with applications to *Pleurobranchaea. Soc. Neurosci. Abs.,* 14, 259 (1988).

[23] M. Dao, M. Habit, J. P. Richard, and D. Tallot, "CABRI, an interactive system for graph manipulation," in G. Tinhofer and G. Schmidt, Eds., *Graph Theoretical Concepts in Computer Science,* Lecture Notes in Comp. Sci., vol. 246, Springer-Verlag, New York, 1987, pp. 58–67.

[24] J. Hynd and P. D. Eades, The types graph editing system—TYGES, *3rd Australasian Conf. on Comp. Graphics, Proc.* Brisbane, Australia pp. 15–19 (1985).

CHAPTER 3
Neural Network Models of Concept Learning

PHILIPPE G. SCHYNS

"Ned," he said, "you are a killer of fish and a very clever fisherman. You have caught a great number of these interesting animals. I would bet, though, that you don't know how to classify them."
"Oh yes I do," the harpooner replied in all seriousness. "They can be classified into fish that can be eaten and fish that can't!"
"That's the distinction of a glutton," rejoined Conseil.
JULES VERNE,
20,000 Leagues Under the Sea.

3.1 INTRODUCTION

Every day we deal successfully with a continuously changing environment. Rarely, if ever, do we observe the same object in exactly the same conditions. However, the variations do not seem to interfere with our recognition of physically different objects. Think, for example, of funnels. Funnels often vary in size, shape, color and they also vary in the material they are made out of. The context in which funnels appear may also change. Funnels are likely to be found in a grange, in a kitchen, or a workshop. Some even find it fashionable to use a funnel for a hat. However singular the context, and whatever the form, color and size of funnels, they are recognized correctly

---

Neural and Intelligent Systems Integration, By Branko Souček and the IRIS Group.
ISBN 0-471-53676-8 ©1991 John Wiley & Sons, Inc.

most of the time. Underlying this phenomenon is the ability to segment the world into classes of equivalence.

The mechanisms by which intelligent systems group physically distinct objects into classes of equivalence are probably among the most fundamental aspects of cognition. Without these mechanisms, every instance of each type of object, events or situations you can possibly think of would appear new every time it was encountered. To get an intuition of the strange mental state that might result if these mechanisms were missing, think of a familiar event, seeing your own face in a mirror. Every morning, due either to a different point of view, the signs of time, or different lighting conditions, your own face would evoke a brand new perception. Without a capacity to abstract something as fundamental as your own face, your mental life would quickly be saturated with a continuous flow of unrelated experiences. The processes and representations involved when we construct abstractions from experiences represent one of the primal mechanisms of knowledge acquisition. Through the use of these processes and representations we make sense of a world that would otherwise appear chaotic. In order to build intelligent computing machines, modelling these processes is of first priority.

Concept learning is the area of cognitive science that deals with the aforementioned mechanisms. A *concept* is the mental representation of a category of objects in the world. For example, the category *dog* could be characterized by the concept consisting of the semantic features {*has-four-legs, has-a-head, has-a-tail, barks,* etc.}. Clearly, if the properties of the concept *dog* have to characterize the entire category, features such as {*has-pointed-ear, is-aggressive, is-tall, has-German-origin*} should not be included in the representation, since they characterize only a subcategory of *dog*. *Categorization* is the operation by which an object is identified as the member of a particular category. It should be clear that categorization and concepts are not independent from one another. Since concepts segment the external world into categories, a large amount of finely tuned concepts will categorize the external world more subtly than a small amount of coarsely tuned concepts.

In this chapter, I will present the major neural network architectures for concept learning. The section has two main aims. I will show how the operations of representing concepts and categorizing exemplars are implemented with neural network techniques, and I will discuss the psychological effects and the psychological characteristics of conceptual memory that have been modelled by such architectures. Supervised models, their virtues, and drawbacks will be explored first. Then, we will turn to unsupervised networks and finally motivate a modular hybrid architecture. This chapter should not be seen as an exhaustive review of concept learning models. I chose the architectures that, to me, presented attractive and general properties, rather than models tuned to specific effects of human conceptual memory.

3.2 SUPERVISED TECHNIQUES

Supervised learning techniques of concept learning share two major characteristics. The first one has been known as *ostensive learning*. Ostensive learning occurs when a particular object, situation or event is experienced while being told what this object, situation or event is. Thus, ostensive learning implies that a teacher provides the correct category label to which the exemplars of a category should be associated. In neural networks, supervised techniques of concept acquisition assume that the correct category name is encoded in a specific format, and provided to the system each time an exemplar is presented. The second characteristic is that semantic memory is implemented as a form of associative memory (see Fig. 3.1). Thus, the knowledge of the system lies in the connections of the associative matrix.

Two major supervised neural network architectures have been used to study concept formation: the linear associator and back propagation.

3.2.1 The Linear Associator

The simplest form of associative memory is certainly the linear associator. Its mathematics have been clearly worked out in Anderson [1], Anderson and Hinton [2], and Kohonen [3]. Knapp and Anderson [4] and McClelland and Rumelhart [5] have presented applications of the linear associator to concept learning. I will present a schematic version of Knapp and Ander-

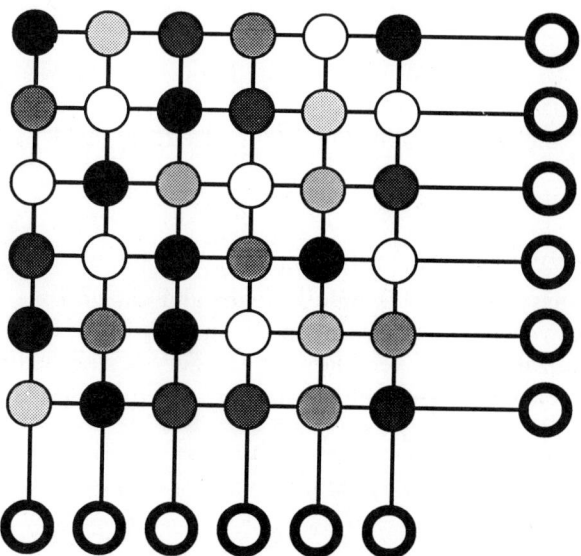

Figure 3.1 *The architecture of the simple, one-layer, associative memory. Input units are directly connected to output units with connection strengths.*

son's model in some detail, since its mathematical development is particularly clear. Although Knapp and Anderson worked out a sophisticated coding of their input patterns, for ease of presentation this aspect will be skipped and I will develop instead a version of their model that keeps the original's major qualitative properties.

Assume that each object of one category is encoded with n real-valued dimensions such as height, weight, and so forth. By doing so, each object is described by a feature vector **i** in n-dimensional space. Suppose also that the object's category name is known and is coded on a vector **o** in m-dimensional space. Given these two assumptions, we can start to build the simplest associative memory by relying on the mathematics of linear algebra (see [4]).

The memory of a single exemplar can be learned with the association equation

$$\mathbf{W} = \alpha \cdot \mathbf{o} \cdot \mathbf{i}^T \tag{3.1}$$

This association implements Hebb's rule: *The strength of each synapse is changed proportionally to the product of the pre- and postsynaptic neurons.* $\mathbf{o} \cdot \mathbf{i}^T$ is the outer product of the two vectors that code respectively the category name and one exemplar of the category. Alpha is the learning constant parameter. **W** is the connection strength matrix resulting from the association between **i** and **o**.

Categorization, the production of the correct category name **o** given exemplar **i** and the associative memory **W**, is achieved by doing some algebraic manipulations of (3.1). If **i** is presented to the associative memory, categorization is

$$\mathbf{W} \cdot \mathbf{i} = \alpha \cdot \mathbf{o} \cdot [\mathbf{i}^T \cdot \mathbf{i}] \tag{3.2}$$

where $[\mathbf{i}^T \cdot \mathbf{i}]$, the inner product of **i** with itself, is equal to 1. So, (3.2) becomes

$$\mathbf{W} \cdot \mathbf{i} = \alpha \cdot \mathbf{o} \tag{3.3}$$

Of course, a system that is able to learn and categorize correctly one exemplar is not very interesting. Due to the linearity of the Hebbian associator, the generalization to n exemplars per category is straightforward. Let $\mathbf{i}_i$ denote the ith exemplar associated with the category name **o**. Learning n different exemplars is simply expressed as

$$\mathbf{W} = \alpha \cdot (\mathbf{o} \cdot \mathbf{i}_1^T + \mathbf{o} \cdot \mathbf{i}_2^T + \cdots + \mathbf{o} \cdot \mathbf{i}_n^T) \tag{3.4}$$

or,

$$\mathbf{W} = \alpha \cdot (\mathbf{W}_1 + \mathbf{W}_2 + \cdots + \mathbf{W}_n)$$

W is simply the sum of the partial matrices that encode the associations between the category name **o** and each exemplar $\mathbf{i}_i$.

In the case of multiple exemplars, the categorization of an exemplar $\mathbf{i}_x$ becomes

$$\mathbf{W} \cdot \mathbf{i}_x = \alpha \cdot (\mathbf{o} \cdot \mathbf{i}_1^T + \mathbf{o} \cdot \mathbf{i}_2^T + \cdots + \mathbf{o} \cdot \mathbf{i}_n^T) \cdot \mathbf{i}_x$$
$$= \alpha \cdot (\mathbf{o} \cdot [\mathbf{i}_1^T \cdot \mathbf{i}_x] + \mathbf{o} \cdot [\mathbf{i}_2^T \cdot \mathbf{i}_x] + \cdots + \mathbf{o} \cdot [\mathbf{i}_n^T \cdot \mathbf{i}_x])$$
$$= \alpha \cdot \mathbf{o} \cdot ([\mathbf{i}_1^T \cdot \mathbf{i}_x] + [\mathbf{i}_2^T \cdot \mathbf{i}_x] + \cdots + [\mathbf{i}_n^T \cdot \mathbf{i}_x]) \quad (3.5)$$

Therefore, the categorization of the exemplar coded by $\mathbf{i}_x$ is the name of the category coded by **o** affected by the sum of the inner products between $\mathbf{i}_x$ and the memorized exemplars $\mathbf{i}_i$. It is important to note that if the $\mathbf{i}_i$'s are normalized (if their length is 1), the inner products can be interpreted as correlation measures. Note also that if the $\mathbf{i}_i$'s are orthogonal to one another, all the $[\mathbf{i}_i^T \cdot \mathbf{i}]$ will be equal to 0 except for the $\mathbf{i}_j$ that is exactly equal to $\mathbf{i}_x$.

The different members of a specific category can be generated in many different ways. As a simple assumption, it seems plausible that the members of the same category should be very similar to one another. In other words, they should be more or less equivalent along a subset of the n dimensions that describe them. Take for example the category *dog*. Most dogs share at least the dimensions {*has-four-legs, has a head, has-a-tail, barks,* etc.}. Mathematically, this amounts to requiring $[\mathbf{i}_i^T \cdot \mathbf{i}_j] \gg 0$ for all $\mathbf{i}_i, \mathbf{i}_j$ members of a specific category. Knapp and Anderson used this assumption to create categories where each exemplar was slightly different from the prototype of the category. In other words,

$$\mathbf{i}_i = \mathbf{p} + \mathbf{d}_i \quad \text{for all } i \quad (3.6)$$

where **p** is the prototype of the category and $\mathbf{d}_i$ is the small distortion vector that singularizes the exemplar $\mathbf{i}_i$. If the distortions are such that they cancel each other when averaged, that is,

$$\sum_i \mathbf{d}_i = 0 \quad (3.7)$$

the network learns

$$\mathbf{W} = \alpha \cdot (\mathbf{o} \cdot \mathbf{i}_1^T + \mathbf{o} \cdot \mathbf{i}_2^T + \cdots + \mathbf{o} \cdot \mathbf{i}_n^T)$$
$$= \alpha \cdot \mathbf{o} \cdot (\mathbf{i}_1^T + \mathbf{i}_2^T + \cdots + \mathbf{i}_n^T) \quad (3.8)$$

where $\mathbf{i}_i^T$ is $(\mathbf{p} + \mathbf{d}_i)^T$
So, (3.8) can be rewritten as

$$\mathbf{W} = \alpha \cdot \mathbf{o} \cdot ((\mathbf{p} + \mathbf{d}_1)^T + (\mathbf{p} + \mathbf{d}_2)^T + \cdots + (\mathbf{p} + \mathbf{d}_n)^T) \quad (3.9)$$

However, by (3.7), the distortions average out and (3.9) becomes

$$\mathbf{W} = \alpha \cdot \mathbf{o} \cdot (\mathbf{p} + \mathbf{p} + \cdots + \mathbf{p})^T \quad (3.10)$$

Thus, the associative memory has really learned n times the prototype of the category rather than n different exemplars. The generalization to m different categories constructed as distortions around m different prototypes is straightforward. If we assume, for the simplicity of presentation, that each category is composed of exactly n exemplars, concept learning in a linear model is described as follows:

$$\mathbf{W} = \alpha \cdot n \cdot (\mathbf{o}_1 \cdot \mathbf{p}_1^T + \mathbf{o}_2 \cdot \mathbf{p}_2^T + \cdots + \mathbf{o}_n \cdot \mathbf{p}_n^T) \quad (3.11)$$

The categorization of an unknown exemplar is simply

$$\begin{aligned}\mathbf{W} \cdot \mathbf{i}_x &= \alpha \cdot n \cdot (\mathbf{o}_1 \cdot \mathbf{p}_1^T + \mathbf{o}_2 \cdot \mathbf{p}_2^T + \cdots + \mathbf{o}_n \cdot \mathbf{p}_n^T) \cdot \mathbf{i}_x \\ &= \alpha \cdot n \cdot (\mathbf{o}_1 \cdot [\mathbf{p}_1^T \cdot \mathbf{i}_x] + \mathbf{o}_2 \cdot [\mathbf{p}_2^T \cdot \mathbf{i}_x] + \cdots + \mathbf{o}_n \cdot [\mathbf{p}_n^T \cdot \mathbf{i}_x])\end{aligned} \quad (3.12)$$

The prototypes of the different categories are the concepts learned by the system (3.11), and the categorization of an unknown exemplar $\mathbf{i}_x$ is a function of the correlation between the unknown exemplar and the known prototypes of the system (3.12). The argument of orthogonality presented earlier still applies here. If the different prototypes are orthogonal to one another, and if the correlations $[\mathbf{p}_j^T \cdot \mathbf{i}_x]$ are close to 0 for all j except when $x = i$, an almost perfect categorization will be achieved. However, if there is a nonzero correlation between the prototypes, and if the $[\mathbf{p}_j^T \cdot \mathbf{i}_x]$ are not close to 0 for some j, interferences will occur. Since the categorizations performed by the system depend essentially on the correlations between the prototypes of the categories and the input exemplars, some interesting properties of the linear associator can be deduced.

Among these properties is the prototype effect. The prototype effect is an interesting phenomenon that has been widely studied in cognitive psychology [4,6,7]. Consider the following task. As a subject to a psychological experiment, you are shown exemplars from one artificial category and you are asked to learn them. Once you have learned these exemplars, you are tested on your knowledge of the category with different sorts of items: the exemplars from the training set, new exemplars from the category that have never been presented to you before, and the prototype of the category that you have never seen. According to the prototype effect, the prototype of the category is the most easily recognized item among the different types of testing stimuli. The intriguing phenomenon is that in the experiment suggested, you have never experienced the prototype directly. Posner and Keele [7] proposed an explanation for this phenomenon. They suggested that categories are represented as prototypes. So, even if you have never experienced the prototype directly, by learning the exemplars of the artificial category, you have extracted its average, the prototype. Consequently, when exposed to different sorts of testing patterns, the prototype is best recognized because it exactly matches the concept formed during the training stage.

In a linear associator, the prototype effect is a by product of the correlational aspect of categorization. Because the output of categorization depends on the correlation of the unknown exemplar with every stored prototype, categorization is best for exemplars that have the highest correlation with one of the prototypes (i.e., the prototypes themselves).

So far, we have seen how a linear associator could lump different exemplars of a category into a prototypical representation [see (3.8) to (3.10)]. However, we may want to preserve in memory the *specificity* of some singular but repeated elements of a particular category. Consider, for example, the category *bird*. Among the properties that characterize the category, we would certainly find {*have feathers, fly,* etc.}. However, a penguin is a bird, but it does not fly. It may therefore be useful to keep a separate representation for *penguin*. A stored association will be distinct if the three following conditions apply: a new category name is associated to the singular exemplar, the representation of *penguin* is sufficiently uncorrelated with the other birds, and the exemplar is presented repeatedly [5].

The linear model of concept learning offers many virtues. The first one is certainly its clear mathematical foundations—straightforward linear algebra. Assuming a correct coding of the categories, the prototype effect occurs as a side effect of the correlational structure of categorization. It should be clear that this model is not solely a categorizer that groups together different exemplars of a category, it is also able to store separately specific exemplars and thereby capture the complex structure of categories. These virtues make the linear associator a nice model to use on simple tasks. Its drawbacks, however, prevent the learning of more complicated associations.

One of the major drawbacks of the Hebbian linear associator is the orthogonality constraint. For learning to occur without interferences, the different patterns have to form an orthogonal set. This constraint can be prevented by using another learning rule, known as the Widrow-Hoff, error-correction, least mean square, or Delta rule.

$$\Delta \mathbf{W} = \alpha \cdot (\mathbf{t} - \mathbf{o}) \cdot \mathbf{i} \qquad (3.13)$$

The error-correction procedure states that the increment of weights $\Delta \mathbf{W}$ is a function of the error between a desired output vector $\mathbf{t}$, also called the teaching output, and the output vector $\mathbf{o}$ produced by the network. Once the association between the input $\mathbf{i}$ and the desired output $\mathbf{t}$ has been learned, the error term is equal to $\mathbf{0}$ and the weights no longer change.

Use of the error-correction rule requires a milder constraint on the set of patterns that can be perfectly learned: the set has to be linearly independent.

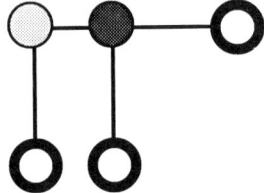

Figure 3.2 *Simple two-input units and one-output unit network.*

Clearly, this constraint is less strong than the orthogonality constraint implied by Hebb's rule, since an orthogonal set must be linearly independent, whereas a linearly independent set is not necessarily orthogonal. However, even with the more powerful error-correction procedure, the network is exposed to another constraint, the linear predictability constraint. The linear predictability constraint states that any output vector must be predicted by a linear combination of any input vector. To illustrate linear predictability, consider the simple network of Figure 3.2 exposed to a linearly predictable and a nonlinearly predictable input set.

In the first problem, we are given two simple categories to learn, *dog* versus *not-dog*. To keep the problem simple, assume we are just using two semantic features to describe exemplars: *has-hairs* and *barks*. If the symbol ¬ is used to denote not, the problem is described as follows.

$$\neg \text{ has-hairs} + \neg \text{ barks} = \neg \text{ dog}$$
$$\neg \text{ has-hairs} + \text{ barks} = \text{ dog}$$
$$\text{has-hairs} + \neg \text{ barks} = \text{ dog}$$
$$\text{has-hairs} + \text{ barks} = \text{ dog}$$

This learning problem can be framed as a logical OR. Is this function learnable using the error-correction procedure with a linear associator? In other words, is OR a linearly separable problem? A solution is found if we can set up a linear discriminant function $g(\mathbf{i}) = \mathbf{W} \cdot \mathbf{i}$ such that $g(\mathbf{i}) < \phi$ means that the exemplar $\mathbf{i}$ is a member of the first category and $g(\mathbf{i}) > \phi$ means that the exemplar is a member of the other category (see Fig. 3.3). Since such a line exists, the linear associator architecture and the error-correction procedure can learn the OR mapping, although not perfectly (the input set is not linearly independent).

Consider now another simple problem of concept learning. We would like a linear associator to learn about *humanness* from combinations of the semantic features *is-male* and *is-female*. The problem is formulated as follows.

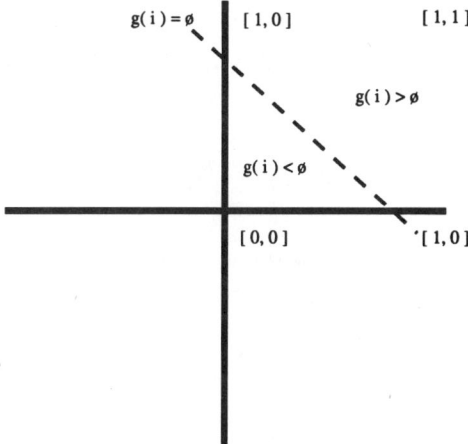

Figure 3.3 *Solution of OR in two dimensions. The linear discriminant function g(i) = **Wi** determines the two decision regions g(i) < 0 and g(i) > 0 that, in turn, delimits category boundaries.*

$$\neg \text{ is-male} + \neg \text{ is-female} = \neg \text{ human}$$
$$\neg \text{ is-male} + \text{ is-female} = \text{ human}$$
$$\text{is-male} + \neg \text{ is-female} = \text{ human}$$
$$\text{is-male} + \text{ is-female} = \neg \text{ human}$$

This problem and XOR are isomorphic. As Figure 3.4 illustrates, any problem isomorph to XOR cannot be learned with the simple kind of network illustrated in Figure 3.2. More formally, a straight decision line cannot be drawn to separate the two decision regions and solve XOR in two dimensions.

To solve XOR, different strategies are available. The simplest one is to recode the input space by adding one dimension. For example, *is-abnormal* and ¬ *is-abnormal*.

$$\neg \text{ is-male} + \neg \text{ is-female} + \neg \text{ is-abnormal} = \neg \text{ human}$$
$$\neg \text{ is-male} + \text{ is-female} + \neg \text{ is-abnormal} = \text{ human}$$
$$\text{is-male} + \neg \text{ is-female} + \neg \text{ is-abnormal} = \text{ human}$$
$$\text{is-male} + \text{ is-female} + \text{ is-abnormal} = \neg \text{ human}$$

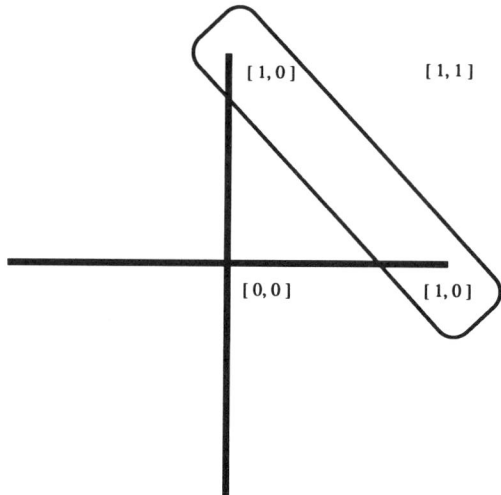

Figure 3.4 *Solution of XOR in two dimensions. It can be readily observed that no linear discriminant function can be computed here.*

The third dimension is a *nonlinear recoding* of the two others—formally a logical AND. If such a coding of the exemplars is allowed, the problem becomes learnable by a simple two-layer network in a three-dimensional space characterized by three semantic features. This problem illustrates that the linear predictability constraint depends on the representation of the set of patterns. Therefore, with the learning power of a linear associator, representational issues will play a major role. A well-chosen representation of the pattern set can transform a hard learning problem into an easy one (see Gluck & Bower, [8]; Corter, Gluck & Bower, [9] LMS procedure for applications of this principle to concept learning in neural networks). There is a conceptual distinction between the logical description of a learning problem and the patterns of activation occurring in the brain. Due to its modular architecture, the cognitive system may, through various steps of preprocessing, gradually transform the initial complexity of a problem. After all, the brain seems to be governed by simple rules implemented in a complex and very intricate architecture. Perhaps this rough intuition could be used as a guiding principle in designing neural systems.

Another strategy to solve XOR is to recode the input space with the network itself. To achieve this goal, an associator needs to be made more complex in two ways: by adding nonlinear units and by allowing a multilayer architecture. The simplest kind of nonlinear neural network has linear threshold output units as expressed in (3.14).

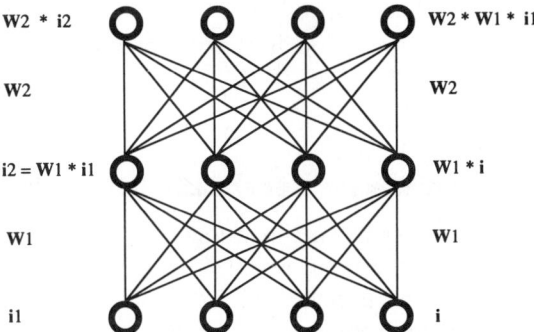

Figure 3.5 *Multilayer, linear network. As shown on the right-hand side, the output of the multilayer linear architecture is equal to* **o** = $W_2 W_1$ = W_n**i**, *a single-layer linear network.*

$$o_i = f(\mathbf{W}_i \cdot \mathbf{i}) \quad \text{where } f(\cdot) = 1 \quad \text{if } \mathbf{W}_i \cdot \mathbf{i} > \phi \text{ and } f(\cdot) = 0 \quad \text{if } \mathbf{W}_i \cdot \mathbf{i} \le \phi.$$
(3.14)

The output of a linear threshold unit is a binary function of its activation state $\mathbf{W}_i \cdot \mathbf{i}$. If the activation state is superior to a threshold ϕ, the output is equal to 1; otherwise, it is equal to 0. If the pattern set is composed only of zeros and ones, this new architecture implements the perceptron [10,11]. A famous theorem is linked to the perceptron: the perceptron convergence theorem. The theorem states that if a set of patterns is learnable by the perceptron, the perceptron learning procedure will find the correct set of weights to solve the learning problem. However, Minsky and Papert [10] showed that XOR isomorphic problems could not be learned by a one-layer perceptron. XOR can be learned by a *multilayer* perceptron, but the convergence theorem does not apply to multilayered perceptrons.

Thus, in order to learn XOR, nonlinearity *and* a multilayer architecture are necessary conditions. As Minsky and Papert showed, a single-layer nonlinear network—the perceptron—cannot learn XOR, and it is easy to show that a multilayer linear network cannot learn XOR either. The argument goes as follows: what can (cannot) be learned by a multilayer linear net can (cannot) be learned by a single-layer linear network. The proof in Figure 3.5 is straightforward.

Much time has been spent discussing the XOR problem carefully because it raises a crucial issue: the problem of building representations through interesting recoding inside a neural network. This theme will become central in the development of the sections to come. I will now turn to an important

82   NEURAL NETWORK MODELS OF CONCEPT LEARNING

method of solving XOR, back propagation. Back propagation is the second and last instance of supervised models of concept learning that will be presented here.

3.2.2 Back Propagation or the Generalized Delta Rule

The major drawback of simple, one-layer linear neural networks, is that no internal representation of the input set is developed by the network itself. The coding of the patterns is achieved by a mechanism external to the learning device. For some problems, like XOR, recoding is necessary, and external recoding just pushes backward the central problem of knowing which mechanisms derive the appropriate coding. This observation lead researchers to develop a learning procedure capable of generating internal representations: back propagation or the generalized Delta rule [12]. In this section, I will present back propagation and show how it successfully captures and represents regularities that characterize the domain of knowledge. For concept learning, the important contribution of back propagation is not only a more powerful learning scheme; it is also that the regularities picked out by the network can be thought of as new, relevant recombinations of features to categorize objects.

The back-propagation learning algorithm is an extension to multilayer nonlinear neural networks of the error-correction procedure developed for linear one-layer networks. The error-correction scheme minimizes an error term defined as the squares of the difference between the teaching output and the actual output of the network summed over all patterns. The error term for pattern p is defined as

$$E_p = \frac{1}{2} \sum_j (\mathbf{t}_{pj} - \mathbf{o}_{pj})^2 \qquad (3.15)$$

The learning rule presented in (3.16) implements gradient descent on a surface in weight space where the height at each point corresponds to the error measure:

$$-\frac{\partial E_p}{\partial \mathbf{W}_{ji}} = \delta_{pi} \mathbf{i}_{pj} \approx \Delta \mathbf{W}_{ji} \qquad (3.16)$$

where $\delta_{pi} = \mathbf{t}_{pi} - \mathbf{o}_{pi}$.

To obtain the generalized Delta rule, we must add a sheet of nonlinear hidden units to the architecture. Thus, a two-layer network is obtained: One layer of weights connects the input units to the hidden units, and one layer of weights connects the hidden units to the output units. The sigmoid function shown in (3.17) is frequently used to implement the nonlinearity at the hidden and output layers.

$$\mathbf{o}_{pj} = f_j(\mathbf{u}_{pj}) \qquad (3.17)$$

where $\mathbf{u}_{pj} = \Sigma_i \mathbf{W}_{ji}\mathbf{u}_{pi}$ and $f(\cdot) = 1/1 + e^{-\beta u_{pj}}$.

Despite the adjunct of nonlinearity, the generalized Delta rule still performs gradient descent in weight space,

$$-\frac{\partial E_p}{\partial \mathbf{W}_{ji}} = \delta_{pi}\mathbf{i}_{pj} \approx \Delta \mathbf{W}_{ji} \qquad (3.18)$$

But δ_{pj} depends on the type of the unit considered. If the unit is an output unit, its error signal is

$$\delta_{pj} = (\mathbf{t}_{pj} - \mathbf{o}_{pj})f'(\mathbf{u}_{pj}) \qquad (3.19)$$

where $f'(\cdot)$ is the derivative of the sigmoid function. The error signal of the units for which there is no teaching output is computed recursively by back-propagating through the weights the error that has been computed in the layer above, either an output or a hidden layer.

$$\delta_{pj} = \sum_k \delta_{pk}\mathbf{W}_{kj}f'(\mathbf{u}_{pj}) \qquad (3.20)$$

A more detailed account of the error-correction procedure for linear one layer and nonlinear multilayer network can be found in Rumelhart et al. [12].

There is a vast amount of literature on the application of back propagation to concept learning. Rather than making an exhaustive presentation, I have selected two works that, to me, illustrate the most interesting features of back propagation, the recoding of the input set in the hidden units.

Hinton [13] provides a very good example of recoding a domain of knowledge. The domain he addressed was family relationships represented by simple propositions (agent relation patient), see Fig. 3.6a. Twenty-four names could stand as agent or patient, and 12 relationships {*father, mother, husband, wife, son, daughter, uncle, aunt, brother, sister, nephew, niece*} were available. For example, the propositions could be

(Christopher husband Penelope)
(Christopher father Victoria)
(Charles uncle Charlotte)

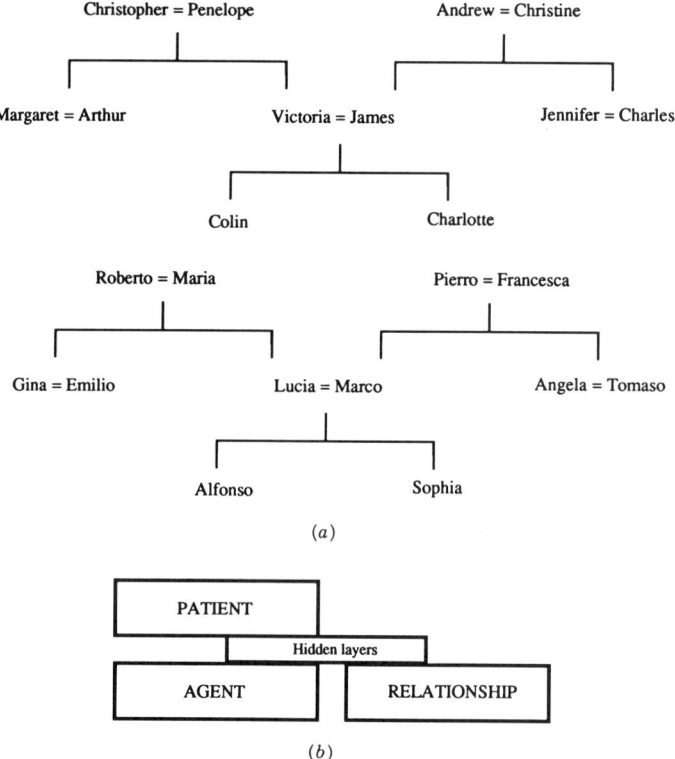

Figure 3.6 (a) The two isomorphic family trees. The first represents the English family, the second the Italian family. (b) A schematic functional representation of the network used by Hinton.

Two sets of relationships were built for two different families, an Italian and an English one. Each set of relationships could be represented by an isomorphic family tree (see Fig. 3.6). During the learning phase, the network was exposed to instances of relationships from two families.

As I already stressed, successful learning in linear models relies on an appropriate encoding of the set of input stimuli. It is assumed that similar instances must have similar intput patterns, and these carefully coded patterns are provided to the network by a device—usually unspecified—external to the learning system. In these models, the network does not translate a neutral representation of the input into its own representation before making an association. By relying on back propagation, Hinton's model does exactly the opposite. Each element of a proposition is represented by a single *neutral* unit, and these units are transformed by the network into its own representation of an agent, a relation, and a patient (see Fig. 3.6b).

After learning, the six hidden units representing the agent of the relationship were analyzed. The first unit could differentiate whether the input was a member of the first or the second family tree—English or Italian. The representation of a member of the first tree was very similar to the representation of a member of the second tree. Thus, the network exploited the structural isomorphism between the two family trees. Unit 2 encoded the generation to which an agent belongs. Three generations were possible, and they corresponded to a particular level of activation of unit 2. Unit 6 learned to recognize the branch of the family of which a person was a member. It should be stressed that the family tree, the offspring, and the branch of the family were all dimensions that adequately describe the domain of knowledge. However, all of them were not *directly* represented on the neutral input pattern. Instead, they were abstracted by the learning procedure. The informational bottleneck formed by the six hidden units has been able to characterize the relevant criteria by which the two family trees could be abstractly characterized. The characterization of the encoding performed by hidden units is a difficult task, and just three out of six hidden units were given a simple straightforward semantic interpretation; the others could not be interpreted at all.

Gorman and Sejnowski [14] provided a nice solution to the problem of interpreting the behavior of hidden units. Moreover, it is a nice application of categorization to a real-world problem. Imagine that, as an engineer, you are asked to design a system that categorizes sonar echoes from explosive mines that lie in waterways and sonar echoes from rocks of the same size. This problem is a typical instance of categorization where complexity arises for two reasons: echoes from the two classes of objects are indistinguishable to the untrained human ear, and echoes from each category are widely different due to a large range of variations in the sonic character of the objects—rocks and mines come in different sizes, shapes, and positions.

Neural network are very good at making sense of noisy signals. Gorman and Sejnowski fed their network with representations of sonar returns collected from a metal cylinder and a rock sitting approximately at 5 ft on a sandy ocean floor. The input units were clamped with the amplitudes of different frequency components of sonar echoes. The echoes were associated with their category labels: (1, 0) represented mine, and (0, 1) represented rock. The sonar returns and category labels were associated by using back propagation. After training, the network could categorize its training exemplars with great accuracy (99.8 percent). It could also categorize new instances of sonar returns with 84.5 percent of accuracy, demonstrating a good generalization capability.

Each signal was represented with 60 different dimensions, where each dimension coded a particular aspect of the stimulus. Collectively, this gives a simultaneous profile of the sonar echoes along 60 dimensions. Now, perhaps one particular profile characterizes the two classes of echoes; or perhaps a collection of dimensions are singly necessary and jointly sufficient to isolate the two profiles and so forth. In other words, categorization in such a network is understood once an account of its mechanisms are given. Unfortunately, due to nonlinear processing the categorizations by backpropagation cannot receive a simple, straightforward, solution as was the case with the linear associator [see (3.11), (3.12)]. This is where Gorman and Sejnowski's work becomes most interesting.

To understand the representations built on the hidden layer, the authors interpreted the patterns of weights from input to hidden by using a technique called hierarchical clustering [15]. Schematically, the procedure goes as follows.

1. For each hidden unit h, compute for each pattern $\mathbf{i}^k$ a weight-state vector: $\mathbf{q}_j^k[h] = [\mathbf{W}_{ij}\mathbf{i}_j^k]$.
2. Compute the Euclidean distance between each pair of weight-state vectors obtained from 1.
3. Apply a hierarchical clustering scheme to the distance matrix obtained in 2. Each obtained cluster is a set of sonar echoes whose weight-state vectors are similar to one another.
4. Compute centroids of the clusters by averaging their members. The centroids can be thought of as the prototypes of the different subspaces segmented by the network.
5. For each hidden unit, rank each centroid according to the activation it engenders in the hidden unit.
6. The hidden unit response to every centroids defines the dimension along with the hidden unit categorizes the input sample.

Conceptually, this scheme provides a method to understand how the hidden units encode the different signals. Each hidden unit is exposed to the center of the subcategories that have been extracted by the network—the centroids. Then the dimension of the stimuli to which a hidden unit responds—the dimension it encodes—is extracted by ranking the centroids as a function of the activation level of the idden unit. When the ranking is obtained, the modeler can figure out the criteria used by the hidden unit for the ranking. This technique is certainly one of the best means to analyze the behavior of hidden units, and it can be applied to many forms of classification that can be achieved through back propagation.

For example, using this technique, Gorman and Sejnowski could characterize the recoding made by the hidden layer along three correlated dimensions, or three correlated signal features: the bandwidth, the onset, and the decay characteristics of the signal. As the bandwidth of the sonar return decreases, the hidden unit's activation level increases. The onset and decay of the signals were also turning on the activation of the hidden units. These three signals are correlated because wide-band signals tend to have a progressive onset and decay while narrow-band signal do not. Thus, the behavior of the hidden units could predict the categorization performed by the network: A cylinder echo—wide bandwidth—would typically turn off all the hidden units, and a rock echo—narrow bandwidth—would tend to turn them on.

It should be worth emphasizing that the three criterial features for classification—bandwidth, onset, and decay—were not *explicitly* represented in the input patterns. They were implicit dimensions of the pattern set that were picked out and coded by the hidden layer. This property is clearly one of the most interesting features of back propagation. As neural network modelers of concept learning, we have to give an account of the emergence of new dimensions along which categorization may occur. Perhaps back propagation is the best supervised system available to account for this phenomenon.

A major drawback of back-propagation-like learning procedures has been known as the *scaling problem* [16]. The scaling problem refers to the fact that difficult learning problems may require thousands of iterations through the pattern set for problems whose solutions require thousands of connections. So, back propagation scales badly. Different strategies have been undertaken to try to speed up the learning procedure: optimizing the parameters dynamically, using a more informed search for a global minimum in weight space, or even start the network with initial learning conditions favorable to the problem at hand. However, even with these learning tricks, some problems still scale badly.

A more promising approach to solve the scaling problem can be referred to as *divide and conquer*. The models I have presented so far were "one-shot" networks in which a large problem had to be solved with one computational module: a single network. However, by taking as analogy the proposal for the "modularity of mind" [17], it may be more interesting to decompose a difficult tasks into subproblems and assign each of them to different expert subnetworks. This design issue will be discussed further at the end of this chapter because it is sensible to think of concept learning in modular terms.

3.3 UNSUPERVISED TECHNIQUES

One of the problems with supervised learning is that its assumption is too constraining as a model of human learning: A fair teacher has to provide feedback about the exemplar's category. In pattern associators, no learning occurs—no concepts are extracted—if the category name and the exemplars are not presented simultaneously to the network. Thus, a supervised model assumes a one-to-one mapping between concepts and category names. Is this a plausible assumption for a human's cognitive architecture? If it were, we would only have as many concepts as we have symbols in our mental lexicon. This is fortunately very unlikely; our knowledge would otherwise be very limited. To make this point intuitively clear, take, for example, the category *chair*. Most of us are able to make fine distinctions among different sorts of chairs: those found in a bar, in a dining room, in a kitchen, in a living room, and so forth. To categorize these objects as members of different kinds, we need a refined conceptual segregation of the domain of knowledge. The number of distinct classifications we can make in the *chair* domain is frequently much larger than the number of labels we have to refer to it. So, concepts and lexical items are in a many-to-one, rather than a one-to-one, relationship. Recent research on infants' conceptual development has shown that category labels may *follow, rather than precede,* the acquisition of concepts [18,19]. In other words, category names seem to be *indicators* of a specific conceptual knowledge rather than the *means* to acquire it. This shows the need to develop techniques where concepts can be learned independently from lexical items.

Unsupervised learning schemes provide a solution to this issue. These methods have been widely studied in Fukishima et al. [20], Grossberg [21–23], Kohonen [24–26], Rumelhart and Zipser [27], and Von der Marlsburg [28]. Abstractly, unsupervised learning is a nonassociative type of learning whose goal is similar to back propagation, building interesting internal representations from the statistics of the input set. Applied to concept learning, these internal representations take the form of high-level features that segregate the pattern set into separate clusters, the categories. In opposition to supervised schemes, no a priori set of categories into which the patterns fall is needed here; the system discovers the structure of the input by developing its own featural representation of the stimuli. Thus, the name *spontaneous learning* was given by Rosenblatt to unsupervised techniques.

3.3.1 Competitive Learning

Competitive learning [21,27] is the first architecture we will explore. Most generally, it is characterized by hierarchically ordered layers of units, with excitatory interlayer connections and inhibitory intralayer connections. Within a layer, the units are organized into a certain number of clusters. Inside a cluster, the units are connected with inhibitory links (see Fig. 3.7).

Learning is based on a winner-take-all, or competitive, procedure. In a cluster, all units compete to respond to the pattern of activation on the layer below. The unit that responds the most wins the competition by sending to the other units of the cluster an inhibition signal proportional to its excitation while sending to itself an excitatory signal proportional to its activation. Thus, after a while one unit in the cluster has its activation value at 1 while the others are at 0 (see Fig. 3.7). The winning units of the different clusters have their afferent weights updated so that they will respond more easily to the pattern they won on. Concretely,

$$\Delta \mathbf{W}_{ij} = \begin{cases} 0 & \text{if unit } j \text{ loses on pattern } k. \\ \alpha \mathbf{u}_{ik}/n_k - \alpha \mathbf{W}_{ij} & \text{otherwise} \end{cases} \quad (3.21)$$

3.3 UNSUPERVISED TECHNIQUES

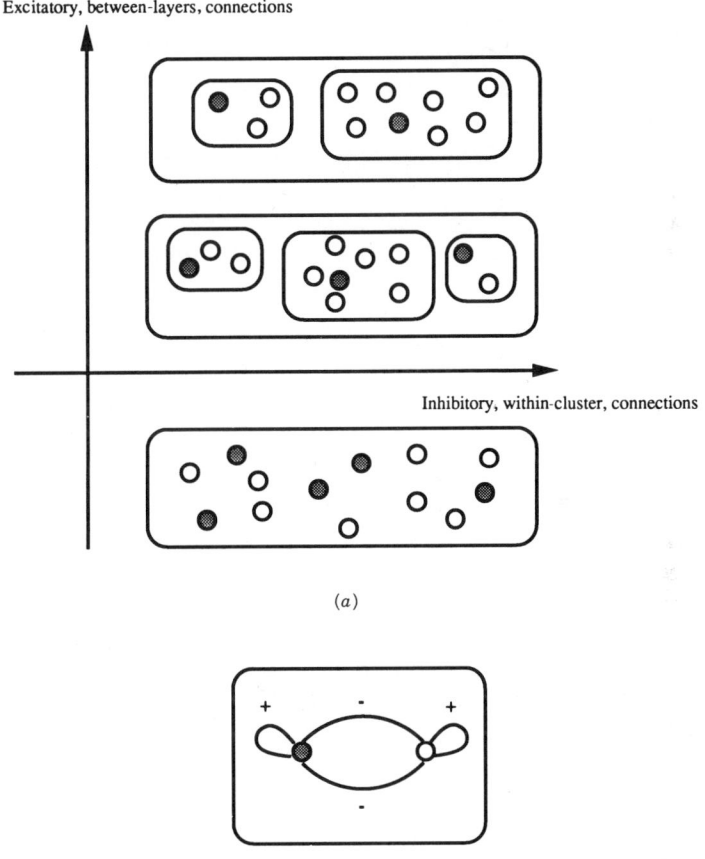

Figure 3.7 (a) A representation of the competitive learning architecture. Note the arrows indicating excitatory between layer connections (vertical arrow) and inhibitory within cluster connections (horizontal arrow). (b) Detail of a cluster. Two units in competition are represented along with their inhibitory and self-excitatory connections.

where W_{ij} is the connection strength between unit i on the lower layer and unit j on the upper layer. $\Sigma_i W_{ij} = 1$ (pseudonormalization), n_k is the number of units turned on (1) in pattern $\mathbf{u}_k$ in the lower layer, and $\mathbf{u}_{ik}$ is the activation value (0 or 1) of the ith unit in pattern $\mathbf{u}_k$. The variable $\mathbf{u}$ can either be an input stimulus or the activation pattern of an intermediate layer. The first component of (3.21) is the important feature of unsupervised learning procedures. It represents the operation of adding a proportion of the input vector to the weight vector afferent to the winning units. Thus, in the long term, if a unit in a cluster wins on a specific pattern, it is likely that the pattern is highly correlated with the weight vector afferent to the unit.

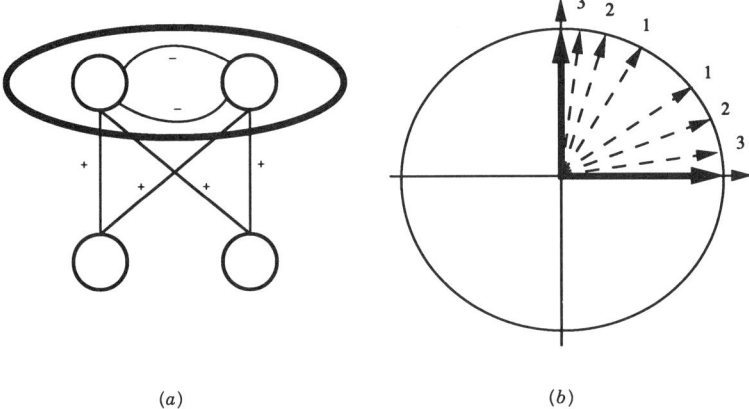

Figure 3.8 (a) the competitive learning architecture applied to the problem of learning to discriminate two categories. The output units forms a single inhibitory cluster. (b) Geometrical representation of the learning procedure. Here, the weight vectors (dashed arrows) are normalized—the length is 1. The input vectors are represented with plain arrows. [1, 0] represents the only exemplar that characterizes category 1, and [0, 1] is the exemplar that characterizes category 2. As we can see, with the number of iterations (1, 2, 3) the weight vectors rotate in weight space in direction of the input vector in input space.

A straightforward geometrical interpretation of the learning procedure is given in Figure 3.8. The example shows the learning of two categories, which, for simplicity, are described along two dimensions—the generalization to n dimensions is mathematically straightforward, but graphically impossible. Each category comprehends a unique exemplar, described by [1, 0] for the first category and [0, 1] for the second one. In the example, the weights are assumed to be strictly normalized (their length is 1).

The illustration given in Figure 3.8 is, of course, oversimplified. Very few categories are characterized by a single exemplar. The example was meant to illustrate the overall effect of the learning procedure, rotating the weight vectors in weight space in direction of relevant subspaces of the input space. After learning, the weight vectors filter the input pattern so that a specific unit in each cluster responds to a different class of patterns.

In the most general case, a cluster with m units can be thought of as an m-ary feature detector in which every stimulus is classified by having one of the possible values of this feature. The learning scheme can be expanded to n exemplars per category. If the patterns are described as variations around a prototype, the weight vectors will pick out the high-level features that describe the most characteristic features of the categories, those of the prototypes themselves.

Thus, competitive learning *discovers* the structure of the input space. What are the limitations on the type of problems a competitive scheme can learn? Is it as severely limited as simple networks analogous to the perceptron? This issue was addressed by Rumelhart and Zipser [27]. They showed that a multilayer competitive learning architecture could learn to discriminate, under certain circumstances, nonlinearly separable classes—vertical and horizontal lines. This characteristic makes competitive learning a powerful learning device.

The major drawback of competitive learning is its instability. The learning procedure can be thought of as driving the system to equilibirum states. At equilibrium, the weights of the system do not change anymore. Thus,

$$\alpha \mathbf{u}_{ik}/n_k = \alpha \mathbf{W}_{ij} \qquad (3.22)$$

Each equilibrium state the system can reach corresponds to a specific set of classifications of the pattern set. When the stimuli population is highly structured, the classifications are highly stable. In other words, the system stays in the same equilibrium state, and the same units in the clusters always respond to the same class of patterns. However, if the external world is poorly structured, the classifications are more variable. The system will meander around in state-space from one stable state to another one, and a particular unit will respond to a given class, then to another one, and so on. If the stimuli do not fall into nice, well-separated clusters, the system undergoes continual evolution.

3.3.2 The Stability-Plasticity Dilemma and ART

Self-organizing networks face a difficult problem. They have no a priori knowledge of the different categories that structure the input space. When a new pattern appears, if it is close to a learned pattern, the system has to decide whether it is an exemplar of an old category or an exemplar of an, as yet, unknown category. If the network is sufficiently plastic, a new exemplar will be incorporated as the first member of a new category. If it is not plastic enough, new categories will not be learned. Clearly, a system cannot be too plastic because it would create a new category for each exemplar it encounters. Abstraction is frequently obtained at the cost of some, but not all, plasticity.

If the amount of resources available to represent the input space is finite, plasticity can lead to another problem. Assuming we have a nonstationary distribution of the input patterns—a distribution that changes with time—old knowledge could well be washed away by new knowledge. This occurs to some extent in humans. However, carried too far, this property would be very annoying. Imagine that every 24 h, your entire conceptual memory was washed away by new concepts; you had better make sure to go home everyday to keep this concept stored. Stability is a desirable property of a knowledge system.

Very few architectures are at the same time stable and plastic. The simple associators, the gradient descent and competitive learning algorithms work well under a stationary input distribution. Controlling the trade-off between stability and plasticity is a difficult issue. At least two sources of control should be considered. First, some knowledge of the input distribution characteristics should be included in the system (its spread, whether categories are unimodal or multimodal, etc.). Second, the input exemplars that are close to one another in the real world should be represented as close to one another in the input space.

Grossberg [21] adaptive resonance theory (ART) handles these issues with an elaborate nonlinear architectural design. The flexibility of the network—its plasticity—is achieved by an acceptance, or "vigilance," parameter, which measures the acceptable distance between the new exemplar and the prototypes of the old categories. If the new exemplar is within the range of acceptability of an old category, it will be learned as an exemplar of this category and eventually update the prototype of the category. In general, a sequential search through the known categories tries to find a matching within the acceptance range. If no match is found, a new unit is assigned to the pattern, and a new exemplar is thereby attached to a new category.

ART's functional architecture has two distinct parts. The first part is similar to the competitive learning scheme presented before. It consists of an input layer and a set of grandmother cell output units. These grandmother cells are nested in a winner-take-all network such that when an input is presented on the input layer, the activation of one output unit depicts the category of which the exemplar is a member. Stability is mainly achieved by top-down connections that go from the grandmother cells to the input layer. In a typical learning epoch, the top layer tries to match the input presented on the input layer using the top-down connections. If the input closely matches the top-down weights, a feedback loop develops and, ultimately, a pattern is formed in the first layer and a single unit is activated on the grandmother cell layer. Resonance refers to this feedback loop that resonates in the network when the input is categorized by one cell on the top layer.

The other component of ART's functional architecture controls the first one. It makes the decisions relative to the acceptance of patterns; it resets the first layer when the architecture tries to recognize the input with another grandmother cell. Because of its grandmother cells top layer, ART has some problems dealing with the relational and dynamical characteristics of conceptual memory. Conceptual memory is organized hierarchically; we know that a German shepherd is a dog and that a dog is a mammal that is itself an animal. These hierarchical dependencies are not handled easily in ART. A different layer of output units is needed to map each level of the conceptual hierarchy. However, since ART is a self-organizing architecture, this structure of the input space—the fact that the categories are hierarchically dependent—is not strictly captured by the learning procedure.

Conceptual memory is dynamic in the sense that concepts are in perpetual evolution. Children start by differentiating *bird* from *cat* from *dog*, with very crude, poorly refined concepts. With experience, as they pick out more and more structure inside a particular category, they make finer discriminations in the domain of knowledge. They can distinguish *sparrow* from *pigeon*, *pinscher* from *German shepherd*. However, the relationships between their initial broad concept and the finely tuned ones are not lost. A German shepherd remains a dog, and a sparrow is still a bird. The mechanisms by which knowledge refinement occurs seem to be related to the mechanisms by which hierarchical dependencies are constructed. Due to its grandmother cell recoding of the input space, ART handles these characteristics of conceptual development with difficulty. The acceptance parameter should be changed at some stage of learning so that patterns initially categorized as members of an old category may form a new category. At the same time, something should keep track of the relationship between general and specific concepts. ART could solve these issues, but the solution might be awkward.

3.4 UNSUPERVISED AND SUPERVISED TECHNIQUES: TOWARD A MODULAR APPROACH OF CONCEPT LEARNING

When a mechanic sees a car, he or she may not only see a car, as a novice would do, but also a particular model of a car, that has a particular kind of engine, that is made by a particular company. In a word, it is said that the car mechanic is an *expert* in a specific domain of knowledge, cars. Rosch et al. [29] suggested that expertise was characterized by a larger amount of low-level concepts that allow experts to deal with the larger amount of details they know about. As alluded to earlier, the mechanisms by which expertise is acquired can be thought of as similar to those by which conceptual hierarchies are constructed. Both cognitive tasks are characterized by a reorganization of conceptual knowledge where crude concepts are refined into more specific ones. As concepts get sharpened, categorization is affected. With more low-level concepts, a domain of knowledge can be classified more finely. According to the account just given, expertise is not an all-or-none state. Every domain of knowledge is conceptualized with some degree of expertise. The threshold of expertise needed to be labeled "expert in X" is more a matter of cultural norms than a normative amount of knowledge.

The development of expertise is an important aspect of knowledge acquisition. Ideally, a system should constantly update its knowledge base as a function of its experiences. Two fundamental operations underlie this development. The first is the representation of concepts, the second is the representation of hierarchical relationships. A definitive solution to these problems would be a breakthrough in knowledge representation. Unfortunately, we must remain modest, and see in actual models only feeble metaphors, not to say guidelines, of how these tasks could be accomplished.

Expertise has really two facets. On one hand, conceptual expertise, on the other hand, lexical expertise. As discussed earlier, labels can be conceived of as indicators of knowledge. Nothing is more true in expertise. Take for example the category *bird*. Most of us are able to categorize distinctively various sorts of birds. In the course of our life, we have come to learn some concepts that allow a segmentation of this category into a couple of subcategories. However, not being zoologists or bird lovers, we may not know the name of each species we are able to classify. This argument was made before to motivate an unsupervised learning scheme.

Since we are able to distinguish different subclasses of birds, we have distinct concepts for these subclasses. Therefore, we should be able to tag the subcategories the concepts refer to with arbitrary symbols such as *eagle, sparrow,* and so on, if these symbols were given to us: In general, it is impossible to guess the arbitrary names that, by linguistic conventions, have been assigned to categories. Thus, the network has to be told what name should be associated with what concept.

For these reasons, a modular architecture seems appropriate to model expertise acquisitions. An unsupervised module that handles concept learning and the development of conceptual hierarchies, and a supervised module that tags the learned concepts with category names. Schyns [30,31] devised a model with these characteristics. The unsupervised module, based on Kohonen's self-organizing feature map, achieved concept learning and categorization of objects [30,32,33]. The tagging of concepts was implemented by the supervised module, an autoassociator using error correction. As we will see, this model is able to refine its conceptual knowledge by constructing hierarchical dependencies. Its overall architecture is presented in Figure 3.9.

Kohonen's self-organizing feature map is somewhat similar to competitive learning. A layer of input units is connected to a layer of output units, or feature detectors. The single layer of weights in competitive learning and in Kohonen maps performs the same function: It recodes the input space. The major differences lie in the structure of the output layer. Rather than being segmented into distinct clusters, the output units of a Kohonen map are organized as a two-dimensional uniform sheet of neurons—the map. The learning rule is also different. Schyns [33,34] developed a learning rule whose functional characteristics were similar to Kohonen's [25].

$$o_w = \text{WTA}_i(\mathbf{w}_i^T \mathbf{i}) \qquad (3.23)$$

$$\Delta \mathbf{w}_i = \begin{cases} \mathbf{i}[(1 - |o_w|) * \text{lc}(o_i, o_w)] & \text{for } o_i, \text{ locally connected to } o_w \\ 0 & \text{otherwise} \end{cases} \qquad (3.24)$$

Similarly to competitive learning, the activation of all units on the map is computed by taking the inner product of the input vector with every weight

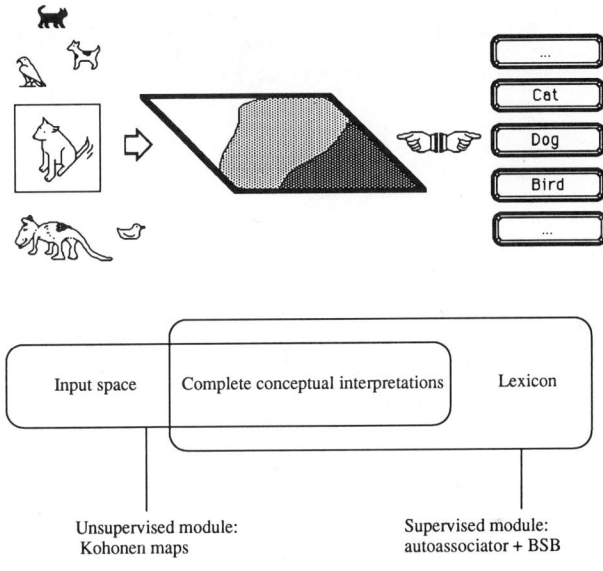

vector afferent to the output units. A winner-take-all scheme is applied to select o_w, the unit of the entire two-dimensional sheet that has the largest activation (3.23). Once the winner is selected, its afferent weights, *as well as those of its neighbors,* are updated (3.24). The updating rule is a bit difficult. The expression in brackets computes a proportion of the input vector. Thus, as in (3.21), the weight vectors are updated by this proportion. So far, I have purposefully left an important detail unexplained, the neighborhood structure. Every unit on the map is locally connected to its neighbors, with Gaussian connections, up to a fixed Euclidean distance. The strength of the connections is a function of distance: the further apart the units, the lower the strength. $lc(o_i, o_w)$ gives the local connections strength between unit i of the map and the winning unit w. So, when the winning unit is known, its weight vector, as well as those of the locally connected units, are updated by a proportion of the input vector. This proportion depends on $lc(o_i, o_w)$ and on the inverse of the activation of the winning unit $(1 - |o_w|)$.

The effect of the learning procedure is complex. First, the weight vectors afferent to a *region* of the map around the winning unit rotate in input space; different regions will respond to different classes of patterns. As the weight vector of a winning unit rotates in direction of a particular class of patterns, the unit's activation values become closer and closer to 1. Thus, $\Delta \mathbf{w}_i$ of neighboring units gets smaller and smaller with learning. Since the $\Delta \mathbf{w}_i$ were

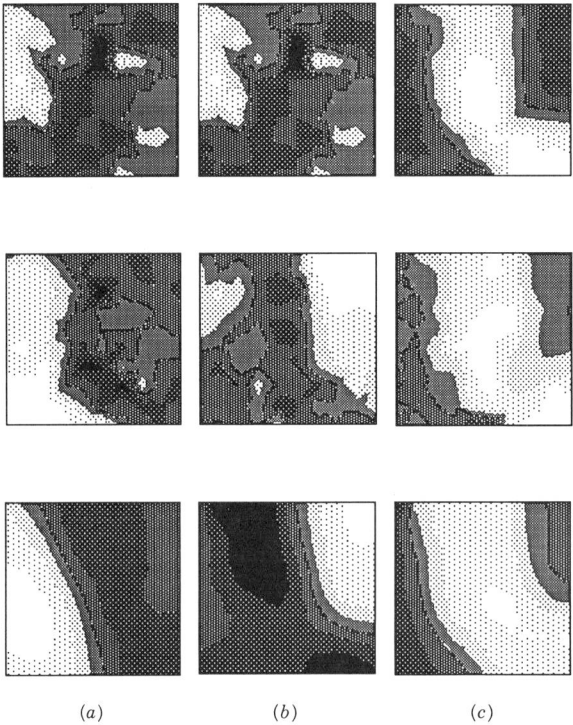

(a)   (b)   (c)

Figure 3.10 *Development of the conceptual map after 10 (a), 50 (b), 1000 (c) iterations of learning. Column a shows the development of the region that responds mostly to category 0, b the region responding mostly to category 1, c the region that encodes category 2.*

already affected by low Gaussian connection strengths, the neighborhood structure tends to shrink dynamically as learning proceeds.

Kohonen has proven that a similar one-dimensional architecture would converge on a solution provided the neighborhood size shrinks with time and the learning constant decreases with time [25]. With the local connections and the learning rule presented here, these two conditions occur as a side effect of the network dynamics—no explicit mechanism, exterior to the learning system, is required [30].

In the following experiment, three categories were constructed around three prototype in a way qualitatively similar to Knapp and Anderson [4]. Exemplars were selected randomly from the three categories and presented to a Kohonen map of 10 × 10 output units. Figure 3.10 shows the organization of the two-dimensional sheet of output units after 10, 50, and 1000 iterations of learning.

A set of units delimits a region that responds with the highest activation to exemplars from a specific category, thus the name *conceptual map* (for a discussion of the virtues of conceptual map as encoder of conceptual infor-

3.4 UNSUPERVISED AND SUPERVISED TECHNIQUES

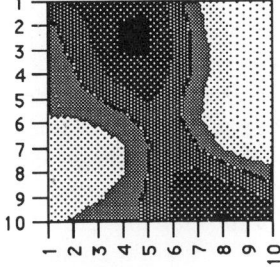

Exemplar 1 from **P1**
Winning unit location: 9 5
Activation value of 9 5: 0.738
Vector_cosine(**Ex**, **P0**) = 0.457
Vector_cosine(**Ex**, **P1**) = 0.732
Vector_cosine(**Ex**, **P2**) = -0.14

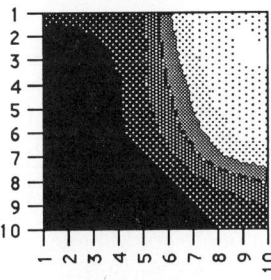

Exemplar 2 from **P1**
Winning unit location: 9 2
Activation value of 9 2: 0.813
Vector_cosine(**Ex**, **P0**) = -0.30
Vector_cosine(**Ex**, **P1**) = 0.771
Vector_cosine(**Ex**, **P2**) = -0.10

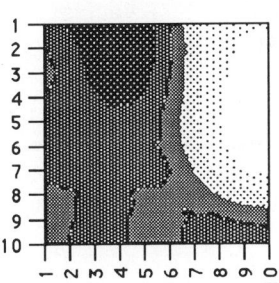

Exemplar 3 from **P1**
Winning unit location: 8 5
Activation value of 8 5: 0.901
Vector_cosine(**Ex**, **P0**) = 0.234
Vector_cosine(**Ex**, **P1**) = 0.855
Vector_cosine(**Ex**, **P2**) = 0.009

mation see Ritter and Kohonen [32] and Schyns [35]). On the maps, we can observe that category 0 is encoded in the lower left corner, category 1 in the upper right corner, and category 2 in the diagonal.

As categories are organized around prototypes, the weight vectors underlying the units of different regions represent the prototypes of the categories. By (3.23), the activation over a region shows a graded response that depends on the similarity, or distance, between the exemplar presented and the prototype. The closer the input exemplar is to the prototype, the higher is the activation over the region that encodes the category. More technically, each region provides a typicality measure—level of activation—where every known concepts—the prototypes—contribute to the evaluation of the input. This is shown in Figure 3.11.

Three exemplars from category 1 are presented to the network. The map has each time a higher activation over the region that encodes category 1, and the activation over the other regions depends on the similarity between the exemplar and the prototypes of each category—indicated on Figure 3.11 by the vector_cosines. Observe, for example, that the exemplar 2 is very similar to the prototype of category 1 (the vector_cosine measure is 0.771), but it is highly dissimilar to the prototype of category 0 (the vector_cosine measure is −0.30). Therefore, the region of the map that encodes category 0 is highly inhibited, reflecting that the exemplar is not a member of category 0. The first exemplar is more ambiguous. It is more similar to category 1 (vector_cosine = 0.732) than to category 0 (vector_cosine = 0.457). Thus, the state of activation over both regions is high, with a slight difference in favor of category 1. Figure 3.11 shows an important property of conceptual maps. There are inherently ambiguous. Since they reflect the interpretation of the exemplar with all the available conceptual knowledge of the system, I will refer to the state of activation of the map as a *complete conceptual interpretation of the exemplar*.

I have presented how an unsupervised module could build complete conceptual interpretations of exemplars. Complete conceptual interpretations are eventually named by the supervised module. To achieve this goal, an autoassociator and error-correction constitute a first, simple, approximation. An autoassociator has the same architecture as the pattern associator, where input and output units are the same state vector **s**. The vector **s** is the concatenation of components **s′** and **s″**, where **s′** contains the complete conceptual interpretation of the exemplar and **s″** is an ASCII representation of the correct category label of the exemplar. Typically, learning to tag complete conceptual interpretations proceeds as follows:

1. Present an exemplar from category A to the unsupervised module.
2. Compute the complete conceptual interpretation of the exemplar.
3. Autoassociate, with error correction, the conceptual interpretation with the label of category A.

Once the names are learned, they can be retrieved with the brain-state-in-a-box (BSB) dynamic search [36–40]. Equation (3.26) presents BSB.

$$\mathbf{s}(t+1) = \lim (\alpha \mathbf{s}(t) + \beta \mathbf{W}\mathbf{s}(t)) \qquad (3.26)$$

To retrieve the category name of an exemplar from its conceptual interpretation, the state vector **s** is filled with its first component **s′** and **s″** is initialized to **0**. BSB attempts, through its dynamic search for stable states, to fill **s″** with the correct category name. The state vector at time t is multiplied by the autoassociative matrix to give **s** at $t + 1$. Eventually, the system reaches an energy minimum, and **s″** is filled. Put differently, BSB disambiguates the ambiguous conceptual interpretation of an exemplar and assigns a

3.4 UNSUPERVISED AND SUPERVISED TECHNIQUES   99

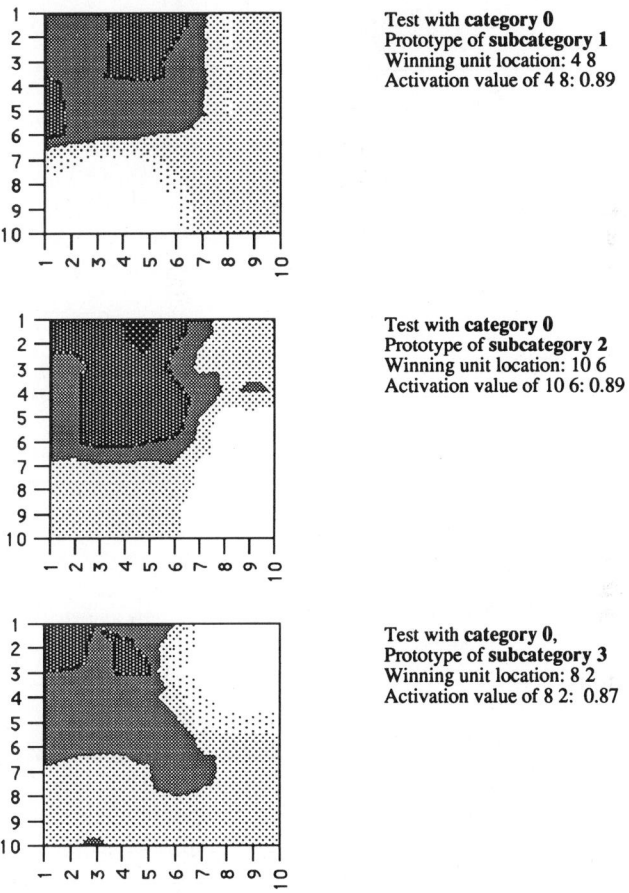

Figure 3.12 *Test of the conceptual map with the prototypes of the three subcategories the system is an expert of.*

name to it. (For a more detailed account of this specific procedure, see [30]. For a formal account of dynamic search, see [36,38,41–43].)

Expertise and hierarchical relationships are natural extensions of the architecture presented. If we make the assumption that an expert sees more frequently his or her category of expertise than the categories for which he or she is a novice, expertise is given a simple account. For simplicity, assume the system experiences four categories with the following probabilities: .65, .2, .1, and .05. Assume also that each category can be decomposed into three subcategories very similar to one another. After learning, typical conceptual interpretations of exemplars look like those of Figure 3.12.

In Figure 3.12, the network is tested with the prototypes of the different subcategories. We can see a large, right-angle-shaped region of activation covering the right and bottom sides of the maps. This high activation indi-

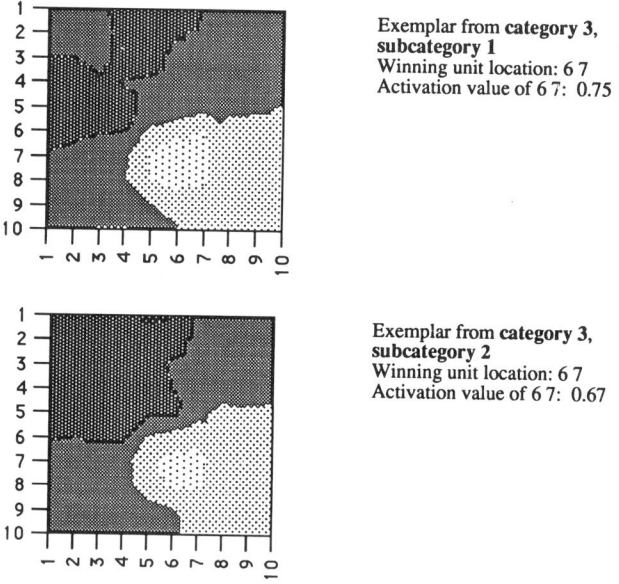

Figure 3.13 Test of the conceptual map with the prototypes of two out of the three subcategories of which the system is a novice.

cates that the exemplar is a member of category 0, the most frequent one. The three maps show an important property: *conceptual nesting*. A specific subregion of the large region responding to category 0 has a higher state of activation, indicated by a white patch on the map. This subregion indicates the subcategory of which the exemplar is a member. We can observe on Figure 3.13 that three different subregions encode the three subcategories of category 0. Thus, the category of expertise has all its structure represented, high-level as well as low-level concepts. What about the other categories? Consider, for example, the least frequent category. The results are shown in Figure 3.13.

The first map shows the conceptual interpretation of an exemplar from category 3, subcategory 1. The output of the unsupervised module for this exemplar can be compared with the output produced with an exemplar from category 3, subcategory 2. No significant difference in the conceptual interpretations of the subcategories is observed. However, a specific region located roughly at the center of the map responds to the category as a whole. Thus, the internal structure of the least frequent category is not reflected on the conceptual interpretations. Figures 3.12 and 3.13 present the conceptual interpretations of two extremes of the probability distribution. The other categories have their internal structure conceptual expertise in different domains of knowledge as alluded to before. Some categories are represented with a large amount of details, and others are represented more crudely.

The explanation for this phenomenon is in the weight vectors underlying the map. The learning procedure rotates the weight vectors in the direction of interesting subspaces of the input space. Since the three subcategories of each category are close to one another, the weight vectors that represent them lie close to one another in different subspaces of a region of the input space. Because the learning procedure updates the weights of a winning units and those of its locally connected neighbors (3.24), interferences can occur in the updating of the weights representing subcategories. The number of output units allocated to a region is a function of the probability of the category represented by the region [34]. So, if a category has plenty of space to be represented—if the category is very frequent—the weight vectors in the region will be able to pick out the internal structure of the category. On the contrary, if the category has little amount of resources to be represented, the weight vectors will not pick out ali the internal structure of the category. The reason for this is the *critical perimeter of interference* [30]. As the weights are updated in a region, in order to have two distinct regions, they have to be sufficiently separated from one another so that the updating of one region will not interfere with what the other region encodes. Thus, the network displays an interesting property. It allocates dynamically representational resources—output units—as a function of the frequency of each category.

If distinct subregions corresponding to different subcategories are formed on the map, conceptual interpretations of exemplars can be easily tagged with hierarchies of labels—for example, *animal, bird, eagle*. On the contrary, if the subregions are not distinguishable on the conceptual interpretations, as is the case for the least experienced categories, the low-level labels will not be tagged to a distinguishable portion of the conceptual interpretations. The psychological implication of these phenomena is straightforward. A category of expertise can be named at all levels of abstraction (for example, *animal, bird, sparrow*), whereas a category for which the system is a novice can only be named at a certain level, indicating that low-level concepts have not been integrated (for example, *animal, bird*). This phenomenon reflects the spectrum of lexical expertise, which depends on the spectrum of conceptual expertise. Thus, this model provides an example of arbitrary symbols—the category names—that stands for indicators or pointers to specific pieces of knowledge.

3.5 CONCLUSIONS

In this chapter, I have discussed the virtues of different neural network modeling strategies in the context of concept learning. Representations were of prime importance. The simple linear associator was presented first; and I stressed how nonlinear category learning problems could be solved provided a suitable recoding of the input space. The major problem was, of course, to justify the providential *external* mechanism that would achieve the necessary preprocessing. Back propagation was presented as a multilayer nonlinear method that could recode its input space in the hidden units and thereby solve hard learning problems through the construction of *internal* representations. We saw that the construction of internal representations through supervised means was not sufficient to account for concept learning. Unsupervised learning schemes that build internal codes without a teacher were presented as an alternative. The major goal of these learning techniques is to extract the structure of the external world by constructing high-level feature detectors that respond preferentially to certain categories of objects. I completed the presentation of the models with a modular architecture in which unsupervised learning was extracting concepts while supervised learning was tagging conceptual interpretations of exemplars with the correct category name.

I think that modular approaches to difficult tasks such as categorization or object recognition have a promising future in neural networks. Modular architectures could be organized hierarchically so that modules sitting at the bottom of the hierarchy, implemented with unsupervised learning techniques, would progressively extract higher- and higher-level features from the environment. At the top of this modular hierarchy would stand a supervised learning scheme that associates the output of that hierarchical preprocessing of the input sample with the appropriate action to undertake. A modular approach has been suggested by several authors as an attractive solution to the scaling problem that pervades current powerful supervised learning techniques such as back propagation [16,44]. The general idea is to progressively preprocess the input sample to construct a good representation that reduces the burden of the supervised module that stands on top of the hierarchy. Another attractive feature of modular architectures is their ability to construct *task-independent* representations. These representations could potentially be used by many modules and many problems would be solved by the same architecture.

A modular design and the divide and conquer approach generate difficult problems. In general, since large and difficult learning tasks are decomposed into many independent subtasks realized by different modules, a crucial issue relates to the segmentation of the task. From an engineering point of view, this problem could be solved with criteria of efficiency and optimality. However, from a cognitive science standpoint, this segmentation poses theoretical difficulties. Functionally distinct operations (functions that characterize the task) may not correspond to structurally independent modules in the brain. Therefore, if one's goal is to model psychological phenomena, the segmentation of the task should have clear empirical and theoretical motivations (see Schyns [30] for a discussion of this issue). The propagation of the flow of information is another important aspect of modular architectures.

Because knowledge is "informationally encapsulated" in each module, a particular module is blind to the other's processes and representations. Since modules communicate only at their input and output stage, the interface between two modules plays an important role in encoding and transmitting relevant information. In the modular model presented earlier, the interface between the modules was the conceptual map. I have shown how it could construct a code in which the hierarchical relationships between concepts, as well as the level of expertise in each category were encoded on the map. I presented in [30,31] how the correct learning of category names is constrained by the code constructed at the interface between the two modules—how category names are grounded on conceptual knowledge. The coding capacities of the channel of communication between modules should be the focus of research in further modular models.

All current research in neural network models of concept learning have assumed that exemplars from the same category were very similar to one another. In other words, the conceptual coherence of categories is achieved by a relationship of similarity between exemplars. The constraint of similarity is also important in human concept learning. However, it does not have the *necessary* character assumed by the models reviewed in this chapter. Murphy and Medin [45] proposed that high-level knowledge could be another important means by which humans would form categories. For example, having a good diet, working out, sleeping enough are exemplars of the category of "things to do to be in good health." Clearly, the exemplars of this category are not perceptually similar to one another as Doberman pinscher and German shepherd are. Modeling this high-level knowledge and the kind of inferences it allows in concept learning should be an important avenue of research in the future.

REFERENCES

[1] J. A. Anderson, A simple neural network generating an interactive memory, *Math. Biosciences,* 14, 197–220 (1972).

[2] J. A. Anderson and G. E. Hinton, "Models of information processing in the brain," in G. E. Hinton and J. A. Anderson, Eds., *Parallel Models of Associative Memory,* Lawrence Erlbaum, Hillsdale, N.J., 1981.

[3] T. Kohonen, *Associative Memory—A System Theoretic Approach,* Springer-Verlag, Berlin, 1977.

[4] A. G. Knapp and J. A. Anderson, Theory of categorization based on distributed memory storage, *J. Experimental Psychology: Learning, Memory and Cognition,* 10, 616–637 (1984).

[5] J. L. McClelland and D. E. Rumelhart, Distributed memory and the representation of general and specific information, *J. Experimental Psychology: General,* 114, 159–188 (1985).

[6] D. Homa and D. Chambliss, The relative contributions of common and distinctive information on the abstraction from ill-defined categories, *J. Experimental Psychology: Human learning and Memory,* 1, 351–359.

[7] M. I. Posner and S. W. Keele, On the genesis of abstract ideas, *J. Experimental Psychology,* 77, 353–363 (1968).

[8] M. A. Gluck and G. H. Bower, From conditioning to category learning: An adaptative network model, *J. Experimental Psychology: General,* 117(3) 227–247 (1988).

[9] J. E. Corter, M. A. Gluck, and G. H. Bower, Basic levels in hierarchically structured categories, *Proceedings of the Tenth Annual Conference of the Cognitive Science Society,* Montreal, Canada, Hillsdale, NJ: Lawrence Earlbaum Associates.

[10] M. Minsky and S. Papert, *Perceptrons: An Introduction to Computational Geometry,* MIT Press, Cambridge, Mass., 1969.

[11] F. Rosenblatt, The perceptron: A probabilistic model for information storage and organization in the brain, *Psychological Review,* 65(6) 386–408 (1958).

[12] D. E. Rumelhart, G. E. Hinton, and R. J. Williams, "Learning Internal Representation by Error Propagation," in J. L. McClelland and D. E. Rumelhart, *Parallel Distributed Processing: Explorations in the Microstructure of Cognition,* Bradford Books, Cambridge, Mass., 1986.

[13] D. E. Rumelhart, G. E. Hinton, and R. J. Williams, "Learning Representations by backpropagating erros, *Nature,* **323,** 533–536, (1986).

[14] R. P. Gorman and T. J. Sejnowski, Analysis of hidden units in a layered networks trained to classify sonar targets, *Neural Networks,* 1, 75–89 (1988).

[15] S. C. Johnson, Hierarchical clustering schemes, *Psychometrika,* 32, 241–153 (1967).

[16] D. E. Rumelhart, "The Architecture of Mind: A Connectionist Approach," in M. I. Posner, Ed., *Foundations of Cognitive Science,* Bradford Books, Cambridge, Mass., 1990, pp. 133–159.

[17] J. A. Fodor, *The Modularity of Mind,* MIT Press, Cambridge, Mass., 1983.

[18] K. L. Chapman, L. B. Leonard, and C. B. Mervis, The effect of feedback on young children's inappropriate word usage. *J. Child Language,* 13, 101–117 (1986).

[19] P. C. Quinn and P. D. Eimas, On categorization in early infancy, *Merrill-Palmer Quarterly,* 32(4) 331–363 (1986).

[20] K. Fukishima, S. Miyake, and T. Ito, Neocognitron: A neural network model for a mechanism of visual pattern recognition, *IEEE Trans. Systems, Man and Cybernetics,* SMC-13, 826–834 (1983).

[21] S. Grossberg, Adaptive pattern classification and universal recoding: I. Parallel development and coding of neural feature detectors, *Biological Cybernetics,* 23, 121–134 (1976).

[22] S. Grossberg, How does the brain build a cognitive code? *Psychological Review,* 87, 1–51 (1980).

[23] S. Grossberg, Competitive learning: From interactive activation to adaptive resonance, *Cognitive Science,* 11, 23–63 (1987).

[24] T. Kohonen, Self-organized formation of topologically correct feature maps, *Biological Cybernetics,* 43, 59–69 (1982).
[25] T. Kohonen, *Self-organization and Associative Memory,* Springer-Verlag, Berlin, 1984.
[26] T. Kohonen, The neural phonetic typewriter, *IEEE Computer,* 21, 11–22 (1988).
[27] D. E. Rumelhart and D. Zipser, Feature discovery by competitive learning, *Cognitive Science,* 9, 75–112 (1985).
[28] C. von der Marlsburg, Self-organization of orientation sensitive cells in the striate cortex, *Kybernetik,* 15, 85–100 (1973).
[29] E. Rosch, C. B. Mervis, W. D. Gray, D. M. Johnson, and P. Boyes-Braem, Basic objects in natural categories, *Cognitive Psychology,* 7, 273–281 (1976).
[30] P. G. Schyns, (1990). A modular neural network of concept and development, *Cognitive Science,* in press.
[31] P. G. Schyns, A modular neural network model of the acquisition of category names, in D. Touretzhi, J. Elman, T. Sejnowski, and G. Hinton (Eds), Proc. 1990 Connectionist Models Summer School, Morgan Kaufmann, San Mateo, CA , 1990.
[32] H. Ritter and T. Kohonen, Self-organizing semantic maps, *Biological Cybernetics,* 62(4) 241–255 (1989).
[33] P. G. Schyns, "Concept Representation with a Self-Organizing Architecture," Cognitive and Linguistic Sciences Department, Brown University, Providence, R.I., 1989.
[34] P. G. Schyns, Expertise acquisition through the refinement of a conceptual representation in a self-organizing architecture, *Proc. Internat. Joint Conference on Neural Networks,* 1990.
[35] P. G. Schyns, On the virtues of two-dimensional self-organized maps as a medium to encode conceptual knowledge, *Neural Network Review,* in-press.
[36] J. A. Anderson, J. W. Silverstein, S. A. Ritz, and R. S. Jones, Distinctive features, categorical perception, and probability learning: Some applications of a neural model, *Psychological Review,* 84, 413–451 (1977).
[37] J. A. Anderson and G. L. Murphy, Psychological concepts in a parallel system, *Physica,* D22, 318–322 (1986).
[38] R. M. Golden, The "brain-state-in-a-box" neural model is a gradient descent algorithm, *J. Math. Psychology,* 30, 73–80 (1986).
[39] Golden, Modelling neural schemata in human memory: a connectionist approach, unpublished Doctoral thesis, Psychology Department, Brown University, 1987.
[40] A. H. Kawamoto, Distributed representations of ambiguous words and their resolution in a connectionist network, in S. Small, G. Cottrell, and M. Tanenhaus, Eds., *Lexical Ambiguity Resolution: Perspectives from Psycholinguistics, Neuropsychology, and Artificial Intelligence,* Morgan Kaufmann, San Mateo, Calif., 1988.
[41] D. H. Ackley, G. E. Hinton, and T. J. Sejnowski, A learning algorithm for Boltzmann machines, *Cognitive Science,* 9, 147–169 (1985).

[42] J. J. Hopfield, Neural networks and physical systems with emergent collective computational capabilities, *Proc. National Acad. Sciences,* 79, 2554–2558 (1982).

[43] J. J. Hopfield, Neurons with graded response response have collective properties like those of two states neurons, *Proc. National Acad. Sciences,* 81, 3088–3092 (1984).

[44] G. E. Hinton and S. Becker, An unsupervised learning procedure that discovers surfaces in random-dot stereograms. *Proc. Internat. Joint Conference on Neural Networks,* **1,** 218–222, Lawrence Earlbaum Associates, Hillsdale, NJ , 1990.

[45] G. L. Murphy and D. L. Medin, The role of theories in conceptual coherence, *Psychological Review,* 92(3) 289–315 (1985).

CHAPTER 4
Fast Algorithms for Training Multilayer Perceptrons

NAZIF TEPEDELENLIOGLU,
ALI REZGUI,
ROBERT SCALERO, AND
RAMONA ROSARIO

4.1 INTRODUCTION

An artificial neural network is an interconnection of computational elements known as neurons or nodes. A type of neural network that has found wide application in many classification problems is the feedforward multilayer perceptron [1–5]. The supervised training of such an artificial neural network is the subject of this chapter. For our purposes the neural net can be viewed as mapping a set of input vectors to another set of desired output vectors. To train the neural net to perform the desired mapping, we apply an input vector to the net and compare the actual output of the net with the desired output (the output vector corresponding to the applied input). The difference between the actual output and the desired output (i.e., the error) is used to update weights and biases associated with every neuron in the network until this difference averaged over every input/output pair is below a specified tolerance. Such training is termed *supervised training* due to the availability of this difference during the training procedure. A net performs the desired mapping when each neuron in the net yields a correct response.

When the net is composed only of a single layer, the correct response of every node is the desired output of the net. This fact makes the single-layer perceptron easy to train.

A single-layer perceptron, however, is limited in performance to the so-called linearly separable problems. If one wants to deal with problems that

---

Neural and Intelligent Systems Integration, By Branko Souček and the IRIS Group.
ISBN 0-471-53676-8 ©1991 John Wiley & Sons, Inc.

are not linearly separable, like the XOR problem for example, more than one layer is needed [6,7]. Thus in the feedforward architecture, the neurons are arranged in consecutive layers, with inputs to neurons in the first layer being the inputs to the net, and outputs of neurons in the last layer being the outputs of the net. Inputs to neurons in the other layers—the hidden layers—are the outputs of neurons in the immediately preceding layer.

The addition of hidden layers, however, presents an obstacle: lack of specification of a correct response for nodes in the hidden layers. Until very recently, this lack prevented the development of effective supervised training algorithms, which resulted in limited application for multilayer nets.

The development that makes the training of a multilayer network possible is the backpropagation algorithm (backprop for short) [6,8–11].

Backprop first uses the known difference between the desired and actual outputs of nodes in the output layer to come up with an error in the output of nodes in the hidden layers. It then employs the LMS algorithm [12,13] to update the weights and biases. The realization that these two functions—estimation of error and updating of weights—are distinct and separate has an important ramification: It allows variations on the weight update mechanism, to improve performance. The error at the hidden nodes can still be estimated by the same method employed by backprop, but the algorithm that uses the error to update the weights and biases need not be the LMS algorithm; it can be any one of a number of more sophisticated descent methods. Replacing the LMS algorithm in this manner can do away with several shortcomings of the standard backprop algorithm: shortcomings such as inconsistent training, getting stuck (i.e., failing to converge for a choice of initial weight vectors), slow convergence (i.e., the necessity of having to present the input vector set an inordinately large number of times before the output error is tolerable). Section 4.5 details one such algorithm: the Kalman algorithm. Whereas the LMS algorithm is a first-order descent method (weight update is governed by the first derivative of the error), the Kalman algorithm is a second-order descent method (weight update is governed by the first and second derivatives of the error), thus bringing about a faster rate of convergence.

Another training algorithm that brings about a faster rate of convergence than standard backprop, but based on a different philosophy from the one already outlined, is the method of variable slopes. Standard backprop updates only the weights and bias of each neuron and leaves the activation function or node nonlinearity unchanged. Performance, however, is sensitive to the slopes of the node nonlinearities as well as the initial set of weights; hence it is expected that changing the slopes as well as the weights and biases during the training procedure will provide improvement in the rate of convergence. This method is derived in Section 4.6. Experimental results verify that not only is the rate of convergence accelerated but the susceptibility to getting stuck is reduced.

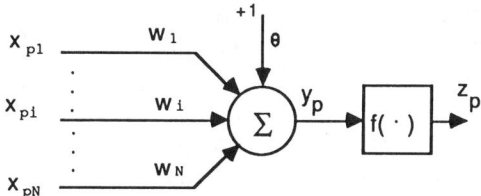

Figure 4.1 A single neuron or node.

4.2 THE NORMAL EQUATION

The basic components of a neural network are the neurons or nodes (Fig. 4.1). Associated with each neuron is a weight vector and a bias. In a feedforward network the neurons are arranged in consecutive layers (Fig. 4.2). The inputs to a neuron in a particular layer are composed solely of the outputs of the neurons in the immediately preceding layers. A neuron yields its output by forming the weighted sum of the outputs of the neurons in the immediately preceding layer, adding a bias, and passing the result through a nonlinear function known as the activation function or the node nonlinearity. A widely used activation function is the sigmoid, given by

$$f(y) = \frac{1 - e^{-sy}}{1 + e^{-sy}} \tag{4.1}$$

which will also be used in this chapter. The parameter s in (4.1) specifies the steepness of the curve.

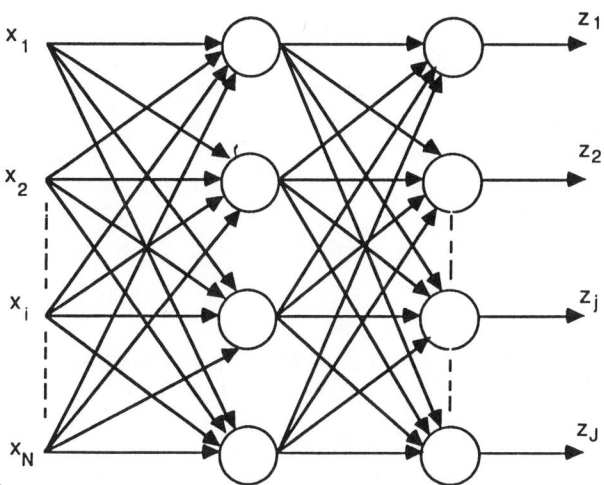

Figure 4.2 A multilayer feedforward perceptron.

When it is necessary to designate a particular neuron, we shall use subscripts to identify the layer and the position within the layer. Thus, for example, the vector $\mathbf{x}_{pl} = [x_{pl1}, x_{pl2}, \ldots, x_{plN}]^T$, where T denotes transpose, is the input vector to any neuron in the lth layer corresponding to the pth pattern presented to the net. Associated with the nth neuron in the lth layer is a weight vector $\mathbf{w}_{ln} = [w_{ln1}, w_{ln2}, \ldots, w_{lnN}]^T$ and a bias θ_{ln}; N is the number of weights in the layer.

The neuron forms the sum $y_{pln} = \mathbf{x}_{pl}^T \mathbf{w}_{ln} + \theta_{ln}$. Augmenting $\mathbf{x}_{pl}$ by 1 and $\mathbf{w}_{ln}$ by θ_{ln} allows this sum to be written in a simpler form as $y_{pln} = \mathbf{x}_{pl}^T \mathbf{w}_{ln}$, where $\mathbf{x}_{pl}$ and $\mathbf{w}_{ln}$ are now $N + 1$ vectors. The output of the neuron is

$$z_{pln} = f(y_{pln})$$

The input to any neuron in the next layer is then

$$\mathbf{x}_{p,l+1} = [x_{p,l+1,1}, x_{p,l+1,2}, \ldots, x_{p,l+1,J}]^T$$
$$= [z_{pl1}, z_{pl2}, \ldots, z_{plJ}]^T$$

where J is the number of neurons in the lth layer. The neurons in the $(l + 1)$st layer generate their outputs, which are the inputs to neurons in the next layer, and this continues to the neurons in the output layer whose outputs are the outputs of the net (Fig. 4.2). When referring to any neuron in general, we shall dispense with the subscripts.

The neural network is trained and will perform the desired mapping when the weight vectors of each neuron are such that the output of the net is within a specified tolerance of the desired output. Training the net, hence, implies training each neuron in the net.

The portion of the neuron (see Fig. 4.1) that generates the weighted sum is known as the linear combiner. Assume that the desired output values of the neuron are known. Hence, the output values of the linear combiner d_p, $p = 1, 2, \ldots, M$, are known, where M is the number of patterns in the input set. Define the error of a node corresponding to pattern p as

$$E_p = \frac{1}{2}(d_p - y_p)^2 \tag{4.2}$$

where $y_p = \mathbf{w}^T \mathbf{x}_p$ is the actual output of the linear combiner.

The total mean squared error of the node is

$$E = \sum_{p=1}^{M} E_p \tag{4.3}$$

To find the optimum weight vector that minimizes E, we take the gradient of E with respect to $\mathbf{w}$ and set it to zero:

$$\nabla E = -\sum_{p=1}^{M} (d_p - \mathbf{w}^T\mathbf{x}_p)\mathbf{x}_p = 0 \tag{4.4}$$

Hence, we have

$$\left(\sum_{p=1}^{M} \mathbf{x}_p\mathbf{x}_p^T\right)\mathbf{w} = \sum_{p=1}^{M} d_p\mathbf{x}_p \tag{4.5}$$

Defining

$$\mathbf{R} = \sum_{p=1}^{M} \mathbf{x}_p\mathbf{x}_p^T \tag{4.6}$$

and

$$\mathbf{p} = \sum_{p=1}^{M} d_p\mathbf{x}_p \tag{4.7}$$

we can write Eq. (4.4) in matrix form as

$$\mathbf{Rw} = \mathbf{p} \tag{4.8}$$

The matrix $\mathbf{R}$ can be interpreted as the correlation matrix of the input set, and the vector $\mathbf{p}$ as the cross-correlation between the training patterns and the corresponding desired responses.

Equation (4.8) is referred to as the deterministic normal equation in the context of adaptive filtering, and the optimum weight vector is the solution to (4.8).

This equation can be solved iteratively by a number of descent methods, such as the method of steepest descent, which results in the so-called Delta rule, conjugate-gradient method, which results in the Delta rule with momentum, and several quasi-Newton methods, which are slightly more complex computationally but yield fast convergence.

Ideally the quantity to be minimized during training is the mean squared error over the entire set given by (4.3). Clearly this requires that the weights be held fixed during an epoch (a presentation of the entire set), and only one update be made per epoch. This is inefficient and may render the iterative methods mentioned useless.

Hence, the general practice is to update the weights in the direction of minimizing (4.2) after each presentation of an input x_p to the net. It is pointed out in the literature that this is a good approximation to minimizing the mean square error over the entire set as long as the amount of update is kept small [6].

4.3 BACKPROPAGATION OF THE ERROR

The iterative methods mentioned in Section 4.2 require knowledge of the error at the output of the linear combiner for each neuron. However, except for the neurons in the output layer, this error is not known. It may even be argued that the concept of error is not defined for the neurons in the hidden layers; for what is the desired response of a neuron in a hidden layer?

To develop a generalized error concept for the neurons in the hidden layers, we begin by considering

$$E_{Tp} = \sum_n E_{pLn} \tag{4.9}$$

where E_{Tp} is the total mean squared error of the net associated with the pth training pattern, L is the number of layers in the network, and the index n is over the neurons in the output layer. Note that since (4.9) defines the error at the input to the nonlinearity, d_{pLn} is the desired response referred to the input of the nonlinearity at the output node n. In other words, if the desired output response of the nth node in the output layer is d'_{pLn}, then the desired input to the nonlinearity d_{pLn}, which is the output of the linear combiner, has to be $f^{-1}(d'_{pLn})$.

We note from (4.9) and (4.2) that the error

$$\delta_{pLn} = d_{pLn} - y_{pLn} \tag{4.10}$$

associated with the pth training pattern at the nth output node can be expressed as

$$\delta_{pLn} = -\frac{\partial E_{Tp}}{\partial y_{pLn}} \tag{4.11}$$

Based on this observation, we now define the error associated with any node—hidden or otherwise—as the derivative of E_{Tp} with respect to the linear combiner output at that node:

$$\delta_{pln} = -\frac{\partial E_{Tp}}{\partial y_{pln}} \tag{4.12}$$

A moment's reflection reveals that this generalization is also intuitively pleasing, for a value of δ_{pln} close to zero indicates that the value of y_{pLn} is just the right value for the node n in the layer l; that is, no weight adjustment in that node is necessary at iteration p.

The task at hand is to calculate δ_{pln} at every node. For the output nodes the desired result is given by (4.10). For the hidden layers (i.e., for $l = 1, 2, \ldots, L - 1$) note that

$$\delta_{pln} = -\frac{\partial E_{Tp}}{\partial y_{pln}} = -\frac{\partial E_{Tp}}{\partial z_{pln}} \frac{\partial z_{pln}}{\partial y_{pln}} \tag{4.13}$$

By the chain rule, the first factor in (4.13) can be expressed as a linear combination of the derivatives of E_{Tp} with respect to the variables associated with the nodes in the next layer $l + 1$ as follows:

$$\frac{\partial E_{Tp}}{\partial z_{pln}} = \sum_r \frac{\partial E_{Tp}}{\partial y_{p,l+1,r}} \frac{\partial y_{p,l+1,r}}{\partial z_{pln}} \tag{4.14}$$

where r is over the nodes in the layer $l + 1$.

Using (4.12) in (4.14), we arrive at the recursion

$$\delta_{pln} = f'(y_{pln}) \sum_r \delta_{p,l+1,r} w_{l+1,r,n} \tag{4.15}$$

where

$$f'(y_{pln}) = \frac{\partial z_{pln}}{\partial y_{pln}} \tag{4.16}$$

by definition.

To summarize, the output errors calculated from (4.10) are backpropagated to the hidden layers by recursively using (4.15).

Once the error is available in a particular node it can be used in a number of different ways to update the weights, as will be demonstrated in what follows.

4.4 THE DELTA RULE

Consider node n in layer l. Let the weight vector at iteration t be $\mathbf{w}_{ln}(t)$. The simplest method of solving (4.8), or, equivalently, minimizing E, is to calculate the gradient of E (or that of E_p) with respect to the weight vector and to adjust $\mathbf{w}_{ln}(t)$ in the direction of steepest descent:

$$\mathbf{w}_{ln}(t + 1) = \mathbf{w}_{ln}(t) + \mu(-\nabla E_p) \tag{4.17}$$

where μ is a suitably small number known as the step size.

Since we have

$$y_{pln} = \mathbf{w}_{ln}^T \mathbf{x}_{pl} \tag{4.18}$$

we conclude that

$$\nabla E_p = \frac{\partial E_{Tp}}{\partial y_{pln}} \cdot \nabla y_{pln} \tag{4.19}$$

Substituting (4.12) in (4.19) and observing from (4.18) that $\nabla y_{pln} = \mathbf{x}_{pl}$, we find

$$\nabla E_p = -\delta_{pln}\mathbf{x}_{pl} \tag{4.20}$$

Thus we have

$$\mathbf{w}_{ln}(t+1) = \mathbf{w}_{ln}(t) + \mu\delta_{pln}\mathbf{x}_{pl} \tag{4.21}$$

Equation (4.21) is variously called the Delta rule, the LMS algorithm, or the method of steepest descent.

When (4.21) is used in conjunction with (4.10) and (4.15), the resulting iterative scheme is referred to as the backpropagation algorithm in the literature [6,8–10].

Up to now we have used the index p to refer to a certain input pattern in the input set, and the index t to refer to the iteration or presentation number. Since the input patterns are being presented one by one to the net (i.e., the pth pattern is presented at iteration t), we shall use the two interchangeably.

Below we give a summary of the backpropagation algorithm:

1. Start with a random set of weights.
2. Calculate y_p by propagating $\mathbf{x}_p$ through the network.
3. Calculate E_p
4. Adjust weights by

$$\mathbf{w}_{ln}(t+1) = \mathbf{w}_{ln}(t) + \mu\delta_{pln}\mathbf{x}_{pl}$$

   where δ_{pln} is given by (4.10) if node n is an output node, and by (4.15) if node n is a hidden node.
5. Repeat by going to step 2.

The backpropagation algorithm was the first, and until recently the only, algorithm to train feedforward multilayer perceptrons. It is still widely used. Furthermore it is computationally simple. These reasons account for its use as a standard against which other algorithms are evaluated. It is customary to talk about how much better, faster, and so on, one's algorithm is compared with the backprop. We shall do the same in what follows.

An alternative to, or an improvement on, the method of steepest descent is the conjugate gradient algorithm [14], in which the weight vector is updated not in the direction of the negative gradient but in a direction that is a linear combination of the current negative gradient and the preceding direction vector.

The algorithm requires the knowledge of the **R** matrix as well as the gradient of the error surface. Under these conditions, the algorithm converges in N steps, where N is the number of weights in the neuron. Of course, in practice, since **R** and the gradient can only be estimated, convergence will take many more iterations than N. Still an improvement over the steepest descent method may be expected with the algorithm. We shall give only the description and refer the reader to [14] for details.

Conjugate Gradient Algorithm

Begin

set $\quad\quad\quad\quad\quad\quad\quad \mathbf{c}_0 = -\nabla E_0$

for each p $\quad\quad\quad \mu_p = -\dfrac{(\nabla E_p)^T \mathbf{c}_p}{\mathbf{c}_p^T \mathbf{R} \mathbf{c}_p}$ $\quad\quad\quad\quad\quad$ (4.22)

$$\alpha_p = -\frac{(\nabla E_{p+1})^T \mathbf{R} \mathbf{c}_p}{\mathbf{c}_p^T \mathbf{R} \mathbf{c}_p} \quad\quad\quad\quad (4.23)$$

$$\mathbf{w}(t+1) = \mathbf{w}(t) + \mu_p \mathbf{c}_p \quad\quad\quad\quad (4.24)$$

$$\mathbf{c}_{p+1} = -\nabla E_{p+1} + \alpha_p \mathbf{c}_p \quad\quad\quad\quad (4.25)$$

If estimating **R** is not feasible, we may choose to set μ_p and α_p as constants: $\mu_p = \mu$ and $\alpha_p = \alpha$. In this case, after substituting (4.25) in (4.24) the rule by which the weight vector is updated becomes

$$\mathbf{w}(t+1) = \mathbf{w}(t) + \mu(-\nabla E_p) + \mu\alpha \mathbf{c}_{p-1} \quad\quad\quad\quad (4.26)$$

but since

$$\mathbf{c}_{p-1} = (1/\mu)(\mathbf{w}(t) - \mathbf{w}(t-1)) \quad\quad\quad\quad (4.27)$$

from (4.24), (4.26) is written as

$$\mathbf{w}(t+1) = \mathbf{w}(t) + \mu(-\nabla E_p) + \alpha(\mathbf{w}(t) - \mathbf{w}(t-1)) \quad\quad\quad\quad (4.28)$$

or substituting the value of ∇E_p from (4.20) and rewriting (4.28) for node n in layer l, we have

$$\mathbf{w}_{ln}(t+1) = \mathbf{w}_{ln}(t) + \mu \delta_{pln} \mathbf{x}_{pl} + \alpha[\mathbf{w}_{ln}(t) - \mathbf{w}_{ln}(t-1)] \quad\quad\quad\quad (4.29)$$

The last term in (4.29) is referred to as the *momentum* in the literature [11] and is usually justified with the heuristic argument of imparting to the algorithm a sense of history; that is, rather than driving the weight vector in the direction of the present gradient, one forms a blend of the present gradient direction and the direction taken in the previous iteration.

Erik Johansson et al. [23] propose a different procedure for applying the conjugate-gradient method to the training of the neural network. This implementation considers the weights of the entire net to be one vector with respect to which the error given by Eq. (4.58) is minimized hence there is only a single μ. Johansson's approach avoids estimating the **R** matrix by using the fact that the correct value of μ at each iteration is that which minimizes the error in the given direction (i.e., the direction which the conjugate-gradient algorithm chooses to follow at that particular iteration). Therefore a one-dimensional line-search is used to yield the value of μ. To avoid the **R** dependence in Eq. (4.23) either the Heistenes–Stiefel formula [24]:

$$\alpha p = \frac{(\nabla E_{p+1})T[\nabla E_{p+1} - \nabla e_p]}{c_p T[\nabla E_{p+1} - \nabla E_p]}$$

or the Polak–Ribiere formula [25]:

$$\alpha p = \frac{(\nabla E_{p+1})T[\nabla E_{p+1} - \nabla e_p]}{\nabla E_p T \nabla E_p}$$

or the Fletcher–Reeves formula [26]:

$$\alpha p = \frac{(\nabla E_{p+1})T \nabla E_{p+1}}{\nabla E_p T \nabla E_p}$$

is used to calculate α.

Although this method seems to be less computationally burdensome than estimating **R** for every layer it nevertheless requires a great number of computations compared to backprop. The parity problem on which it has been tested shows that it performs better than the backprop but not as well as the Kalman algorithm.

4.5 THE KALMAN ALGORITHM

Although widely used, the classical backprop has several shortcomings. It is extremely sensitive to the initial weight set: There are many choices of initial weight vectors for which the algorithm will not converge. Even when it does converge, sometimes it trains very slowly—the mean square error diminishes only after a great number of iterations. Another problem of the backprop is the inconsistency of training—the mean squared error might stay the same for many iterations and then suddenly plunge to a lower value.

The algorithm presented here does not suffer from these disadvantages. It is not as sensitive to initial weight settings, and it converges quickly and consistently. This algorithm is obtained by modifying the weight update mechanism. It uses the Kalman algorithm [13] to solve iteratively the deterministic normal equation (4.8), derived in Section 4.1 and repeated here for convenience:

$$\mathbf{Rw} = \mathbf{p} \tag{4.8}$$

Recall that the matrix

$$\mathbf{R} = \sum_p \mathbf{x}_p \mathbf{x}_p^T \tag{4.6}$$

is the correlation matrix of the inputs to a node, and the vector

$$\mathbf{p} = \sum_p d_p \mathbf{x}_p \tag{4.7}$$

is the cross-correlation between the inputs to a node and the corresponding desired responses.

Equation (4.8) has to be solved at every node. If the desired response d_p at every node were known, then the optimum weights could be found in a straightforward manner: Compute $\mathbf{R}_1$, the correlation matrix of the input training set. Compute $\mathbf{p}_{1n}$ for every node n in the first layer using the desired response of the node and the input training patterns [Eqs. (4.6) and (4.7), respectively]. Compute the weight vector for each node n in the first layer as the solution to the equation:

$$\mathbf{R}_1 \mathbf{w}_{1n} = \mathbf{p}_{1n}$$

Once $\mathbf{w}_{1n}$ is known, the inputs to the nodes in the next layer are known and hence $\mathbf{R}_2$ and $\mathbf{p}_{2n}$ are known. The foregoing procedure can thus be repeated to find the weight vectors at nodes in the second layer, and this process continues from one layer to the next up to the output layer. The weights in the net will then be at their optimum values. Note that this is a one-pass procedure involving no iterations.

As mentioned in Section 4.3, however, the desired responses of nodes in the hidden layers are not known, so the preceding procedure must be altered in the following way: All weights in the net are first set randomly, and the difference between the desired and actual output of the net, the error, is back-propagated by using the rule outlined in Section 4.3, to nodes in the hidden layers. These errors then yield an estimate for $\mathbf{p}$ that is, in turn used in (4.8) to find a new weight vector. The weights obtained in this way are hopefully closer to the optimum weights than the initial randomly selected values. The error is back-propagated over the new weight vectors again, and **p** is estimated again, the weights are recalculated and this process continues until the error at the output of the net is below a specified value. This procedure involves many passes, with the error being propagated backward from the output layer to the input layer, and the weights being adjusted forward from the input layer to the output layer.

Any algorithm to train a multilayer perceptron has to invoke this procedure, including the classical backprop and the algorithm that we shall soon present. Where these two algorithms differ is in the method of solving the weight equation. Backprop uses the LMS algorithm, and the new algorithm uses the Kalman algorithm.

Equations (4.6) and (4.7) imply that all the training patterns must be run through the network before the estimates **R** and **p** are determined and each weight-vector update is performed. This poses no problem when the training set consists of a small number of patterns. However, for large training sets, the enormous amount of time required to pass the entire training set through the network before each weight-vector update is performed would render the procedure useless. Hence, Eqs. (4.6) and (4.7) must be modified to make it possible to form a running estimate of the correlations **R** and **p** with every presentation of an input to the net. To this end we introduce

$$\mathbf{R}(t) = \sum_{k=1}^{t} \mathbf{x}(k)\mathbf{x}^{\mathrm{T}}(k) \qquad (4.30)$$

and

$$\mathbf{p}(t) = \sum_{k=1}^{t} d(k)\mathbf{x}(k) \qquad (4.31)$$

as the estimates of the correlation matrix and correlation vector, respectively. The weight vector at each node can now be updated with the presentation of every training pattern, since one does not have to wait for M training patterns for these estimates to be available.

In doing this, a new problem is created, which can be easily fixed. With the exception of the first layer, estimates of one layer are based upon data received from the preceding layer. Since the preceding layer is untrained at the beginning, the correlation estimates are not as good as they will be at or near the end of the training period. These inaccurate older estimates, however, are still strongly included in the estimation process later on when the network has become smarter. This would, at best, increase the training time of the network.

To remedy this problem, we insert a positive forgetting factor $b < 1$ into the correlations and rewrite (4.30) as

$$\mathbf{R}(t) = \sum_{k=1}^{t} b^{t-k}\mathbf{x}(k)\mathbf{x}^{\mathrm{T}}(k) \qquad (4.32)$$

and (4.31) as

$$\mathbf{p}(t) = \sum_{k=1}^{t} b^{t-k}d(k)\mathbf{x}(k) \qquad (4.33)$$

The forgetting factor b will alow the new information to dominate while the correlation estimates from earlier training will have negligible effect on the current estimates.

If b is too small, the network will forget too much of the past training information and will only train to the most recent patterns. Too large a value of b will cause the network to remain trained to information that is no longer valid. The value for b used in simulations was 0.99.

We now turn our attention to the task of solving (4.8). The algorithm introduced in this section solves (4.8) by making use of the Kalman algorithm [13,15–18] to estimate $\mathbf{R}^{-1}$ in the following equation, which is merely (4.8) rearranged:

$$\mathbf{w} = \mathbf{R}^{-1}\mathbf{p} \qquad (4.34)$$

To derive a recursive equation for the estimate of $\mathbf{R}^{-1}$, we begin by writing (4.32) and (4.33) in their recursive forms as

$$\mathbf{R}(t) = b\mathbf{R}(t-1) + \mathbf{x}_p\mathbf{x}_p^{\mathrm{T}} \qquad (4.35)$$

and

$$\mathbf{p}(t) = b\mathbf{p}(t-1) + d_p\mathbf{x}_p \qquad (4.36)$$

We now invoke the matrix inversion lemma [13,19], which states that if an $n \times n$ matrix $\mathbf{A}$ can be written in the form

$$\mathbf{A} = \mathbf{B}^{-1} + \mathbf{C}\mathbf{D}^{-1}\mathbf{C}^{\mathrm{T}} \qquad (4.37)$$

its inverse can be expressed as

$$\mathbf{A}^{-1} = \mathbf{B} - \mathbf{B}\mathbf{C}(\mathbf{D} + \mathbf{C}^{\mathrm{T}}\mathbf{B}\mathbf{C})^{-1}\mathbf{C}^{\mathrm{T}}\mathbf{B} \qquad (4.38)$$

where T denotes transpose, and matrices **B**, **C**, and **D** are of suitable dimension so that the product $\mathbf{C}\mathbf{D}^{-1}\mathbf{C}^T$ is an $n \times n$ matrix.

Equation (4.35) is in the form of Eq. (4.37) if we define $\mathbf{A} = \mathbf{R}(t)$, $\mathbf{B}^{-1} = b\,\mathbf{R}(t-1)$, $\mathbf{C} = \mathbf{x}_p = \mathbf{x}(t)$, and $\mathbf{D}^{-1} = \mathbf{I}$. The inverse autocorrelation matrix $\mathbf{R}^{-1}$ can now be expressed, through Eq. (4.38), as

$$\mathbf{R}^{-1}(t) = b^{-1}\mathbf{R}^{-1}(t-1) - \frac{b^{-1}\mathbf{R}^{-1}(t-1)\mathbf{x}(t)\mathbf{x}^T(t)b^{-1}\mathbf{R}^{-1}(t-1)}{1 + \mathbf{x}^T(t)b^{-1}\mathbf{R}^{-1}(t-1)\mathbf{x}(t)} \quad (4.39)$$

By defining the Kalman gain $\mathbf{k}(t)$ as

$$\mathbf{k}(t) = \frac{\mathbf{R}^{-1}(t-1)\mathbf{x}(t)}{b + \mathbf{x}^T(t)\mathbf{R}^{-1}(t-1)\mathbf{x}(t)} \quad (4.40)$$

we can write Eq. (4.39) as

$$\mathbf{R}^{-1}(t) = [\mathbf{R}^{-1}(t-1) - \mathbf{k}(t)\mathbf{x}^T(t)\mathbf{R}^{-1}(t-1)]b^{-1} \quad (4.41)$$

Substituting (4.36) and (4.41) into Eq. (4.34) yields

$$\begin{aligned}\mathbf{w}(t) &= [\mathbf{R}^{-1}(t-1) - \mathbf{k}(t)\mathbf{x}^T(t)\mathbf{R}^{-1}(t-1)]\mathbf{p}(t-1) + d(t)\mathbf{R}^{-1}(t)\mathbf{x}(t) \\ &= \mathbf{R}^{-1}(t-1)\mathbf{p}(t-1) - \mathbf{k}(t)\mathbf{x}^T(t)\mathbf{R}^{-1}(t-1)\mathbf{p}(t-1) \\ &\quad + d(t)\mathbf{R}^{-1}(t)\mathbf{x}(t).\end{aligned} \quad (4.42)$$

After we replace $\mathbf{R}^{-1}(t-1)\mathbf{p}(t-1)$ in (4.42) with $\mathbf{w}(t-1)$, (4.42) becomes

$$\mathbf{w}(t) = \mathbf{w}(t-1) - \mathbf{k}(t)\mathbf{x}^T(t)\mathbf{w}(t-1) + d(t)\mathbf{R}^{-1}(t)\mathbf{x}(t) \quad (4.43)$$

Expanding the last term in (4.43), we obtain

$$d(t)\mathbf{R}^{-1}(t)\mathbf{x}(t) = b^{-1}d(t)[\mathbf{R}^{-1}(t-1) - \mathbf{k}(t)\mathbf{x}^T(t)\mathbf{R}^{-1}(t-1)]\mathbf{x}(t) \quad (4.44)$$

Replacing $\mathbf{k}(t)$ in (4.44) with the right side of (4.40) produces

$$\begin{aligned}d(t)\mathbf{R}^{-1}(t)\mathbf{x}(t) &= b^{-1}d(t)\mathbf{R}^{-1}(t-1)\mathbf{x}(t) \\ &\quad - \frac{b^{-1}d(t)\mathbf{R}^{-1}(t-1)\mathbf{x}(t)\mathbf{x}^T(t)\mathbf{R}^{-1}(t-1)\mathbf{x}(t)}{b + \mathbf{x}^T(t)\mathbf{R}^{-1}(t-1)\mathbf{x}(t)}\end{aligned} \quad (4.45)$$

Using $b + \mathbf{x}^T(t)\mathbf{R}^{-1}(t-1)\mathbf{x}(t)$ as the common denominator and canceling terms, allows us to express (4.45) as

$$d(t)\mathbf{R}^{-1}(t)\mathbf{x}(t) = \frac{d(t)\mathbf{R}^{-1}(t-1)\mathbf{x}(t)}{b + \mathbf{x}^T(t)\mathbf{R}^{-1}(t-1)\mathbf{x}(t)} \quad (4.46)$$

or

$$d(t)\mathbf{R}^{-1}(t)\mathbf{x}(t) = d(t)\mathbf{k}(t). \tag{4.47}$$

The weight update (4.43) can now be written as

$$\mathbf{w}(t) = \mathbf{w}(t-1) + \mathbf{k}(t)[d(t) - \mathbf{x}^T(t)\mathbf{w}(t-1)] \tag{4.48}$$

Note that $\mathbf{x}^T(t)\mathbf{w}(t-1)$ in (4.48) can be replaced by the scalar $y(t)$, and from (4.11) for the hidden layers, $d(t) - y(t)$ can be replaced by $\mu\delta$. Therefore, the weight-update equation for nodes in the output layer is

$$\mathbf{w}(t) = \mathbf{w}(t-1) + \mathbf{k}(t)[d(t) - y(t)] \tag{4.49}$$

and for nodes in the hidden layers it is

$$\mathbf{w}(t) = \mathbf{w}(t-1) + \mu\delta\mathbf{k}(t) \tag{4.50}$$

Thus far in the derivation, we have focused our attention on a single node and have omitted node and layer indices. At this point, we will rewrite the Kalman filter equations to reference a specific node and layer. Hence, for the lth layer we have the following equations:

The Kalman gain vector:

$$\mathbf{k}_l(t) = \frac{\mathbf{R}_l^{-1}(t-1)\,\mathbf{x}_l(t)}{b + \mathbf{x}_l^T(t)\mathbf{R}_l^{-1}(t-1)\,\mathbf{x}_l(t)} \tag{4.51}$$

the update equation for the inverse matrix:

$$\mathbf{R}_l^{-1}(t) = [\mathbf{R}_l^{-1}(t-1) - \mathbf{k}_l(t)\,\mathbf{x}_l^T(t)\,\mathbf{R}_l^{-1}(t-1)]b^{-1} \tag{4.52}$$

the update equation for the weight vectors in the output layer:

$$\mathbf{w}_{Ln}(t) = \mathbf{w}_{Ln}(t-1) + \mathbf{k}_L(t)(d_n - y_{Ln}) \tag{4.53}$$

the update equation for the weight vectors in the hidden layers:

$$\mathbf{w}_{ln}(t) = \mathbf{w}_{ln}(t-1) + \mu\mathbf{k}_l(t)\delta_{ln}(t) \tag{4.54}$$

where t is the present iteration number and n is the node in layer l. Constants b and μ are the forgetting factor and back-propagation step size, respectively.

The Kalman algorithm as presented in (4.51) through (4.54) requires $(N + 1)^2 + 2(N + 1)$ multiplications for the Kalman gain in (4.51) and $3(N + 1)^2$

multiplications for the inverse matrix update in (4.52). The number of computations may be reduced [13] by taking advantage of the symmetry of $\mathbf{R}_l^{-1}(t-1)$ and defining

$$\mathbf{a}_l(t) = \mathbf{R}_l^{-1}(t-1)\mathbf{x}_l(t) \tag{4.55}$$

Equations (4.51) and (4.52) now become

$$\mathbf{k}_l(t) = \frac{\mathbf{a}_l(t)}{b + \mathbf{x}_l^T(t)\mathbf{a}_l(t)} \tag{4.56}$$

and

$$\mathbf{R}_l^{-1}(t) = [\mathbf{R}_l^{-1}(t-1) - \mathbf{k}_l(t)\mathbf{a}_l^T(t)]b^{-1} \tag{4.57}$$

Be rewriting the Kalman equations in this way, the number of multiplications for the $\mathbf{a}_l$ vector, Kalman gain, and inverse matrix update equations become $(N+1)^2$, $2(N+1)$, and $2(N+1)^2$, respectively. Thus, we have eliminated a costly $(N+1)^2$ term.

Here is a summary of the algorithm:

1. Initialize.
   Equate all node offsets x_{l0} to some nonzero constant for layers $l = 1$ through L.
   Randomize all weights in the network.
   Initialize the inverse matrix $\mathbf{R}^{-1}$.
2. Select training pattern.
   Randomly select an input/output pair $\mathbf{x}_{p1}$ and $\mathbf{d}_p$ to present to the network.
3. Run selected pattern through the network.
   For each layer l from 1 through L, calculate the summation output

   $$y_{pln} = \mathbf{x}_{pl}^T \mathbf{w}_{ln}$$

   and the function output

   $$f(y_{pln}) = \frac{1 - e^{-sy_{pln}}}{1 + e^{-sy_{pln}}}$$

   for every node n. The constant s is the sigmoid slope.
4. Invoke Kalman filter equations.
   For each layer l from 1 through L, calculate

   $$\mathbf{a}_l(t) = \mathbf{R}_l^{-1}(t-1)\mathbf{x}_l(t)$$

calculate the Kalman gain

$$\mathbf{k}_l(t) = \frac{\mathbf{a}_l(t)}{b + \mathbf{x}_l^T(t)\mathbf{a}_l(t)}$$

and update the inverse matrix

$$\mathbf{R}_l^{-1}(t) = [\mathbf{R}_l^{-1}(t - 1) - \mathbf{k}_l(t)\mathbf{a}_l^T(t)]b^{-1}.$$

5. Backpropagate error signals.
    Compute the derivative of $f(y_{pln})$, using

    $$f'(y_{pln}) = (s/2)(1 - f^2(y_{pln}))$$

    Calculate error signals in the output layer, where $l = L$, by evaluating

    $$\delta_{pLn} = d_{pn} - y_{pLn}$$

    for every node n.
    For the hidden layers, starting at layer $l = L - 1$ and decrementing through $l = 1$, find error signals by solving

    $$\delta_{pln} = f'(y_{pln}) \sum_r \delta_{p,l+1,r} w_{l+1,r,n}$$

    for every node in the lth layer, where r runs over nodes in the $(l + 1)$st layer.

6. Find the desired summation output.
    Calculate the desired summation output at the Lth layer by using the inverse function

    $$d_{pLn} = f^{-1}(d'_{pLn}).$$
    $$= \frac{1}{s} \ln \frac{1 + d'_{pLn}}{1 - d'_{pLn}}$$

    for every nth node in the output layer.

7. Update the weights.
    The weight vectors in the output layer L are updated by

    $$\mathbf{w}_{Ln}(t) = \mathbf{w}_{Ln}(t - 1) + \mathbf{k}_L(t)(d_n - y_{Ln})$$

    for every nth node.

For each hidden layer $l = 1$ through $L - 1$, the weight vectors are updated by

$$\mathbf{w}_{ln}(t) = \mathbf{w}_{ln}(t - 1) + \mu \mathbf{k}_l(t)\delta_{ln}(t)$$

for every nth node.
8. Test for completion.

At this point, we can either use the mean squared error of the network output as a convergence test, or run the algorithm for a fixed number of iterations. Either way, if completion is not reached, go back to step 2.

4.6 VARIABLE SLOPES

As mentioned earlier, the back propagation algorithm suffers from a major drawback: The convergence of the algorithm is very sensitive to the initial set of weights. Very often, one finds that the algorithm fails to converge. This phenomenon is usually explained as the algorithm's "getting stuck at a local minimum" of the error surface [6,20]. Although this may be the correct explanation in some cases, there is another important phenomenon that prevents convergence: It is common that the input vector and the weights of a node are such that the output of the linear combiner falls where the nonlinearity saturates and hence has a small derivative. The weight-update term in (4.21) is proportional to this derivative and thus may get so small that the weights update very slowly, if at all, which causes the algorithm to fail to converge or to converge only after a large number of iterations [21,22].

The likelihood of this occurring is increased if the value of s in (4.1) is high, since s determines the steepness of the slope. Too low a value for s, however, has two undesirable effects: The value of the derivative decreases even in the unsaturated regions, thus slowing convergence, and the output of the linear combiner almost always falls in the linear region of the sigmoid. The latter is undesirable because it thwarts the effect of the nonlinearities, thus depriving the net of its multilayered nature (since any number of layers having linear activation functions can be replaced by a single layer).

There is thus an optimum value for s lying somewhere between these two extremes. This value is not necessarily the same for every node. The complexity of the net makes it impossible to predetermine this optimal slope set, so an adaptive means of finding this set is required. Specifically, we have to determine adaptively the slope of the nonlinearity s_{ln} associated with each neuron (node) as well as the weight vector w_{ln} so as to minimize

$$E_{Tp} = \frac{1}{2} \sum_n (d_{pLn} - z_{pLn})^2 \qquad (4.58)$$

that is, the total mean squared error associated with the pth training pattern, where L is the number of layers in the network, and the index n is over the neurons in the output layer. Equation (4.58) differs from (4.9) in that d_{pLn} is now considered to be the desired output referred to the output of the nonlinearity instead of to the input of the nonlinearity, so y_{pLn} in (4.9) is replaced by z_{pLn} in (4.58).

We begin by viewing (4.1) as a function of y and s:

$$f(y,s) = \frac{1 - e^{-sy}}{1 + e^{-sy}} \qquad (4.59)$$

The partial derivatives of $f(y,s)$ with respect to y and s are

$$f_y(y,s) = (s/2)(1 - f^2(y,s)) \qquad (4.60)$$

and

$$f_s(y,s) = (y/2)(1 - f^2(y,s)) \qquad (4.61)$$

respectively.

To minimize E_{Tp} with respect to the slopes s_{ln}, we must then evaluate $\partial E_{Tp}/\partial s_{ln}$. Using the chain rule, we have

$$\frac{\partial E_{Tp}}{\partial s_{ln}} = \frac{\partial E_{Tp}}{\partial y_{pln}} \frac{\partial y_{pln}}{\partial z_{pln}} \frac{\partial z_{pln}}{\partial s_{ln}} \qquad (4.62)$$

Since

$$\frac{\partial z_{pln}}{\partial s_{ln}} = f_s(y_{pln}, s_{ln}) \qquad (4.63)$$

$$\frac{\partial y_{pln}}{\partial z_{pln}} = \frac{1}{f_y(y_{pln}, s_{ln})} \qquad (4.64)$$

and

$$\frac{\partial E_{Tp}}{\partial y_{pln}} = -\delta_{pln}$$

we have

$$\frac{\partial E_{Tp}}{\partial s_{ln}} = -\delta_{pln} \frac{f_s(y_{pln}, s_{ln})}{f_y(y_{pln}, s_{ln})} \qquad (4.65)$$

where δ_{pln} is given by (4.10) or (4.15), depending on where the node is.

The slopes of the activation function s_{ln}'s are updated according to the Delta rule by

$$s_{ln}(t+1) = s_{ln}(t) - \beta \frac{\partial E_{Tp}}{\partial s_{ln}} \qquad (4.66)$$

or

$$s_{ln}(t+1) = s_{ln(t)} + \beta \, \Delta s_{ln} \qquad (4.67)$$

where

$$\Delta s_{ln} = (d_{ln} - y_{pln}) \frac{f_s(y_{pln}, s_{ln})}{f_y(y_{pln}, s_{ln})} \qquad (4.68)$$

if the neuron is in the output layer, and

$$\Delta s_{ln} = f_s(y_{pln}, s_{ln}) \sum_r \delta_{p,l+1,r} w_{l+1,r,n} \qquad (4.69)$$

if the neuron is in a hidden layer.

The range of β is between 0 and 1. A momentum term to consider past slope changes can be added, and (4.67) can be rewritten as

$$s_{ln}(t+1) = s_{ln}(t) + \beta \, \Delta s_{ln} + \rho(s_{ln}(t) - s_{ln}(t-1)) \qquad (4.70)$$

where ρ is the momentum term and its range is between 0 and 1.

Equation (4.70) is the only step added to the classical algorithm. It can be seen in (4.70) that the slopes are updated in a very similar manner to the way the weights are updated in (4.21). The new algorithm has only two more terms than the classical algorithm: Δs_{ln} and $f_s(y_{pln}, s_{ln})$. The term $f_s(y_{pln}, s_{ln})$ is the derivative of the nonlinearity with respect to the slope and has the same computational complexity as the term $f_s(y_{pln}, s_{ln})$. The term Δs_{ln}, as seen in (4.69), is related to δ_{pln}, which is used to update the weights. Therefore, the new algorithm does not have any more computational burden than the classical one; the δ_{pln}'s, used to update the weights, are also used to update the slopes.

We summarize the algorithm.

1. Start with a random set of weights and a random set of slopes.
2. Calculate y_p by propagating $\mathbf{x}_p$ through the network.
3. Calculate E_p.
4. Adjust slopes according to

$$s_{ln}(t+1) = s_{ln}(t) + \beta \, \Delta s_{ln} + \rho(s_{ln}(t) - s_{ln}(t-1))$$

where $s_{ln}(t)$ is the slope at time t belonging to node n in the lth layer, β is the step size, ρ is the momentum constant, and

$$\Delta s_{ln} = (d_{pln} - z_{pln}) \cdot f_s(y_{pln}, s_{ln})$$

if n refers to an output node and

$$\Delta s_{ln} = \left(\sum_r \delta_{p.l+1.r} w_{l+1.r.n} \right) \cdot f_s(y_{pln}, s_{ln})$$

if n refers to a hidden node. This step is the only addition to the classical algorithm.

5. If $s_{ln}(t+1) < s_{min}$, then $s_{ln}(t+1) = s_{min}$, where s_{min} is a small positive number used to prevent the slopes from taking very small or negative values.
6. Adjust weights according to

$$\mathbf{w}_{ln}(t+1) = \mathbf{w}_{ln}(t) + \mu \delta_{pln} \mathbf{x}_{pl} + \alpha [\mathbf{w}_{ln}(t) - \mathbf{w}_{ln}(t-1)]$$

where δ_{pln} is given by (4.10) if node n is an output node and by (4.15) if node n is a hidden node.
7. Repeat by going to step 2.

4.7 EXPERIMENTAL RESULTS

In this section we present a number of simulations of two bench-mark problems to compare the performance of the two new algorithms with that of backprop.

We first examine the so-called circle in a square problem, which can be described as follows. Determine if a set of coordinates is inside or outside a circle of a given radius. The network is a feedforward multilayer perceptron consisting of two inputs (x and y coordinates) and one output. The input coordinates are selected randomly, with values ranging uniformly from -0.5 through $+0.5$. If the distance of the training point from the origin is less than 0.35, the desired output is assigned the value -0.5, indicating that the point is inside the circle. A distance greater than 0.35 means that the point is outside the circle, and the desired output becomes $+0.5$. Training patterns are presented to the network alternating between the two classes (inside and outside the circle). The network has one hidden layer containing nine nodes.

Figure 4.3 shows the mean squared error versus the iteration number for both Kalman and backprop algorithms during training. Ten different training runs (ensemble members) are averaged together. The initial weights are changed from one run to the next, but both algorithms start each run with identical weights (values ranging from -20 through $+20$) and both are pre-

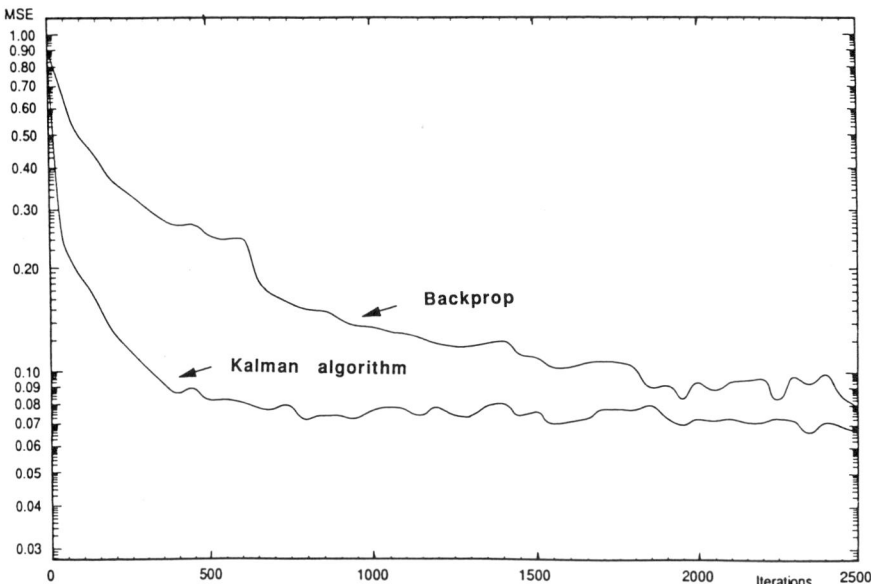

Figure 4.3 Mean squared error versus iterations: backprop and Kalman algorithm.

sented with the same training patterns, selected in exactly the same order. The sigmoid constant s is set to 0.2 for all nonlinear function blocks in both algorithms. For the back propagation algorithm, the step size μ and momentum term α for both layers are 0.8 and 0.9, respectively. The Kalman algorithm uses forgetting factors of 0.99 for all layers and a step size of 40 for the hidden layer. The new algorithm remains below a mean squared error of 0.10 after 320 iterations, as opposed to the back propagation algorithm at 1850 iterations.

The output of the network when passed through a hard limiter is shown at various iteration numbers in Figures 4.4a through 4.4d for the new algorithm, and Figures 4.5a through 4.5d for the back propagation algorithm. It is seen that the new algorithm forms a circle in 200 to 300 iterations, whereas the back propagation algorithm takes 1800 iterations. The new algorithm remains below a mean squared error of 0.10 after 320 iterations, as opposed to the backpropagation algorithm at 1850 iterations.

These results show that for the circle in a square problem the Kalman algorithm converges faster than the backpropagation algorithm. The difference in the convergence rate is especially more pronounced at the beginning of training. On the average, the Kalman algorithm converges to a mean squared error of 0.10 in approximately 12 percent of the iterations required by the back propagation algorithm. This is an improvement ratio of better than 8 to 1, provided that both algorithms are run on a parallel machine.

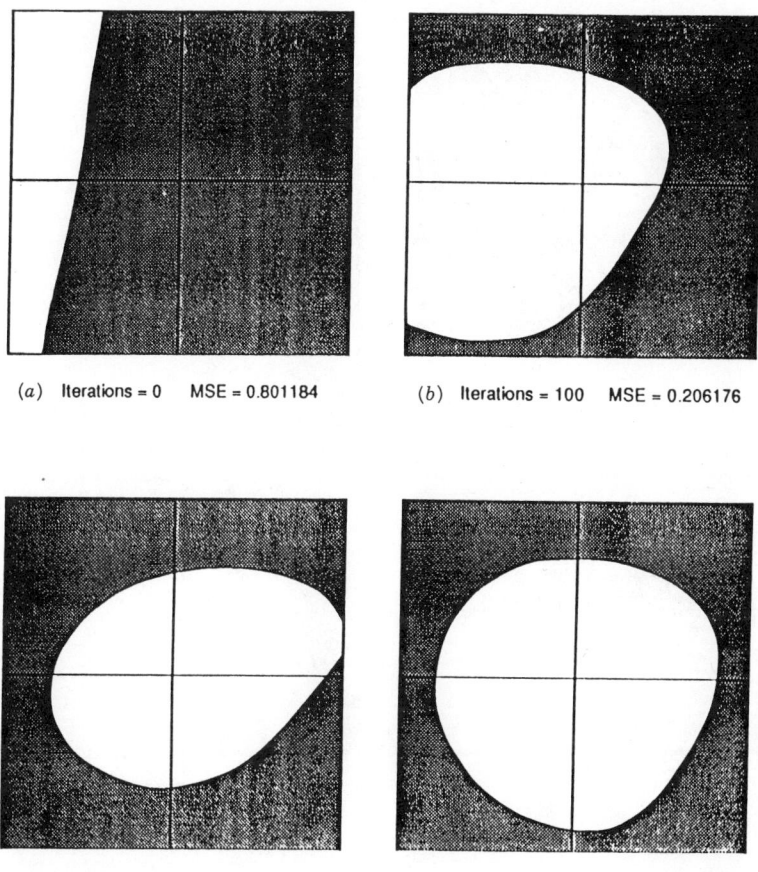

Figure 4.4 Two-dimensional circle plots for the Kalman algorithm: (a) iterations = 0, MSE = 0.801184; (b) iterations = 100, MSE = 0.206176; (c) iterations = 200, MSE = 0.115151; (d) iterations = 300, MSE = 0.092668.

Figure 4.6 shows the mean squared error versus the iteration number for the variable slopes and backprop during training. The curves shown are the averages of five different training runs. For backprop $s = 0.8$, and the same value was used for the initial slope of the variable-slopes algorithm. Both algorithms were used with $\mu = 0.1$ and $\alpha = 0.2$. The modified algorithm has additional parameters $\beta = 0.001$, $\alpha = 0.001$, and $s_{min} = 0.2$. The variable-slopes algorithm reaches a mean squared error of 0.1 in about 2500 iterations, whereas the backprop reaches the same error in about 3500 iterations.

The second problem we consider is the so-called parity problem [3,6]. The problem is a generalized XOR problem. The neural network is presented with an n-bit word as the input and is required to give an output of 1 when the input has an odd number of ones and -1 otherwise. Note that the input

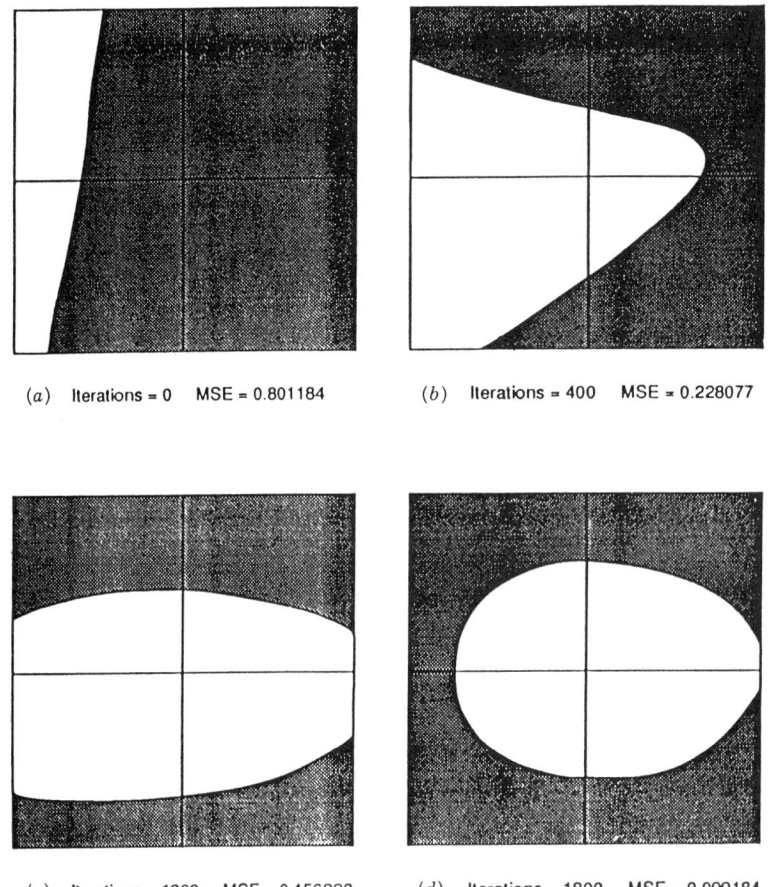

Figure 4.5 Two-dimensional circle plots for the backprop: (a) iterations = 0, MSE = 0.801184; (b) iterations = 400, MSE = 0.228077; (c) iterations = 1300, MSE = 0.156898; (d) iterations = 1800, MSE = 0.099184.

patterns that differ by one bit require different outputs. This makes the parity problem difficult.

We present experimental results with 4-bit patterns. The net has one hidden layer with four nodes and a single output node. The four nodes in the hidden layer is the minimum required for 4-bit patterns [6]. With this architecture it is very difficult for the net to respond correctly to all 16 inputs. What often happens is that the net trains to 15 patterns and responds incorrectly to one pattern (usually the all-ones pattern) and cannot train any further, suggesting the existence of a local minimum.

Table 4.1 shows the results of two typical runs for the back propagation algorithm and the Kalman algorithm. Each algorithm is fine-tuned to its best performance. For the backprop we have $s = 0.2$, $\mu = 0.8$, $\alpha = 0.9$. Weights

4.7 EXPERIMENTAL RESULTS

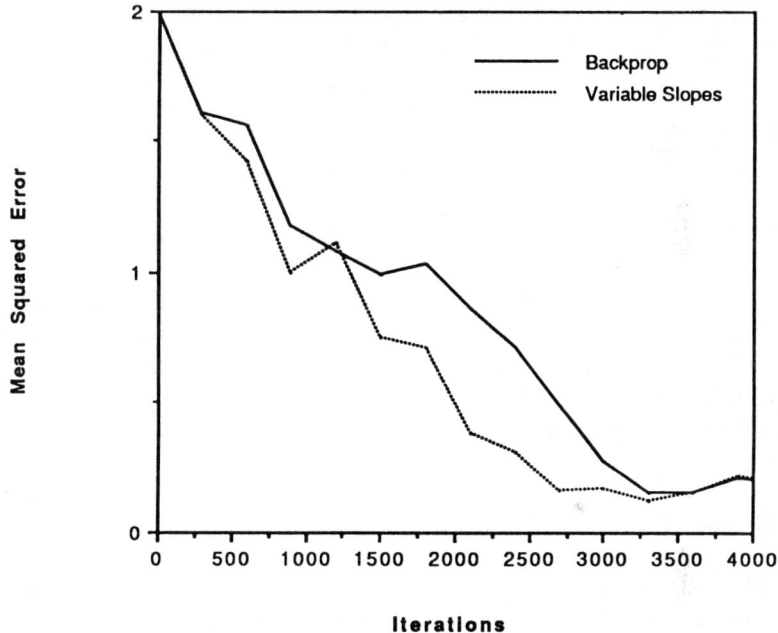

Figure 4.6 *Mean squared error versus iterations: backprop and variable slopes.*

are chosen randomly from a uniform distribution between -20 and $+20$. For the Kalman algorithm we have $b = 0.99$, $s = 1.0$, and $\mu = 100$ for the hidden layer and $\mu = 1$ for the output layer. Input levels are $+0.5$ and -0.5, and desired outputs are likewise $+0.5$ and -0.5.

It is seen that the Kalman algorithm clearly outperforms the backprop. Out of the 11 runs for which both algorithms converged, the average number of iterations is 20,455 for the backprop and 4800 for the Kalman algorithm.

Next we present the performance of the variable-slopes algorithm. The parameters used in this simulation are different from those already given because the two simulations were done by different people. Parameters in this case are not fine-tuned.

Table 4.2 shows the results of two typical runs. For both algorithms $s = 0.5$ (this is the initial value for the variable slopes algorithm), $\mu = 0.1$, and

TABLE 4.1 Comparison between the Backpropagation Algorithm and the Kalman Algorithm for the Parity Problem

	Number of Iterations to Converge	
Simulation Number	Backprop	Kalman Algorithm
1	10,000	6,200
2	62,000	2,000

TABLE 4.2 Comparison between the Backpropagation Algorithm and the Variable-Slopes Algorithm for the Parity Problem

	Number of Iterations to Converge	
Simulation Number	Backprop	Variable Slopes
1	56,000	32,000
2	64,000	16,400

$\alpha = 0.2$; in addition, for the variable-slopes algorithm $\beta = 0.001$ and $\rho = .001$. The weights are chosen randomly from a uniform distribution between -10 and $+10$. Input levels are $+1$ and -1, and desired outputs are $+1$ and -1 also. We note from the table that the variable-slopes algorithm also shows a superior performance.

REFERENCES

[1] F. Rosenblatt, The perceptron: A probabilistic model for information storage and organization in the brain, *Psychological Review* 65, 386–408 (1958).

[2] F. Rosenblatt, *Principles of Neurodynamics: Perceptrons and the Theory of Brain Mechanisms*, Spartan Books, New York, 1962.

[3] M. A. Minsky and S. A. Papert, *Perceptrons*, MIT Press, Cambridge, Mass., 1969.

[4] M. A. Minsky and S. A. Papert, *Perceptrons—Expanded Edition*, MIT Press, Cambridge, Mass., 1988.

[5] B. Souček and M. Souček, *Neural and Massively Parallel Computers*, Wiley, New York, 1988.

[6] D. E. Rumelhart and J. L. McClelland, *Parallel Distributed Processing*, Vol. 1, MIT Press, Cambridge, Mass. 1986.

[7] B. Widrow, R. G. Winter, and R. A. Baxter, Layered Neural Nets for Pattern Recognition, *IEEE Trans. Acous., Speech, Signal Processing*, 36(7), pp. 1109–1118, (1988).

[8] P. Werbos, "Beyond Regression," Doctoral dissertation Harvard University, Cambridge, Mass. 1974.

[9] D. B. Parker, "Learning Logic," Tech. Rep. TR-47, MIT, April 1985.

[10] R. Hecht-Nielson, "Theory of the backpropagation neural network," *Proc. of the Int. Joint Conf. on Neural Networks*, I, pp. 593–605, IEEE Press, New York, June 1989.

[11] R. P. Lippman, An Introduction to Computing with Neural Nets, *IEEE ASSP Magazine*, 4(2), pp. 4–22, (1987).

[12] B. Widrow and S. D. Stearns, *Adaptive Signal Processing*, Prentice-Hall, Englewood Cliffs, N.J., 1985.

[13] S. S. Haykin, *Adaptive Filter Theory*, Prentice-Hall, Englewood Cliffs, N.J., 1986.

[14] D. G. Luenberger, *Introduction to Linear and Nonlinear Programming*, Addison-Wesley, Reading, Mass., 1973.

[15] J. G. Proakis, *Digital Communications*, McGraw-Hill, New York, 1983.

[16] R. Scalero, "A Fast New Algorithm for Training Feed-Forward Neural Networks," Doctoral dissertation, Florida Institute of Technology, Melbourne, 1989.

[17] R. Scalero and N. Tepedelenlioglu, "A fast training algorithm for neural networks," *Proc. of the Int. Joint Conf. on Neural Networks*, I, pp. 715–718, Lawrence Erlbaum Associates, Hillsdale, N.J., January 1990.

[18] R. Scalero and N. Tepedelenlioglu, A fast new algorithm for training feed-forward neural networks, *IEEE Trans. Acous., Speech, Signal Processing*, in press.

[19] B. D. O. Anderson and J. B. Moore, *Optimal Filtering*, Prentice-Hall, Englewood Cliffs, N.J., 1979.

[20] R. Hecht-Nielsen, *Neurocomputing*, Addison-Wesley, Reading, Mass., 1990.

[21] A. Rezgui, "The Effect of the Slope of the Activation Function on the Performance of the Backpropagation Algorithm," Doctoral dissertation, Florida Institute of Technology, Melbourne, 1990.

[22] A. Rezgui and N. Tepedelenlioglu, "The effect of the slope of the activation function on the performance of the backpropagation algorithm," *Proc. of the Int. Joint Conf. on Neural Networks*, I, pp. 707–710, Lawrence Earlbaum Associates, Hillsdale, N.J., January 1990.

[23] E. M. Johansson, F. U. Dowla and D. M. Goodcan, "Backpropagation learning for multi-layer feed-forward neural networks using the conjugate gradient method," *Lawrence Livermore National Laboratory*, Report #UCRL-JC-104850.

[24] M. R. Hestenes and E. L. Stiefel, "Methods of conjugate gradients for solving linear systems," *J. Research NBS*, 49(6), pp. 409–436, December 1952.

[25] R. Fletcher and C. M. Reeves, "Function minimization by conjugate gradients," *Computer Journal*, 7, pp. 149–154, 1964.

[26] E. Polak and G. Ribiere, "Note sur la Convergence de Methods de Directions Conjures, *Revue Francaise Information Recherche Operationnelle*, 16, pp. 35–43, 1969.

CHAPTER 5

Teaching Network Connections for Real-Time Object Recognition

STEPHEN S. WILSON

5.1 INTRODUCTION

This chapter describes a method using simulated annealing to automatically train a neural network. Complex objects are recognized by the network using an industrial machine vision system. In this method, the objective of neural training is to find an optimum connectivity pattern for a fixed number of inputs that have fixed weights, rather than the usual technique of finding the optimum weights for a fixed connectivity [1]. The recognition model uses a two-layer artificial neural network, where the first layer consists of edge vectors that have both magnitude and direction. Each neuron in the second layer has a fixed number of connections that connect only to those first-layer features that are best for distinguishing the object from a confusing background. Simulated annealing is used to find the best parameters for defining edges in the first layer and the optimum connection pattern to the second layer. Weights of the second layer connections are ± 1, so that multiplications are avoided and the system speed is considerably enhanced. This method has been found to work very well on objects that have reasonably distinctive features, but can otherwise have large differences due to nonuniform illumination, missing pieces, occluded parts, or uneven discoloration.

When an artificial neural network is used for object recognition in the image domain, there is, in principle, at least one neural cell for each image pixel in each layer. There can also be a large number of inputs to each cell. In

Portions of this chapter will be published in the *Proceedings of the IEEE.*

Neural and Intelligent Systems Integration, By Branko Souček and the IRIS Group.
ISBN 0-471-53676-8 ©1991 John Wiley & Sons, Inc.

a feedforward network, each cell will have input connections from a neighborhood of cells in the previous layer, where the size of the neighborhood can be as large as the object to be recognized. In general, all possible connections within the neighborhood must be considered, since it is not known beforehand which connections are useful for locating the object, and which are not.

There are various forms of Hebbian learning [2]. One involves incrementally changing the weights according to a Delta rule, on a fixed set of connections by applying a training data set to the inputs and comparing the output with a desired response. There are a large number of neural models (see Rumelhart and McClelland [3] for references and examples) that use various types of incremental learning. In a sense, the training algorithm discovers useful connections by virtue of the magnitudes of the final trained weights. Since an image is the same when translated over a Euclidean spatial domain, the resulting large number of weight values can be reduced by assuming translation invariance of both the weights and connectivity over the image. Networks of this type might be called *iconic* networks [4], since the structure has a one-to-one mapping to the image domain. Each cell in the same layer will have an identical pattern of connectivity and weights. But this assumption will not reduce the large number of computations in the network. For an N^2 image and an M^2 object size, the number of computations is roughly N^2M^2. Hinton [5] has investigated the learning of translation invariance of patterns in networks, but without an explicit assumption that the weights and connections are restricted to have translation invariance by the neural model.

The net of a cell is defined as the sum of the inputs multiplied by the weights. The net for cell j receiving signals s_i from cells i is generally given [3] by

$$\text{net}_j = \sum_i w_{ij} s_i$$

Each index i and j ranges over the two-dimensional image. Because an iconic network is translation-invariant, the weights w_{ij} for a given layer cannot be a function of the image coordinates j. Thus the net must be restricted and redefined for translation invariance. The image coordinates are more conveniently indexed by two component spatial vectors $\mathbf{x}$ and $\mathbf{u}$:

$$\text{net}_\mathbf{x} = \sum_\mathbf{u} w_{\mathbf{x}-\mathbf{u}} s_\mathbf{u}$$

The computational problem is easier if connections with large weights are set to 1, and connections with small weights are 0. Large inhibitory weights would all be set to -1. With this simplification there is not much loss in robustness because it is reasonable to give important features the same degree of importance by virtue of the unit excitatory or inhibitory weights.

This simplification has been verified in practice. Furthermore, there is no point in considering weights that are zero to have any connections. A better representation for a sparsely connected iconic network, is provided by using the commutative property of the net computation, which is similar to a convolution. The connectivity is given by a set $C = \{\mathbf{c}\}$, where vectors $\mathbf{c}$ indicate the relative coordinate positions of the connections to any given cell from a neighborhood of cells in the previous layer. The summation is over the smaller set C:

$$\text{net}_\mathbf{x} = \sum_{\mathbf{c} \in C} w_\mathbf{c} s_{\mathbf{x}-\mathbf{c}}$$

Since the weights are constant (± 1), the net can be further simplified by splitting the set $C = E \cup I$ into an excitatory set E and an inhibitory set I:

$$\text{net}_\mathbf{x} = W_E \sum_{\mathbf{c} \in E} s_{\mathbf{x}-\mathbf{c}} - W_I \sum_{\mathbf{c} \in I} s_{\mathbf{x}-\mathbf{c}} \qquad (5.1)$$

where W_E and W_I are nominally 1, but can be different when it is desirable to weight excitation and inhibition differently.

The primary reason for this type of network architecture is to cut down on the computational expense of two aspects that are common in neural networks. First, multiplications that are costly on fine-grain massively parallel hardware are avoided. Second, the number of nonzero, constant weights are limited to a specified, fixed number. Thus, for weights defined in this manner, an automatic learning method cannot use Delta rule training techniques where the best set of weights for a fixed connectivity are found by using small incremental adjustments of weights. A new learning method must be defined that involves choosing the best places to connect cells to the previous layer by using a limited number of connections, all with constant weights. In short, this concept of neural training must optimize the connectivity of inputs that have fixed weights, rather than finding the optimum weights for a fixed connectivity. The idea of generating new connections and removing connections with small weights has been studied by Honavar and Uhr [6].

The connectivity is easily and quickly trained automatically by using simulated annealing [7,8]. In this method the locations of a fixed number of connections to representative feature points are like a fixed number of molecules that can move during each time interval by an amount proportional to a temperature. During simulated annealing, the energy of the system gradually lowers. Thus the energy of a particular connection scheme must be defined in such a way that annealing will allow quick convergence to a good set of connections. In this case, an optimal connection scheme is one that will cause one neural cell, or a very small neighborhood of cells representing the object location, to fire strongly and all other cell outputs to remain small. It will be shown that a good definition of energy is the value of the largest resulting neural output other than that at the location of the object being trained. These other outputs can be considered as potential false recognitions, and a major goal is to minimize false recognitions. At high temperatures, the system can globally seek energy minima by radically changing the connection schemes. As the temperature is lowered, the connections move less, and the system will approach a local minimum energy, so that false recognitions are at a minimum.

Simulated annealing has been used by Patarnello and Carnevali [9] to train Boolean networks. Their method is very successful at finding good solutions to problems involving logic decisions. They use an energy definition that is the average number of wrong output bits during training. The temperature is used in determining the probability of energy transitions during annealing, but the temperature does not play a role in "molecular" movement.

The remainder of this chapter will cover details of the definitions of energy, temperature, and molecules. Then some real-world applications will be discussed in detail.

5.2 DESCRIPTION OF THE NETWORK

The network is composed of two layers. The input to the first layer is an 8-bit gray-level image. The first layer provides local edge detection with four separate directions. The second layer has a sparse connectivity to the first layer. The output of the second layer provides an indication of the presence and location of trained objects.

5.2.1 First Layer

The first layer is composed of four separate noninteracting *sublayers,* so that there are four neurons per pixel in the first layer. A sublayer is defined to be an iconic translation invariant group of cells, where each sublayer within a layer has a unique connectivity only to the previous layer and sends outputs only to the next layer. The connectivity and weights of each of the sublayers of the first layer are given by the difference of offset Gaussians (DOOG), obtained by convolving the image with a Gaussian function and then computing finite differences, x_0 and y_0, in the vertical and horizontal directions. The DOOG is given by

$$D(x,y) = e^{-((x-x_0)^2+(y-y_0)^2)/\sigma^2} - e^{-(x^2+y^2)/\sigma^2} \tag{5.2}$$

Figure 5.1 shows an example of a discrete DOOG function with $\sigma = 0.7$ and an offset of 2 pixels. These weights are simple to compute and are qualitatively similar to Gabor functions, which are often used in vision preprocessing since they simulate the early stages of biological systems [10].

5.2 DESCRIPTION OF THE NETWORK

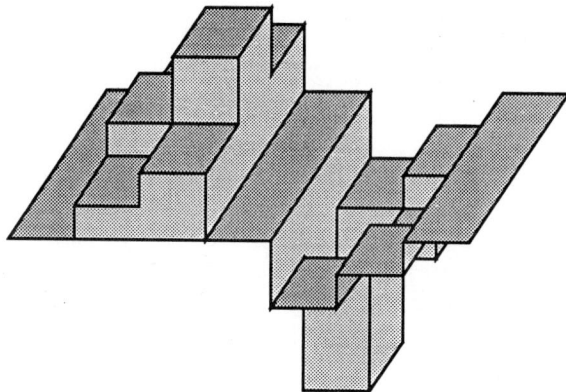

Figure 5.1 *A simple DOOG function.*

Figure 5.2 shows the connectivity pattern and weights for the types of cells in two of the four sublayers in the first layer: One cell is sensitive to edge transitions in the north direction, and the other is sensitive to edges in the west direction. Offsets given by Eq. (5.2) are $(x_0,y_0) = (0,1)$ for the north direction, and $(x_0,y_0) = (-1,0)$ for the west direction. The other two sublayers are sensitive to the south and east directions, with offsets $(0,-1)$ and $(1,0)$, respectively. There are N^2 cells in each sublayer. Each sublayer undergoes a binary threshold operation, where the same threshold is applied to each output. Although these operations can be thought of as a set of thresholded convolutions on a local and compact neighborhood, they still follow the definition of an artificial neural network—in particular, an iconic network. The outputs of the first layers are binary digits, where 1 represents edges that exceed the threshold. There are four summing operations, one for each of the four directions, but only two are independent. The weights associated with the south direction are the negative of those associated with the north direction. It is easiest to compute the DOOG and then apply a positive threshold for the north direction and a negative threshold for the south direction. A similar simplification applies to the east and west directions. Because the weights of the first layers are constrained to be DOOG operations, there are only three parameters that need to be trained: the Gaussian sigma, the offset distance, and the single threshold common to all directions.

Inhibition signals are also defined in the first layer. Computations are simplified by inverting the sign of the signals rather than using negative weights. An inhibition signal is not a separate computation; but is derived as the union of the other four sublayers, as shown in Figure 5.2. A logical complement of the output of the OR gate (NOR) is equivalent to the negation of the signal. The Boolean nature of the inhibition layer allows a very high speed computation on a computer system with a fine-grained massively parallel architecture.

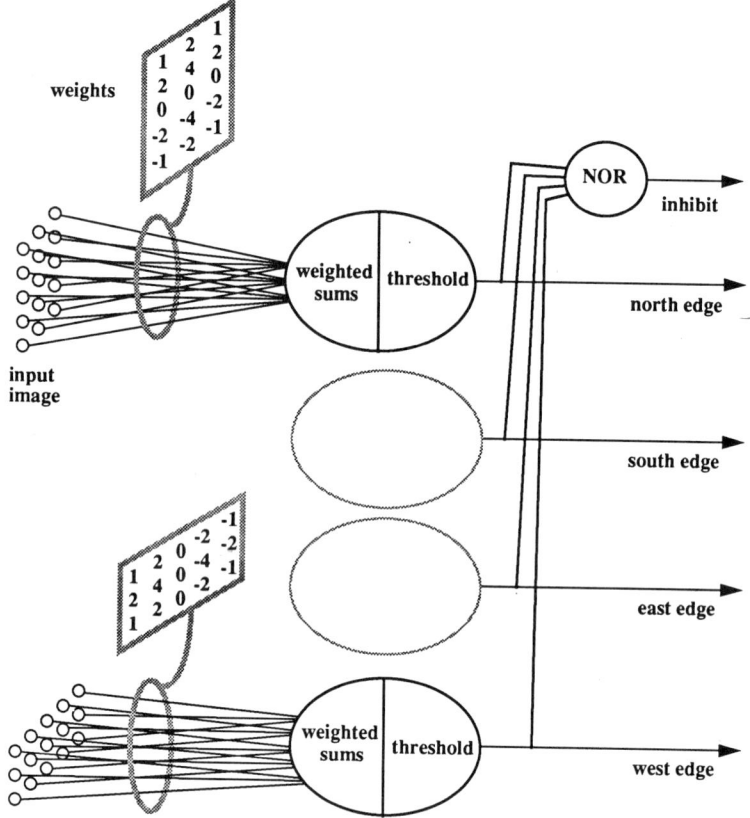

Figure 5.2 *Network configuration for the first sublayers.*

5.2.2 Second Layer

The binary outputs of the first four sublayers and the inhibition signal connect to the second layer (Fig. 5.3), and are constrained as follows. There are a fixed number of connections from each of the first sublayers, where each of the four direction signals will have the same number of connections, and, in general, there will be a different but fixed number of inhibition connections. Whereas the first layer consists of local connections that provide edge detection, the connections to the second layer are global, have weights that are all one, and span the dimension of the object. All outputs of the first layer can be spatially "dilated" slightly (not shown in Fig. 5.3) before being connected to the second layer. The result of a dilation is that a small neighborhood of neurons surrounding any firing neuron will be forced to fire so that detection of objects that are rotated or scaled slightly from the training images will not be lost. Dilation also improves detection when there are occasional pixel dropouts due to noise. Dilation of signals between layers can technically be defined as another layer, as is done in the Neocognitron [11].

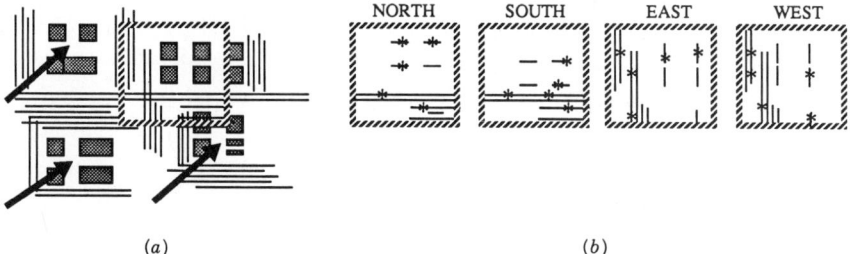

Figure 5.4 *(a) Image: pattern to be located. (b) Edges: quantum states (lines) and molecules (asterisks).*

any of these features to a second-layer neuron located at the center of the box would give a maximum output from that neuron for the object pattern centered in the box. These sets of features that define the object of interest are called spatial *quantum states*. During training, a number M of candidate connections, called *molecules*, are chosen from these spatial quantum states to ensure that valid features are always used. The M molecular positions are defined by the set of vectors $C = \{c\}$ in Eq. (5.1) and can all be represented as a single coordinate position in an M-dimensional *phase space*. The number of molecules will generally be much lower than the number of quantum states. Typical molecules are illustrated as asterisks in Figure 5.4b.

Neurons not located at the center of the pattern of interest all have the same connectivity due to the translation-invariant nature of the network. Not all image features of those neurons would line up with the connections, and thus those neurons would ordinarily not fire. The objective of training is to choose connections that will minimize firing of neurons from any other area in the image, especially areas shown by the arrows in Figure 5.4a. Thus, the major goal is to find connections chosen from the quantum states that will give the lowest false recognition rate.

5.3.2 Definition of Temperature and Energy

During training, various trial candidate connections, or molecules, are chosen. A particular set of molecules are related to those in the previous trial in the sense that each molecule has moved somewhere in the neighborhood of the corresponding molecule in the previous trial. The size of that neighborhood of movement from which new trial molecules are chosen is called the *temperature*. The distance measure of the neighborhood is not important; so the city block metric is used because it is easiest to compute. The relation of temperature to this definition of molecular movement from one time increment to the next results in a simple physical picture: The degree of movement is analogous to the average kinetic energy of the system, which is related to the physical temperature. Thus, a very high temperature means that molecules are chosen almost at random from the quantum states, and

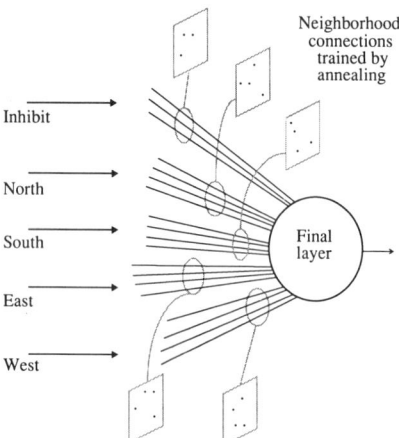

Figure 5.3 *Second layer.*

The second layer has one neural cell per pixel. During training, specific connections from the outputs of the first layer edges to the second layer are evaluated by using the simulated annealing process. A specific neuron firing in the second layer indicates the coordinates of an object of interest. In general, a small compact neighborhood "blob" of neural cells will fire because the edges in the first layer will be a few pixels wide, depending on the parameters of the DOOG computation and because of the dilation of these edges. The centroid of the blob has been found to give the location of the object to around 0.25-pixel accuracy.

5.3 THE PHYSICAL MODEL

In order to provide a method of training connections using simulated annealing, a relation between the neural model and a physical thermodynamic model must be developed. This relation is provided by explicitly defining temperature and energy in an intuitively reasonable manner, so that the methods of thermodynamics can be meaningfully applied.

5.3.1 Definition of Molecules and Quantum States

A particularly difficult object to locate is a local pattern in the midst of a larger surrounding similar pattern such as a specific area on an integrated circuit chip. Figure 5.4*a* shows an example of a pattern of interest enclosed in the heavy box. Note that there are three other similar patterns, indicated by heavy arrows, that can be confused with the desired pattern. Figure 5.4*b* shows thresholded edges in the four directions. Any of these edge pixels could be used as features for locating the object. That is, connections from

5.3 THE PHYSICAL MODEL   143

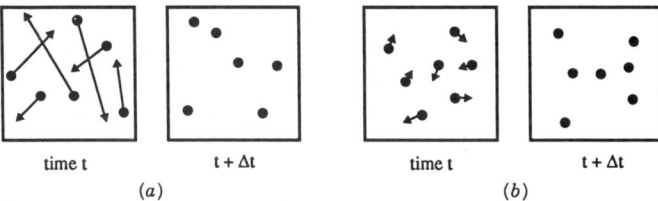

Figure 5.5 Movement of connections: (a) hot, (b) cold.

corresponds to a global scan over the molecular phase space. A low temperature means that there is only a slight difference from the previous configuration, and corresponds to a scan about a local area. Figure 5.5a shows a simple example of molecules chosen from quantum states while at a high temperature, and Figure 5.5b shows molecules at a low temperature.

The potential *energy* of the system can be defined as the net value (the sum of products of signals times weights) of the neuron that has the largest number of inputs that are firing, other than those neurons corresponding to the object being trained. These other neurons correspond to false recognitions, and the objective to minimize false recognitions is realized by minimizing the energy during training. Figure 5.6a shows an example of an image cross section of net values of neurons in the output layer. In this example there are 20 input connections. The cross section includes a peak at the object being trained (value, 20) along with a few false near recognitions. The largest false recognition has a value 16, and there are six neurons with that value. Thus the system energy has the value 16. Figure 5.6b represents a different molecular state that also has an energy value 16, but only two cells have that value. The definition of energy must be changed to reflect the fact that the molecular state resulting in Figure 5.6b is a better choice because there is less chance of a false recognition.

The revised definition of potential energy will involve a 2-byte word where 1 byte is an integer and the other byte is a fractional part. The integer byte is the largest false net as previously defined. The fraction byte is the

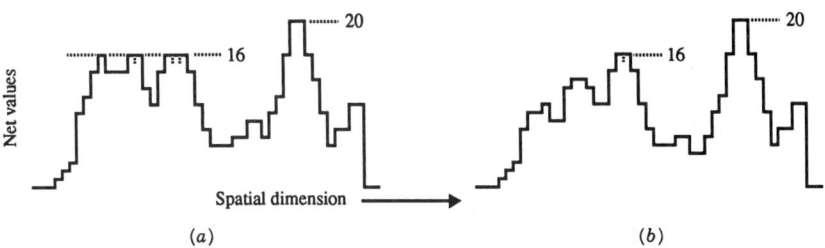

Figure 5.6 Net values: (a) configuration with six near-false recognitions: (b) two near-false recognitions.

number of cells that fire at that net value divided by 256. The fraction byte is truncated to 255/256 to prevent overflow of the fraction part into the integer part. Thus, a molecular configuration with the lowest energy has the lowest false recognition net value with the fewest cells at that value.

5.4 TRAINING BY SIMULATED ANNEALING

The methods of training the first and second network layers are quite different because of the nature of their definitions. In the first layer, DOOG parameters must be found. In the second layer, the best "molecular" configuration must be found. In training the second layer, the DOOG parameters remain fixed during the annealing of the second layer.

5.4.1 Annealing the Second Layer

With the foregoing definition of temperature and energy, the cycles of simulated annealing are very straightforward. According to the usual method for simulated annealing [7], we start with a random molecular configuration at a high temperature. The energy for that state is computed. Next a new molecular state is randomly derived from the previous state according to the temperature, and the new energy is computed. A transition to the new state is made with a probability P according to the temperature T, the energy difference ΔE compared to the previous state, and according to the Boltzmann probability distribution

$$P \sim \exp(\Delta E/kT)$$

where k is the Boltzmann constant, which is determined empirically, as described later. If the new state has a lower energy, then a transition to the new state occurs. If the new state has a larger energy, then the probability of a transition is according to the preceding probability, as in the method given by Metropolis [7]. Since the temperature is initially large, the molecules can move large distances, the energy more readily makes transitions, and there is a good chance that the system will reach a state near the global minimum. The temperature is gradually lowered according to an annealing schedule that is slow enough to provide thermodynamic equilibrium. The molecules have less freedom to move and approach a state that gives a local minimum energy. Annealing is analogous to doing a random walk in phase space with a gravitational effect.

Several different annealing schedules have been tested, and it was found that energy transitions seldom occur at intermediate temperatures. In this application of simulated annealing, the best overall annealing schedule for different types of objects requires that about 20 percent of the time be spent at a very high temperature such that molecular motion is essentially random.

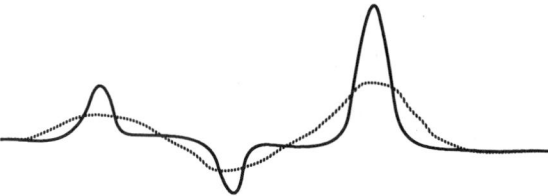

Figure 5.7 *Dynamic threshold. The dotted line is a data-dependent threshold.*

The remainder of the annealing requires an immediate drop to 4° and then equal time spent at the temperatures 4°, 3°, 2°, and 1°. There seem to be many configurations with roughly the same energy minimum, where the area around each of these minima seems to be about 4°. Temperatures above 4° do not seem to push the configuration to a lower minimum any better than a random molecular motion would.

5.4.2 Training the First Layer

Training the best DOOG parameters is straightforward. For the Gaussian sigma, and the offset, there is actually very little choice. In most industrial applications a sigma close to zero generally results in too much noise. Sigmas above 1.0 often unnecessarily reduce image detail, but are useful for increasing rotational independence several degrees. Thus, the best value most often is $\sigma = 0.7$. A DOOG offset of 2 pixels works best for most images. These parameters are not trained, but fixed during annealing.

The first-layer threshold value, which determines the minimum strength of edges that will be output, must be trained for each new application. The use of edges as inputs to the second layer results in a large degree of illumination independence. However, for the widest illumination range, the threshold must be carefully chosen. The threshold represents an optimum trade-off between signal and noise. A threshold that is too low will result in an excessive number of signals from the first layer that are triggered by noise. The probability of triggering a second layer output would then be increased. Weak edges will be lost if the threshold is too high.

Dynamic thresholds give the best results for a wide illumination range. One simple dynamic method is to compute the average value in the neighborhood around each of the four edge layers by using either a Gaussian or uniform convolution kernel. This will result in an image that has broader and lower amplitude edges, as shown by the dotted line in Figure 5.7. The lower amplitude image will act as a threshold. This threshold is roughly a constant fraction of the edge height and will automatically track edges as illumination changes over the scene. A north edge, for example, is defined to be the region where the DOOG is larger than the dynamic threshold and both functions are positive. A south edge is where the DOOG is smaller and both functions are negative. A region where the DOOG and dynamic threshold have opposite signs represents the absence of an edge. A noise squelch is needed to prevent an output response from cells that have small values. Dynamic thresholds require slightly more computer time, but are generally not needed for images where the illumination level does not change by more than a factor of 8 over the image.

The simplest, but perhaps longest, way to train a fixed threshold is to start by setting it to a relatively high value and then running the annealing cycles. The threshold is lowered to three quarters of the previous value, and then the annealing cycles are repeated. The threshold is continually lowered in this manner until the threshold corresponding to the lowest energy is found. The DOOG parameters can be trained in a similar manner if optimal values are not already known.

The threshold used to determine the set of quantum states is chosen to be twice as high as the threshold used during annealing and running. The reason is that the quantum states should be derived from relatively strong edges so that if the illumination level should drop by a factor of 2 while the system is running, then the trained edges will still be present. Experiments show that edges are still valid and noise is acceptably low if the illumination increases by a factor of 4 or more.

5.5 PROPERTIES OF THE ENERGY SPACE

A number of techniques were used to study the energy space in order to evaluate convergence, determine the best annealing schedule, evaluate the Boltzmann constant, and reduce the training time.

5.5.1 Energy Histograms

For 1000 quantum states and 20 molecules, there are 10^{60} possible molecular configurations; so there is no chance of finding the global minimum energy. The nature of the energy states can be explored by looking at a histogram of just a few of these states, say 1000 configurations chosen at random. Only the integer part of the energy is plotted in the histogram because there is no good scale for comparing the fractional part. Typical histograms for a complex image are shown in Figure 5.8. The goal is to find the state that leads to the lowest energy, namely that located at the far left on the tail of the histogram. Because of the limited number of states used in the histogram, the tails appear truncated. It might be possible that a histogram with a larger number of random states, say 10,000, would show points at the next incremental energy further down the tail; or possibly there is a longer tail that leads to a lower energy in a histogram for the same training object but for different first-layer parameters. This could be the case even though the average energy is higher, as shown in the shaded histogram in Figure 5.8.

5.5 PROPERTIES OF THE ENERGY SPACE

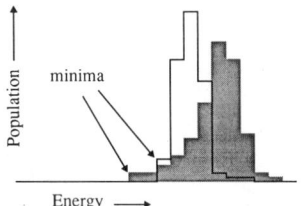

However, the actual existence of cases with long shallow tails to lower energy states is rather hypothetical and not useful since, in practice, the information is limited to a relatively small statistical sampling that can be gathered in a reasonable amount of time. Thus very low energy states with small populations will most likely not be found, especially since there are many local minima, all with roughly the same value.

To increase the speed of learning the first-layer parameters, it is not necessary to actually perform the annealing cycles for every choice of first-layer parameters, but it is sufficient to look at the energy histograms and make a decision based on the average energy and width of the histograms, and then choose the first-layer parameters on the basis of the nature of the histograms. This method will not tell which choice of parameters will converge to the lowest possible energy, but it will tell which choice will most quickly lead to the lowest energy, given a limited number of annealing cycles. A procedure that is very successful in quickly predicting the lowest energy on the basis of the energy histogram is given in a later section on character recognition.

5.5.2 Properties of the Area around a Local Minimum

Histograms can also be used to visualize the nature of the area around a local minimum. First, a molecular configuration is selected, either at random or using annealing. If a point is selected at random, then a few low-temperature annealing cycles are needed to ensure that the starting configuration refers to a local minimum. Next the energy of a large number of random configurations near that minimum are computed. Typical histograms of these energies are shown in Figure 5.9a,b. Figure 5.9a shows that there are very few points at the lowest energy, so that this configuration refers to a sharp local minimum. In Figure 5.9b, there are a large number of points at or near the minimum, so that either the region about the minimum is rather broad or the energy space may not be at a minimum but is in an area that is very rough, possibly where many shallow canyons lead away from the starting point. It was found that the latter is often the case, because during many further cycles of low-temperature annealing molecules will significantly depart from the starting configuration and the energy will very gradually decrease to slightly lower values.

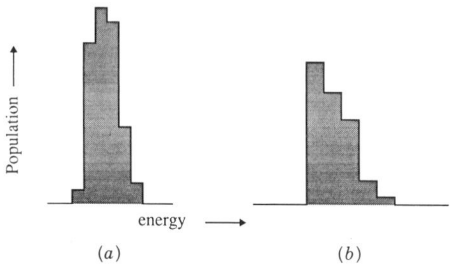

Figure 5.9 Energy histograms near local minima: (a) at local minimum, (b) at false minimum.

5.5.3 The Boltzmann Constant

Unlike physical systems, there is no relation between temperature and energy as defined in this annealing model. Thus, the Boltzmann constant, which relates units of temperature to that of energy, must be determined experimentally. One simple way to investigate this relation is to explore histograms of a large number of random molecular configurations at various temperatures, as done in the previous section. The standard deviation of the temperature is compared to the standard deviation of the integer part of the energy. These standard deviations give a rough measure of ΔE as a function of ΔT. For T near zero, $\Delta T = T - 0$. If we equate the Boltzmann constant to $k = \Delta E/T$, then k is roughly the ratio of the standard deviations. With several different types of training objects and over several ranges of low temperatures, it was found that k computed in this manner varies from 0.5 to 1.0. Generally, in the literature, k is assumed to be 1. The fact that k is experimentally close to 1 here is coincidental and strongly depends on the definition of energy.

With the Boltzmann constant set to 0.5, hundreds of annealing cycles are required to maintain equilibrium during the annealing process and allow convergence to a reasonable solution. Convergence is several times faster if k is set to zero. In this case, if the energy of a new state is less than or equal to the energy for the current state, then there is always a transition to the new state, and if the new state energy is greater than that for the current state then no transitions takes place. Although annealing with $k = 0.5$ was found to lead to a better solution (lower energy) than annealing with $k = 0$, annealing with $k = 0$ leads to a better solution under a stringent time constraint.

5.6 THE SYSTEM HARDWARE

This application was programmed to run on a single-board Applied Intelligent Systems Inc. AIS-3000 computer [12] with 64 fine-grained single-in-

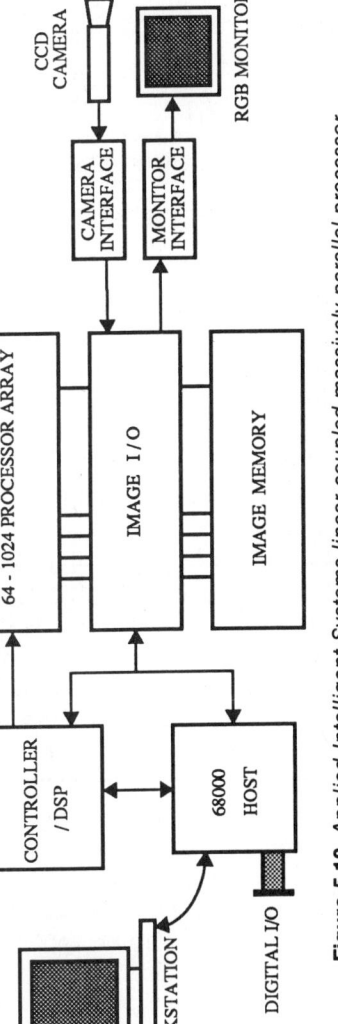

Figure 5.10 *Applied Intelligent Systems linear coupled massively parallel processor.*

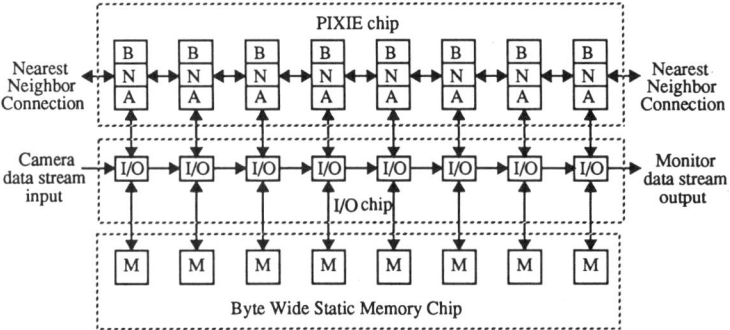

Figure 5.11 *One processing chip coupled to a memory chip*

struction, multiple-data (SIMD) processing elements, where a single instruction is broadcast to all processing elements in the array during each clock cycle. The system shown in Figure 5.10 is designed for low-cost industrial machine vision applications and contains a single board with a host computer, controller, video I/O, and a massively parallel SIMD array of processors called PIXIE chips [13]. There are from 64 to 1024 processing elements in the array, depending on the specific computer model in the family of products. The controller broadcasts instructions to all PIXIE processing elements; then the processors cycle through all pixels in the region of interest for each new instruction. The processors are connected in a linear chain, where each processing element communicates with its nearest neighbor. Figure 5.11 is a block diagram of the PIXIE chip, where there are three types of operations available: Boolean (B), neighborhood (N), and arithmetic (A). The neighborhood operators receive data for the east and west parts of the neighborhood from adjacent processors. There are eight processors per chip. The processor chip is coupled to a bytewide I/O chip and a bytewide static memory. A number of these chip triplets make up a large array. Each processor in the array operates on the data in a column of the image. However, in the programming model for this system, there is one virtual processing element for each pixel in the image. Most function calls in the programming language cause a single operator to be applied to the entire region of interest, which is often the entire image.

The SIMD array is ideal for running both layers of the network. The DOOG first layer is composed of near-neighbor operations that are a part of the image processing library, and they are very efficient. Although the second layer requires more global communication, the operations are translation-invariant and run with high efficiency on the SIMD architecture. Figure 5.12 illustrates how a SIMD machine can process the second-layer sums in Eq. (5.1). One of the first sublayers, say the north edge, is stored in a single bit plane N. The summation over the connections is the net and is stored in byte plane Σ, as shown in Figure 5.12a. The Σ plane is initialized to zero.

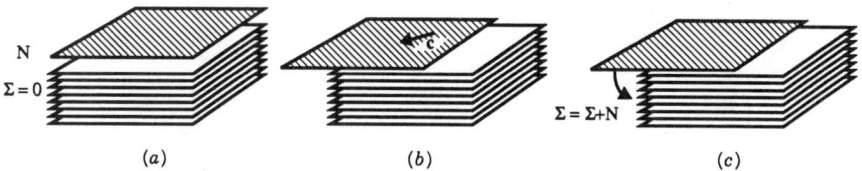

Figure 5.12 *Parallel computation of second layer: (a) starting position, (b) shifting operation, (c) summation operation.*

There are only two instructions in the inner loop of the algorithm. The first instruction (Fig. 5.12b) is a shifting operation that translates the N plane by a vector **c** that corresponds to one of the connection vectors in Eq. (5.1). The second instruction (Figure 5.12c) sums the entire N plane into the corresponding pixels of the Σ byte plane. These two cycles are repeated for all connections in all sublayers. The order of shifting operations is chosen so that unnecessary shifting is avoided by shifting to the nearer connections first. After the signals from all connections are processed, a final thresholding operation is applied. All cells in the Σ plane are compared to a constant threshold T by subtracting Σ from T. The sign bit is copied to an output plane. Wherever the value in a cell in Σ exceeds T, a binary 1 is output.

Processing time for object recognition for the two layers is around 70 ms. Only 20 ms processing time is required in the second layer, since operations consist of simple additions. Image capture requires an additional 30 ms. Simulated annealing training time is roughly 45 s.

5.7 EXAMPLES AND APPLICATIONS

Objects that can be trained by this method include integrated circuit wafer patterns, marks on printed circuit boards, objects to be located by robots, and alphanumeric characters to be recognized or verified. Three examples of training will be given.

5.7.1 Computer-Generated Patterns

A simple example of training involves computer-generated images of a square with other shapes in the background that are similar [1]. In the first example, the square in Figure 5.13a is the object being trained within a background context of other squares with missing corners that are the primary source of high-energy false-alarm states. In this case, after training, perfect squares give the maximum output; squares with missing corners exhibit a low net output; and squares with missing edges that were not in the training image exhibit a rather high false-alarm signal. In Figure 13b, the square being trained is in the context of other squares with missing edges. In this case, squares with missing corners have a high false-alarm signal. When

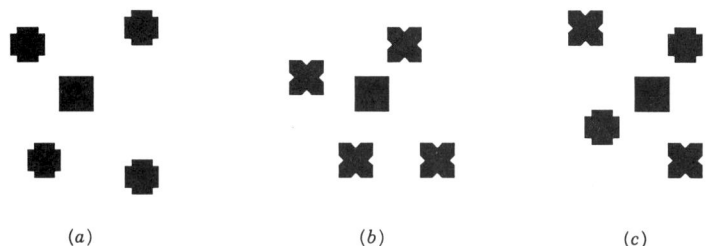

Figure 5.13 *(a–c) Training a square in context with other similar shapes.*

squares with missing corners and edges are in the background of the training image (Fig. 5.13c), then both have a low false-alarm signal, although not quite as low as the two previous cases. Table 5.1 shows the net output signals for these three conditions, where it is apparent from Figure 5.13c that the network can be trained to ignore squares with both types of defects as long as they are present during training. Sixteen connections and 700 annealing cycles are used in this example.

5.7.2 Integrated Circuit Die Surface

The next example involves finding the location and rotation of an integrated circuit (IC) die. Two small and widely separated areas on the chip are located. The die angle and position can be determined from those two locations. The system can reliably detect areas on ICs that are not easily distinguishable by the human eye. The concepts have been tested with hundreds of thousands of parts. There are no parameters for the user to set. In training, the user specifies a window about the area to be learned, and then starts the system. The system informs the user if the region is not easily distinguishable from the background (final energy is too high). This can occur, for example, on some repetitive circuit patterns that are not unique in the context of the background. An example of a difficult pattern is shown in Figure 5.14a, where the pattern in the middle is trained in the context of patterns on the right and left that have only slight differences in the area being trained. To make the problem more difficult, the area being trained does not include the upper edge of the dark rectangle, which is an obvious discriminating

TABLE 5.1 Net Outputs After Training Objects in Figure 5.13

See Figure	13a	13b	13c
Squares	16	16	16
Missing corners	8	14	10
Missing edges	15	9	10

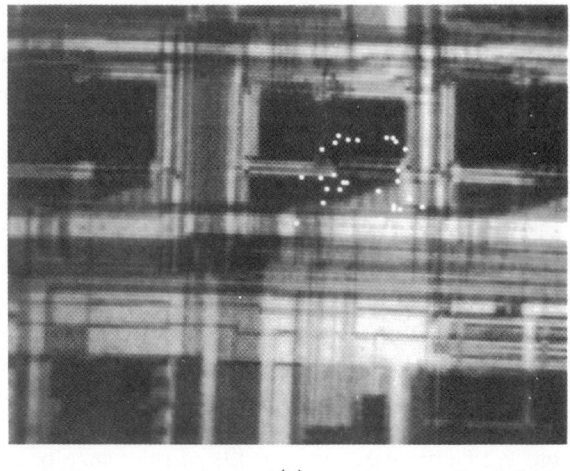

Figure 5.14 *Training an IC pattern: (a) IC surface; (b) second layer output after annealing.*

feature. The white dots indicate locations of typical molecules during annealing. An iconic map of the neural outputs for a random configuration where no annealing has occurred will result in rather high outputs for the three similar patterns. Results after 100 cycles of annealing are shown in Figure 5.14b. The output in the region of the background pattern is considerably suppressed.

It was found that the usage of inhibition signals from the first layer did not always lead to lower-energy states on IC patterns. A number of experiments

TABLE 5.2 Effect of Inhibition of Final Annealed Energy

Circuit Region	Energies with Inhibition	Energies with No Inhibition
A	20.27	20.54
B	17.23	19.23
C	19.44	18.53

were run to test the effect of inhibition. Three areas on the IC surface were tested with and without inhibition. In those cases that used inhibition, eight connections were inhibitory and five connections were used in each of the four edge planes. In cases with no inhibition, seven connections to each of the direction planes were used. Thus, in all cases, a total of 28 connections were used. Table 5.2 summarizes the results of the usage of inhibition. The final energies are averages from a number of trials, each with 200 cycles of annealing.

Table 5.2 shows that inhibitory connections have little effect on the performance of the system when complex images such as IC patterns are trained. The pattern in circuit region B in Table 5.2 shows the most improvement when inhibition is used. An inhibition connection will help lower the energy if it is connected to a feature of a nearby similar pattern that is not contained in the trained pattern. In practice, inhibition is not used in this type of application because the operator will avoid training the system on patterns that are not unique.

5.7.3 Character Recognition

Character recognition requires a more complex network configuration: The second layer will have one sublayer for each character to be recognized. The neural cells associated with each pixel are shown in Figure 5.15, where each cell is responsible for the identification of a single character. This type of configuration is often called a local memory, where knowledge of a character is localized to a single cell at a pixel site. It may be possible to use a distributed memory concept by more efficiently encoding the second layer. The number of cells could be as low as $\log_2(z)$ per pixel, where z is the number of characters. Knowledge for each character is distributed over several cells at each pixel. With distributed memory, the system would be much more difficult to train using simulated annealing.

Each of the z sublayers must be separately trained on its associated character, with all of the other characters in the background of the image. If a sublayer strongly responds to a wrong character, then that response is a false alarm that will result in a large energy. Inhibition is very important for character recognition. For example, the strokes of an F are included in the strokes of an E. Thus, without inhibition, the cell trained for an F will also

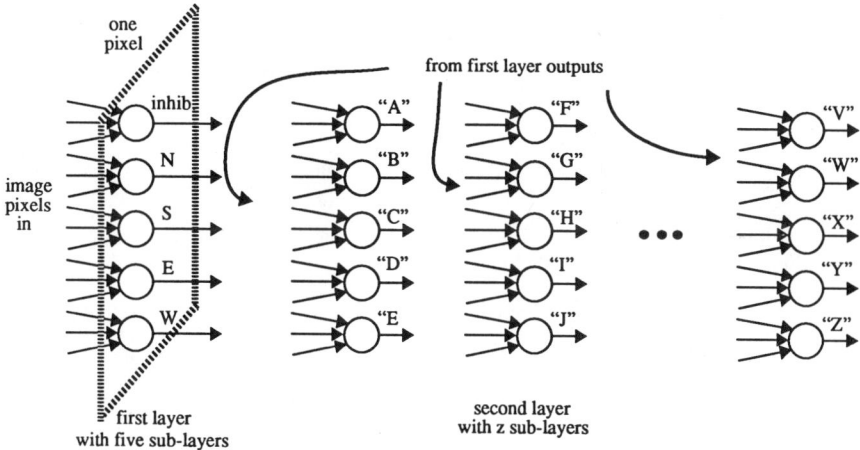

Figure 5.15 *Neural cells associated with a single pixel.*

respond to an E. If inhibitory connections are made to the F cells from the region of the lowest horizontal stroke, then an F will still be recognized, since it does not contain that lower stroke, but an E will be inhibited from responding in F cells. Without inhibition, an F cell will also strongly respond to P, R, and B, for the same reasons. Figure 5.16 illustrates the training of an F in the context of eight other confusing characters. Twelve excitatory and 16 inhibitory connections are used as inputs to the F cell. After 500 cycles of annealing, the energy is 22 + 15/256 (abbreviated as 22.15), which means that the net of the largest responding false alarm cells is 22 (80 percent of maximum), and there are 15 cells in that layer with that response.

The connections that were found in this example are shown in Figure 5.17a, where edge connections and directions are indicated by N, S, E, and W, and I indicates inhibition. Final energies after a much slower annealing with 5000 cycles are shown in Figure 5.17b, where the final energy is 22.04. The final energy and connection configurations are very similar to those corresponding to the faster annealing schedule. Thus, an annealing schedule of 500 cycles is adequately near equilibrium. Inhibition connections cluster

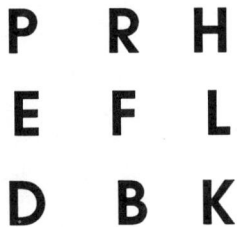

Figure 5.16 *Training an F in a background context of other characters.*

156 TEACHING NETWORK CONNECTIONS FOR REAL-TIME OBJECT RECOGNITION

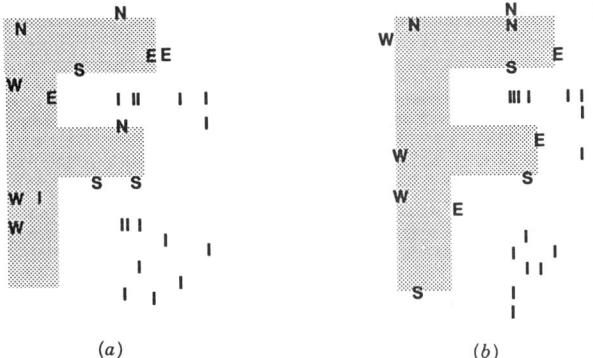

Figure 5.17 *Connections: N = north, S = south, E = east, W = west, and I = inhibit; (a) connections after 500 annealing cycles; (b) connections after 5000 annealing cycles.*

most strongly around areas that prevent confusion with the characters P, R, E, and B. In some cases, two or three inhibition connections are identical and can be interpreted to indicate negative weights that have values of −2 or −3. After 5000 cycles of annealing, the net outputs of the cells in the F sublayer are shown in Figure 5.18. That sublayer responds in a single compact blob for the correct character. Response to other characters is rather uniform and diffuse. The degraded and rotated F characters shown in Figure 5.19 result in a net output shown in Figure 5.20. These degraded characters still show a larger net value than any other characters in the image.

During training with simulated annealing, the system will tend to improve the performance with respect to those characters that give the largest false alarms. False characters that give a lower net response will be ignored since

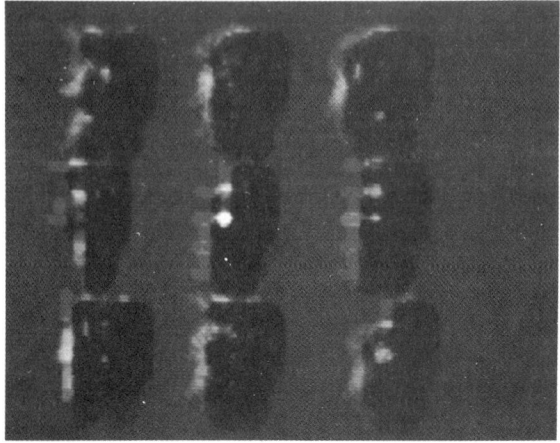

Figure 5.18 *Net outputs from the F sublayer.*

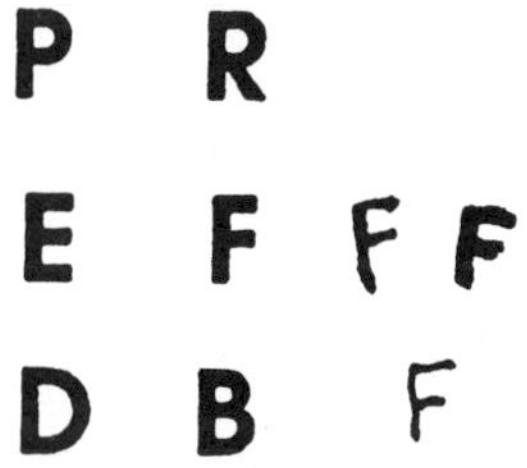

P R

E F F F

D B F

Figure 5.19 *Degraded characters.*

they do not directly play a part in the energy function. That is, the energy function is highly nonlinear, and it makes no difference to the energy function if the net of a possible false alarm is slightly smaller or very much smaller than the net of the largest false alarm. Thus, simulated annealing tends to improve the performance of high-false-alarm characters at the expense of lower-false-alarm characters. This property is also true for learning in other types of network architectures.

Performance is a function of the relative number of excitatory and inhibitory connections. Table 5.3 gives the energy as a function of various numbers of inputs for various characters. Boldface values in Table 5.3 indicate the configuration with the best energy. It is clear that the best balance of excitation and inhibition depends on the character. Recognition of characters that are fairly distinct from others in the set, such as R, D, and K works best with no inhibition. Characters such as F and E have strokes in common with several characters and are best recognized with a substantial number of inhibition connections.

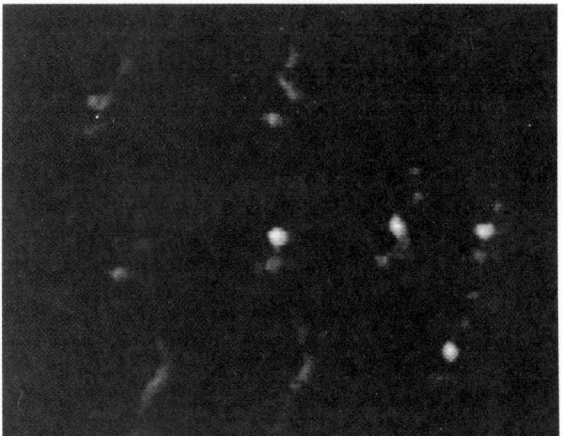

Figure 5.20 *Net output from degraded characters.*

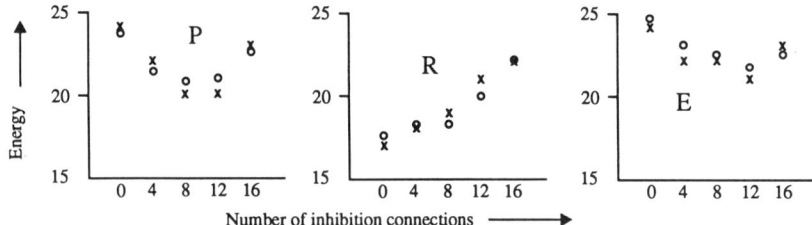

Figure 5.21 Best energy (x) and predicted energy (o) for characters P, R, and E.

It is desirable to have a way of predicting the best numbers of inhibition and excitation connections rather than running an annealing loop for each possibility. It turns out that the best energy, E_p, can be predicted on the basis of the mean value and standard deviation of the energy histogram, and is given by the empirical formula E_p = mean − 3.47 × STD. Figure 5.21 shows the best energies after annealing (x) and the predicted energy E_p (o) for characters that require a small, medium, and large number of inhibition points. The predicted energies are accurate within one energy unit. A more important observation is that the predicted energy is a minimum at the same points as the best annealed energies. This search procedure is an order of magnitude faster than annealing since only a relatively small number of energy points are needed to determine the mean and standard deviation.

The approximate processing time for an AIS-3500 computer with 128 fine-grained processing elements to recognize an alphabet of 10 characters by this method is around 150 m. Thirty-six characters would require around 450 m. The system runs just as quickly if there are multiple occurrences of the same character in the field of view since the network is translation-invariant.

TABLE 5.3 Energy as a Function of Inhibition

Number of excitation points	28	24	20	16	12
Number of inhibition points	0	4	8	12	16
Best energy for F	26.04	25.03	23.07	**22.01**	22.15
Best energy for P	24.06	22.01	**20.07**	20.12	23.02
Best energy for R	**17.12**	18.07	19.03	21.03	22.02
Best energy for H	18.03	**16.03**	17.09	17.01	19.05
Best energy for E	24.04	22.05	22.02	**21.07**	23.02
Best energy for L	23.10	21.02	**20.01**	20.09	21.33
Best energy for D	**16.01**	16.05	16.07	17.08	19.07
Best energy for B	20.02	**19.13**	21.01	22.07	22.13
Best energy for K	**17.01**	17.01	18.02	18.08	21.15

5.8 CONCLUSIONS

The idea of training the connectivity of a network with fixed weights is very successful for the location of a pattern to subpixel accuracy, discrimination of an object among very similar objects, and classification. A two-layer translation invariant network can be used, where the first layer consists of thresholded edges and the second layer is trained by simulated annealing. A detailed study of the characteristics of this type of network on various objects under various conditions leads to several conclusions:

- During annealing, a relationship is defined between the potential energy and the largest network outputs that correspond to false recognitions. This results in convergence to solutions that are meaningful in a practical application.
- Degraded objects with several degrees of rotation are easily discriminated.
- Objects with a large potential for false recognitions are trained at the expense of other objects that have a low potential for false recognitions.
- The Boltzmann constant is experimentally found to be around 0.5 to 1.0 for a wide variety of cases.
- Under a stringent time constraint (45 s), annealing with a Boltzmann constant set to zero leads to the best solution. If the annealing time is 5 min (500 annealing cycles), then better results are obtained if the Boltzmann constant is 0.5.
- Knowledge of the mean and standard deviation of an energy histogram is a good predictor of the final annealed energy. Predictions are useful for finding those parameters, such as thresholds and the number of inhibition connections, that give the lowest energies. Obtaining the histogram is an order of magnitude faster than going through annealing cycles.
- A fine-grained parallel processor quickly trains and runs the network at speeds and costs that meet the requirements of real-time industrial applications.

REFERENCES

[1] S. S. Wilson, Teaching network connectivity using simulated annealing on a massively parallel processor, *Proc. IEEE,* 7(4), (April 1991).

[2] D. O. Hebb, *The Organization of Behavior,* Wiley, New York, 1949.

[3] D. E. Rumelhart and J. L. McClelland, *Parallel Distributed Processing,* A Bradford Book, MIT Press, Cambridge, Mass., 1986.

[4] S. S. Wilson, Vector morphology and iconic neural networks, *IEEE Trans. on Syst., Man, and Cybern.*, SMC-19(6) (Nov/Dec, 1989).

[5] G. E. Hinton, "Learning Translation Invariant Recognition in a Massively Parallel Network" in G. Goos and J. Hartmanis, Eds., *PARLE: Parallel Architectures and Languages, Europe,* Lecture Notes in Computer Science, Springer-Verlag, Berlin, 1–13 1987.

[6] V. Honavar and L. Uhr, "A Network of Neuron-like Units That Learns to Perceive by Generation as Well as Reweighting of Its Links," in D. Touretzky, G. Hinton, and T. Sejnowski, Eds., *The Proceedings of the 1988 Connectionists Models Summer School,* Morgan Kaufmann, San Mateo, Calif., 1988.

[7] N. Metropolis, A. Rosenbluth, M. Rosenbluth, A. Teller, and E. Teller, Equations of state calculations by fast computing machines, *J. Chem. Phys.*, 21, 1087–1091 (1953).

[8] S. Kirkpatrick, C. D. Gelatt, Jr., and M. P. Vecchi, Optimization by simulated annealing, *Science,* 220 (4598), 671–680 (1983).

[9] S. Patarnello and P. Carnevali, "Learning Capabilities of Boolean Networks," in I. Aleksander, Ed., *Neural Computing Architectures,* MIT Press, Cambridge, Mass., 1989.

[10] J. G. Daugman, Uncertainty relation for resolution in space, spatial frequency, and orientation optimized by two-dimensional visual cortical filters, *J. Optical Soc. America,* A 2, 1160–1169 (1985).

[11] K. Fukushima, Neocognitron: A hierarchical neural network capable of visual pattern recognition, *Neural Networks,* 1, 119–130 (1988).

[12] S. S. Wilson, A single board computer with 64 parallel processors, *Electronic Imaging '87,* Institute for Graphic Communication, Inc., pp: 470–473, Boston, Nov. 2–5, 1987.

[13] L. A. Schmitt and S. S. Wilson, The AIS-5000 parallel processor, *IEEE Tran. Pattern Anal. Machine. Intell.*, 10(3), 320–330 (May 1988).

CHAPTER 6

The Discrete Neuronal Model and the Probabilistic Discrete Neuronal Model

RON SUN

6.1 INTRODUCTION

This chapter presents a neural network formalism previously proposed in Sun et al. [1,2,15]: the discrete neuronal (DN) model and the probabilistic discrete neuronal (PDN) model, and related learning algorithms.

The DN and PDN models have been applied to some biological neural networks (e.g., the stomatogastric ganglion in lobsters) and have demonstrated their generality and usefulness in simulating information processing capabilities of real neural networks given their exact or inexact connectivity patterns and endogenous firing patterns. (For details, see Sun et al. [1].) Another application of this model is in building rule-based inference systems. The advantages of these types of models are that they can handle variable bindings easily, and that they can incorporate certainly factors propagation mechanisms into inference systems. The detailed description of the system and some descriptions of the neuronal models are contained in Chapter 14 (see also Sun [2,3]). Systems for sequential processing based on this model have been proposed. Aleksander and colleagues [4] used similar models and applied them to other domains.

Because of their usefulness and interesting properties we need to understand the models better, especially in learning aspects. But, because of their

---

This work was supported in part by the Defense Advanced Research Projects Agency, administered by the U.S. Air Force Office of Scientific Research under contract F49620-88-C-0058.

Neural and Intelligent Systems Integration, By Branko Souček and the IRIS Group.
ISBN 0-471-53676-8 ©1991 John Wiley & Sons, Inc.

generality and complexity, it is difficult to perform learning in the same way as conventional parallel distribution processing (PDP) models did. This chapter describes an initial step toward a better understanding of learning issues in this type of networks.

Neural network models did not start with weighted-sum models. Although the most frequently seen model is the weighted-sum (thresholding) model, its biological validity and computational sufficiency are questionable. The popularity of this model mainly stems from its simplicity. Many alternative neuronal models may provide better computational properties and biological plausibility.

To a large extent, finding alternative neuronal models is an area yet to be explored. Devising more complicated models seems to be an inevitable step toward useful, efficient, and powerful networks. In this endeavor we could look into some past efforts in this.

McCulloch and Pitts [11] first proposed a binary element as a model of neurons, and networks of them were constructed to show their emergent properties. Many used this kind of model to do useful things. One of the earliest models was Rosenblatt's perceptron [11], a simple weighted-sum model. Another is Widrow's model for adaptive pattern recognition [11]. Although simple, it has many desirable computational properties, such as limited rotation-invariance. Many biological models were proposed in the 1960s and 1970s, which employed first-order differential equations. Grossberg [12] studied the computational properties of these models based on mathematical analysis and proposed several classes of models based on their computational characteristics, such as additive models, shunting models, feedforward, and recurrent competitive models.

Other researchers provided different ideas. Klopf [5] proposed the idea of self-interested "hedonistic" neurons. Barto and his colleagues have worked on models based on this and other ideas [6]. Feldman and his colleagues [13] worked on their own models, which differentiate inputs from different sites. Aleksander and his colleagues [4] have been working on models they call "probabilistic logic neurons," which are similar to our model, although derived differently. Discrete neuronal models were proposed in Sun [2] and Sun et al. [1], in which a set of discrete states is used instead of continuous activation functions.

6.2 DN AND PDN

An automation-theoretical description follows. Basically, a discrete neuronal model (DN) is a 2-tuple

$$W = \langle N, M \rangle$$

where $N = \{\langle S, O, A, I, B, T, C \rangle\}$, is a set of neurons
- S = set of all possible states of a neuron,
- A = set of all actions (output symbols),
- O = set of all outputs,
- B = set of all input symbols,
- I = set of all inputs,
- T = state transition function: $S \times I \to S$,
- C = action function: $S \times I \to O$,
- M = connectivity among neurons in the set N: $M = \{(i,j,k) | i$ and j are indices of neurons and k is a label unique for a particular synapse$\}$.

In this model a set of discrete states is explicitly specified instead of a continuous activation function. By introducing the state variable, we hope to be able to model more aspects of the working of real neurons.

A way to understand the model is to think of a neuron as having a set of inputs and a set of outputs:

$$i = (i_1, i_2, \ldots, i_n) \in \Pi B = B'$$
$$o = (o_1, o_2, \ldots, o_m) \in \Pi A = A'$$

The state transition and action functions are the following mappings:

$$C: S \times B' \to A'$$
$$T: S \times B' \to S$$

A network of DN neurons is a directed graph with (multiple) edges between pairs of nodes. An added constraint is if there is a directed link from node i to node j, then $A'_i = B'_j$.

The DN units can be implemented with the simple PDP neuron types. A linear neural network model can be expressed as

$$\frac{d}{dt} X = AX + BI$$

where X is the activation vector and I is the input vector. A determines the contribution of current X in deciding the next X, and B that of input I.* It is easy to show that it can simulate a finite-state automaton (FSA) and, therefore, the DN model. We can apply standard system theory to study if a

---

* In case of semilinear systems, we can easily replace X on the right-hand side of the equation by a function $h(X)$.

particular FSA is implementable. For applications mentioned, the implementation is straightforward.

The PDN model is an extension of this. It assumes that the state transition function and the action function are probabilistic instead of deterministic. Mathematically, it can be described as

$$W = \langle N, M \rangle$$

where $N = \{\langle S, A, B, I, O, T, C, P_1, P_2 \rangle\}$,
    S = set of all possible states of a neuron,
    A = set of all actions,
    B = set of all input symbols,
    I = inputs,
    O = outputs,
    T = state transition function: $S \times I \to S$; P_1,
    C = action function: $S \times I \to O$; P_2,
    M = connectivity among neurons in N, as before.

Here P_1 and P_2 are sets of probabilities associated with each state transition or each action. The state transition function is a mapping to a statistical distribution of states and the action function is a mapping to a statistical distribution of actions:

$$T: S \times \Pi B \to D(S)$$

$$C: S \times \Pi B \to \Pi D(A)$$

where $D(\cdot)$ denotes a statistical distribution of the argument. For simplicity, sometimes we can assume T is fixed (i.e., P_1 is a set of 1's, or, equivalently, $D(S) = S$) and simply adjust P_2 in C.

Biophysiological justifications can usually be argued but cannot be conclusive at all, considering the state of the art. We simply do not know enough at this time to fully describe the operation of the neuron. Generally speaking, the probabilistic nature of the model can be attributed to the unreliability and noise in the cell, statistical nature of membrane operations (channel opening and closing), and random environmental influences.

Notice that in the foregoing definitions there is no provision regarding the form of functions (specifically the action function and the state transition function). They could be (1) a fixed function with adjustable parameters, such as the weighted-sum function, or (2) a set of functions, in which case learning amounts to the selection of one appropriate function out of the set, or (3) a table specifying a mapping, in which case learning is the precise specification of table entries. In the following discussion the third choice is assumed, due to its generality and simplicity.

6.3 LEARNING ALGORITHMS

Learning is the major advantage of neural network (connectionist) models. Devising better, more powerful learning algorithms is the main thrust of this field of research. We present a set of simple learning algorithms and describe experiments performed on them. Then we extend these algorithms by incorporating techniques from other related computational models. It is hypothesized that the integration of these techniques may result in very general and very powerful models for a large variety of domains.

Learning algorithms for PDN can be classified into three categories:

1. *Learning Only Action Functions:* There are two possible ways of dealing with states in the formalism: (1) ignoring them—that is, assuming there is only one state and it never changes, or (2) setting up a fixed sequence of states to go through for each node, useful for dealing with temporal sequences that are generated in systems that consistently go through a fixed sequence of states.
2. *Learning Only State Transitions:* With fixed action probability, the learning of state transitions is easier, and we can apply some known algorithms.
3. *Learning both Action Functions and State Transition Functions:* This is a combination of the first two cases. There are, however, some extra complications that will be discussed later.

6.3.1 Simple Algorithms and Experiments

A learning algorithm for a network of PDN nodes can be easily devised based on the idea of reinforcement learning [5]. We assume that for each input-state combination in a node, there is a table with entries specifying all possible next states with their respective probabilities. If these probabilities do not sum to 1, the remaining probability is assumed to be uniformly distributed over all remaining possible states. It is the same for the action function.

SCHEME 6.1: *If a node fires correctly (according to an external teacher or internal critic, error signal, or reinforcement signal), then increase the firing probability of the corresponding entry in case that entry exists, or create an entry for that action if the corresponding entry does not exist. If it fires incorrectly, then decrease the probability.*

The preceding algorithm addresses the issue of training a single neuron, with input, output, and a reinforcement signal. In case of a connected network of such neurons, we have a simple algorithm that is an extension of the previous scheme and can tune the network into performing the required input/output mapping.

SCHEME 6.2 [Pessimistic: Learning Action Functions Only] *Clamp the input nodes and look at the output nodes. If all the output nodes are correct or all are wrong, then apply Scheme 6.1 to each node, assuming that every node in the network is correct (or wrong). Otherwise, apply Scheme 6.1 to the correct output nodes for positive reinforcement, and then, assuming the rest of the nodes are all wrong, apply it for negative reinforcement.*

An alternative is to reduce the number of negatively reinforced nodes as much as possible, trying to preserve the part of the network that produced the correct results (if we negatively reinforce all of the network nodes except those that produce correct results, we may inadvertently reverse the correct subnetworks). So we arrived at the next scheme.

SCHEME 6.3 [Optimistic: Learning Action Functions Only] *Clamp the input nodes and look at the output nodes. If all the output nodes are correct or all are wrong, then apply Scheme 6.1 to each node, assuming that every node in the network is correct (or wrong). Otherwise, apply Scheme 6.2 to the incorrect output nodes for negative reinforcement, and then, assuming the rest of the nodes are all correct, apply it for positive reinforcement (or, assuming neutrality regarding correctness, do nothing with these nodes).*

This algorithm has a different problem, namely if there are two few negatively reinforced nodes, we may never get the optimal or near optimal solution, because we lose many degrees of freedom due to our decision to give positive reinforcement to the rest of the network.

The rate of change made in each node has a profound impact on the performance of the networks. It can help to decide if a network will converge, how fast it will converge, and if it will converge to a correct value.* A simple scheme is to add a fixed amount to a probability in case of reward and to subtract a fixed amount from a probability in case of penalty; that is,

$$p_j = p_j + r_{increment} \quad \text{if } j = k$$

or

$$p_j = p_j - r_{decrement} \quad \text{if } j \neq k$$

where j is the action taken and k is the desired output. Another scheme, an expedient scheme, is the linear reward-penalty algorithm:

$$f_j = ap_j$$
$$g_j = b(1/(r-1) - p_j)$$

---

* The formal notions are expedient, absolute expedient, ϵ-optimal, and optimal. See Narendra and Thathahar [8]

where f is the amount of increment to the probability of an action if positively reinforced and g is the amount of decrement if negatively reinforced. r is the total number of different actions. Suppose the correct output is k. Then

$$p_j = p_j + f_j \quad \text{if } j = k$$
$$p_j = p_j + g_j \quad \text{if } j \neq k$$

We can turn the algorithm into the linear reward-inaction algorithm, which is absolutely expedient, by simply ignoring the second equation. Many other schemes have various properties and performances (see Williams [7]).

Such a network can be viewed as a confederation of units that face difficulties in optimizing their own performance because of the lack of knowledge of the behavior of the other units due to distributivity, limited communication, and inability to access centralized control information (as suggested by Barto [6]). Each of them wants to maximize an evaluation function to its own internal goal (cf. Klopf [5]). A learning algorithm is a way that each of them can learn to cooperate with one another and adapt itself so that its own decisions fit into the overall situation, including not only environmental factors but also the decision making of its peers. This situation is generally referred to as distributed decision making. (cf. Ho [14])

The idea behind these algorithms is simple: If things are going well, then keep doing the same thing or even do more of it; otherwise, stop doing the same thing and find an alternative. The problem is that, if not all units are observable (or there may be no reinforcement signal for some of them), then how we decide if they performed correctly, or in other words, how we assign credit and blame. The previous two algorithms try to tackle that problem (though a little simplistic and without utilizing domain knowledge). We need more formal analyses and empirical studies to better understand this type of situation.

The foregoing algorithms are stateless, namely, the state variables in the generic definition of DN and PDN are not used at all. On one hand, this greatly simplifies the problem of learning in PDN networks, but, on the other hand, this results in losing some computational power and generality. In order to incorporate states, we can extend the reward-penalty algorithm for both actions and states, using the same reinforcement signal.

SCHEME 6.4 [Learning both State Transition and Action Functions] *Run the network as before. For each individual node, if the node deserves reward, give it to both state transitions and actions; if the node deserves a penalty, give it to (1) the action only, (2) the state transition only, or (3) both the action and the state transition. The reinforcement signal can be generated externally or by straight comparisons between the desired output and the actual output. For hidden units, we can either adopt the pessimistic strategy*

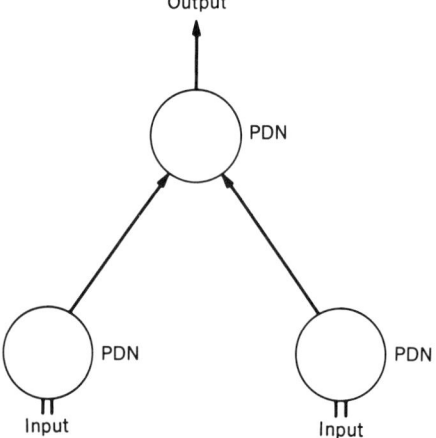

Figure 6.1 *A network structure for the parity problem.*

(assuming that all are incorrect, if there are any incorrect output nodes), the passive strategy (doing nothing), or the optimistic strategy (assuming that all are correct).

The following experiment was conducted: We constructed a three-node two-layer feedforward network with two input nodes and one output node (see Figure 6.1). Because in DN/PDN models the adaptive parameters are within an individual node itself, we have in this network two layers of adaptable nodes or, in other words, the two input nodes are actually two "hidden" nodes (i.e., they are adaptable but there is no known correct output pattern for them). So this network is equivalent to a three-layer feedforward PDP network in topology.

Input	Output
0 0 0 0	0
0 0 0 1	1
0 0 1 0	1
0 0 1 1	0
0 1 0 0	1
0 1 0 1	0
0 1 1 0	0
0 1 1 1	1
1 0 0 0	1
1 0 0 1	0
1 0 1 0	0
1 0 1 1	1
1 1 0 0	0
1 1 0 1	1
1 1 1 0	1
1 1 1 1	0

Figure 6.2 *The parity problem.*

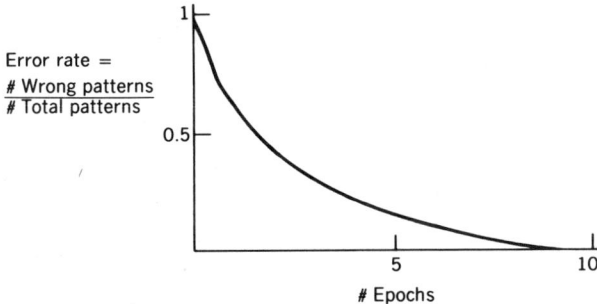

Figure 6.3 *Curve of the error rate going to zero.*

We first apply this network to the parity problem with four inputs and one output detecting oddity (see Figure 6.2). We assume that the nodes are stateless; that is, there is only one state in each node. The pessimistic strategy is adopted for the two input nodes. The adjustable parameters in the network include the rate of increment when rewarded and the rate of decrement when punished. When the parameter setting is $r_{increment} = 0.25$ and $r_{decrement} = 0.35$, the network converges to zero usually within 10 epochs. A typical curve of error rates is shown in Figure 6.3.

Some comparison is in order here: with back-propagations, the task takes on average more than 10,000 epochs to converge; with the algorithm in Aleksander [4], it takes on average 32 epochs to converge. With our algorithm, the average number of training epochs needed is 8.7, based on 10,000 experiments. In all the 10,000 experiments, the algorithm converged. From this data, it is evident that this algorithm provides a viable alternative to conventional PDP learning algorithms, though more comparisons are needed to fully characterize the algorithm.

The equations can be summarized as

$$p_j = p_j + r_{increment}$$

or

$$p_j = p_j - r_{decrement}$$

This way a probability value may go above 1 or below 0. We allow this to happen and interpret it as being more firmly on one side or the other. And later changes toward the other direction will have to take a lot more effort (more data presentations) than the probability just being 1 or 0. This can help to prevent too much oscillation (i.e., error rates go up and down).

We tried some other variants, too. We tested the optimistic strategy and found that it does not work as well as the pessimistic strategy, probably due to insufficient degrees of freedom (only punishing the output node). We also

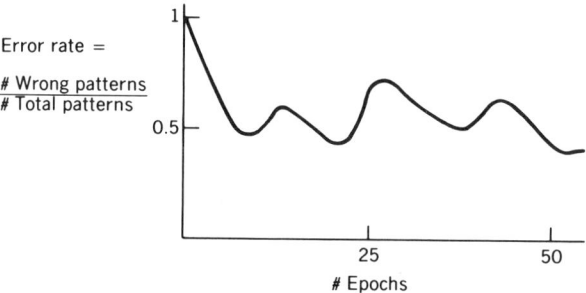

Figure 6.4 *A case of oscillating error rates.*

tested linear reward-penalty algorithms (including reward-inaction variants). They generally oscillate a lot (i.e., error rates go up and down) and rarely converge to zero (see Figure 6.4 for a typical case). Even when it converges, it takes several hundred training epochs. With linear associative algorithms, it is generally better to have *a* greater than *b* (the rate of reward and the rate of penalty, respectively, as defined earlier).

We extend the model to two states in each node, and we apply the network to the same parity problem. Most of the time, it converges to zero within 10 epochs. This result shows that for simple problems with no need for states, adding multiple states into the model will not do much harm, whereas with states the model can tackle more difficult problems, as will be explained later. We tried all three variants regarding penalty for actions, states, or both.

We also apply this network to sequence learning problems, in which we do need state variables and multiple states in the model, because sometimes outputs depend on past activities, which are represented by the state. We use the same network of three nodes. What we aim at is the following: Given a fixed sequence of inputs, the model should produce a fixed sequence of desired outputs.* We tested the network on a set of sequences as in Figure 6.5, and the system learns them rapidly, with parameter settings $r_{increment} = 0.4$ and $r_{decrement} = 0.45$.

The difficulty comes in when there are multiple, contradictory† sequences that have to be accounted for. We tested the network on a difficult sequence where there are contradictory items, such as items 2 and 3 in Figure 6.6. In this set of sequences, the output at time *t* cannot be determined solely by the input at time *t*. It also depends on the input at time $t - 1$. So in this case the

---

* This formulation of problems can be justified as follows: suppose we have a device to control, and we know several possible sequences of situations the device can go through. When a particular situation in a particular sequence comes up, the controller is required to send a particular control signal to the device, so that the controller can receive a zero reinforcement signal.

† For example, different outputs are required when the same inputs are given at different times.

Seq#	Input	Output
1	0 0 0 0	0
	0 0 1 1	0
	0 0 0 1	0
	0 0 1 0	0
	0 0 0 1	0
2	0 0 0 0	0
	0 0 1 0	0
	0 0 1 1	0
	0 0 0 1	0
	0 0 1 0	0
3	1 1 1 1	1
	0 1 1 1	1
	0 1 1 1	1
	0 1 1 0	1
	1 0 0 1	1
4	0 0 0 0	0
	0 0 0 0	0
	0 0 1 0	0
	0 0 1 1	0
	0 0 1 1	0
5	1 1 1 1	1
	0 1 0 1	1
	0 1 0 0	1
	0 1 0 0	1
	0 1 0 1	1

Figure 6.5 *A set of sequences to be learned.*

state that represents the previous activities is very important. We call these sequences state-dependent sequences. We applied the previous algorithm, with penalty to both state transitions and actions, and the system sometimes converged to zero. The parameters were $r_{increment} = 0.05$ and $r_{decrement} = 0.7$. From the experiment, we can conclude that the system will converge (i.e., the error rate drops to zero) probabilistically for state-dependent sequences.* However, there is no analytic result concerning the probability of convergence in any of the foregoing cases.

6.3.2 Applying Other Related Algorithms

To fully appreciate the difficulty of developing a general learning algorithm, we can compare DN/PDN models with some known computational models, namely, hidden Markov Models (HMM), learning automata, team decision theory, probabilistic logic neurons (PLN) (see Aleksander [4]), and so on. We can apply the knowledge and insight we have of other models to DN and PDN, integrating the useful aspects of each related field and taking advan-

---

* This is similar to the TD method (see Sutton [9]) in a way: taking into account past activities in making current predictions, by utilizing the state variables that summarize all past activities in this case. Although this method seems to be more powerful, the convergence is only probabilistic and empirical results indicate that the probability of convergence is very low.

Seq#	Input	Output
1	0 0 0 0	0
	0 0 1 1	0
	0 0 0 1	0
	0 0 1 0	0
	0 0 0 1	0
2	0 0 0 0	0
	0 0 1 0	0
	0 0 1 1	0
	0 0 0 1	0
	0 0 1 0	0
3	1 1 1 1	1
	0 1 1 1	1
	0 1 1 1	0
	0 1 1 0	1
	1 0 0 1	1
4	0 0 0 0	0
	0 0 0 0	0
	0 0 1 0	0
	0 0 1 1	0
	0 0 1 1	0
5	1 1 1 1	1
	0 1 0 1	1
	0 1 0 0	1
	0 1 0 0	1
	0 1 0 1	1

Figure 6.6 *A more difficult set of sequences.*

tage of the known algorithms for more restricted domains or subproblems. The integration and combination of these techniques may result in some new algorithms that are both general and efficient.

First, it is not difficult to see that a DN node is an FSA. It is certainly limited in its computational power, compared with Turing machines. But if we hook up a network of them, DN network can simulate a Turing machine. As a matter of fact, a linear array of DN nodes of sufficient lengths can be equivalent to the computational power of Turing machine.

Furthermore, if we extend DN models to PDN models as described earlier, we end up with something resembling an HMM. But HMM does not have inputs, which are extra complications of PDN models.

The PDN can be viewed as a learning automaton (as in Narendra and Thathachar [6]). Thus a PDN network can be viewed as a set of corporating learning automata, trying to optimize their performances according to a reinforcement signal generated by a noisy environment.

In state transition processes, we can analogize the sequential transition with a trajectory through a state-space, each dimension of which is the state of an individual node. An important thing that can be used to determine the optimal trajectory is the energy landscape, which is over the entire state-space. While each node wants to maximize its own gain ("hedonistic neurons"), the overall system has to reach its global maximum. This leads to the team decision theory [14], which studies the ways that the strategy of each node to achieve its own maximum gain can help to guide the whole system into a global maximum. Global cooperation is achieved through the choice of local individual strategies. With an appropriate strategy for each node (usually the same for all), the system will be able to follow an appropriate trajectory to settle into an optimal state in the state-space. The optimal state to settle into is determined by a global energy function, and the state each node should be in depends on the states of other nodes and cannot be decided in isolation. The mapping of terminology is simple: An *energy function* is equivalent to a *payoff matrix* in team theory, a *synaptic input* is the same as an *observation,* an *action* is a *decision variable,* and an *updating rule* is a *strategy.* The only difference is that our model allows communications between nodes, whereas the team theory usually assumes there is no communication of any sort.

By allowing communication through synaptic links, each node knows the previous actions of a subset of nodes that have synapses with it. Then it must try to predict what that subset of nodes will do next and choose an action that maximizes the energy. We call it the *predictive actions* algorithm, which uses ideas of team decision theory.

SCHEME 6.5 [Predictive Actions: Learning Action Functions Only] *Each node should choose an action, at each monent, that achieves*

$$\max \left(\sum_i P_i E_{ij} \right)$$

and modify its C function accordingly (by incrementing the probability of the chosen action by a certain small amount), where P_i is the probability of being in the ith combination of inputs next (which depends on the current input combination), and E_{ij} is the local energy of choosing the jth action in the face of the ith input combination (measuring what the action ought to be, given the actions of other nodes at the same moment), assuming the global energy is a monotone increasing function of local energy values.

We will call the *i*th input combination the global state *i* relative to a particular node. For each node, P_i can be expressed as the probability of being in the global state *i* (i.e., the *i*th combination of input to the node) at time $t + 1$, given that it is in the global state k at t. So

$$P_i(k) = \prod_{\text{for all related nodes } l} P^{(l)}(i/k)$$

Here, $P^{(l)}(i/k)$ is the probability of the *l*th node being in a state that is part of a global state *i* at time $t + 1$, given that it is in a state that is part of the global state k at time t. These statistics are unknown a priori and have to be collected during the run.

And

$$E_{ij} = \text{Expected}-\text{Reinforcement}(i,j)$$

In other words, E_{ij} is equal to the expected reinforcement when the global state is i and the output is j. It can also be easily calculated according to collected statistics. So, overall the algorithm is easy to implement.

By introducing multiple states into the model, we have a more powerful model and thus need more complex learning algorithms. From analogy to HMM, we can extend the well-known Viterbi algorithm to learn an optimal state-space trajectory, optimal in the sense that a state is most likely individually at a particular time for each time period. After calculating this trajectory, we can use the linear reward-penalty algorithm (from learning automata theory) to modify the system. This algorithm performs supervised learning, since a sequence of desired outputs is needed. With this algorithm, oscillation is possible, because of mutual dependency among connected nodes.

SCHEME 6.6 [Viterbi/L-RP: Learning State Transitions Only] *Use the Viterbi algorithm to calculate the most likely state at each moment, given a sequence of desired outputs. Run the network and use the linear reward-penalty algorithm to modify parameters.*

The Viterbi algorithm can be expressed as the calculation of the likelihood of a state at time t for an output sequence:

$$\gamma_t(i) = \frac{\alpha_t(i)\beta_t(i)}{\text{Prob}(Os/Is)}$$

where i is a state, Prob(Os/Is) is the probability of the desired output sequence Os given the input sequence Is, $\alpha_t(i)$ is the probability of the partial output sequence until time t and being in state i at time t, and $\beta_t(i)$ is the probability of the partial output sequence from time $t + 1$ until the end and being in state i at time t. Both can be calculated easily given C and T, inputs at each moment, and the initial state (see Rabiner and Juang [10]). So is Prob(Os/Is). In case there are multiple sequences, as happens most of the time, we calculate different likelihood values for different sequences, and we use one or the other based on which one is closest to the sequence in question. When ambiguity cannot be resolved, we add all potentially possible likelihood values to form the overall likelihood, and use it for the moment. When more and more inputs are given, ambiguities can decrease. We may even resolve ambiguity right at the beginning, if the first input uniquely identifies a sequence.

When we want to learn both state transition and action functions at the same time, we need to consider the interaction between these two functions.

For example, we could be punishing a state transition and an action at the same time: While we are gradually reducing the probability of that particular state transition to near zero, we are messing up the action function in the target state at the same time unnecessarily. With this in mind, we can combine the predictive algorithm for action functions and Viterbi/L-RP algorithm for state transitions together in the following way:

SCHEME 6.7 [Combining Viterbi/L-RP and Predictive Action Algorithm: Learning both Action and State Transition Functions] *We use the Viterbi/ L-RP and predictive action algorithms together for state transition and action functions respectively. When both deserve rewards, give both rewards. When both deserve penalties, penalize only the state transition. When the state transition deserves a reward but the action deserves a penalty, give them correspondingly. When the state transition deserves a penalty but the action deserves a reward, give only the penalty to the state transition.*

More precisely, given desired input/output mappings, we first initialize the T and C to be random distributions (uniform or otherwise); then we run the network and apply the Viterbi algorithm to find the optimal state trajectory while calculating E_{ij} and P_i for action functions. We run the network again, applying L-RP to T based on the otimal state trajectory found, apply the predictive action algorithm to C, and use Scheme 6.7 for maintaining consistency. In the meantime, we will collect and update all the statistics in the Viterbi and predictive action algorithms.

It will also work if we replace the predictive algorithm with straight comparisons of the desired output and the actual output, if available. It is also possible to devise an algorithm without using the viterbi algorithm so that there is only one reinforcement signal for both state transitions and actions, which was presented as Scheme 6.4.

Another interesting algorithm, which is more or less retrospective instead of predictive, is the Baum-Welch algorithm from HMM theory (see Rabiner and Juang [10]).* In this scheme, a set of extended HMM processes, hooked up in a network fashion, work together through mutual communication to achieve cooperative computations. This algorithm is absolutely expedient. It learns both state transition and action probabilities.

This discussion shows that the integration of these above ideas as ingredients can result in some very powerful and useful algorithms. But the ideas in this subsection are not meant to be fully tested, working algorithms. They are informal, in the sense that it is not clear how well they will work under

---

* Here we have to extend the HMM, as defined in [10], so that the state transition and action functions are contingent on the input at the moment and the current state, whereas in [10] they depend only on the current state. Another extension is that we deal with multiple sequences here.

6.4 DISCUSSION

Once we remove the requirement that a unit has to be very simple in terms of computational resources, we get into a brand new type of networks, which has many interesting properties that are not present in conventional PDP models. There is a strong possibility that this type of networks may solve many problems difficult for conventional PDP models. For example, it may be used to solve problems of variable bindings, symbol manipulations, and sequence generation and recognition.

In building symbolic inference systems (see Sun [2,3]), variable bindings are better handled by this model than any other schemes. Learning correct bindings can be done without internal states, only input/output mappings. To fully develop symbolic manipulation capability in connectionist models, we have to take into account past activities and current contexts, which can be represented by state variables.

In sequence generation and recognition, state variables are also important in providing the flexibility of handling multiple sequences, so that one state can represent one particular sequence or one particular segment of a sequence. In recurrent networks, we might do without the state variable, but the learning will be slow, because the global state has to be considered instead of the local state (as represented by the state variable).

This type of network was also used for modeling biological neural networks, especially local neural networks, in which individual firing patterns and connectivities can be clearly identified and expressed in DN models. Sun et al. [1] presents one such application.

As a matter of fact, this type of situation—that is, self-interested units trying to maximize their own rewards work together to maximize the overall "rewards" by devising and following some general rules—is ubiquitous: It exists in economical systems, ecological systems, and in societies in general. The model can be used for simulating individual cognitive agents, a society of such agents, and their interactions with the environment.

We can compare our algorithms (Scheme 6.4 in particular) with Aleksander [4] qualitatively too (we did some quantitative performance comparisons). In Aleksander [4] a learning algorithm is proposed for a neural network model called the probabilistic logic node (PLN), which is similar to PDN, except that it has no internal states. Basically, it presets all entries of output functions to be random (uniformly distributed) and runs the network until the output matches the desired output; then sets all current mappings (of all entries involved) to be deterministic. If in later trials, one entry is ever found consistently mismatching the desired output, reset the entry to be random.

Compared with the PLN learning algorithm, the algorithms proposed here (especially Scheme 6.4) have the following advantages:

- They allow contradictory information and noises, and thus work in ill-structured domains, because they are based on statistical sampling.
- During learning, new information does not wipe out old ones. Rather, the change is gradual and cumulative.
- After learning, they still have versatility due to their probabilistic nature and variability of behavior due to their capabilities for further learning.

6.5. CONCLUDING REMARKS

In this chapter a connectionist network formalism and its learning issues were discussed. These models are generalizations of conventional PDP models but much more powerful, exhibiting many interesting properties. In the models each neuron is purposive self-interested units, with a set of input/output mappings and internal states. This leads to some interesting similarities to HMM, learning automata, PLN models, reinforcement learning paradigms, and team decision theory. Some learning algorithms for the PDN network models were devised. The differences and similarities of these learning algorithms, compared with existing algorithms for similar models, were discussed. These algorithms can be used for various purposes in different domains.

REFERENCES

[1] R. Sun, E. Marder, and D. Waltz, Model local neural networks in the lobster stomatogastric ganglion, *Proc. Internat. Joint Conference on Neural Networks*, Washington, D.C., IEEE Press, II-604, 1989.

[2] R. Sun, A discrete neural network models for conceptual representation and reasoning, *Proc. of 11th Cognitive Science Society Conference,* 1989.

[3] R. Sun, Integrating Rules and Connectionism, Proc. of AAAI workshop on Integrating Neural and Symbolic Processes, Boston, 1990.

[4] I. Aleksander, "The Logic of Connectionist System," in *Neural Computing Architectures,* MIT Press, Cambridge, Mass., 1989.

[5] R. Williams, Reinforcement learning in connectionist networks, TR 8605, ICS, University of California, San Diego, 1986.

[6] K. Narendra and M. Thathachar, Learning automata, *IEEE Trans. Systems, Man and Cybernetics,* 13 323–334 (1974).

[7] A. Barto, From Chemotaxis to Cooperativity, COINS TR 88-65, Tech. Rep. University of Massachusetts, Amherst, 1988.
[8] A. Klopf, The Hedonistic Neuron, *Hemisphere,* 1982.
[9] R. Sutton, The temporal difference method, *Machine Learning,* 1989.
[10] L. Rabiner and B. Juang, Introduction to Hidden Markov Model, *IEEE ASSP,* 4–16, Jan. 1986.
[11] J. Anderson and E. Rosenfeld, eds. *Neurocomputing Foundation of Research,* MIT Press, Cambridge, Mass, 1988.
[12] S. Gooseberg, *The Adaptive Brain,* North-Holland, Netherlands, 1987.
[13] J. Feldman and D. Ballard, Connectionist models and their properties. *Cognitive Science,* pp. 205–254, July 1982.
[14] Y. Ho, Decision theory and information structures, *Proc. IEEE,* 68, 644–654 (1980).
[15] R. Sun, The descrete neuronal model and probabilistic discrete neuronal models, *Proc. of INNC-Paris,* 902–907, 1990.

CHAPTER 7

Experimental Analysis of the Temporal Supervised Learning Algorithm with Minkowski-r Power Metrics

RYOTARO KAMIMURA

7.1 INTRODUCTION

This chapter presents a method to accelerate the temporal supervised learning algorithm of Williams and Zipser [1,2]. Our method consists mainly in the replacement of the squared error function with the Minkowski-r power metrics. Using our modified version of the temporal supervised learning algorithm (TSLA), we attempt to show that fully recurrent neural networks can recognize the characteristics of sequences more clearly than the simple recurrent neural network developed by Elman [3] to examine temporal sequences. The simple recurrent neural network is only a modified version of the feedforward network.

Feedforward networks with a back-propagation learning method have been very popular in the community of neural networks. Multiple applications have been reported, showing the efficiency of the back-propagation method. However, the architecture of the feedforward network is certainly not plausible for complex systems like the human brain in that the architecture is too restrictive. The fully recurrent neural network is general enough to take into account the characteristics of the complex structures. Multiple architectures can be obtained by imposing appropriate restrictions on the fully recurrent neural network. The feedforward network is only one of the possible configurations obtained by imposing such restrictions.

Neural and Intelligent Systems Integration, By Branko Souček and the IRIS Group.
ISBN 0-471-53676-8 ©1991 John Wiley & Sons, Inc.

Recently, the back-propagation method has been generalized to the fully recurrent network by Pineda [4]. Then, this recurrent back propagation of Pineda has been further extended to the generation and the recognition of the dynamic behaviors of the systems. For example, see Williams and Zipser [1,2] and Pearlmutter [5,6]. In a method developed by Pearlmutter, the full back-propagation method is incorporated into the learning algorithm, which can reduce the computational requirement. On the other hand, a method developed by Williams and Zipser, which we call *TSLA*, does not use the full back propagation and requires, thus, a great deal of numerical complexity when the network size is relatively large. For further details concerning the comparison between two algorithms for time-dependent neural networks, see Pineda [9,10]. However, TSLA is very simple and general in designing the network. It need not be unfolded in time and it can be used in the real-time mode. The computational requirement of TSLA is, however, on the order of n^4 and the required memory is on the order of n^3, where n is the number of units in the network. To overcome this shortcoming, Zipser [11] has recently proposed a subgrouping method that consists mainly of the reduction of network size by considering an original fully recurrent network as a group of small self-sufficient recurrent networks. By this technique, the number of units in the small networks can be reduced, which may improve the learning time.

This chapter is concerned with a new acceleration method, in which Minkowski-r power metrics are used instead of the ordinary squared error function. The squared error function has been extensively used to train the network. Hanson and Burr [12] introduced Minkowski-r power metrics into the back-propagation learning method of the feedforward network. Minkowski-r power metrics are defined by the equation $(1/r)\Sigma_k|e_k|^r$, where e_k are the errors between the target values and actual outputs. They have applied the network with Minkowski-r power metrics to a problem of noise reduction—for example, the recovery of the shape surface from random noises. They also point out that by decreasing r values the convergence may significantly be improved with little change of the nature of the solution.

Following the suggestion of Hanson and Burr, we employ Minkowski-r power metrics for the temporal supervised learning algorithm of Williams and Zipser [7,8]. In TSLA, Minkowski-r power metrics are defined by the equation $(1/r)\Sigma_k|e_k(t)|^r$, where $e_k(t)$ are the errors between the targets and observed outputs at time t. Changing the value of r, we have confirmed that the network can converge to an acceptable point significantly faster. Particularly when the learning is advanced, and errors become small, the ordinary squared error function is insensitive to errors. On the other hand, the network with an appropriate small Minkowski-r continues to be sensitive to small errors, which enables the network to converge to an acceptable point significantly faster than the network with the squared error function.

In addition to the aforementioned acceleration method for the temporal supervised learning algorithm, we attempt to show to what extent some characteristics of the sequences of natural language can be clarified with the recurrent neural networks. Natural language is one of the most challenging subjects in the fields of artificial intelligence, cognitive science, and neural networks. Several attempts to apply the neural network to the analysis of natural language have been made [3,13]. However, in previous studies, the network architectures used in the experiments were all related to the feedforward neural network. One of the most successful networks is called the *simple recurrent neural network,* because the simple network uses the feedforward network with some recurrent connections. The simple recurrent neural network employs the pattern of activation of a set of the hidden units at time step $t - 1$ with the inputs at time step t to predict the output at time step $t + 1$. Using this simple recurrent neural network, Elman [3] tried to discover the notion of "word" by observing errors between the target outputs and observed outputs. The network shows higher errors when it must estimate the first letter of the "word." However, as can be seen in the architecture of the simple recurrent network, it is expected that the simple network is not good at estimating a sequence with long-distance dependence like natural language. Poor results have already been reported concerning estimation of the sequence with long-distance correlation like embedded sentences. Cleeremans et al. [13] tried to evaluate the performance of the simple recurrent neural network, when it was applied to the processing of the embedded sequences. They reported that the simple recurrent neural network had difficulty in estimating a sequence whose last letter was identical to the first letter. On the other hand, the fully recurrent neural network with TSLA showed good performance in processing embedded sentences, which was reported by Smith and Zipser [14].

Moreover, the architecture of the feedforward neural network is too restrictive to be applied to the analysis of complex systems like natural language in that it is not plausible that the structure of natural language can be the feedforward type. On the other hand, the fully recurrent network has a lot of potential in that the network architecture is as general as possible (in the fully recurrent network, the units are all connected). Thus, it is possible to impose any kind of restriction so as to obtain an appropriate architecture for an unknown structure like natural language. Our experimental results show that the fully recurrent neural network can recognize several syntactic characteristics, for example, the "phrase" and "sentence" in addition to the boundary of the "word." In a previous study by Elman [3], only the category "word" could be recognized by the simple recurrent neural network, which is a modified version of the ordinary feedforward network. Thus, the fully recurrent network is more powerful than the simple recurrent neural network.

In Section 7.2, we present the standard TSLA of Williams and Zipser and a modified version with Minkowski-r power metrics. The teacher-forced learning algorithm, which is a modified version of the TSLA, is also presented. This algorithm has been introduced in order to accelerate the TSLA. Finally, the actual computational methods to update the weight change are briefly presented.

In Section 7.3, we compare the performance of the standard TSLA with that of the TSLA equipped with Minkowski-r power metrics, changing the Minkowski-r value. We show that the network with Minkowski-r power metrics (especially $r = 1.5$) can converge to an acceptable point significantly faster than the network with the ordinary squared error function. In addition, we show that the fully recurrent neural network can converge significantly faster if an appropriate number of hidden units are chosen. Finally, the minimum number of hidden units, necessary for the generation of natural language, is experimentally determined.

In Section 7.4, four experimental results are discussed concerning the generation of the natural language. In all these experiments, we aim to show the good performance of the fully recurrent network with our modified version of TSLA. First, the generation of a periodic sequence with a relatively long period is discussed in order to show that the fully recurrent network can recognize the characteristics of the target sequences clearly. Second, we discuss the experimental results regarding the generation of natural language with letters. We try to show that the fully recurrent neural network can recognize the hierarchical structure found in the sequence of natural language such as "phrase," and "sentence," and so on. Third, we discuss the generation of natural language with grammatical categories (parts of speech). It is found that the network can recognize the "sentence," though the sentence boundaries are not explicitly given. Fourth, the generation of the whole text with words is discussed. Finally, in Section 7.5, we summarize and conclude the chapter.

7.2 THEORY AND COMPUTATIONAL METHODS

7.2.1 Temporal Supervised Learning Algorithm

Standard Temporal Supervised Learning Algorithm In this section, we briefly present the temporal supervised learning algorithm formulated by Williams and Zipser [1,2]. Let $x_i(t)$ denote an activity of the ith unit at time t, and let w_{ij} represent connection strength from the jth unit to the ith unit. Then an activity of the ith unit at time $t + 1$ is given by

$$x_i(t + 1) = f(s_i(t)) \tag{7.1}$$

where f is a logistic function:

$$f(s_i(t)) = 1/(1 + e^{-s_i(t)}) \tag{7.2}$$

and

$$s_i(t) = \sum_j w_{ij} x_j(t) \tag{7.3}$$

To simplify the following formulation, we use a set function introduced by Pineda [4]. The function is defined by

$$\Theta_{i,\Phi} = \begin{cases} 1 & \text{if } i \in \Phi \\ 0 & \text{otherwise} \end{cases} \qquad (7.4)$$

Some of the units can be taken as a set of visible units V, which should have their own target values. The network is trained to learn the target values by minimizing the difference between target and actual values.

Consider the difference at time t,

$$e_k(t) = [d_k(t) - x_k(t)]\Theta_{k,V} \qquad (7.5)$$

where $d_k(t)$ show target values, assigned to visible units.

The overall network error at time t is defined as

$$J(t) = \frac{1}{2} \sum_k [e_k(t)]^2 \qquad (7.6)$$

The weight change for any particular weight at time t is defined by

$$\Delta w_{ij}(t) = -\eta \frac{\partial J(t)}{\partial w_{ij}} = \eta \sum_k e_k(t) \frac{\partial x_k(t)}{\partial w_{ij}} \qquad (7.7)$$

where η is a learning rate. The derivative $\partial x_k(t)/\partial w_{ij}$ can be computed from the equation

$$\frac{\partial x_k(t+1)}{\partial w_{ij}} = \varepsilon f'(s_k(t)) \left[\sum_l w_{kl} \frac{\partial x_l(t)}{\partial w_{ij}} + \delta_{ik} x_j(t) \right] \qquad (7.8)$$

where ε is a constant necessary for the convergence, and with initial conditions

$$\frac{\partial x_k(t_0)}{\partial w_{ij}} = 0 \qquad (7.9)$$

Finally, the overall weight change is simply represented by

$$\Delta w_{ij} = \sum_t \Delta w_{ij}(t) \qquad (7.10)$$

Formulation using Minkowski-r Power Metrics Minkowski-r power metrics have been introduced by Hanson and Burr [12] for the back-propagation learning method of the feedforward networks. In this chapter, we employ Minkowski-r power metrics for the TSLA of Williams and Zipser to accelerate temporal learning.

In the formulation using Minkowski-r power metrics, instead of the squared error function defined by Eq (7.6), we use a generalized version defined by the equation

$$J(t) = \frac{1}{r} \sum_k |e_k(t)|^r \qquad (7.11)$$

Differentiating this equation with respect to w_{ij}, we have the weight change

$$\Delta w_{ij}(t) = -\eta \frac{\partial J(t)}{\partial w_{ij}}$$

$$= \eta \sum_k |e_k(t)|^{r-1} \frac{\partial x_k(t)}{\partial w_{ij}} \operatorname{sgn}(e_k(t)) \qquad (7.12)$$

Thus, the ordinary squared error function is a special case of the functions with Minkowski-r power metrics.

Teacher-Forced Learning Algorithm The teacher-forced learning algorithm, a version of the original algorithm has been very powerful for many problems, especially those such as the stable oscillation. The basic idea consists of replacing actual outputs $x_k(t)$ by teacher signals $d_k(t)$. To simplify the explanation, we introduce new variables $z_k(t)$ that are defined by

$$z_k(t) = d_k(t)\Theta_{k,V} + x_k(t)\Theta_{k,H} \qquad (7.13)$$

where H denotes a set of hidden units. The network dynamics is described by Eq. (7.1) using new variables. Differentiating both sides of the equation, we have

$$\frac{\partial z_k(t+1)}{\partial w_{ij}} = \Theta_{k,H} f'(s_k(t)) \left[\sum_{l \in H} w_{kl} \frac{\partial x_l(t)}{\partial w_{ij}} + \delta_{ik} z_j(t) \right] \qquad (7.14)$$

We have observed that without this teacher-forced learning algorithm, almost all complex tasks cannot be accomplished in a reasonably short time. The same method was used with good results to train the recurrent neural network for learning in autoassociative memory. For further details, see Pineda [4].

Computational Methods In this section, we briefly present two computational methods we used in the experiments. A detailed explanation regarding other minor methods necessary for rapid convergence is found in Fahlman [15]. The first method concerns weight change. The weight update is performed by the equation

$$\Delta w_{ij}^{(n)} = -\eta \left[\alpha \sum_{t} \left[\frac{\partial J(t)}{\partial w_{ij}} \right]^{(n)} - (1 - \alpha) \Delta w_{ij}^{(n-1)} \right] \quad (7.15)$$

where n is the number of iterations in the learning process and $\Delta w_{ij}^{(n-1)}$ is a weight change without modification at the $(n - 1)$st iteration. The parameter α is set to 0.1 through all our experiments discussed in the following sections.

Second, we change η according to the increase of the mean absolute value of error. The method can be summarized as follows. Once the mean absolute error increases, the value of η is reduced by a factor of 2. When the mean absolute error ceases to increase, the value of η is reset to a previous value or 0.5 if the previous value is below 0.5. This method has been introduced to stabilize the learning process.

7.3 ACCELERATION OF THE TEMPORAL SUPERVISED LEARNING

7.3.1 Overview of the Experimental Methods

In this section, we present the experimental results concerning the performance of a fully recurrent neural network with variable Minkowski-r. In the experiments, we used a sequence of natural language, which will appear in Section 7.4.3. The data were encoded in the binary numbering system with 5 bits. Table 7.1 shows an example of data represented in the binary numbering system. The network was composed of five visible units and several hidden units. The learning was considered to be finished when the Hamming distance between the target sequence and the observed sequence was zero. All the computations were performed on SX-1. As discussed in the introduction, we used a fully recurrent neural network for all the experiments to be presented. In this section, we explain a fully recurrent network and how to train the network.

For simplicity, suppose that a fully recurrent neural network is composed of four units, in which three units are employed as hidden units and the other is used as visible unit. Figure 7.1 shows a fully recurrent neural network with four units all interconnected. The visible unit must have the target, but the hidden units do not need targets. The network is trained to minimize errors between the target and the observed output at the visible unit. The fully recurrent neural network is very general compared with the feedforward network, which is only one possible configuration of the recurrent neural network. The recurrent network is, however, expensive with respect to the computational requirement. The generality of the recurrent network is necessary in that the knowledge of the target object to be analyzed by the network can be incorporated into the network by means of appropriate restrictions on the network architecture.

TABLE 7.1 An Example of Data, Composed of Letters and Represented in the Binary Numbering System

M	I	S	S		M	A	R	T	H	A
0	0	1	1	1	0	0	1	1	0	0
1	1	0	0	1	1	0	0	1	1	0
1	0	0	0	0	1	0	0	1	0	0
0	0	1	1	1	0	0	1	0	0	0
1	1	1	1	1	1	1	0	0	0	1

Let us explain the learning process of a periodic sequence

010101010101010101010101

At an initial state of the learning process, activity values are all set to equal values, and weights are set to some random values. At time step 1, the visible unit is trained to produce 0 and at time step 2 the network must generate 1, and so on. See Figure 7.2. The network is trained until it can generate an exact periodic sequence above a training interval. As can be seen in the figures, the recurrent neural networks used in our experiments have no input units. Thus, our model of the recurrent neural network can be used for simulating spontaneous behaviors that are frequently observed in humans.

7.3.2 Minkowski-r Power Metrics

In this section, we evaluate the performance of the network with Minkowski-r power metrics with different values of r. Table 7.2 shows the

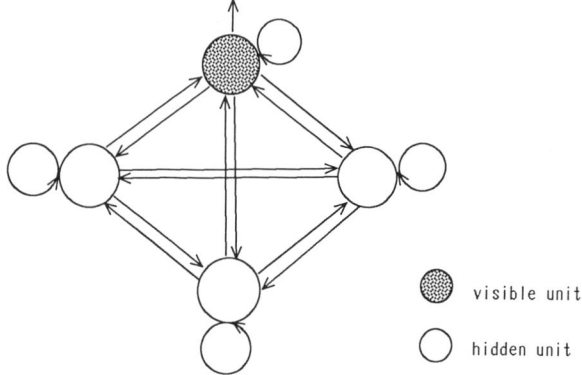

Figure 7.1 *An example of a fully recurrent neural network.*

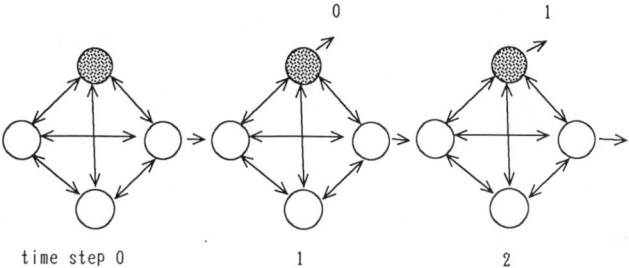

Figure 7.2 The learning process of a fully recurrent network.

number of learning cycles necessary for a target sequence of natural language, when the values of r range from 1.1 to 2.0. Hanson and Burr [12] have suggested that a Minkowski-r network with $r = 1.5$ gives good performance. Thus, in Table 7.2, we present the ratios of the number of learning cycles needed to converge for the network with Minkowski-r power metrics for $r \neq 1.5$, to the number of learning cycles necessary for a network with $r = 1.5$. Figure 7.3 depicts the aforementioned ratios as a function of Minkowski-r. As can be seen in the table and the figure, in almost all cases the networks show the best performance when r is 1.5. This tendency seems to be salient, as the length of the sequence is longer. Note that the learning

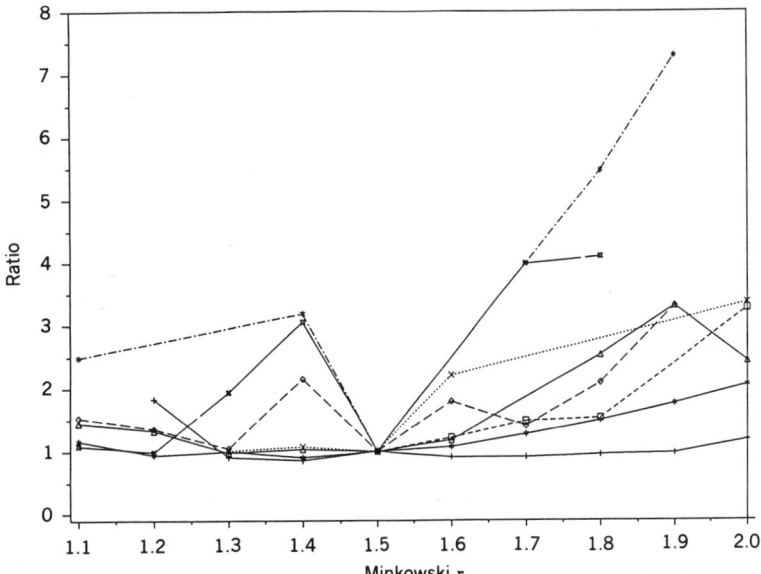

Figure 7.3 The ratio of the number of the learning cycles necessary for the network with Minkowski-r power metrics with $r \neq 1.5$ to that of the learning cycles for the one with $r = 1.5$.

TABLE 7.2 The Number of Learning Cycles for Variable Minkowski-r Values and the Ratios of the Number of Cycles Necessary for Minkowski-r Power Metrics with $r \neq 1.5$ to the Number of Cycles for Minkowski-r with $r = 1.5$

Length			10	20	30	40	50	60	70	80	90
ε			1.0	0.5	0.5	0.5	0.5	0.5	0.3	0.3	0.3
Hidden Units			1	4	6	7	10	13	14	14	14
	1.1	Cycle	117	*	*	5324	*	*	3840	5688	4557
		Ratio	1.17	*	*	2.49	*	*	1.53	1.45	1.09
	1.2	Cycle	94	8056	*	*	*	*	3418	5220	4154
		Ratio	0.94	1.83	*	*	*	*	1.36	1.33	0.99
	1.3	Cycle	100	4009	3190	*	*	*	2636	3842	8124
		Ratio	1.00	0.91	1.01	*	*	*	1.05	0.98	1.94
	1.4	Cycle	91	3759	3404	6815	1888	*	5408	4069	12825
		Ratio	0.91	0.86	1.08	3.19	1.11	*	2.15	1.04	3.06
	1.5	Cycle	100	4394	3152	2137	1699	1297	2513	3917	4188
	1.6	Cycle	107	3988	6967	*	1636	1586	4487	4639	*
		Ratio	1.07	0.91	2.21	*	0.96	1.22	1.79	1.18	*
	1.7	Cycle	127	3979	*	*	3947	1923	3508	*	16660
		Ratio	1.27	0.91	*	*	2.32	1.48	1.40	*	3.98
	1.8	Cycle	149	4181	*	11660	*	1985	5249	9905	17124
		Ratio	1.49	0.95	*	5.46	*	1.53	2.09	2.53	4.09
	1.9	Cycle	176	4278	*	15587	5024	*	8298	12918	*
		Ratio	1.76	0.97	*	7.29	2.96	*	3.30	3.30	*
	2.0	Cycle	205	5175	10566	*	10026	4231	*	9519	*
		Ratio	2.05	1.18	3.35	*	5.90	3.26	*	2.43	*

Asterisks denote that the learning failed. Note that several parameters and the learning rates were adjusted, based mainly on the case when $r = 1.5$.

cycles in the table were computed, mainly based on the Minkowski-r network with $r = 1.5$. Thus, if the learning rate is smaller and the number of hidden units is larger, learning may succeed even in cases where it had failed.

Let us examine why the network with a small Minkowski-r (especially 1.5) can converge to an acceptable point significantly faster. Figure 7.4 shows mean absolute errors (MAE) as a function of the number of iterations (learning cycles). Sequence length (English sentences) was 50 letters. The

7.3 ACCELERATION OF THE TEMPORAL SUPERVISED LEARNING

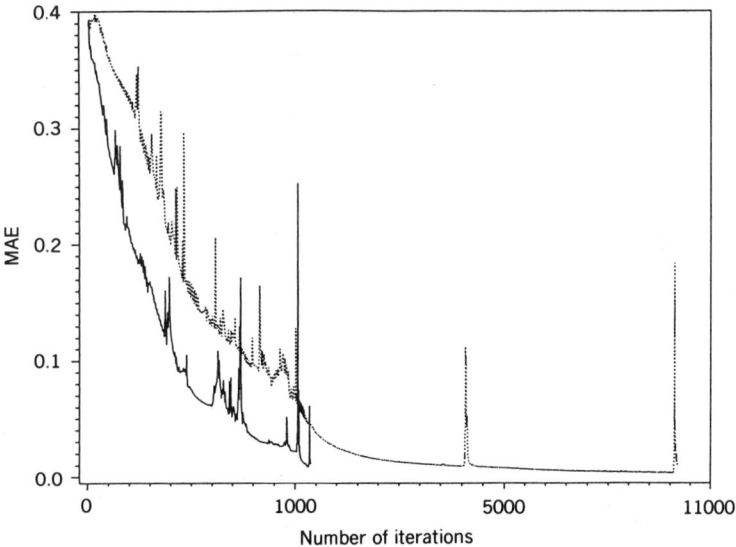

Figure 7.4 *Mean absolute errors as a function of learning cycles (iterations). A solid line shows a case where r is 1.5. A dotted line is for r = 2.0, which means the ordinary squared error function. The length of sequence (English sentences) is 50 letters.*

network was composed of 5 visible units and 10 hidden units. Although it took 1699 iterations for the network with $r = 1.5$ to reach a final point (zero Hamming rate), it took 10,026 iterations for the network with $r = 2.0$ to produce an original sequence. Thus, the network with Minkowski-r power metric ($r = 1.5$) can converge six times faster than the one with $r = 2.0$, which is a network with the ordinary squared error function. As the figure clearly shows, the network with the squared error function becomes insensitive to small errors as learning is advanced.

Figure 7.5 also shows the mean absolute errors as a function of learning cycles. The length of the sequence was 60 letters, and 13 hidden units were employed. It took 4231 iterations for the network with the square error function ($r = 2$) to finish the learning, but only 1297 learning cycles for the network with $r = 1.5$ to converge to an acceptable point. When the learning is advanced, the squared error function is not so sensitive to small errors, as can be seen in the figure. On the other hand, the network with Minkowski-r power metrics can be sensitive to small errors up to a final point.

The reason why the network with Minkowski-r power metric ($r = 1.5$) is so rapid in convergence is clear. As can be seen in the figures, the mean absolute errors for the network with the ordinary squared error function decrease extremely slowly as errors become small. The MAE cannot easily reach a final point because of this slow decrease of the errors. The reason is that the ordinary error function is not so sensitive to small errors. Minkowski-r power metrics with small r can be significantly sensitive to

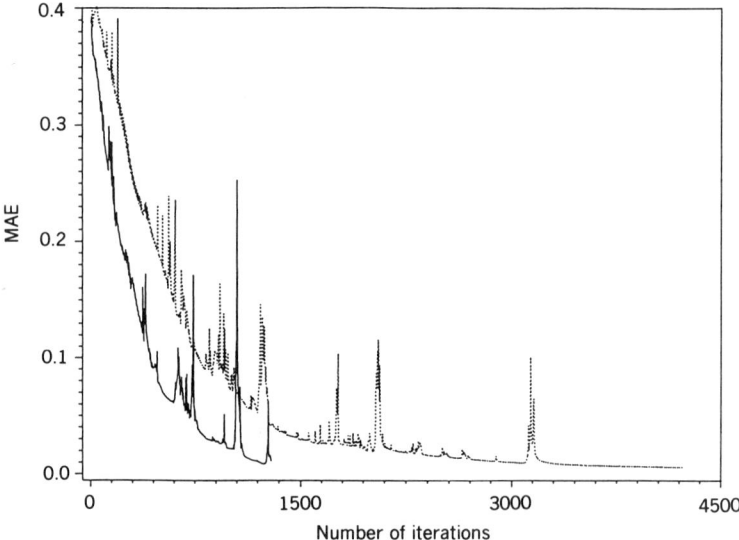

Figure 7.5 *Mean absolute errors as a function of learning cycles (iterations). A solid line shows a case where* r *is 1.5. A dotted line is for* r = 2.0, *which means the ordinary squared error function. The length of sequence (English sentences) is 60 letters.*

small errors in that small errors are overestimated and large errors are underestimated.

Finally, we have observed that the network with the ordinary squared error function cannot sometimes finish the learning, especially if the sequence is long. In this case, the MAE calculated with the teacher-forced learning algorithm is close to zero. Since the check for the convergence is performed with a free-running version of the TSLA (original version), we think that a solution for the teacher-forced version is different from that for the free-running version. A network with ordinary error function definitely has difficulty coping with long complex sequences.

7.3.3 Influence of the Number of Hidden Units

In this section, we consider the influence of the number of hidden units on learning speed. Our results suggest increasing the number of hidden units significantly reduces the number of iterations, and CPU time is reduced if an appropriate number of hidden units are chosen. We discuss two typical cases where the lengths of sequences are 30 and 50 letters, respectively. In all of the experiments to be discussed, we use networks with Minkowski-r power metrics with $r = 1.5$.

Figure 7.6 presents CPU time as a function of the number of hidden units when the length of a sequence is 30 letters. The minimum number of hidden units necessary for a given sequence was six. The figure shows that the best CPU time (1.22 min) is obtained when the number of hidden units is seven. It

7.3 ACCELERATION OF THE TEMPORAL SUPERVISED LEARNING

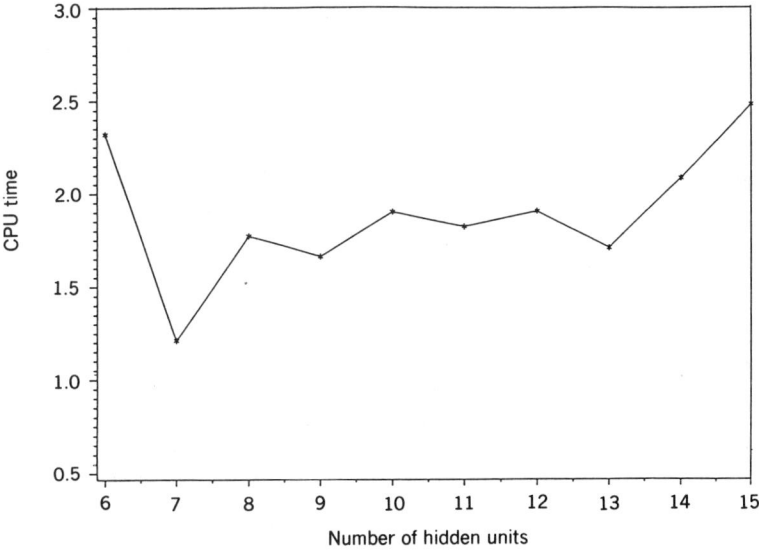

Figure 7.6 *CPU time as a function of the number of hidden units. The length of sequence is 30 letters. CPU time is measured in minutes.*

has been confirmed that this tendency is strictly observed especially when the length of sequence is relatively small.

Figure 7.7 shows CPU time as a function of the number of hidden units when sequence length is 50. The minimum CPU time is obtained when the number of hidden units is 15. We have observed that the appropriate in-

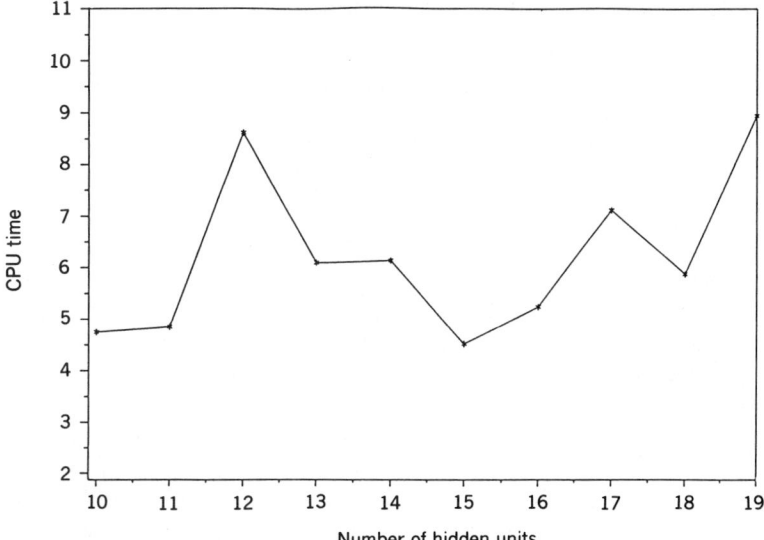

Figure 7.7 *CPU time as a function of the number of hidden units. The length of sequence is 50 letters. CPU time is measured in minutes.*

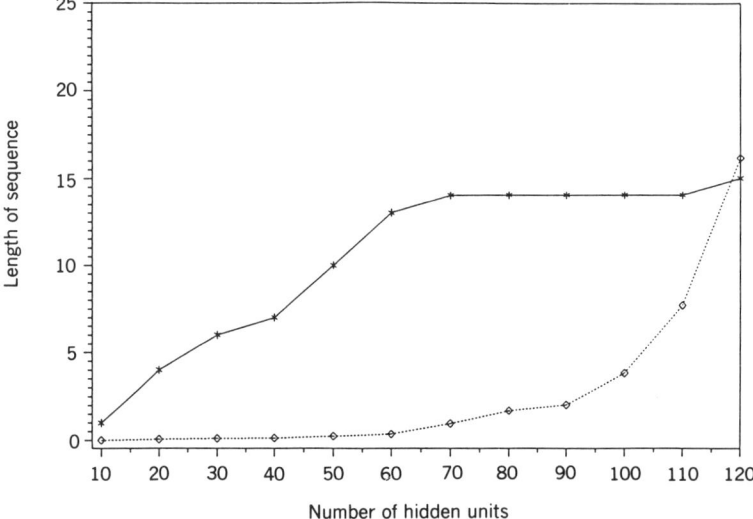

Figure 7.8 *The minimum number of hidden units and CPU time as a function of the length of training sequences. CPU time is measured in minutes and multiplied by 0.05. The minimum number of hidden units is represented by a solid line, CPU time by a dotted line.*

crease of hidden units can greatly improve the learning time when the sequence length is relatively small. When the length is large, the tendency has not been strictly observed.

7.3.4 The Minimum Number of Hidden Units

In this section, we determine the minimum number of hidden units for given sequences. As discussed in the introduction, the computational requirement of the TSLA increases on the order of n^4, where n is the number of units in the network. Thus, it is necessary to confirm that only a relatively small number of hidden units is necessary for recognition of temporal patterns. This procedure is also related to the problem of compressing long sequential data into the relatively small number of hidden units. Note that networks with Minkowski-r power metrics ($r = 1.5$) are used.

Table 7.3 and Figure 7.8 show the minimum number of hidden units for a given sequence as a function of sequence length. Several trials were performed for sequences with different lengths. We could obtain approximately the same results for any trial. As the table and the figure show, the number of hidden units does not necessarily increase in proportion to the sequence length. Until the sequence length is 70 letters, the number of hidden units increases gradually in proportion to the sequence length. Since the sequence length becomes 70, the number of hidden units ceases to increase and the number of hidden units remains 14 units until the length is 110. When the

TABLE 7.3 The Minimum Number of Hidden Units as a Function of the Sequence Length

Length	10	20	30	40
Hidden units	1	4	6	7
CPU time	0.05	1.32	2.32	2.62
Cycles	100	4394	3152	2137
ε	1.0	0.5	0.5	0.5

Length	50	60	70	80
Hidden units	10	13	14	14
CPU time	4.75	7.18	18.97	33.75
Cycles	1699	1297	2513	3917
ε	0.5	0.5	0.3	0.3

Length	90	100	110	120
Hidden units	14	14	14	15
CPU time	40.57	76.38	153.88	323.03
Cycles	4188	7098	13004	21715
ε	0.3	0.3	0.2	0.2

The length and CPU time are measured in letters and minutes, respectively.

length is 120 letters, the number of hidden units increases slightly to 15. These results show that the number of hidden units necessary for a given sequence does not necessarily increase in proportion to the length of the sequences, but remains small compared with sequence length.

7.4 GENERATION OF NATURAL LANGUAGES

7.4.1 Overview of the Results

Several attempts to apply the recurrent neural network to estimating the sequence of natural language have been already made (Cleeremans et al. [13] and Elman [3]). However, the simple recurrent neural network was used in these studies. Actually, this simple recurrent network is the same as the ordinary feedforward network. Experimental results have already been reported that the fully recurrent neural network is superior to the simple recurrent neural network in complex problems (Smith and Zipser [14]).

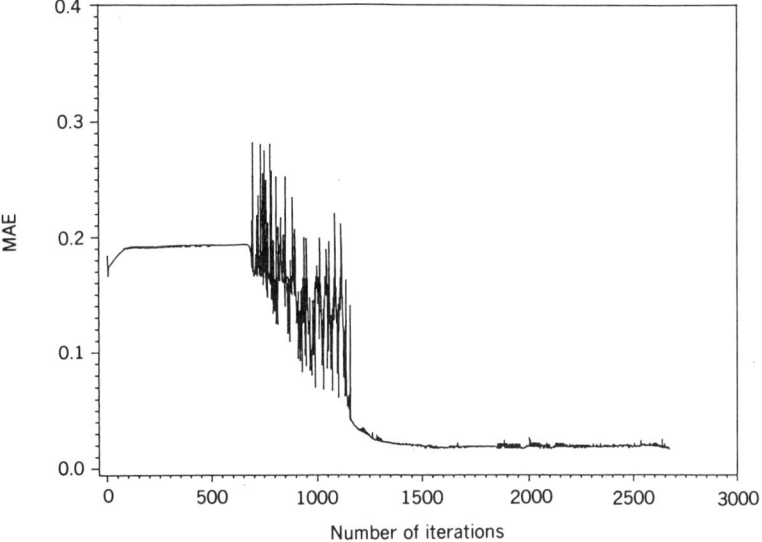

Figure 7.9 *Mean absolute errors as a function of learning cycles (iterations) for a periodic sequence.*

In this section, we present four experimental results. First, the generation of a periodic sequence is presented to show that the recurrent network can clearly recognize the characteristics of the sequence. Second, we attempt to generate the sequence of natural language whose fundamental unit is the letter. Third, the generation of natural language with grammatical categories is discussed. Finally, we attempt to restore the whole text with words.

7.4.2 Generation of a Periodic Sequence

In this section, we attempt to generate a periodic sequence to evaluate the performance of the recurrent network. The periodic sequence we used is

$$100000100000100000100000100000$$

The network architecture we used in the experiments was composed of three hidden units (minimum number) and one visible unit.

Figure 7.9 shows the variation of the MAEs as a function of the learning cycles. The MAEs continue to be stable up to the 700th learning cycle. The MAEs then, oscillate greatly. Finally, they recover stability and reach the final point (zero Hamming distance). A weight matrix obtained after finishing the learning did not show clear characteristics. However, the error between target and observed output at the visible unit showed clear periodic characteristics. This error analysis was developed by Elman [3] to understand the characteristics of target sequences. Figure 7.10 shows the variation of er-

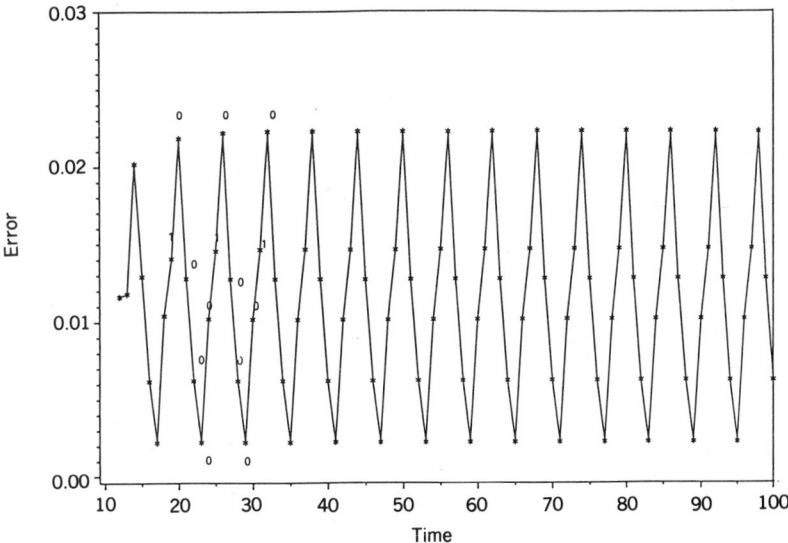

Figure 7.10 *The sum of absolute errors between the target periodic sequence and the observed outputs at the visible unit.*

rors. As can be seen clearly, the error between observed output and target output is high when the network must estimate the symbol following 1. It has been confirmed that the fully recurrent network can recognize and generate the symbol sequences with longer periods if the appropriate number of hidden units is prepared.

7.4.3 Generation of Natural Language with Letters

Overview of the Experiments The network used in the experiments was fully connected and composed of 5 visible units and 15 hidden units. The length of a sequence was 120 letters. This sequence was encoded in the binary numbering system with 5 bits. The learning was considered to be finished when the Hamming distance between target and observed sequences was zero. The learning method used the variable learning rate and a Minkowski-r power metric with $r = 1.5$, which was discussed in Section 7.2.

First, the learning process of the network is discussed to show how the network can learn a target sequence. Then we attempt to generate a sequence of natural language above a training sequence. Finally, error analysis is applied to see to what extent the network can recognize the structures in the sequence.

Learning Process In this section, we present three sequences generated by the network at three stages of the learning process, to show how the network learns a target sequence. Figure 7.11 shows the variation of the

196 EXPERIMENTAL ANALYSIS OF THE TEMPORAL SUPERVISED LEARNING ALGORITHM

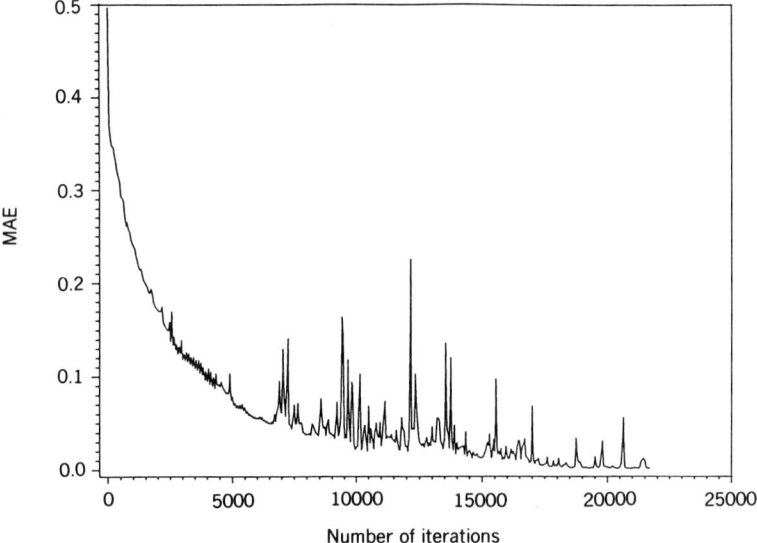

Figure 7.11 *Mean absolute errors as a function of the number of learning cycles (iterations). The sequence length is 120 letters.*

MAEs as a function of the number of learning cycles (iterations). It took 21,715 iterations (about 5 h) to reach this final point. As can be seen in the figure, the variation of errors remains large up to the final point. This shows that Minkowski-r power metrics are effective up to the final point.

After 1000 learning cycles, we had the sequence

```
MISWLERGI    GLQVDIS    MYVLISW    HASLISW-
LERGI WLEREY GLQVL ISVMIWLISWLAREI WLEREI
GL QVEISGM WLISWHASLMQSMISWL EREI WLAVEI
GLQVEISW.
```

Obviously, there are no meaningful words in the sequence. After 17,500 learning cycles (iterations), we had the sequence

```
MISS     MARTHA    MEACHAM    KEPT    TMDS
*ET EHS EHCG*HT*DMYS FMXU* E EKC JERTXAETJTYE
JEXIOWZE*S EKTXOMSSUHCWTI RSUHS UPKTHE* MAR-
TIE STET UHE DAUXEMPQTHT EPQ.
```

At this stage, several English words appear. The network generated the following original sequence perfectly after 21,715 learning cycles:

```
MISS MARTHA MEACHAM KEPT THE LITTLE BAKERY
ON THE CORNER THE ONE WHERE YOU GO UP THREE
STEPS AND THE BELL SOUNDS WHEN YO
```

Automatic Generation We produce a regenerated sequence above a training interval. Our objective is to confirm whether the network really acquires some kind of structure in the sequence. We reproduce three sequences with 30, 50, and 120 letters in the training sequences. The regenerated sequence above a training section is emphasized.

When the length of the training sequence was 30 letters, we obtained the sequence

MISS MARTHA MEACHAM KEPT THE L*EPT THE LEPT THQ KGSLARDHAESHAHEIMSS MARTHA* KEQ MAR-TISSALAIUIG*DH*DSX EIEMISS MEPTYWLARTHA IEQ EAEI*.

In this case, the network was composed of five visible units and six hidden units. It took 3152 iterations to reach a final state. The network could not reproduce grammatical sentences above a training interval. The only English words are THE and MARTHA.

When the length of the training sequence increased to 50 letters, a typical regenerated sequence was

MISS MARTHA MEACHAM KEPT THE LITTLE BAKERY ON THE L*ADPLMNIT*D V DHE LIURMAK KSXKOVYUK-ERT VNER*YELPTYT LESZMMVYQKERP NNDPLISNEPQ KORY ON THE LADPLMNIT*T D*.

The number of hidden units of the network increased to 10, and it took 1699 iterations to converge. In this case, we cannot see any grammatical sentences. The preposition ON and the article THE were generated by the network.

Finally, when sequence length increased to 120 letters, the generated sequence was

MISS MARTHA MEACHAM KEPT THE LITTLE BAKERY ON THE CORNER THE ONE WHERE YOU GO UP THREE STEPS AND THE BELL SOUNDS WHEN Y*OU GO UP THREE STEPS AND THE BELL SOUNDS WHEN YOU GO UP THREE STEPS AND THE BELL SOUNDS WHEN YOU GO UP THREE STEPS AND THE BELL SOUNDS*.

At this stage, the network could reproduce infinitely the grammatical sequence WHEN YOU GO UP THREE STEPS AND THE BELL SOUNDS. The network estimated the sequence following WHEN YO to be a sequence U GO UP THREE STEPS. The correlation among words can span about 10 words.

198 EXPERIMENTAL ANALYSIS OF THE TEMPORAL SUPERVISED LEARNING ALGORITHM

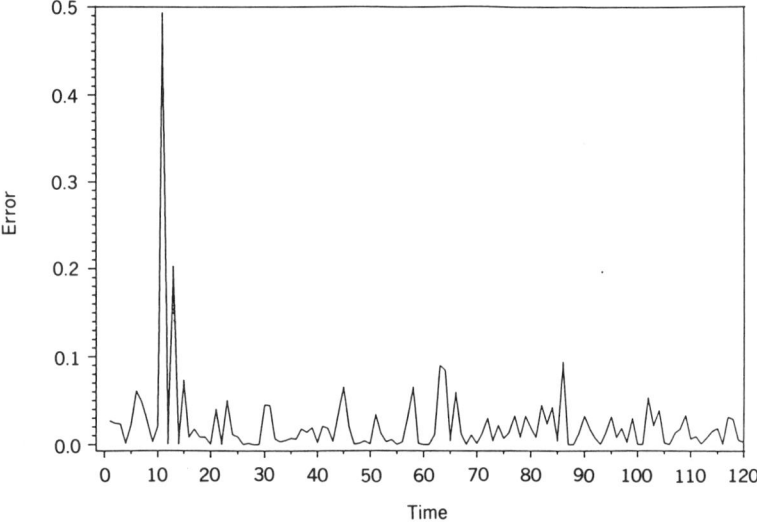

Figure 7.12 *The sum of the absolute errors between the target letters and outputs of five visible units as a function of time step (the order of occurrence of letters). The length of the sequence is 120 letters.*

Error Analysis In this section, we attempt to show where the network has difficulty inferring the next letter. Our main results are simply summarized by three points. First, the network produced the high error values when it attempted to estimate new words. Second, the word with high frequency could be easily estimated. Third, the network demonstrated difficulty when it tried to infer the first part of a new sentence or a new phrase.

Figure 7.12 shows the sum of the absolute errors over all outputs of the visible units. The interval (time interval) ranged from the first letter M to the 120th letter O. The original data are given in the Section 7.4.3. As can be seen clearly, the first part of the sequence produces extremely high error values because the network must estimate a sequence of a proper noun, that is, a name of a woman, which should not be used in ordinary English sentences.

Figure 7.13 also depicts the sum of the absolute errors between the target letters and the outputs of visible units. The time interval is restricted to that ranging between the 26th and 61st letters to show clearly the behaviors of the network. Four salient peaks in the figure are observed at time steps 30, 45, 51, and 58. Time steps 30 and 51 are assigned to the first letters as the words LITTLE and CORNER. We can also see a high peak at the beginning of the word BAKERY at time step 37. These results shows that the network has difficulty estimating the first letter of a new word. In other words, the information contained in the first letter is higher than any other letter in a word. This is completely reasonable from the information-theoretical point of

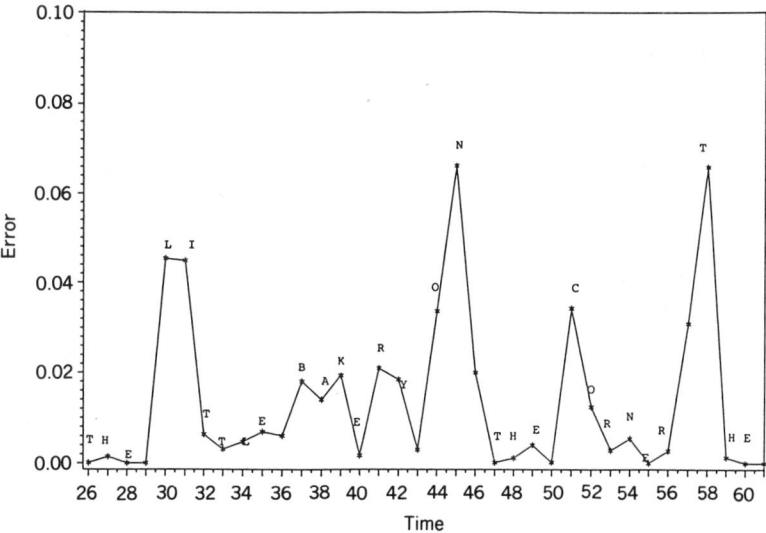

Figure 7.13 *The sum of the absolute errors between target letters and outputs of five visible units as a function of time step. The time interval ranges from the 26th to 61st letters, that is, THE LITTLE BAKERY ON THE CORNER THE.*

view. For experimental results regarding the information contained in words from the information-theoretical point of view, see Petrova et al. [13]. We think that the high peaks at the 44th and 45th positions of the sequence show the beginning of the prepositional phrase ON THE CORNER.

At time step 58, we also see an error peak for the first letter of THE. This peak occurs because a new construction begins at this time step. On the other hand, the word THE at the beginning of the graph, that is, from time step 26 to 28, does not produce a large error, because THE is a word with the highest frequency in the sequence and the network can easily estimate words with a high frequency.

7.4.4 Generation of Natural Language with Grammatical Categories

Learning Process In this section, we present the experimental results concerning the generation of natural language with grammatical categories (parts of speech). The data are in Section 7.4.5. To simplify the experiments and to show the characteristics of the sequences easily, we use only eight categories: noun, pronoun, verb, determiner, adjective, adverb, preposition and conjunction. We used eight input units, corresponding to the categories. In addition to 8 input units, 15 hidden units were employed. The local representation was used to obtain a clear understanding of the hidden units. For

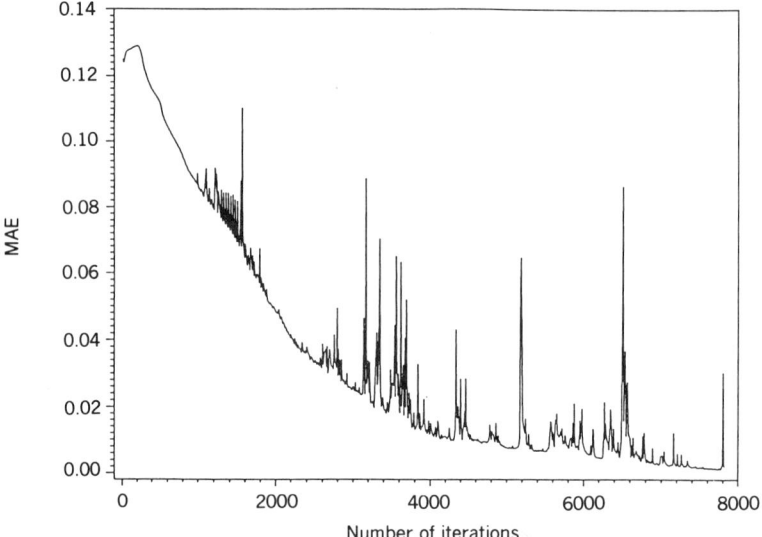

Figure 7.14 *Mean absolute errors as a function of the learning cycles for the network with grammatical categories.*

example, the sentence

MR WHITE FOUND WORK IN A SMALL TOWN

was encoded as

N N V N Prep Det Adj N.

The purpose of the experiments is to confirm that the network can recognize a hierarchical structure, like phrase and sentence, in the sequence.

Figure 7.14 depicts the variation of the mean squared errors as a function of the learning cycles. The values of the mean absolute errors are relatively small at the first part of the cycles, because local representation is employed in encoding the data. On the other hand, the binary numbering system is employed for other experiments.

Error Analysis Figure 7.15 shows the sum of the absolute errors between the target grammatical categories and actual outputs at the visible units. The network clearly shows higher errors when it must estimate a conjunction like AND. For example, at time step 19, the high errors are observed in the figure. In addition, the network has difficulty when it must estimate the first word of the sentence. For example, at time steps 26 and 41 higher errors are observed clearly. In these time steps, the topics of the sentence change.

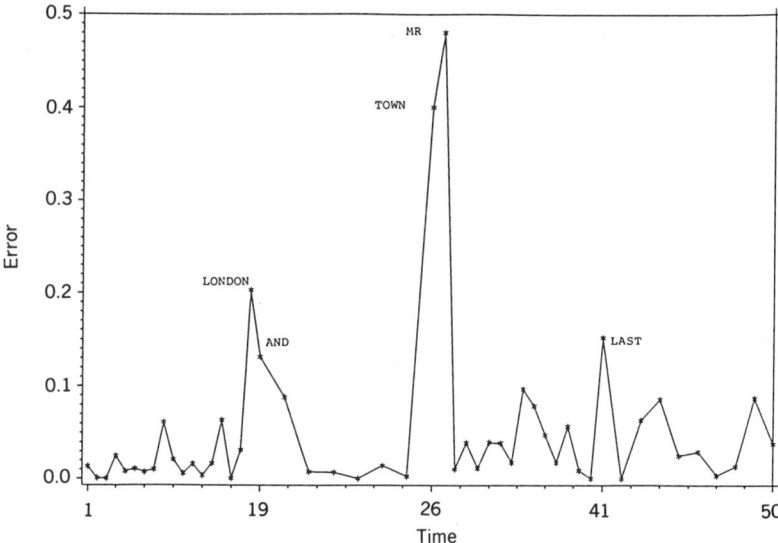

Figure 7.15 *The sum of the absolute errors between the target letters and the outputs of eight visible units as a function of time.*

Thus, the network can recognize the change of topics in the text. These results suggest that the network tries to recognize the so-called hierarchical structure such as "sentence" and the change of topics in a text.

7.4.5 Generation of Natural Language with Words

In this section, we reproduce the regenerated sequences of natural language whose fundamental unit is the word. The network architecture used in the experiment was composed of 25 hidden units and 7 visible (output) units. There were no input units in the network. The length of the text used in the experiments was 129 words. In the text, 75 distinct words were observed. These words were encoded in the binary numbering system with 7 bits. See Table 7.4. The learning was considered to be finished when the Hamming distance between target and observed sequences was zero.

After 3009 learning cycles, we had the sequence

```
MR WHITE OF WHITE HIS AND HE AND HOUSE AND A
WITH YOU NEW HURTING HOUSE AND OF TOWN MR
NEW MY IN WIFE WIFE DO WITH WIFE OR NEW HIM MR
OF LONDON IN.
```

At this stage, we cannot see any meaningful sequence of words except MR WHITE. Thus, only 30 words are reproduced. After 8520 learning cycles,

TABLE 7.4 An Example of Data, Composed of Words and Represented in the Binary Numbering System

MR	WHITE	FOUND	WORK	IN	A	SMALL	TOWN
0	0	1	0	0	0	0	0
0	0	0	0	0	0	1	0
0	0	0	0	0	0	0	1
0	0	0	1	0	1	1	0
1	0	0	1	1	0	1	0
1	1	1	1	1	0	1	0
0	1	0	1	1	0	1	1

the regenerated sequence was

MR WHITE FOUND WORK IN A SMALL TOWN AND HE AND HIS WIFE SOLD THEIR HOUSE IN LONDON AND BOUGHT ONE IN THE NEW TOWN MR WHITE WENT TO WORK BY BUS EVERY DAY WITH SOME OF HIS NEW NEIGHBOURS LAST MONTH HE WAS ON THE BUS AND ONE OF HIS NEW NEIGHBOURS WAS IN FRONT OF HIM *IN WHITE HIM YESTERDAY WHITE DAY YOUR WHERE SOLD HIS WHERE HURT WORK AND ME AND MY HIM AND.*

Now the network produced a perfect original sequence up to the 58th word. After that stage, a random sequence was produced by the network. Note that the random part of the sequence is emphasized.

Finally, after 9056 learning cycles, we had

MR WHITE FOUND WORK IN A SMALL TOWN AND HE AND HIS WIFE SOLD THEIR HOUSE IN LONDON AND BOUGHT ONE IN THE NEW TOWN MR WHITE WENT TO WORK BY BUS EVERY DAY WITH SOME OF HIS NEW NEIGHBOURS LAST MONTH HE WAS ON THE BUS AND ONE OF HIS NEW NEIGHBOURS WAS IN FRONT OF HIM MR WHITE SAID TO HIM ONE OF MY TEETH IS HURTING VERY BADLY IS THERE A GOOD DENTIST HERE YOU DO NOT WANT A DENTIST ANSWERED THE NEIGHBOUR ONE OF MY TEETH HURT A LOT YESTERDAY TOO BUT MY WIFE PUT HER ARMS ROUND ME AND KISSED ME AND THE TOOTH STOPPED HURTING GO AND TRY THAT TOO MR WHITE STOOD UP GOOD HE SAID HAPPILY WHERE IS YOUR WIFE NOW.

This is a perfect original text. Since the network can produce the whole text, we can now arrive at the problem of the self-organization of the syntactical structure of natural language. We think that the self-organization of the grammatical structure can be obtained by the appropriate restriction upon the network architecture.

7.5 CONCLUSION

This chapter is concerned with the analysis of the performance of an acceleration method with Minkowski-r power metrics for the temporal supervised learning algorithm of Williams and Zipser. First, it has been shown that the network with an appropriate Minkowski-r can converge to an acceptable point significantly faster than a network with ordinary squared error function. We have then shown that the recurrent network can recognize some important characteristics of symbol sequences more clearly than the feedforward network.

Concerning Minkowski-r power metrics, experimental results have shown that the network with Minkowski-r converges significantly faster than the network with the ordinary squared error function. This tendency is salient as the sequence length increases. We think that this rapid convergence is due to the fact that the network with a small Minkowski-r is significantly sensitive to small errors. This sensitivity is greatly effective as learning is advanced, and errors become significantly small because small errors are overestimated and large errors are underestimated. On the other hand, the squared error function becomes insensitive to small errors, when the learning is advanced. This prevents the network with the squared error function from converging faster.

In addition to Minkowski-r power metrics, the appropriate number of hidden units can contribute to the rapid convergence of learning. Finally, it has been shown that the number of hidden units necessary for generating a sequence of natural language is not in direct proportion to the length of a sequence, but remains small compared with the length of the sequence.

We have also demonstrated the performance of a recurrent neural network with the temporal supervised learning algorithm when applied to the generation of symbol sequences. Four experimental results have been discussed.

First, we have demonstrated that the recurrent neural network can recognize and generate a periodic sequence, even if a period of the sequence is long. The periodic characteristics of the sequence can be clarified by using error analysis between the target and observed output at the visible unit after finishing the learning. This suggests that error analysis is a good method for clarifying the characteristics of the sequence.

Second, we have tried to reproduce a sequence of natural language whose fundamental unit is the letter. It has been observed that the recurrent network can produce the sequence of natural language even above a training interval of the natural language. Then, error analysis has also shown some characteristics of the sequence of natural language. For example, the error tends to be higher when the network must estimate the first position of the word or phrase or sentence. This suggests that the recurrent network does not learn a sequence simply but tries to recognize the so-called syntactical structure of the sequence. These results have not been observed when using the simple recurrent neural network [3].

Third, the generation of natural language with grammatical categories has been discussed. It has been observed that the network tries to discover characteristics of the sequence such as "sentence."

Fourth, we have tried to generate the entire text with words. The network can regenerate the whole text after a long training. Since the network can produce the entire text, the information contained in the text must be compressed into the hidden units or weights of the recurrent neural network. Thus, it is necessary to develop a method to extract the information hidden in the connections.

REFERENCES

[1] R. J. Williams and D. Zipser, Experimental analysis of the real-time recurrent learning algorithm, *Connection Science*, 1(1), 87–111 (1989).

[2] R. J. Williams and D. Zipser, A learning algorithm for continually running fully recurrent neural networks, *Neural computation*, 1, 270–280 (1989).

[3] J. L. Elman, Finding Structure in Time, CRL Technical Report, No. 8801, University of California, San Diego, 1988.

[4] F. J. Pineda, Dynamics and architecture for neural computation, *J. Complexity*, 4, 216–245 (1988).

[5] B. A. Pearlmutter, "Learning State Space Trajectories in Recurrent Neural Networks," in *Proceedings of the 1988 Connectionist Models Summer School*, Carnegie Mellon University, Pittsburgh, Pa., 1988, pp. 113–117.

[6] B. A. Pearlmutter, Learning state space trajectories in recurrent neural networks, *Neural Computation*, 1(2), 263–269 (1989).

[7] R. Kamimura, "Application of Temporal Supervised Learning Algorithm to Generation of Natural Language," *Proc. Internat. Joint Conference on Neural Network*, Vol. 1, San Diego, IEEE Neural Networks Council, New York, 1990, pp. 201–207.

[8] R. Kamimura, "Experimental Analysis of Performance of Temporal Supervised Learning Algorithm, Applied to a Long and Complex Sequence," *Proceedings of International Neural Network Conference*, Vol. 2, Paris, Kluwev Academic Publishers, The Netherlands, 1990, pp. 753–756.

[9] F. J. Pineda, Recurrent backpropagation and the dynamical approach to adaptive neural computation, *Neural Computation*, 1(2), 161–172 (1989).

[10] F. J. Pineda, "Time Dependent Adaptive Neural Networks," in D. S. Touretzky, Ed., *Advances in Neural Information Processing Systems*, Vol. 2, Morgan Kaufmann, 1990, pp. 710–718.

[11] D. Zipser, A subgrouping strategy that reduces complexity and speeds up learning in recurrent networks, *Neural Computation,* 1, 552–558 (1989).

[12] S. J. Hanson and D. J. Burr, "Minkowski-r back-propagation: learning in connectionist models with non-Euclidian signals," in D. Z. Anderson, Ed., *Neural Information Processing Systems,* American Institute of Physics, New York, 1988, pp. 348–357.

[13] A. Cleeremans, D. Servan-Schreiber, and J. L. McClelland, Finite state automata and simple recurrent networks, *Neural Computation,* 1(3), 372–381 (1989).

[14] A. W. Smith and D. Zipser, Learning sequential structure with the real-time recurrent learning algorithm, ICS Technical Report, No. 8903, University of California, San Diego, 1989.

[15] S. E. Fahlman, "Faster-Learning Variations on Back-Propagation: An Empirical Study," in *Proc. 1988 Connectionist Models Summer School,* Morgan Kautmann, California, 1988, pp. 38–51.

[16] N. Petrova, R. Piotrowski, and P. Guiraud, Charactéristique informationelles du mot français, *Bull. de la société de Linguistique de Paris,* 65, 14–28, 1971.

CHAPTER 8

Genetic Programming: Building Artificial Nervous Systems with Genetically Programmed Neural Network Modules

HUGO DE GARIS

8.1 INTRODUCTION

One of the author's interests is studying future computing technologies and then considering the conceptual problems they will pose. The major concern in this regard is called the *complexity problem;* in other words, how will computer scientists over the next 20 years or so cope with technologies that will provide them with the ability to build machines containing roughly the same number of artificial neurons as there are in the human brain. In case this seems a bit farfetched, consider four of these future technologies, namely wafer scale integration (WSI), molecular electronics, nanotechnology, and quantum computing. WSI is already with us. It is the process of using the full surface of a slice of a silicon crystal to contain one huge very large scale integrated (VLSI) circuit. It is estimated by the mid-1990s [1] that it will be possible to place several million artificial neurons on a WSI circuit. Molecular electronics [2] is the attempt to use molecules as computational devices, thus increasing computing speeds and allowing machines to be built with an Avogadro number of components. Nanotechnology is even more ambitious [3,4], aiming at nothing less than mechanical chemistry—that is, building nanoscopic assemblers capable of picking up an atom here and putting it there. We already know that nanotechnology is possible, because we have the existence proof of biochemistry. Nanoscopic assemblers could build any substance, including copies of themselves. Quantum computing

---

Neural and Intelligent Systems Integration, By Branko Souček and the IRIS Group.
ISBN 0-471-53676-8 ©1991 John Wiley & Sons, Inc.

promises to use subatomic scale (quantum) phenomena in order to compute [5], thus allowing the construction of machines with the order of 10^{30} components.

On the assumption that most of these technologies will be well developed within a human generation, how are future computer scientists to cope with the truly gargantuan complexity of systems containing 10^{30} components? This problem is not new. The biological world is confronted with this massive design problem every day. The solution taken by nature, of course, is Darwinian evolution, using techniques and strategies such as genetics, sex, death, reproduction, mutation, and so on. The traditional approach to computer architecture (i.e., explicit preplanned design of machines) will probably become increasingly impossible. It is likely that machine designers of the future will probably be forced to take an evolutionary approach as well. Components will be treated more and more as black boxes, whose internal workings are considered to be too complex to be understood and analyzed. Instead, all that will matter will be the performances of the black boxes. By coding the structures of components in a linear "chromosome"-like way, successful structures (i.e., those that perform well on some behavioral test) may see their corresponding chromosomes survive into the next generation with a higher probability. These linear structure codes (e.g., DNA in the biological world) can then be mutated by making random changes to them. If by chance an improvement in performance results, the mutated chromosome will gradually squeeze out its rivals in the population. Over many generations, the average performance level of the population will increase.

This chapter is based on evolutionary philosophy. It is concerned with the evolution of neural network modules that control some process, say stick legs that are taught to walk, or an artificial creature with a repertoire of behaviors. Before launching into a description as to how this is done, readers who are unfamiliar with either the genetic algorithm (GA) [6] or neural networks [7] are given a brief overview of these topics. This is followed by a description as to how the GA can be used to evolve neural network modules (called GenNets) that have a desired behavior. By putting GenNets together into more complex structures, one can build hierarchical control systems, artificial nervous systems, and other systems.

Three examples of these ideas are presented in relative detail, based on some of the author's papers. The final sections provide ideas as to how genetic programming can play an important part in the new field of artificial life [8] by showing how artificial nervous systems can be constructed and how these ideas might be put directly into silicon and into robots.

8.2 NEURAL NETWORKS

For the last 20 years, the dominant paradigm in artificial intelligence (AI) has been symbolism—using symbols to represent concepts. This approach has only been moderately successful, which probably explains why so many AI researchers abandoned the symbolist camp for the neuronist camp (alias parallel distributed processing, connectionism, or neural networks after the neuronist revival in the 1980s. This was certainly true for the author, who always had the feeling when interpreting the symbols he was using that somehow he was cheating. He, and not the machine, was giving the symbols their interpretation. At the back of his mind was always the impression that what he was doing was not "real" AI. Real AI to him was trying to model neural systems, and secretly he was reading neurophysiology, hoping to get into brain simulation in one form or another. After the first International Neural Networks Conference [9] was held, and the author got hold of the proceedings, there was no turning back. Similar experiences probably occurred to many people, because suddenly there were thousands of neuronists and six monthly neural network world conferences.

In the light of the complexity problem introduced in the introduction, one soon realizes that one does not have to put many artificial neurons together to create a complicated whole, whose dynamics and analysis quickly transcend the scope of present-day techniques. It seems probable that the intelligence shown by biological creatures such as mammals is due to the enormous complexity of their brains (i.e., literally billions of neurons. It seems also not unreasonable that to produce machines with comparable intelligence levels, we may have to give them an equivalent number of artificial neurons and to connect them together in ways similar to nature. But how are we ever to know how to do that, when we are talking about a trillion neurons in the human brain, with an astronomical number of possible connections?

Considering these weighty questions led the author to the idea that a marriage of artificial neural networks with artificial evolution would be a good thing—the basic idea being that one might be able to design functional neural networks without fully understanding their dynamics and connections. This is, after all, the way nature has designed its brains. The author's view was that an artificially evolved neural net (which came to be called a GenNet) should be considered a black box. What was important was not *how* it functioned (although it is always nice to have an analytical understanding of the dynamics) but that it functioned at all. The art of designing GenNets and putting them together (an undertaking called genetic programming) was thus to be highly empirical, a "let's try it and see" method. To insist that one should have a theoretical justification for one's approach before proceeding seemed to be unnecessarily restrictive, and would only hinder the rate of progress in designing artificial neural nets and artificial nervous systems.

There are other pragmatic arguments to justify this empirical approach. The author's belief is that technological progress with conventional computer design in terms of speed and memory capacities will allow neural net simulations of a complexity well beyond what the mathematicians are capable of analyzing. One can expect the simulations to hit upon new qualitative

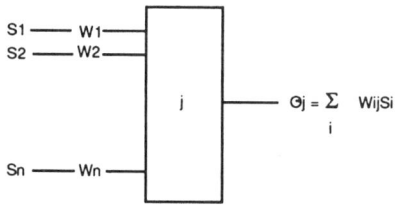

Figure 8.1 *An artificial neuron.*

phenomena that the analysts can subsequently attempt to understand. This chapter is essentially such an investigation. Several such qualitative phenomena have been discovered, as will be explained in later sections.

The foregoing has been more a justification for, rather than an explanation of, genetic programming (GP). Since GP is a new approach, simply explaining what GP is without attempting to justify why it is would seem to provide an incomplete picture of what GP is all about. Further justification of the approach are the results already obtained and that will be discussed. This section, on neural networks, makes no pretence at being complete. There are plenty of good textbooks now on the basic principles of neural nets [7]. However, enough needs to be explained in order for the reader to understand how the genetic algorithm is applied to the evolution of neural nets. This minimum now follows.

Artificial neural networks are constructed by connecting artificial neurons together, whose function consists essentially of accepting incoming signals, weighting them appropriately, and then emitting an output signal to other neurons, depending on the weighted sum of the incoming signals. Figure 8.1 shows the basic idea.

The incoming signals have strengths S_i. Each incoming signal S_i is weighted by a factor W_i, so that the total incoming signal strength is the dot product of the signal vector $\{S_1, S_2, S_3, \ldots\}$ and the weight vector $\{W_1, W_2, W_3, \ldots\}$. This dot product is then usually fed through a nonlinear sigmoid function, shown in Figure 8.2.

There are binary versions of these ideas, where the input signals are binary, the weights are reals, and the output is binary, depending on the sign of the dot product. For further details, see any of the neuronist textbooks [7]. However, for GP, the incoming signals are assumed to be reals, as are the weights and output signals.

Now that the functioning of a single neuron has been discussed, we need to consider how a group of neurons are usually connected to form networks. After all, the very title of this section is neural networks. Broadly speaking, there are two major categories of networks used by neuronists, known as *layered feedforward* and *fully connected*. Both types have been used in the

8.2 NEURAL NETWORKS

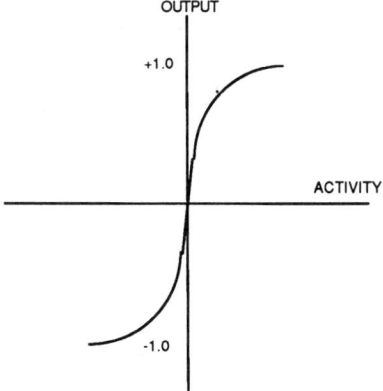

OUTPUT = -1 + 2/(1 + e-ACTIVITY)

Figure 8.2 *The neural output function.*

author's GP experiments. Figure 8.3 shows a layered feedforward network. The use of the word *layered* is obvious. The usual number of layers is 3, namely an input layer, a hidden layer, and an output layer, as shown in the figure. The word *feedforward* is used because no output signals are fed back to previous layers. For given weights, a given input vector applied to the input layer will result in a definite output vector. Learning algorithms exist that can modify the weights, allowing a desired output vector for a given input vector. The best-known such algorithm is called the *backpropagation* (backprop) algorithm [7].

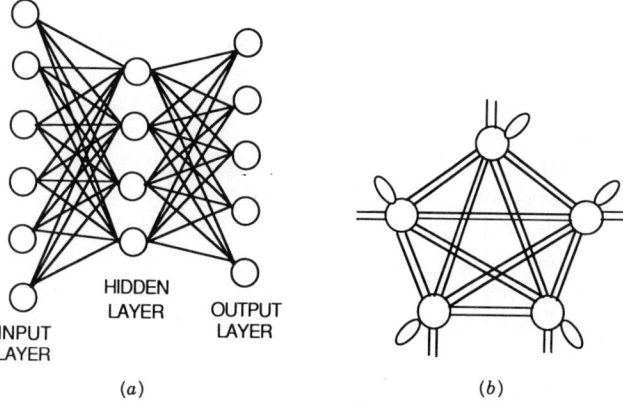

Figure 8.3 *Neural network architectures: (a)* feedforward network; *(b) fully connected network.*

The other common network type is called *fully connected,* as shown also in Figure 8.3. The dynamics of fully connected networks are a lot more complicated, which explains why many neuronist theorists have preferred to work with feedforward networks. As we shall see, the level of complexity of the dynamics of the network is usually irrelevant to the genetic algorithm, which responds only to the quality of the results of the dynamics and not to how the dynamics are produced. A fully connected network obviously includes feedback, and the outputs depend not only on the initial inputs but on the history of the internal signal strengths. This greater complexity may be a disadvantage is terms of its analyzability, but its greater number of degrees of freedom may actually be useful for the functioning of GenNets (e.g., a greater robustness, or giving the GA a larger search space to play with).

Imagine as little as a dozen such artificial neurons all connected together, where the outputs of some of the neurons are used as parameter values in certain calculations. If the results of these calculations are fed back as external inputs to some of the neurons, and the process is repeated for hundreds of cycles or more, one has an extremely complicated dynamical system. However, a computer simulation of such a system is quite within the state of the art, even on relatively modest workstations.

In order to be able to evolve the behavior of such a system described in the preceding section, one needs to understand a little about the genetic algorithm, which is the technique used to control the artificial evolution of this behavior. Hence, a brief description of the genetic algorithm follows.

8.3 THE GENETIC ALGORITHM

The genetic algorithm, as already mentioned, is a form of artificial evolution. But evolution of what? There are a few key concepts that need to be introduced to be able to understand how the GA can be employed to evolve neural network behaviors. Probably the most important idea is that of a linear mapping of the structure of the system one is dealing with, into a string of symbols, usually a binary string. For example, imagine you were trying to find the numerical values of 100 parameters used to specify some complicated nonlinear control process. You could convert these values into binary form and concatenate the results into a long binary string. Alternatively, you could do the reverse and take an "arbitrary" binary string and extract (or map) the 100 numerical values from it.

Imagine forming a population of 50 of these binary strings, simply by using a random number generator to decide the binary value at each bit position. Each of these bit strings (of appropriate length to agree with the numerical representation chosen) will specify the values of the 100 parameters. Imagine now that you have a means of measuring the quality or performance level (or "fitness," in Darwinian terms) of your (control) process. The next step in the discussion is to actually measure the performance levels

```
0101010101010  10101010101  010101010101
1111000011110  00011110000  111100001111
```

Figure 8.4 *Crossover of chromosomal pairs.*

of each process as specified by its corresponding bit string. These bit strings are called *chromosomes* by GA specialists, for reasons that will soon become apparent. Let us now reproduce these chromosomes in proportion to their fitnesses; that is, chromosomes with high scores relative to the other members of the population will reproduce themselves more.

If we now assume that the initial population of chromosomes is to be replaced by their offspring, and that the population size is fixed, then we will inevitably have a competition for survival of the chromosomes among themselves into the next generation. Poor fitness chromosomes may not even survive. We have a Darwinian "survival-of-the-fittest" situation. To increase the closeness of the analogy between biological chromosomes and our binary chromosomes, let us introduce what the GA specialists call *genetic operators,* such as mutation, crossover, and so on, and apply them with appropriate probabilities to the offspring.

Mutation in this case is simply flipping a bit, with a given (usually very small) probability. Crossover is (usually) taking two chromosomes, cutting them both at (the same) two positions along the length of the bit string, and then exchanging the two portions thus cut out, as shown in Figure 8.4.

If by chance, the application of one or more of these genetic operators causes the fitness of the resulting chromosome to be slightly superior to that of the other members of the population, it will have more offspring in the following generation. Over many generations, the average fitness of the population increases. This increase in fitness is due to favorable mutations and the possible combination of more than one favorable mutation into a single offspring due to the effect of crossover. Crossover is synonymous with sex. Biological evolution hit on the efficacy of sex very early in its history. With crossover, two separate favorable mutations, one in each of two parents, can be combined in the same chromosome of the offspring, thus accelerating significantly the adaptation rate (which is vital for species survival when environments are rapidly changing).

In order to understand some of the details mentioned in later sections, some further GA concepts need to be introduced. One concerns the techniques used to select the number of offspring for each member of the current generation. One of the simplest techniques is called *roulette wheel,* which slices up a pie (or roulette wheel) with as many sectors as members of the population (assumed fixed over the whole evolution). The angle subtended by the sector (slice of pie) is proportional to the fitness value. High-fitness chromosomes will have slices of pie with large sector angles. To select the next generation (before the genetic operators are applied), one simply "spins" the roulette wheel a number of times equal to the number of chromosomes in the population. If the roulette "ball" lands in a given sector (as simulated with a random number generator), the corresponding chromosome is selected. Since the probability of landing in a given sector is proportional to its angle (and hence fitness), the fitter chromosomes are more likely to survive into the next generation, hence the Darwinian survival of the fittest, and hence the very title of the algorithm, the genetic algorithm.

Another concept is called *scaling,* which is used to linearly transform the fitness scores before selection of the next generation. This is often done for two reasons. One is to avoid premature convergence. Imagine that one chromosome had a fitness value well above the others in the early generations. It would quickly squeeze out the other members of the population if the fitness scores were not transformed. So choose a scaling such that the largest (scaled) fitness is some small factor (e.g., 2) times the average (scaled) fitness. Similarly, if there were no scaling, toward the end of the evolution the average (unscaled) fitness value often approaches the largest (unscaled) fitness, because the GA is hitting against the natural (local) "optimum behavior" of the system being evolved. If the fitness scores differ by too little, no real selection pressure remains, and the search for superior chromosomes becomes increasingly random. To continue the selective pressure, the increasingly small differences between the unscaled average fitness and the largest fitness are magnified by using the same linear scaling employed to prevent premature convergence. This techniques is used when evolving GenNets.

There are many forms of crossover in the GA literature [6]. One fairly simple form is as follows. Randomly pair off the chromosomes selected for the next generation and apply crossover to each pair with a user-specified probability (i.e., the "crossprob" value, usually about 0.6). If ("two-point") crossover is to be applied, each chromosome is cut at two random positions (as in Figure 8.4) and the inner portions of the two chromosomes swapped.

A fairly typical cycle of the GA is shown in Figure 8.5, where initialization usually means the creation of random binary strings. Popqual translates the binary strings into the structures whose performances are to be measured and measures the fitnesses of these performances. Often the chromosome with the highest fitness is saved at this point, to be inserted directly into the next generation once the genetic operators have been applied. Scaling linearly scales the fitness values before selection occurs. Selection is using a technique such as roulette wheel to select the next generation from the present generation of chromosomes. Crossover swaps over portions of chromosome pairs. Mutation flips the bits of each chromosome with a low probability (typically 0.001 or so) per bit. Elite simply inserts the best chromosome met so far (as stored in the Popqual phase) directly into the next generation. This can be useful, because the stochastic nature of the GA can sometimes eliminate high fitness chromosomes. The elitist strategy prevents this.

Figure 8.5 is far from complete. However, it contains the essence of the subject. For a more detailed analysis, particularly a mathematical justifica-

```
0101010101010  00011110000  010101010101
1111000011110  10101010101  111100001111
```

```
              INITIALIZATION
     LOOP :   POPQUAL
              SCALING
              SELECTION
              CROSSOVER
              MUTATION
              ELITE
```

Figure 8.5 *Typical cycle of the genetic algorithm.*

tion as to why crossover works so well, see the several texts available [6], especially [10]. The GA is a flourishing research domain in its own right, and is as much a mix of empirical exploration and mathematical analysis as are neural networks.

Now that the essential principles of both neural networks and the GA have been presented, a description as to how the GA can be used to evolve behaviors in neural nets can follow.

8.4 GENNETS

A GenNet (i.e., a genetically programmed neural net) is a neural net that has been evolved by the GA to perform some function or behavior, where the fitness is the quality measure of the behavior. For example, imagine we are trying to evolve a GenNet that controls the angles of a pair of stick legs over time, such that the stick legs move to the right across a computer screen. The fitness might simply be the distance covered over a given number of cycles. Most of the work the author has done so far has been involved with fully connected GenNets. Remember the GA is indifferent to the complexity of the dynamics of the GenNet, and this complexity might prove useful in the evolution of interesting behavior.

This section will be devoted essentially to representing the structure of a GenNet by its corresponding bit string chromosome. The representation chosen by the author is quite simple. Label the N neurons in the GenNet from 1 to N. (The number of neurons in the GenNet is a user specified parameter, typically about a dozen neurons per GenNet.) If the GenNet contains N neurons, there will be N^2 interconnections (including a connection looping directly back to its own neuron as shown in Figure 8.3). Each connection is specified by its sign (where a positive value represents an excitatory synapse, and a negative value represents an inhibitory synapse), and its weight value. The user decides the number (P) of binary places used to specify the magnitude of the weights. (Typically this number P is 6, 7, or 8). The weights were chosen to have a modulus always less than 1.0 Assuming one bit for the sign (a 0 indicating a positive weight, a 1 a negative weight), the number of bits needed to specify the signed weight per connection is $P + 1$. Hence for N^2 connections, the number of bits is $N^2(P + 1)$.

The signed weight connecting the ith (where $i = 0 \rightarrow N - 1$) to the jth neuron is the $(iN + j)$th group of $P + 1$ bits in the chromosome (reading from left to right). From this representation, one can construct a GenNet from its corresponding chromosome. In the Popqual phase, the GenNet is "built" and "run;" that is, appropriate initial (user chosen) signal values are input to the input neurons (the input layer of a layered feedforward network, or simply to a subset of the fully connected neurons that are said to be input neurons).

GenNets are assumed to be *clocked* (or synchronous); that is, between two ticks of an imaginary clock (a period called a *cycle*), all the neurons calculate their outputs from their inputs. These outputs become the inputs for the next cycle. There may be several hundred cycles in a typical GenNet behavioral fitness measurement. The outputs (from the output layer in a feedforward net, or from a subset of fully connected neurons that are designated the output neurons) are used to control or specify some process or behavior. Three concrete examples will follow. The function or behavior specified by this output (whether time dependent or independent) is measured for its fitness or quality, and this fitness is then used to select the next generation of GenNet chromosomes.

8.5 GENETIC PROGRAMMING

Now that the techniques used for building a GenNet are understood, we can begin thinking about how to build networks of networks—how to use GenNets to control other GenNets—thus allowing hierarchical, modular construction of GenNet circuits. Genetic programming consists of two major phases. The first is to evolve a family of separate GenNets, each with its own behavior. The second is to put these components together such that the whole functions as desired. In genetic programming, the GenNets are of two kinds, either behavioral (i.e., functional) or control. Control GenNets send their outputs to the inputs of behavioral GenNets and are used to control or direct them. This type of control is also of two kinds, either direct or indirect. Examples of both types of control will be shown in the following sections.

Direct control is when the signal output values are fed directly into the inputs of the receiving neurons (usually with intervening weights of value

+1.0). The receiving GenNets need not necessarily be behavioral. They may be control GenNets as well. (Perhaps "logic" GenNets might be a more appropriate term.) Indirect control may occur when an incoming signal has a value greater than a specified threshold. Transcending this threshold can act as a trigger to switch off one behavioral GenNet and to switch on another. Just how this occurs, and how the output signal values of the GenNet that switches off influence the input values of the GenNet that switches on will depend on the application. For a concrete example, see Section 8.8, which describes an artificial creature called LIZZY.

For direct control, the evolution of the behavioral and control GenNets can take place in two phases. First, the behavioral GenNets are evolved. Once they perform as well as desired, their signs and weights are frozen. Then the control GenNets are evolved, with the outputs of the control GenNets connected to the inputs of the behavioral GenNets (whose weights are frozen). The weights of the control GenNets are evolved such that the control is optimum according to what the control/behavioral–GenNet–complex is supposed to do. See Section 8.6 for a concrete example of direct control. Once the control GenNets have been evolved, their weights can be frozen and the control/behavioral GenNets can be considered a unit. Other such units can be combined and controlled by further control GenNets. This process can be extended indefinitely, thus allowing the construction of hierarchical, modular GenNet control systems or as the author prefers to look at them, artificial nervous systems. Note that there is still plenty of unexplored potential in this idea. The three examples that follow really only scratch the surface of what might be possible with genetic programming.

8.6 EXAMPLE 1: POINTER

Pointer is an example of time-independent evolution, meaning that the outputs of GenNets are not fed back to the inputs. The inputs are conventionally clamped (fixed) until the outputs stabilize. Figure 8.6 shows the setup used in this example. It is the two-eye two-joint robot arm positioning (or pointing) problem. The aim of the task is to move the robot arm from its vertical start position X to the goal position Y. J1 and J2 are the joints, E1 and E2 are the eye positions, and JA1, JA2, EA1, and EA2 are the joint and eye angles of the point Y.

Pointer is an example of the modular hierarchical approach of genetic programming. Two different GenNet modules will be evolved. The first, called the *joint module,* controls the angle JA that a given joint opens to, for an input control signal of a given strength; the second, called the *control module,* receives inputs EA1 and EA2 from the two eyes and sends control signals to the joints J1 and J2 to open to angles of JA1 and JA2.

Figure 8.7 shows the basic circuit design that the GA uses to find the joint and control modules. The joint modules (two identical copies) are evolved

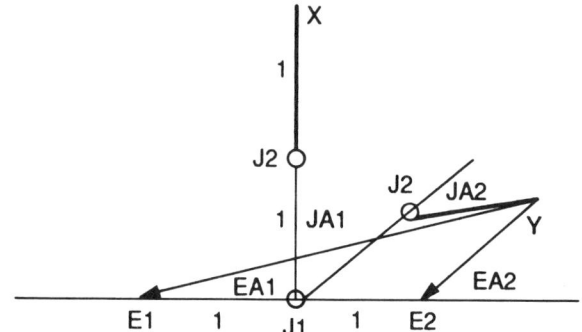

Figure 8.6 Setup for the "pointer" experiment.

first and are later placed under the control of the control module. Each module (containing a user-specified number of neurons) is fully connected, including connections from each neuron to itself. Between any two neurons are two connections in opposite direction, each with a corresponding (signed) weight. The input and output neurons also have "in" and "out" links but these have fixed weights of 1 unit. The outputs of the control module are the inputs of the joint modules as shown in Figure 8.7.

The aim of the exercise is to use the GA to choose the values of the signs and weights of the various modules, such that the overall circuit performs as desired. Since this is done in a modular fashion, the weights of the joint module are found first. These weights are then frozen, and the weights of the control circuit found so that the arm moves as close as possible to any specified goal point Y.

With both weights and transfer functions restricted to the $+1$ to -1 range, output values stabilized (usually after about 50 cycles or so for 1 percent accuracy). In each cycle, the outputs are calculated from the inputs (which

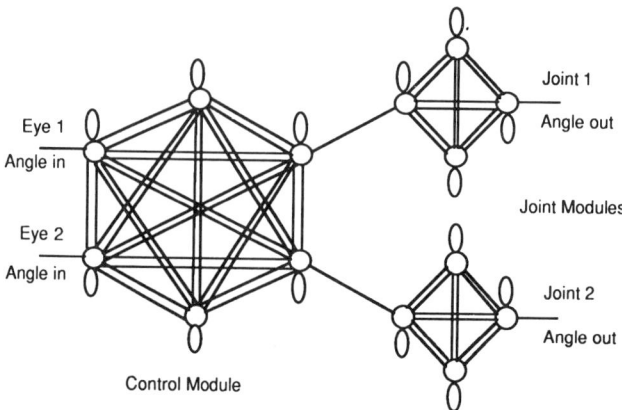

Figure 8.7 GenNet modules for the "pointer" experiment.

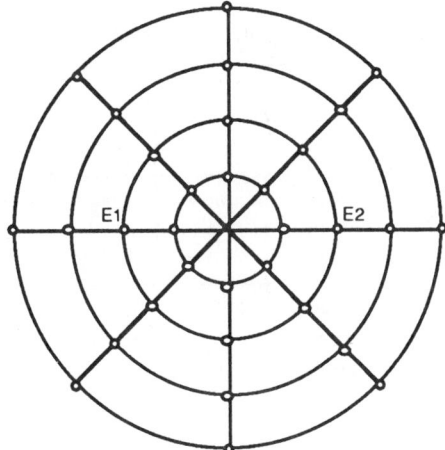

Figure 8.8 *Training points for the "pointer" experiment.*

were calculated from the previous cycle). These outputs become the input values for the neurons that the outputs connect to. The GA is then used to choose the values of the weights, such that the actual output is as close as possible to the desired output.

The function of the joint module is effectively to act as a "halver" circuit; that is, the output is to be half the value of the input. This may seem somewhat artificial, but the point of Pointer is merely to illustrate the principles of modular hierarchical GenNet design (genetic programming). To evolve the joint module, 21 training input values ranging from +1 to −1 in steps of −0.1, were used. The desired output values thus ranged from 0.5 to −0.5, where 0.5 is interpreted to be half a turn of the joint, that is, a joint angle of 180° (a positive angle is clockwise).

The GA parameters used in this experiment were crossover probability, 0.6; mutation rate, 0.001; scaling factor, 2.0; population, 50; number of generations, several hundred. The fitness or quality measure used in the evolution of the joint module was the inverse of the sum of the squares of the differences between the desired and the actual output values.

Figure 8.8 shows the set of training points Y used to evolve the control module. These points all lie within a radius of 2 units, because the length of each arm is 1 unit. For each of these 32 points, a pair of "eye angles," EA1 and EA2, is calculated and converted to a fraction of one half turn. These values are then used as inputs to the control circuit. The resulting 2 joint angle output values JA1 and JA2, are then used to calculate the actual position of the arm Y', using simple trigonometry. The quality of the chromosome that codes for the signs and weights of the control module is the inverse of the sum of the squares of the distances between the 32 pairs of actual positions Y' and the corresponding desired positions Y.

WEIGHTS	FROM NEURON				
		0	1	2	3
TO NEURON	0	-0.96875	0.84375	-0.875	0.90625
	1	-0.96875	-0.78125	-0.875	-0.78125
	2	-0.9375	-0.71875	-0.75	-0.28125
	3	-0.15625	-0.9375	0.3125	-0.65625

Figure 8.9 Weight values of joint module.

Figure 8.9 shows an example solution for the 16 weight values of the joint module. With these weights (obtained after roughly 100 generations with the GA), the actual output values differed from the desired values by less than 1 percent. Similar solutions exist for the 36 weights of the control module, which also gave distance errors less than 1 percent of the length of an arm (1 unit). What was interesting was that every time a set of weights was found for each module, a different answer was obtained, yet each was fully functional. The impression is that there may be a large number of possible adequate solutions, which gives genetic programming a certain flexibility and power.

Also interesting in this result was the generalization of vector mapping that occurred. The training set consisted of 32 fixed points. However, if two arbitrary eye angles were input, the corresponding two actual output joint angles were very close to the desired joint angles.

The above experiment is obviously static in its nature. In Figure 8.7 there are no dynamics in terms of output values from the output neurons fed back to the inputs of the input neurons. However, what one does see clearly is the idea of hierarchical control, where control GenNets are used to command functional GenNets. This idea will be explored more extensively in the LIZZY experiment in Section 8.8. The contents of this section are based on a paper by the author [11].

8.7 EXAMPLE 2: WALKER

Having considered the time-independent process in the previous section, it is fascinating to question whether GenNets can be used to control time-dependent systems, where the fitness would be the result of several hundred cycles of some essentially dynamic process in which feedback would feature. The great advantage of the GenNet approach is that even if the dynamics are complicated, the GA may be able to handle it (although there is

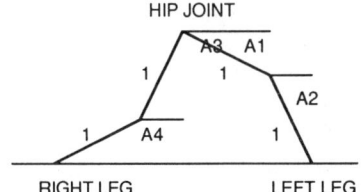

Figure 8.10 Setup for "walker" experiment.

no guarantee; see the final section on evolvability). The GA is usually indifferent to such complexity, because all it cares about is the fitness value. How this value is obtained is irrelevant to the GA. Good chromosomes survive; that is, those that specify a high fitness performance will reproduce more offspring in the next generation.

As a vehicle to test this idea, the author chose to try to get a GenNet to send time-dependent control signals to a pair of stick legs to teach them to walk. The setup is shown in Figure 8.10.

The output values of the GenNet are interpreted to be the angular accelerations of the four components of the legs. Knowing the values of the angular accelerations [assumed constant over one cycle—where a cycle is the time period over which the neurons calculate (synchronously) their outputs from their inputs] and the values of the angles and angular velocities at the beginning of a cycle, one can calculate the values of the angles and the angular velocities at the end of that cycle. As input to the GenNet (control module), the angles and the angular velocities were chosen. Figure 8.11 shows how this feedback works.

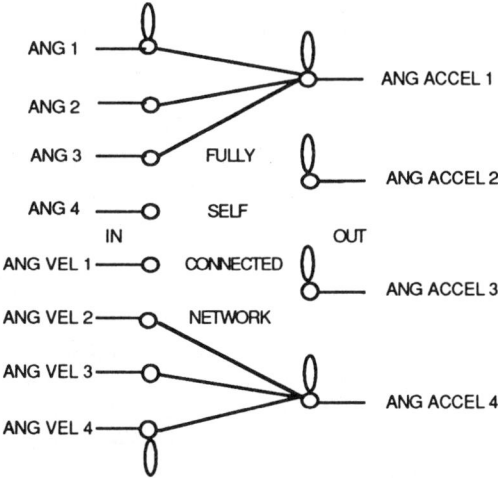

Figure 8.11 The "walker" GenNet.

Knowing the angles, one can readily calculate the positions of the two "feet" of the stick legs. Whenever one of the feet becomes lower than the other, that foot is said to be "on the ground," and the distance (whether positive or negative) between the positions of the newly grounded foot and the previously grounded foot is calculated.

The aim of the exercise is to evolve a GenNet that makes the stick legs move as far as possible to the right in the user-specified number of cycles and cycle time. The GenNet used here consists of 12 neurons: 8 input neurons and 4 output neurons (no hidden neurons). The eight input neurons have as inputs the values of the four angles and the four angular velocities. The input angles range from -1 to $+1$, where $+1$ means one half turn (i.e., 180°). The initial (start) angles are chosen appropriately, as are the initial angular velocities (usually zero), which also range from -1 to $+1$, where $+1$ means a half turn per second. The activity of a neuron is calculated in the usual way, namely the sum of the products of its inputs and its weights, where weight values range from -1 to $+1$. The output of a neuron is calculated from this sum, using the antisymmetric sigmoid function of Figure 8.2 The outputs of neurons are restricted to have absolute values of less than 1 so as to avoid the risk of explosive positive feedback. The chromosomes used have the same representation as described in Section 8.4.

The selection technique used was "roulette wheel," and the quality measure (fitness) used for selecting the next generation was (mostly) the total distance covered by the stick legs in the total time T, where $T = C \times$ cycle time for the user-specified number C of cycles and cycle time. The fitness is thus the velocity of the stick legs moving to the right. Right distances are nonnegative. Stick legs that move to the left scored zero and were eliminated after the first generation.

Using the foregoing parameter values, we undertook a series of experiments. In the first experiment, no constraints were imposed on the motion of the stick legs (except for a selection to move right across the screen). The resulting motion was most unlifelike. It consisted of a curious mixture of windmilling of the legs and strange contortions of the hip and knee joints. However, it certainly moved well to the right, starting at given angles and angular velocities. As the distance covered increased, the speed of the motion increased as well and became more "efficient"; for example, windmilling was squashed to a swimmer's stroke. See Figure 8.12 for snapshots.

In the second experiment, the stick legs had to move such that the hip joint remained above the floor (a line drawn on the screen). During the evolution, if the hip joint did hit the floor, evolution ceased, the total distance covered was frozen, and no further cycles were executed for that chromosome. After every cycle, the coordinates of the two feet were calculated and a check made to see if the hip joint did not lie below both feet. This time the evolution was slower, presumably because it was harder to find new weights that led to a motion satisfying the constraints. The resulting motion was almost as unlifelike as in the first experiment, again with windmilling and contortions, but at least the hip joint remained above the floor.

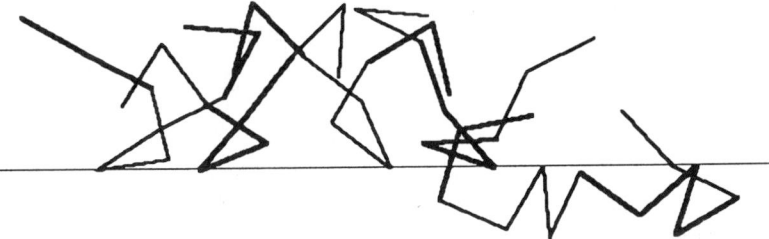

Figure 8.12 Snapshots of "walker" with no constraints.

In the third experiment, a full set of constraints was imposed to ensure a lifelike walking motion (e.g., hip, knees, and toes above the floor, knees and toes below the hip, knee angles less than 180°, etc). The result was that the stick legs moved so as to take as long a single step as possible, and "did the splits" with the two legs as extended as possible and the hip joint just above the floor. From this position, it was impossible to move any further. Evolution ceased. It looked as though the GA had found a local maximum. This was a valuable lesson and focused attention upon the important concept of evolvability (i.e., the capacity for further evolution). A change in approach was needed.

This led to the serendipitous discovery of the concept of behavioral memory—the tendency of a behavior evolved in an earlier phase to persist in a later phase. For example, in phase 1 evolve a GenNet for behavior A. Take the weights and signs resulting from this evolution as initial values for a second phase that evolves behavior B. One then notices that behavior B contains a vestige of behavior A. This form of "sequential evolution" can be very useful in genetic programming, and was used to teach the stick legs to walk (i.e., to move to the right in a steplike manner), in three evolutionary phases.

In the first phase, a GenNet was evolved over a short time period (e.g., 100 cycles) that took the stick legs from an initial configuration of left leg in front of right leg to the reverse configuration. The angular velocities were all zero at the start and at the finish. By then allowing the resulting GenNet to run for a longer time period (e.g., 200 cycles) the resulting motion was "steplike"; that is, one foot moved in front of and behind the foot on the floor, but did not touch the floor. The quality measure (fitness) for this GenNet was the inverse of the sum of the squares of the differences between the desired and the actual output vector components (treating the final values as an eight-component state vector of angles and angular velocities).

This GenNet was then taken as input to a second (sequential evolutionary) phase, which was to get the stick legs to take short steps. The fitness this time was the product of the number of net positive steps taken to the right, and the distance. The result was a definite stepping motion to the right, but distance covered was very small. The resulting GenNet was used in a third phase in which the fitness was simply the distance covered. This time

Figure 8.13 Snapshots of "walker" with full constraints.

the motion was not only a very definite stepping motion, but the strides taken were long. The stick legs were walking. See Figure 8.13 for snapshots.

The preceding experiments were all performed in tens to a few hundred generations of the GA. Typical GA parameter values were as follows: population size, 50; crossover probability, 0.6; mutation probability, 0.001; scaling factor, 2.0; cycle time, 0.03; number of cycles, 200 to 400. A video of the results of the experiments has been made. To obtain a real feel for the evolution of the motion of the stick legs, one really needs to see it. The effect can be quite emotional, as proven by the reaction of audiences at several world conferences. It is really quite amazing that the angular coordination of four lines can be evolved to the point of causing stick legs to walk.

The success of walker had such a powerful effect on the author that he became convinced that it would be possible to use the basic idea that one can evolve behaviors (the essential idea of genetic programming) to build artificial nervous systems by evolving many GenNet behaviors (one behavior per GenNet) and then switching them on and off using control GenNets. A more concrete discussion of this exciting prospect follows in the next section.

8.8 EXAMPLE 3: LIZZY

The section reports on work still under way, so only partial results can be mentioned. The LIZZY project is quite ambitious and hopes to make a considerable contribution to the emerging new field of artificial life [8]. This project aims to show that genetic programming can be used to build a (simulated) artificial nervous system using GenNets. This process is called *brain building* (or, more accurately, *nanobrain building*). The term *nanobrain* is appropriate, considering that the human brain contains nearly a trillion neurons and that a nanobrain would therefore contain the order of hundreds of neurons. Simulating several hundred neurons is roughly the limit of present-day technology, but this will change.

This section is concerned principally with the description of the simulation of just such a nanobrain. The point of the exercise is to show that this kind of thing can be done, and if it can be done successfully with a mere dozen or so GenNets as building blocks, it will later be possible to design artificial nervous systems with GenNets numbering in the hundreds, thousands, and up. One can imagine in the near future, whole teams of human

Figure 8.14 LIZZY, the artificial lizard.

genetic programmers devoted to the task of building quite sophisticated nano (micro?) brains, capable of an elaborate behavioral range.

To show that this is no pipe dream, a concrete proposal will be presented showing how GenNets can be combined to form a functioning (simulated) artificial nervous system. The implementation of this proposal has not yet been completed (at the time of writing), but progress so far inspires confidence that the project will be completed successfully. The "vehicle" chosen to illustrate this endeavor is shown in Figure 8.14.

This lizardlike creature (called LIZZY) consists of a rectangular wireframe body, four two-part legs, and a fixed antenna in the form of a V. LIZZY is capable of reacting to three kinds of creature in its environment, namely, mates, predators, and prey. These three categories are represented by appropriate symbols on the simulator screen. Each category emits a sinusoidal signal of a characteristic frequency. The amplitudes of all these signals decrease inversely as a function of distance. Prey emit a high frequency, mates a middle frequency, and predators a low frequency. The antenna picks up the signal, and can detect its frequency and average strength. Once the signal strength becomes large enough (a value called the *attention* threshold), LIZZY executes an appropriate sequence of actions, depending on the frequency detection decision.

If the object is a prey, LIZZY rotates toward it, moves in the direction of the object until the signal strength is a maximum, stops, and pecks at the prey like a hen (by pushing the front of its body up and down with its front legs), and after a while wanders off. If the object is a predator, LIZZY rotates away from it and flees until the signal strength is below attention threshold. If the object is a mate, LIZZY rotates toward it, moves in the direction of the object until the signal strength is a maximum, stops, and mates (by pushing the back of its body up and down with its back legs), and after a while wanders off.

This description merely sketches LIZZY's behavioral repertoire. In order to allow LIZZY to execute these behaviors, a detailed circuit of GenNets

226 GENETIC PROGRAMMING: BUILDING ARTIFICIAL NERVOUS SYSTEMS

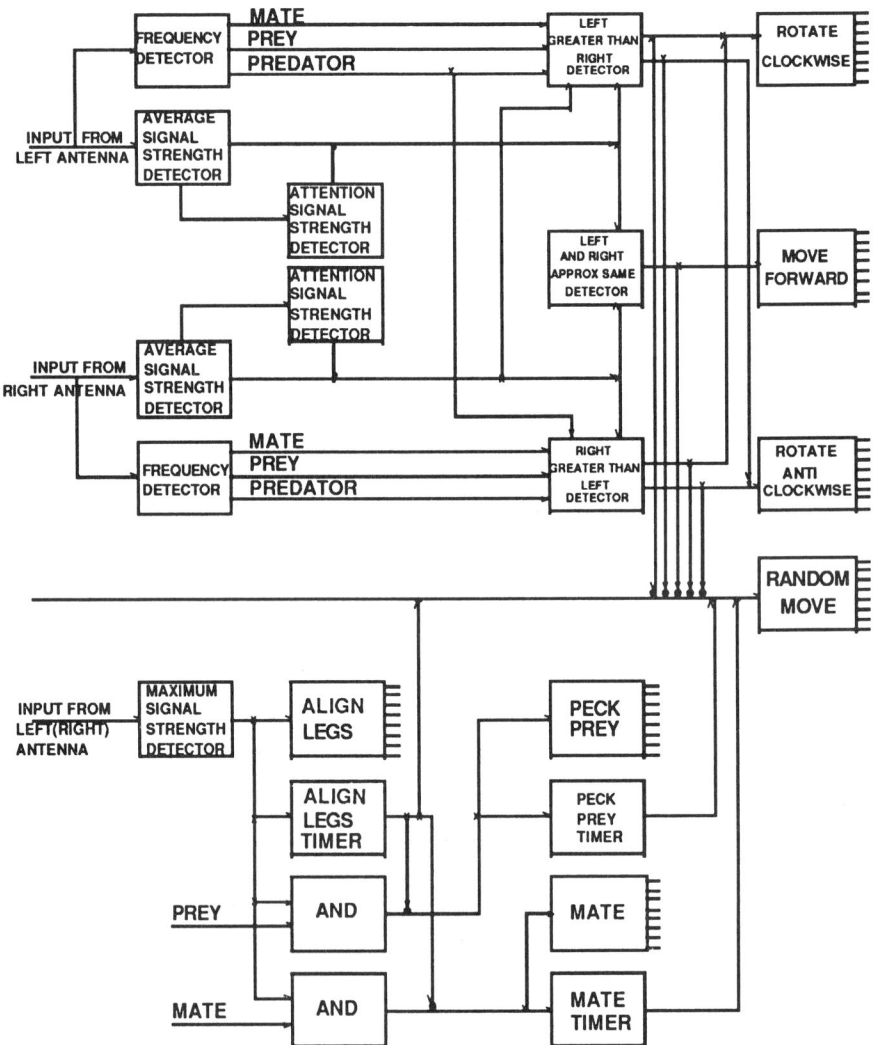

Figure 8.15 The LIZZY GenNet circuit.

and their connections needs to be designed. Figure 8.15 shows an initial attempt at designing such a GenNet circuit. There are seven different motion GenNets, of which "random move" is the default option. When any of the other six motions is switched on, the random move is switched off. Provisionally each motion GenNet consists of eight output neurons, whose output values (as in the walker GenNet) are interpreted as angular accelerations. The position of the upper (and lower) part of each leg is defined by the angle that the leg line takes on a cone whose axis of symmetry is equidistant from each of the XYZ axes defined by the body frame. The leg line is confined to rotate around the surface of this cone as shown in Figure 8.16.

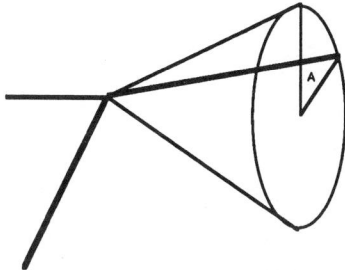

Figure 8.16 *Conical rotation of LIZZY's legs.*

This approach was chosen so as to limit the number of degrees of freedom of the leg angles. If each leg part had two instead of one degree of freedom, and if one wanted to keep the GenNet fully connected, with 16 outputs and 32 inputs (angles and angular velocities)—that is, 48 neurons per GenNet, hence 48 squared connections, the resulting chromosome would have been huge.

The power and the fun of GenNets is that one can evolve a motion without specifying in detail how the motion is to be performed. For example, the fitness for the "move forward" GenNet will be a simple function of the total distance covered in a given number of cycles (modified by a measure of its total rotation and its drift away from the desired straight-ahead direction). Similar simple fitness functions can be defined for the two rotation GenNets. (In fact, only one needs to be evolved. One can then simply swap body side connections for symmetry reasons.)

Understanding the GenNet circuit of Figure 8.15 is fairly straightforward, except perhaps for the mechanism of rotating toward or away from the object. The antenna is fixed on the body frame of LIZZY. If the signal strength on the left antenna is larger than that on the right, and if the object is a prey or mate, then LIZZY is to turn toward it by rotating anticlockwise, or away from it by rotating clockwise if the object is a predator. Eventually the two signal strengths will be equidistant from the object. Once this happens, LIZZY moves forward.

The following paragraphs describe each of the GenNets used in Figure 8.15 and give their quality (fitness) functions. Since this section reports on work still in progress, not all the GenNets have been completed at the time of writing. Those remaining to be evolved are indicated by TBE.

8.8.1 Detector GenNets

- Frequency detector (TBE)
- Average-signal-strength detector (TBE)
- Maximum-signal-strength detector (TBE)
- Left-greater-than-right detector

- Right-greater-than-left detector
- Left-and-right-about-same detector
- Attention-signal-strength detector

Initial attempts at frequency detection using GenNets have shown that this will not be a trivial business, indicating that genetic programming is still very much an art.

Left-Greater-than-Right Detector This GenNet was evolved using supervised learning. It had two input neurons, one output neuron, and seven hidden neurons, 10 in all. The desired output value was to be greater than 0.1 if the "left" input value was greater than the "right" input value by 0.1, and negative otherwise. Each input value ranged from -1 to $+1$ in steps of 0.05, giving 21×21 combinations. The fitness was defined as the inverse of the sum of the squares of the errors; for example, if the inputs were (left 0.60) and (right 0.45) and the output was 0.045, then the error was $(0.1 - 0.045)$. With clamped input values, output values usually stabilize after a few dozen cycles. The output values were measured after a fixed number of cycles (e.g., 20). This GenNet worked fine with very high quality fitness values.

Note that whether the rotate clockwise or counterclockwise motion GenNet is to be switched on depends upon the category of the detected creature. One could either have an $L > R$ detector per creature category or one could try to evolve a multifunction GenNet with extra input(s), to indicate prey/mate or predator. If the predator input value is high, and $L > R$, then the rotate clockwise GenNet should switch on. Experience shows that multifunction GenNets (which have different behaviors for different input "switching" values) are harder to evolve. Sometimes they work, sometimes they do not. (For example, the author was able to get walker to move (with no constraints) either right or left across the screen, depending on whether the input value on an extra input "switching" neuron was fixed to $+0.5$ or -0.5.) Again, GP remains an art at the moment. Mathematical criteria for evolvability do not yet exist unfortunately.

Right-Greater-than-Left Detector This is a trivial modification of the preceding GenNet. One merely switches the input connections.

Left-and-Right-about-Same Detector This GenNet is similar to the previous two. The input training values are the same (21×21 combinations). If the two input values differed by less than 0.1, then the output value was to be greater than 0.1, otherwise negative. The fitness was the same inverse of the sum of the squares of the errors. This GenNet evolved easily.

Attention-Signal-Strength Detector This GenNet takes the output of the average signal strength detector and checks if that average value lies above a given threshold. If so, it gives an output value greater than 0.1, negative otherwise. The training input values ranged in steps of 0.01 over the same ±1 range (201 possibilities). If the input was greater than 0.6 (an arbitrary threshold value), then the output was supposed to be 0.1, otherwise negative. The same fitness definition was used. No problems here.

8.8.2 Logic GenNets

AND Gate This GenNet has two input, one output, and five hidden neurons. An input value is interpreted to be high if it is greater than 0.1 If both are high, the output value is supposed to be greater than 0.1, otherwise negative. The input training values for each input neuron ranged over ±1 in steps of 0.1 (21 × 21 combinations). The fitness was as before.

8.8.3 Effector (Behavioral/Motion) GenNets

- Straight-Ahead Walk
- Rotate clockwise
- Rotate counterclockwise
- Align legs (TBE)
- Timers
- Peck
- Mate
- Random move

Straight-ahead Walk. This GenNet contained 24 neurons, (16 input, 8 output, no hidden). (How does one choose these numbers? There are no formulas. Once again, GP is an art.) The input values are calculated from the output values as described in Section 8.7. The motion definition of LIZZY chosen is as follows. This motion occurs in a cycle. The first step is to find the position and orientation of the body frame resulting from the completion of the previous cycle. The input (conical) angle values are used to position the legs relative to the body frame (where for computational convenience, the body frame is momentarily assumed to lie in the XY plane). LIZZY's body is then given the same orientation and position as at the end of the previous motion cycle.

The next step is to position the legs relative to the floor. LIZZY is lowered or raised vertically so that the lowest leg touches the floor (the XY plane) at the point F1. LIZZY is then rotated about F1 in a vertical plane that includes the second lowest foot until that foot touches the floor at F2. LIZZY is then rotated about the axis F1–F2 until the third lowest foot touches the floor.

To get LIZZY to walk, a further component of the motion definition was introduced. The front two feet were assumed to be "sticky" and the back two feet merely "slid" on the floor. Either the left or right front leg was defined as the "pushleg"—the front leg on the floor that did the pushing of

LIZZY. Whenever the current pushleg lifted off the floor, the other front leg became the pushleg and its position on the floor fixed. Hence the motion definition contains three components—positioning the legs relative to the wire frame body, positioning the body and legs relative to the floor, and then displacing the legs on the floor.

This motion definition seems rather artificial with hindsight, but was only an initial attempt to illustrate the principle of GP nervous system design. The fitness function for getting LIZZY to move straight ahead was simply the component of the distance covered along the straight-ahead direction. LIZZY did walk reasonably well.

Rotate Clockwise The same GenNet architecture as before was used. The motion definition differed slightly in that to get a rotation, the moment that one of the front legs became the pushleg, the orientation of the vector between the pushleg body corner position and the fixed pushleg "foot-on-the-floor" position was fixed. LIZZY had to rotate so as to keep this vector fixed as the orientation of the leg relative to the body changed. The fitness was defined to be $1.0 - \cos(\text{angle rotated})$. LIZZY could do a 180° turn in about 120 cycles.

Rotate Counterclockwise As before, but counterclockwise. In order to measure the sign of the angle rotated, the sine rather than the cosine was used initially, and the cosine calculated from it. Rotations in the wrong direction were given zero fitness and eliminated after the first generation.

Align Legs (TBE) It is debatable whether such a GenNet is needed. The idea is just to take LIZZY from any leg configuration and reorientate it to a standard position with the leg angles taking given (symmetric) values in order to allow LIZZY to peck or mate. For further discussion on this point, see the next section.

Timers These GenNets are used to specify how long the pecking and mating are to go on. The output goes from a high value (e.g., >0.6) to a low value (e.g., <0.1) after a given number of cycles (e.g., 50 cycles). The fitness was the usual inverse of the sum of the squares of the errors between desired and actual output values.

Peck This GenNet was rather trivial to construct. The back legs were frozen in an elevated symmetrical position. The value from any one (and only one) of the angle inputs was fed to all four front leg components. This inevitably led to a pecking motion by LIZZY's head.

Mate This GenNet was very similar to the peck GenNet. This time the front legs were fixed in a symmetric position, and the four back legs were fed one (and only one) of the values of the angle input neurons. This led inevitably to a mating motion.

Random Move Any reasonable motion GenNet will do here, to get LIZZY to move randomly (e.g., a half-evolved GenNet for rotation, or straight-ahead motion).

8.9 DISCUSSION AND FUTURE IDEAS

As is obvious from the previous section, a lot remains to be done before LIZZY is fully functional. Not only are there GenNets that have not yet been evolved, LIZZY's GenNets have yet to be integrated into the LIZZY GenNet circuit. Initially, switching on and off GenNets will use a simple reset; that is, LIZZY's leg positions will be reset to those taken at the beginning of the evolution of the particular motion whose GenNet is being switched on. However, at a later stage, a form of limit cycle behavior may be possible. Instead of evolving a behavior from only one set of starting parameters, one evolves the same behavior over an ensemble of sets, where the fitness is the average performance over the ensemble. If the GenNet performance evolves into a form of limit cycle, the behavior may occur independently of the starting conditions. This is an interesting idea to pursue, but, depending on the motion definition, may require considerable computing power or lots of time for the evolution. If behavioral limit cycling is possible, then transitions between behaviors can be smooth. No resets will be necessary because no matter what the final state of the legs are when one GenNet switches off, the new behavior will be generated from that (starting) state by the next GenNet switching on.

A related question concerns whether GenNets can be evolved which have a form of "behavioral inertia"; that is, they continue to behave in the desired way even when the number of executed cycles goes beyond the number used to evolve the behavior. If behavioral limit cycling is not occurring, this behavioral inertia cannot be taken for granted.

If the ideas discussed in Section 8.8 prove to be successful, several steps will need to be taken in order to make genetic programming a generally applicable methodology to computer science in general and to artificial life in particular. There are at least two major directions of development for future research. One concerns the speed at which artificial evolution can be implemented. It would be nice if these techniques could be accelerated by putting them directly into hardware (e.g., VLSI accelerator chips for GenNet development). Such hardware versions of GenNet generators have been called *Darwin machines* [11–13]. They will be an essential component in future large-scale GP projects. One can imagine teams of (human) genetic programmers devoted to the task of evolving large numbers of GenNets to be later incorporated into increasingly complex artificial nervous systems. If several thousand GenNets need to be developed for example, then evolution speed will be essential. Traditional serial computers are hopelessly slow for this kind of work.

The second major direction for future GP research development is to put GenNets into real-world insect robots—to genetically program them. The alternative to this second direction would be to exclusively simulate increasingly sophisticated nervous systems. It is the first option that the author intends to take, because if one spends a lot of time and effort in creating elaborate simulations, at the end of the day one does not have anything as valuable as a functioning real-world robot for a comparable effort. However, simulation will remain important. One does not build an insect robot without being reasonably sure that it will work. This assurance usually comes from a successful simulation.

The definition of motion given in the LIZZY section is not real-world compatible. It will therefore be changed to one that is, namely moving one leg at a time, and adjusting the orientation of the body frame to each change. The GA can be used to decide the order of moving the legs. This motion definition is much closer to how one would get a real insect robot to move. One can then allow the robot to do its own real-world evolution. If these robots can be made cheaply, many of them could evolve simultaneously, reporting their fitnesses to a central computer that selects the chromosomes of the next generation and passes them back to the robots. This is the vision that the author has for genetically programmed insect robots for the near future.

Building insect robots like this would presumably mean storing GenNet weights in read-only memories (ROMs) and connecting them up to control real-world detectors and effectors such that the robot is given an artificial nervous system. However, there is a problem with evolving GenNet robots, and that concerns the time needed to measure their fitnesses in the real world. The great advantage of computer simulation is computer nanosecond speed. However, if one sends control (ROM) GenNet control signals to real insect robot effectors, they will perform at real-world mechanical speeds (i.e., in milliseconds at best and more likely in seconds). Hence measuring the fitnesses of a population will inevitably take time. There are several approaches one can take to this problem. One approach is to simulate the dynamics of the robots as accurately as possible, to get a ballpark estimate of the fitness, and then use the resulting GenNet weights as starting values in a real-world evolution.

Either one robot could be used sequentially many times, where the results of each chromosome's (dynamical) "expression" would be stored for each generation in order to be able to select the next generation, or, many such robots could evolve in parallel, as mentioned before. As microtechnology is gradually replaced by nanotechnology, the size and cost of these robots may decrease dramatically. Nanorobots ("nanots") could be made in huge numbers and evolved in parallel.

There are other ideas which need to be considered in the future of genetic programming. The artificial nervous system discussed in Section 8.8 was entirely "instinctual"; that is, it was incapable of learning. The LIZZY

GenNet circuit of Figure 8.15 is prewired. There is no adaptation based on LIZZY's experience. Nervous systems are more efficient and useful to the bodies they serve if they are capable of improving their performance on the basis of previous experience. Future GenNets capable of learning will be essential. This is an important short-term research goal of GP.

Another idea that needs to be explored is the possibility of getting the GA (rather than a human genetic programmer) to combine GenNets to build artificial nervous systems. This raises the question as to which GenNets should be combined and how. It is difficult enough to evolve individual GenNets, so evolving artificial nervous systems would be orders of magnitude more difficult and more computationally expensive. However, if it were done incrementally, some progress in this direction might be possible. Modification of already successful artificial nervous systems may lead to more successful variants and thus be amenable to GA treatment.

Another aspect of GP that needs serious consideration concerns the concept of evolvability (i.e., the capacity of a system to evolve rapidly enough to be interesting). From experience, the author knows that if the choice of representation of the dynamics of the system or the choice of the fitness function is inappropriate, evolution can be slow to infinitesimal. For example, if a tiny change in the starting conditions causes major changes in the fitness result, then the "fitness landscape" (in the parameter space of solutions) may be "spiky" instead of just "hilly," making the GA search virtually random and thus very slow. This may not be such a problem if the parameter space is "solution-dense", but if not, then isolated solution spikes may be very hard to find. Criteria for evolvability is a research topic worthy of the GA community. GenNet chromosomes can be very long (thousands of bits), especially when artificial nervous systems are being built, because they may have to code for large numbers of parameter values. The dynamic behavior resulting from many time dependent parameters may be very sensitive to changes in the initial values of these parameters or to the GenNet weights.

One final idea concerns the possibility that elaboration of the middle "logic" layer (i.e., between the detectors and the effectors) in artificial nervous systems may become sophisticated enough to be capable of handling "symbolic reasoning." If one looks at the development of biological life on earth, one sees that instinctual behavior came first, and only later did symbol manipulation evolve. The nervous systems capable of manipulating symbols were essentially superstructures built on an instinctual behavioral base. It is not too difficult to imagine that more elaborate GenNet circuits capable of handling sequences of production rules (of the form "If A&B&C $\rightarrow$ D") may form an elementary basis for symbolic representation and manipulation. If logic GenNets can be used in such a way, we may see the beginnings of a grand synthesis between traditional symbolic artificial intelligence, genetic algorithms, neural networks, and artificial life. Many "lifers" (artificial life researchers) hope that their bottom-up behavioral approach and the top-down disembodied symbolic approach of AI researchers will one day meet somewhere in the middle.

REFERENCES

[1] M. Rudnick and D. Hammerstrom, "An Interconnect Structure for Wafer Scale Neurocomputers," in D. Touretzky, G. Hinton, and T. Sejnowski, Eds., *Proceedings of the 1988 Connectionist Models Summer School,* Morgan Kaufmann, San Mateo, Calif., 1989, pp. 498–512.

[2] M. A. Reed, "Quantum Semiconductor Devices," in F. L. Carter, R. E. Siatkowski, and H. Wohltjen, Eds. *Molecular Electronic Devices,* North Holland, Amsterdam, 1988.

[3] K. E. Drexler, *Engines of Creation: The Coming Era of Nanotechnology,* Doubleday, New York, 1986.

[4] C. Schneiker, "*Nano Technology with Feynman Machines: Scanning, Tunneling, Engineering, and Artificial Life*", in C.G. Langton, Ed., *Artificial Life,* Addison-Wesley, Reading, Mass., 1989, pp. 443–500.

[5] R. P. Feynman, Quantum mechanical computers, *Optics News,* 11–20, (February 1985).

[6] D. E. Goldberg, *Genetic Algorithms in Search, Optimization, and Machine Learning,* Addison-Wesley, Reading, Mass., 1989.

[7] D. E. Rumelhart and J. L. McClelland, *Parallel Distributed Processing,* Vols. 1 and 2, MIT Press, Cambridge, Mass., 1986.

[8] C. G. Langton, Ed., "*Artificial Life*", Addison-Wesley, Reading, Mass., 1989.

[9] M. Caudil and C. Butler, eds., *Proceedings of the 1st International Conference on Neural Networks,* IEEE Catalog no. 87 TH0191-7, SOS Printing, San Diego, 1987.

[10] J. H. Holland, *Adaptation in Natural and Artificial Systems,* Univ. of Michigan Press, Ann Arbor, Mich., 1975.

[11] H. de Garis, "Genetic Programming: Modular Neural Evolution for Darwin Machines," *Proceedings International Joint Conference on Neural Networks,* January 1990, Washington D.C., vol. 1, pp. 194–197, Maureen Caudil, ed., Lawrence Erlbaum, Hillsdale, N.J., 1990.

[12] H. de Garis, "Genetic Programming: Building Nanobrains with Genetically Programmed Neural Network Modules," *Proceedings International Joint Conference on Neural Networks,* Vol. 3, June 1990, San Diego, pp. 511–516.

[13] H. de Garis, "Genetic Programming: Building Artificial Nervous Systems Using Genetically Programmed Neural Network Modules," *Proceedings 7th International Conference on Machine Learning,* B. W. Porter and R. J. Mooney, eds. Morgan Kaufmann, San Mateo, Calif., 1990, pp. 132–139.

CHAPTER 9

Neural Networks on Parallel Computers

**HYUNSOO YOON
JONG H. NANG AND
SEUNG R. MAENG**

9.1 INTRODUCTION

Recently, extensive research efforts are being devoted to the theory, implementation, and application of neural networks, as the computational model based on neural networks are conceived as a promising alternative solution to complex AI problem such as pattern recognition, vision, and speech recognition [1]. However, the long computational time, required in most real-world application problems, has been a critical problem in neural network researches.

To resolve it, there have been mainly two approaches to the implementation of neural networks. One is *direct implementation,* also called a *special-purpose neurocomputer* [2], in which there is a physical processing element for each neuron in a neural network. Even though it can potentially provide a very high performance, it can only support a specific neural network model because it is fixed in hardware. The other is *virtual implementation,* also called a *general-purpose neurocomputer,* in which a processing element takes charge of multiple neurons and simulates them in a time-sharing fashion. As it can generally support several neural network models via programming, it is more widely used than the former approach [3].

The general-purpose neurocomputer can be divided into three approaches according to the underlying hardware architecture: *coprocessor* style, *SIMD* style, and *MIMD* style. The coprocessor style, the simplest approach, is to

---

Neural and Intelligent Systems Integration, By Branko Souček and the IRIS Group.
ISBN 0-471-53676-8 ©1991 John Wiley & Sons, Inc.

use the floating-point boards to accelerate the *sum-of-product computations* in neural network simulations. Typically they plug into the back plane of the host machine, such as an IBM PC, or workstation together with a large memory for implementing the large neurons and interconnections. It, however, has a limitation on the speed and capacity even though it uses a specialized architecture.

Since neural network operations are inherently parallel, a more powerful and natural approach is to use the parallel processing technique. Because a neurocomputer based on a parallel computer usually provides acceptable speed and flexibility without great programming effort, it may be a reasonable compromise between the computational power of direct implementations and flexibility and ease of programming of a coprocessor-style neurocomputer. Implementing a neural network on a parallel computer, however, is not a trivial problem. Major issues such as parallelizing the neural network learning or recalling algorithm, mapping the parallelized algorithm onto the parallel architecture, and interprocessor communication or synchronization need to be investigated.

The parallelizing methods used for neural network simulation can be classified as model-specific and general. Model-specific parallelizing methods are only for a specific neural network model, such as back propagation or the Kohonen feature map. Since they are optimized for the specific network structure and learning algorithm, they can produce maximum performance for the underlying model. However, they cannot be applied to other models. On the other hand, the general parallelizing methods, which do not assume the structure and learning algorithm of a particular neural network, can be applied to any neural network model.

In this chapter we survey the main research efforts on the parallelizing and virtual implementation methods used for neural network computation. In Section 9.2, we introduce the basic neural network model and learning algorithm. The main issues of parallelizing neural networks and a classification of the various parallelization methods are presented in Section 9.3. In Section 9.4, the implementation methods such as the ones based on coprocessor board, systolic array, SIMD, and MIMD are studied separately in each subsection. A parallel back propagation algorithm for a multilayered neural network on distributed-memory multiprocessor is presented in Section 9.5 as a case study of parallelizing method. A summary and future research areas for developing a general-purpose parallel neurocomputer are given in Section 9.6.

9.2 OVERVIEW OF BASIC NEURAL NETWORK MODELS

An artificial neural network, or simply a neural network, is a computational paradigm to mimic the computations of the brain. Individual neural networks are specified by a few key properties: the *neuron characteristics,* the output

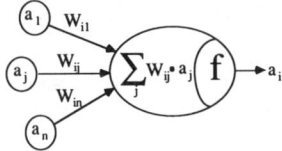

Figure 9.1 *Basic structure of a neuron.*

function of the neuron according to the input; the *network topology,* the interconnection pattern of the neurons; the *training/learning phase,* which establishes the values of the weights [4]. In this section, we briefly describe the basic structure, learning algorithms, and main computations in the neural network simulation. More detailed descriptions can be found in [4,5].

9.2.1 Basic Structure of Neural Networks

An artificial neural network model typically consists of many simple processing elements, called *neurons,* that interact using weighted connections. Each unit is associated with a *state* or an *activation level,* which is determined by the input received from other neurons in the network. Figure 9.1 shows the basic structure of a neuron, in which the activation level of a neuron is usually determined by applying a function f, called an *activation function,* to the weighted sum of received inputs (called *net input*). It can be a linear, sigmoid, or hard-limiter function [4].

In an artificial neural network, neurons are generally configured in regular and highly interconnected topologies. In the Hopfield model, for instance, a single layer of neurons forms the topology. The output from each neuron feeds back to all of its neighbors. In models like the Boltzmann machine and back propagation, the network consists of one or more layers between the input and output neurons. In models like the self-organizing map, the network connects a vector of input neurons to a two-dimensional grid of output neurons.

9.2.2 Learning Algorithms

A neural network is learned so that the application of a set of inputs produces the desired (or at least consistent) set of outputs. Learning is accomplished by sequentially applying input patterns while adjusting network weights according to a predetermined procedure. During training, the network weights gradually converge to values such that each input pattern produces the desired output pattern.

There are mainly two classes of learning algorithms; *unsupervised learning* and *supervised learning.* The supervised learning algorithm modifies the weights to produce the desired output pattern, and the unsupervised learning modifies the weights to produce output patterns that are consistent. The

TABLE 9.1 The Neural Network Model and Characteristics [2]

Neural Network Model	Primary Applications	Strengths	Weakness
Hopfield model	Retrieval of data/images from fragments	Implementable on large scale	Weights must be set
Perceptron	Typed character recognition	Oldest neural network	Cannot recognize complex patterns; sensitive to changes
Delta rule	Pattern recognition	Simple network, more general than the perceptron	Cannot recognize complex patterns
Back-propagation model	Wide range; speech synthesis to loan application scoring	Most popular network, works well & simple to learn	Requires supervised training with abundant example
Boltzmann machine	Pattern recognition (e.g., radar, sonar)	Simple network, uses noise function for global minima	Requires long training time
Self-organizing map	Maps one geometrical region onto another	Outperforms many algorithmic techniques	Requires extensive training time

back propagation algorithm, Hopfield network, and Boltzmann machine are typical examples of the supervised learning algorithm. A major example of an unsupervised learning algorithm is the Kohonen self-organizing feature map. Table 9.1 summarizes the main neural network models and their characteristics.

9.3 PARALLELIZING METHODS OF NEURAL NETWORK COMPUTATIONS

The neural network model is inherently parallel; that is, each neuron can compute its activation level independently if the states of other neurons are available. And the SIMD- and MIMD-style parallel machines can offer the tremendous speed improvements that are needed to neural network computation. However, these machines will not be able to exploit the parallelism available in neural network computation unless new parallel algorithms are developed that are well suited to parallel computer environment. The *bridge* between the powerful parallel computers and the parallelisms available in neural network models is a parallelizing method, a topic we present in this section.

Broadly speaking, there are two kinds of parallelisms obtained when a neural network is executed by multiple processors. A *spatial parallelism* (also called a network partitioning method [6]) is obtained when the whole network is divided into a set of subnetworks and each subnetwork is simulated independently by each processor. A *training parallelism* (also called a data partitioning method [6]) is obtained when a set of training input/output pairs are distributed to each processor. In this scheme, each processor trains the neural network using a subset of training pairs, and they are combined to produce the global one. Since the parallelizing methods based on training parallelism may not converge, and they are less general than the methods based on spatial parallelism, spatial parallelism is the main source of parallel implementation.

Let us discuss the issues encountered when parallelizing the neural network computation with spatial parallelism. In the virtual implementation with spatial parallelism, several neurons are mapped onto a processing element that simulates the function of neurons in a time-sharing fashion. The number of neurons mapped onto a processing element could be the whole network or a subset of it. If the whole network is mapped onto one processing element (i.e., no spatial parallelism), the function of each neuron is sequentially simulated by the single processing element one by one. In this extreme mapping scheme, a pipeline techniques within a processor can be applied.

On the other hand, if a subset of neural network is mapped onto a processing element (i.e., using spatial parallelism in neural network computation), several problems, such as how to divide the neural network, how to map the subnet onto the processing element, and how to parallelize the learning or recalling algorithm, should be resolved. If the topology and learning algorithm of a neural network are given and it has a regular structure, it is relatively easy to parallelize the neural network computation. Some nice heuristic partitioning and mapping strategies are already developed for a specific neural network model such as back propagation and the Kohonen map. However, if the topology and learning algorithm are irregular and have a nondeterministic nature, it is difficult to parallelize them. Partitioning a general neural network is a kind of *min-cut problem,* which is known as NP-complete, and mapping the partitioned neural network onto a processor is a kind of *process scheduling or load balancing.* There is little published (and nice) partitioning and mapping that can be applied to an arbitrary neural network model yet.

In this section we review the published parallelizing methods for a specific model, and fundamental parallelizing and mapping schemes for an arbitrary neural network model.

9.3.1 Model-Specific Parallelizing Methods

On the Back-Propagation Model Since the back-propagation model [7] has been used in a wide range of applications [1], several parallelizing methods have been already developed for it. In these works, it is usually assumed that the structure of a neural network is fully connected and multilayered;

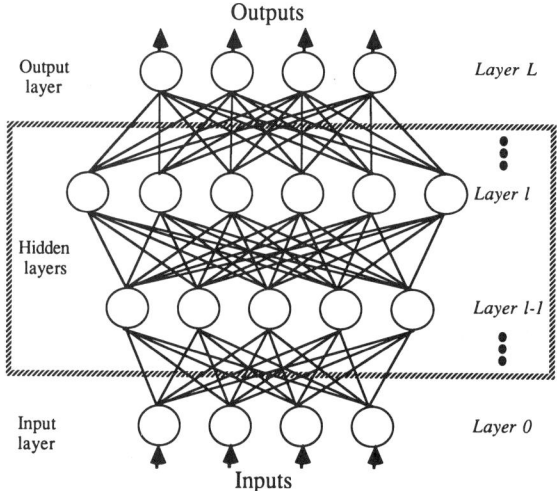

Figure 9.2 *Fully connected multilayered neural network.*

that is, it consists of multiple layers, and each neuron in a layer is connected to all neurons in the next layer as shown in Figure 9.2.

The parallel back propagation learning algorithms for a multilayered neural network model can be classified into the spatial parallel algorithms, training parallel algorithms, and systolic algorithms.

Spatial Parallel Algorithms In the spatial parallel scheme, the neurons in each layer are partitioned and mapped onto the processors as shown in Figure 9.3.

The well-known works on this scheme include [6,8–10]. In these works, a processor typically takes charge of a subset of neurons in each layer, and it performs the required computations for the mapped neurons. The data required in the computation for a mapped neuron are the activation values, the error values of the connected neurons, and the weight values of these connections. Since a fully connected neural network is assumed, all the activation or error values of neurons in the adjacent layer are required in the forward or backward phases of back propagation algorithm. The performance of simulation is heavily dependent on the data distribution strategy.

We now explain data distribution methods used in the spatial parallel algorithm for the back-propagation model. In the partitioning scheme used in [9], each processor keeps either the output or the input weight values of the mapped neurons. If a processor keeps only the input weight values, the forward pass can be executed easily with the communications for the activation value of neurons in the lower layer. However, to compute the error value of the mapped neuron in the backward pass, the processor should know the weight value of output connections. It causes an excessive in-

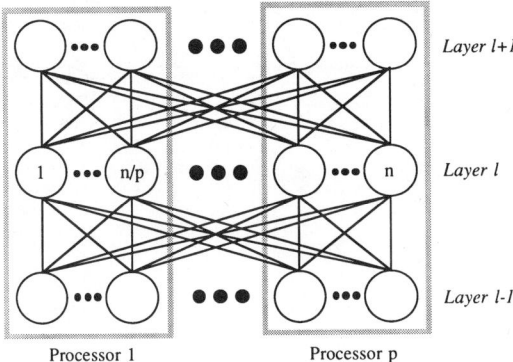

Figure 9.3 *Neural network partitioning used in spatial parallel algorithms.*

terprocessor communications ($O(n^2)$), where n is the number of neurons in a layer since they are stored in different processors.

One way to avoid this communication overhead is the *recomputation method* [10], in which each processor keeps both the input and output weight values and adjusts them independently with additional broadcasting of related data. Because of the recomputation overhead, the p-processor speedup is about $\frac{3}{4}p$ as shown in [10]. Although this method avoids the heavy communication overhead, it still doubles the storage and computations. (More complete discussion of the *recomputation method* used in parallel back propagation algorithm is presented in Section 9.5.)

Panda et al. have resolved this problem by using a special broadcasting hardware in a multiple-bus structure, named the layer bus controller (LBC) [8]. They assume that one processor is designated per neuron, and weight values are duplicated among these processors to provide full parallelism in the forward and backward computations. The consistencies of two values stored in two different processors are automatically maintained by the LBC by copying the weight values from one processor to the others during the computation for the lower layer.

Although the recomputation could be removed by using a special broadcasting hardware as in [8], it also doubles the storage for weight values. Another method is the *partial summing method* used in Warp [6], in which only output weight values for the mapped neurons are stored in a processor, and a processor computes and sends the partial sum of product for mapped neurons to a neighbor processor to produce the total sum of products. However, since the computations for mapped neurons are distributed over several processors, it is not a real spatial parallel algorithm.

Training Parallel Algorithms The back propagation algorithm is a supervised learning algorithm, which is a procedure to find a set of weights of a neural network according to training input/output patterns so that, given

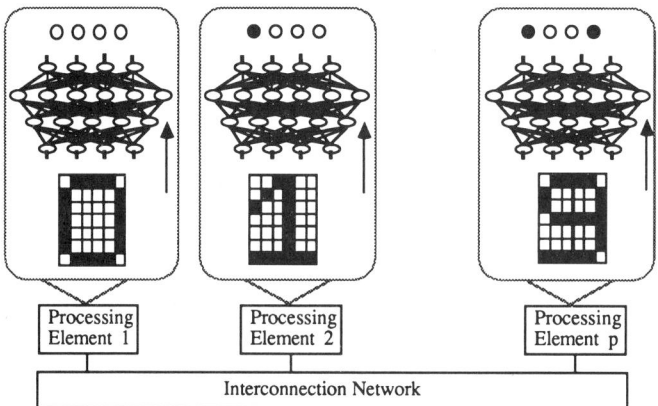

Figure 9.4 *Neural network parallel simulation using training parallelism.*

each input pattern, the output pattern produced by the network is sufficiently close to the desired output pattern. In the works based on the training parallelism [6,11,12], the training input/output patterns are divided into p sections, where p is the number of processors. Each processor conceptually contains a complete copy of the neural network and trains using a subset of the training patterns as shown in Figure 9.4.

In order for the individual processors to collectively work on a solution, they need to operate with the same weight values. Otherwise, each processor would work toward a solution for a peculiar subset of training patterns, not the global solution. This coherency requirement causes the training to become *epoch* training; that is, several training patterns are presented to the network before weight changes are made. After each processor computes the local weight changes, it sends them to a special processor that averages the all of local weight changes to make a global one. Then, the global weight changes are retransmitted to all processors so that they can make a new local weight change. The main difference in the researches based on the training parallelism is the way to communicate each other and the way to make global weight changes. They are heavily dependent on the underlying hardware architecture, for example, a shared-memory multiprocessor such as Encore Multimax [11], a distributed-memory multiprocessor such as Hypercube [12], or a linear array of processor such as Warp [6].

The main drawback of these approaches is that it may not converge sometimes because the weight changes for the training patterns are averaged. The communication bottleneck in the special processor, which averages and retransmits the global weight changes to make the global one, is another drawback.

Systolic Algorithms Since the computation required in neural networks including the back-propagation model is inherently regular and recursive

9.3 PARALLELIZING METHODS OF NEURAL NETWORK COMPUTATIONS

(actually, a matrix-vector multiplication), the systolic algorithms may be an attractive solution. The main characteristic of these approaches is that the computation for a neuron is distributed over the whole processing elements so that they form a pipelined array.

A systolic algorithm for a VLSI implementation of back-propagation model has been proposed in [13], in which one systolic ring array is assigned to each layer to maximize the parallelism. Another systolic algorithm for the back-propagation model has been proposed in [14]. It may be viewed as a method based on spatial parallelism. However, since the computation required for a neuron is distributed over several processors, it can be classified as a systolic one.

Another notable work on the systolic algorithm is [15]. There, a single ring-of-processors array is used to simulate the neurons at the same column in all layers.

Others A different approach to parallelize the back-propagation model is to assign the processing elements to both of the connections and neurons. Since the number of connections is very much larger than the number of neurons in a neural network, it potentially has a chance to produce good performance. In works based on this method, it is usually assumed that the underlying architecture is a massively parallel fine-grained SIMD machine such as CM (connection machine) as in [16,17], or AAP-2 as in [18].

In [16], one processor is assigned to each neuron, and two processors in CM are allocated for each weight. The processor for each neuron is immediately followed by all the processors for its output connections, and immediately preceded by all of its input connections. In their parallel algorithm, the CM-specific operations such as *copy-scan* and *plus-scan* are used. Copy-scan copies the value from the beginning or end of a segment of contiguous processors to all the other processors in the segment. Plus-scan is used to sum all the values in a segment of processors and leave the result in either the first or last processor of the segment. With these operations, the forward and backward computations are easily implemented on the CM. However, this scheme relies heavily on the sophisticated routing hardware of the CM [17].

Another parallelizing method based on the CM can be found in [17], in which a kind of back-propagation model called *recurrent back propagation* is implemented. In this work, two fundamentally different methods are examined: one is to assign processing elements to connections; the other is to assign processing elements to neurons. The former method is to extend the parallelizing method used in [16] in such a way that one of the output weights of a bias node feeds back to the input weight of it. This modification is required to meet the characteristic of the recurrent network. A method to allocate the processing element to neurons is also proposed. It is based on the Tomboulian method to map the graph to an arbitrary SIMD architecture, and its algorithm does not depend on the capabilities of the CM. Performance results show that the former method works well for a highly connected network, whereas Tomboulian's scheme fails for a dense graph.

Parallelizing the back-propagation model on the AAP-2 [18] is another work to assign a processing element to the connection. The architecture of AAP-2 is similar to the CM, and the parallelizing method is almost the same as the work in [16] except that a *table look-up method* is used to obtain the activation value of a neuron.

Cook and Gilbert [19] proposed a data structure and data partitioning scheme of implementing the back-propagation model on the Hypercube architecture. They assume that a matrix representing the connectivity of neural network is usually sparse, so that a linked list data structure can be used in the parallel simulation to reduce the storage and required computations. However, they did not mention the exact parallel algorithm on the Hypercube.

Another work presented in [20] analyzes the speedups of the true gradient method for the back-propagation model on the distributed-memory multiprocessors interconnected by the torus, mesh, and ring. In the true gradient method, since the errors are computed after all training patterns are shown, more parallelisms can be achieved. They partition the weight matrix into p submatrices, where p is the number of processors in the system, and each processor computes the partial sum of product and shifts it to its neighbor. It can be seen as a mixed parallelizing method of the partial summing and training parallelism.

Kohonen Map Model The self-organizing feature Map [21], or Kohonen map, has a vector of input neurons connected to a two-dimensional grid of output neurons. Output nodes are extensively interconnected with many local connections. Continuous input values are applied to a neural network, and it is trained with unsupervised learning. The learning algorithm requires, for each neuron, a definition of a neighborhood that slowly decreases in size with time. It is based on checking the most active neuron and updating its weights.

A parallel algorithm of the Kohonen map is desired because of the large cost of determining the *Euclidean distance* for each neuron in order to find the winning one, and the large number of iterations required to reach a stable state [22].

There have been some works [22,23] on the parallelizing Kohonen map. The work in [22] is a parallel algorithm based on the training parallelism, whereas the work in [23] is based on the spatial parallelism. The former work shows that if the number of training patterns assigned to a processor is small, it appears to have similar performance to the exact algorithm. However, because of the effect of training parallelism that averages the change of weight, the self-organization process occasionally fails, as in the back-propagation model.

In [23] the neurons in the input and output layers are divided equally among the processors, and each processor is responsible for updating the weights connecting the input neurons with the corresponding set of output neurons. The analytic speedup ratios of its parallel algorithm on the ring and mesh are obtained via simulation. However, they did not mention the exact algorithm to find the winning cell and the strategy to update the weights of connections on the multiprocessor system.

Boltzmann Machine The Boltzmann machine [24] is a connectionist architecture that employs a very simple learning procedure for the automatic assignment of connection weights. It is a network of neuronlike processing units, whose state (on or off) is determined by a stochastic function of the states of the connected units, weighted by relative connection strengths. All units in the Boltzmann machine can change their state in parallel. Although it may lead to an erroneously calculated energy difference (ΔE), it does not affect the asymptotic convergence properties of the *simulated annealing technique* because the probability that two (or more) neighboring units change their state simultaneously becomes very small [25].

There is a lot of research on the parallel simulated annealing algorithm because, though it takes an extremely long time to execute, it provides the way to get a near-optimal solution of a combinatorial optimization problem. An excellent survey can be found in [26].

Iazzetta et al. [27] have proposed a transputer implementation of the Boltzmann machine in which there are OCCAM language processes per neurons and connections. The process for a connection contains the weight and the activation values of two connected neurons. The additional processes are required to prevent deadlock conditions in their implementation. To parallelize the Boltzmann machine computations in their implementation, the processes which take charge of the subset of Boltzmann machine are allocated to the available processing elements. However, they did not mention how to allocate them.

9.3.2 General Parallelizing Methods

The parallelizing methods discussed in Section 9.3.1 are only for a specific neural network model. Since they are optimized for the given neural network model, it is hard to apply them to other models or when network structures, such as the number of layers and the connection topology in the model, are changed. To be a general-purpose parallel neurocomputer, a general parallelizing method, which is applicable to an arbitrary neural network model with reasonable performance, is required.

Although there are several proposals for a general-purpose parallel neurocomputer, they are only focused on the hardware structure and do not mention the general parallelizing methods used in their system. In this subsection, we briefly review some published parallelizing methods used in the parallel neural network simulation system.

Rochester Simulator The Rochester simulator [28] is a neural network development tool including a general simulator and a graphic interface developed on the Sun workstation. It is also implemented on the Butterfly multiprocessor, which consists of up to 256 nodes.

The general parallelizing method used in the Rochester simulator implemented on the Butterfly is a kind of spatial parallelism. The default partitioning scheme used is to assign the neurons to the processors sequentially such that each processor has the same number of neurons. Actually, a general simulator, called a *worker,* runs on a processor in the Butterfly. Each worker allocates and maintains an output array for the neurons mapped on the processor. To refer the activation value or weight of the connection, the output array is shared via parallel programming libraries in the BBN Butterfly.

This partitioning strategy may produce the poor simulation performance because it does not use the informations resident in the network structure such as the topology of the network. For example, if the heavily interconnected neurons are allocated to the different processors, it causes a heavy communication overhead.

Ghosh's Work J. Ghosh and K. Hwang have proposed a scheme [3] to map a large-scale neural network with selective interconnection on the message-passing multiprocessors, and have analyzed the communication bandwidth requirements. They assume that the neurons in a neural network can be distinguished by their connectivity; the *core group* which are relatively high interconnected among themselves, the *influence region* with a set of cores whose constituent cells have a substantial number of intercore connection, and the *remote region,* which is the rest of the cells.

Followings are the two heuristics used in [3] when partitioning and mapping the neural network model onto MIMD:

- *Partitioning Principle:* Prefer neurons belonging to the same core for inclusion in the same processor. If a core is larger than the size of local memory of the processor, then this policy tends to minimize the number of processors to which the neurons in a core are distributed. Otherwise, it tends to minimize the number of cores mapped onto the same processor.

- *Mapping Principle:* If processor k takes charge of the neurons in a core group, and the neurons in the influence region are mapped onto processor l, then the distance $d_{k,l}$ should be as small as possible.

Even though these heuristics can be used in partitioning and mapping the neural network, an exact algorithm applicable to an arbitrary neural network is still required.

9.4 IMPLEMENTATION OF NEURAL NETWORKS: NEUROCOMPUTERS

The neurocomputers refer to the all implementation methods designed to optimize the computation of artificial neural networks. The *general-purpose neurocomputer* is a generalized and programmable neurocomputer for emulating a range of neural network models, and the *special-purpose neurocomputer* is a specialized neural network hardware implementation dedicated to a specific neural network model and therefore potentially has a very high performance. To solve a real-word or real-time application problem, the special-purpose neurocomputer may be an attractive method. However, when developing a new neural network model or testing the new applications, the general-purpose neurocomuter is more attractive because of its flexibility. Furthermore, it can also be used in the real-word application if its underlying architecture is a parallel computer powerful enough to fulfil the computational requirement. Figure 9.5 presents the spectrum of neurocomputer architectures with respect to speed and flexibility.

In this section, we first present the classification and overview on the neurocomputers*, and briefly explain them focused on the main design philosophies, characteristics, and developing environments. For the sake of convenience, *neurocomputer* refers to the general-purpose neurocomputer in the rest of this chapter.

9.4.1 Classification

To fulfill the computational and space requirements of neural network computation, numerous commercial or experimental neurocomputers have been developed. The basic architecture of a neurocomputer is either a conventional microprocessor, such as the Motorola 68020 or Intel 80286, or a specialized processor architecture suited for neural network computation, or an existing parallel computer such as CM and a network of transputers. Some of them are equipped with environments such as a specification language, network compiler, neural network libraries for some well-known models, and user interface with graphic facilities to show the network states. On the

---

* In neurocomputer research, a lot of existing commercial microprocessors as well as parallel computers are used as their implementation hardware. Although their primary target is not a neural network simulation, we also call them neurocomputers in this chapter because they can perform the required neural network computation efficiently.

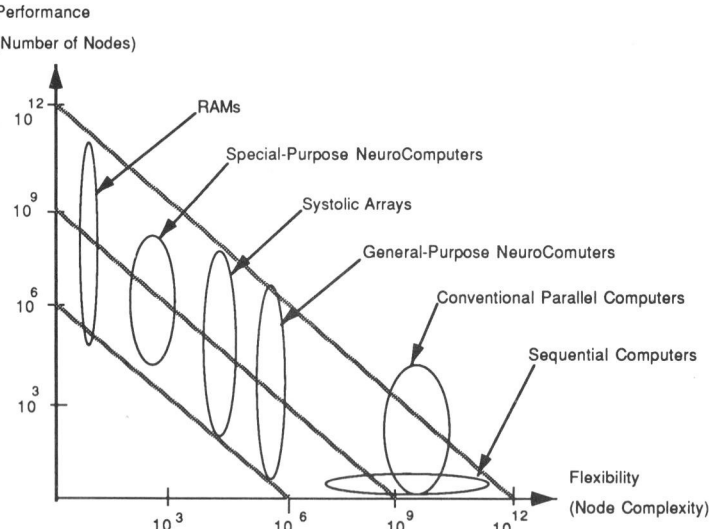

Figure 9.5 *Spectrum of neurocomputer architecture [29].*

other hand, some of them are designed for a specific neural network model without the neural network developing environments.

As to the underlying hardware architecture on which neurocomputers are based, they can be classified as coprocessor-style, SIMD-style, and MIMD-style neurocomputers. Table 9.2 presents the main characteristics of commercial or experimental neurocomputers.

9.4.2 Coprocessor Neurocomputers

The coprocessor-style neurocomputers may be the simplest and most cost-effective, in which floating-point boards to accelerate the *sum-of-product computations* in neural networks are used. Either the conventional digital signal processing (DSP) chip or a specialized microprocessor chip is used as the basic architecture. Typically they plug into the back plane of the host machine, such as an IBM PC or workstation, together with a large memory for implementing the large neurons and interconnections.

We briefly review the well-known coprocessor-style neurocomputers: the ANZA developed by HNC, and the DELTA-SIGMA. The ANZA is a general-purpose neurocomputing coprocessor installed on an IBM PC/AT or compatible computer to support any neural network algorithm. The boards are based on a Motorola MC68020 microprocessor plus a Motorola MC68881 floating-point coprocessor [31]. It was designed to facilitate neurocomputing integration into existing software environments. Interaction between user software and the ANZA is accomplished through a set of callable subroutines that constitute the user interface subroutine library (UISL). The UISL

9.4 IMPLEMENTATION OF NEURAL NETWORKS: NEUROCOMPUTERS

TABLE 9.2 The Main Characteristics of Neurocomputers

Neurocomputer (company)	System Architecture	Environment (supporting models)	Status	Ref.
ANZA-PLUS (HNC)	AT coprocessor board (with 68020+68881)	UISL, AXON (General)	Commercial	30
DELTA-SIGMA (SAIC)	Workstation with DELTA FPP Coprocessor board	ANSim ANSpec (an OOP lang.) (general)	Commercial	31 32 33
Connection Machine	10-D Hypercube interconnected 64K PEs	(back propagation)	Commercial	17 16
AAP-2 (NTT LSI Lab.)	Torus interconnected 64K PEs	AAPL (Back propagation)	Prototype	18
Multimax (MMSS)	Multimax with 18 PEs	(Back propagation)	Experimental	11
Mark III (TRW)	VME bus interconnected 15 PEs (68020+68881)	ANSE (General)	Commercial	34
NEP (IBM Sci. Center)	NEP bus interconnected NEP board (TMS320)	CONE including GNL, NETSPEC, IXP (general)	Experimental	35 36
NNETS (NASA)	A network of transputers	Frame or script (general)	Experimental	37
GRIFFIN (TI)	3-D Array of NETSIM cards	Turbo C (general)	Experimental	38 39
Warp (CMU)	Ring interconnected 10 PEs	(Back propagation, Kohonen style)	Experimental	6 22
Rochester Simulator	BBN butterfly	Sunview (general)	Experimental	28
Panda's Work (USC)	Multiple bus (for VLSI)	(Back propagation)	Under development	8
TLA (Sharp)	TLA with 17 transputers	(Hopfield, back propagation)	Under Development	40
NERV (Univ. of Heidelberg)	VME bus interconnected 320 PEs (68020)	Hypercard of Macintosh (general)	Under development	22
Neural RISC (UCL)	Ring interconnected VLSI RISC PEs	NC (general)	Under development	29
Sandy/8 (Fujitsu)	Ring interconnected 256 TMS320C30	(Back propagation)	Under development	15
GCN (Sony)	Mesh interconnected 80860s with 4M	(Back propagation)	Will make a single chip	41
Ariel (TI)	Hierarchical bus (TMS320C30+68030)	(General)	Under design	42
PANSi (KAIST)	Ring interconnected 32 transputers	(Back propagation)	Toy prototype	9 10

provides access to all of the ANZA's network data structure and functions. This allows the user to quickly and easily add neural network capability to new or existing software. The basic library contains five classic neural network algorithms in a parameterized specification that can be configured for a specific user application. The user can only change the number of neurons, their initial state, weight value, and learning rate.

The SAIC SIGMA neurocomputer workstation is a PC/AT plus a DELTA floating-point processor board, together with a neural programming system comprising ANSim and ANSpec. The DELTA floating-point processor is a high-speed floating-point engine optimized for the neural network computation. The processor has a pipelined Harvard floating-point architecture with one program memory and two data memories. The ANSim library is menu driven and contains 13 well-known user-configurable neural network models. The ANSpec is an object-oriented neural network specification language suitable for implementation of complex neural networks, beyond ANSim. Recently, a new SAIC neurocomputer, called DELTA-II, has been developed, which is also a PC/AT coprocessor board with a specialized RISC processor and large memory capacity [33].

The coprocessor-style neurocomputers, however, have a limitation on speed and capacity even though a specialized processor architecture is used.

9.4.3 SIMD Neurocomputers

The SIMD-style neurocomputers can be characterized by their simple processing elements and regular communication patterns as well as the single instruction stream. Let us present the implementation of neural network computations on SIMD parallel machines CM and AAP-2.

The AAP-2, developed by NTT, is a massively parallel cellular array processor based on very large scale integration (VLSI) technology (Fig. 9.6). It consists of 64K processing elements. All processors are controlled by a single instruction stream broadcast from an array control unit, and are connected to their nearest-neighbor processing elements (torus structure). The AAP-2 has useful operations such as ripple summation, broadcast to all processors, broadcast along the processing element (PE) row (column), and global OR. A back propagation algorithm on the AAP-2 using these useful operations is described in [18].

Although the design goal of the CM was not for neural network computations, its massively parallel characteristic permits high-performance neural network simulations. There are two noticeable works on the CM: [16] and [17]. In these works, the CM-specific operations such as copy-scan and plus-scan are used to implement the back propagation algorithm.

Since the computations in neural network simulation are regular and highly parallel, the SIMD architecture may be the most suitable implementation method. However, if the computation and communication patterns in a neural network simulation are not well defined or irregular (e.g., when devel-

9.4 IMPLEMENTATION OF NEURAL NETWORKS: NEUROCOMPUTERS

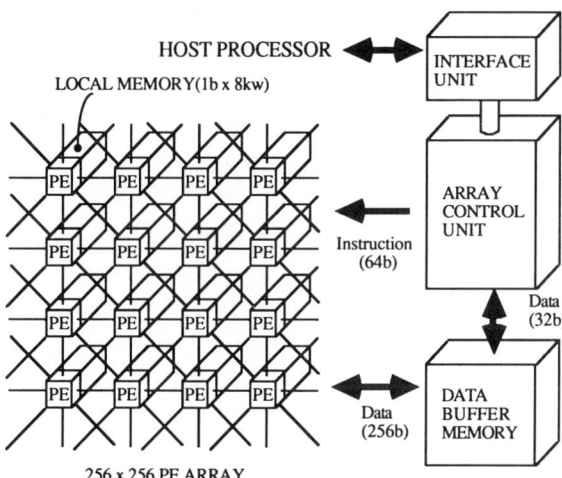

Figure 9.6 *System configuration of AAP-2 [18].*

oping a new neural network model), to extract a regular computation pattern and to devise a SIMD algorithm from the arbitrary neural network is not a trivial problem. Therefore, although it can be used as a simulation tool for a well-defined and regular neural network model, it is hard to use as a general-purpose parallel neurocomputer.

9.4.4 MIMD Neurocomputers

The MIMD-style neurocomputer is the most powerful and widely used implementation vehicle because of its scalability and generality. It can be divided into a shared-memory multiprocessor and a distributed-memory multiprocessor. However, since partitioning and mapping of neural network computations onto a multiprocessor are difficult tasks, almost all MIMD-style neurocomputer can support a limited number of neural network models.

Multimax A parallel back propagation algorithm on the 18-PE encore multimax, a shared-memory multiprocessor, is published in [11]. In their system, a copy of the weight matrix is stored in the shared memory. Each processor computes the error on the individual training pattern and then updates the weights in controlled fashion. Figure 9.7 represents the parallelization approach and procedure in the multimax.

Mark III The TRW neurocomputer family [34] includes the Mark II software simulator, the Mark III, which is a parallel processor array, and the Mark IV, which is a single high-speed, pipelined processor. All share the

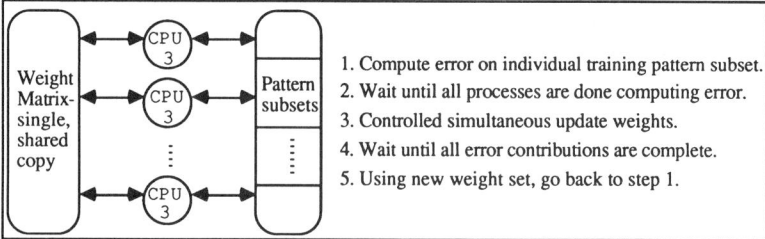

Figure 9.7 *A Parallel back propagation algorithm on multimax [11].*

common artificial neural system environment (ANSE) user programming environment.

The Mark III consists of up to 15 physical processors, each built from a Motorola MC68020 and a MC68881 floating-point coprocessor, all connected to a common VME bus (Fig. 9.8). The data for a neural network to be processed are distributed across the local memories of the PEs. This minimizes the bus traffic because the information required in computations for the mapped subnetwork, such as activation values and connection weights, is in the PE's local memory.

Network Emulation Processor The network emulation processor (NEP) [35] is an IBM-developed parallel processor. Its execution model is based on the flow-of-activation networks (FAN), which is an abstraction of neural networks. A FAN net processes information through the collective parallel computation of a large number of simple processors (*nodes*) communicating with each other through directed point-to-point channels (*links*).

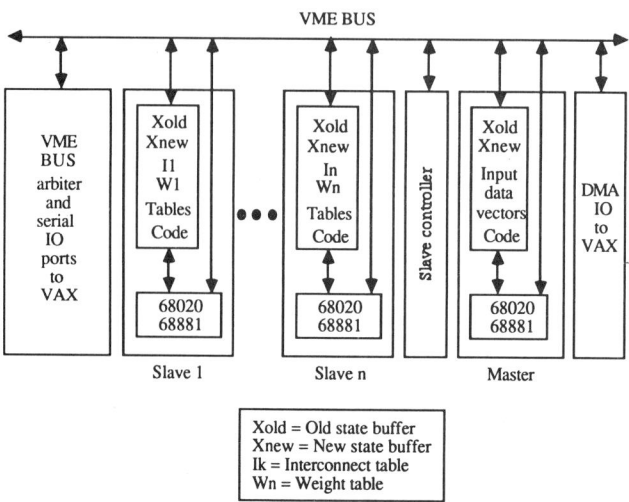

Figure 9.8 *The architecture of Mark III [31].*

9.4 IMPLEMENTATION OF NEURAL NETWORKS: NEUROCOMPUTERS

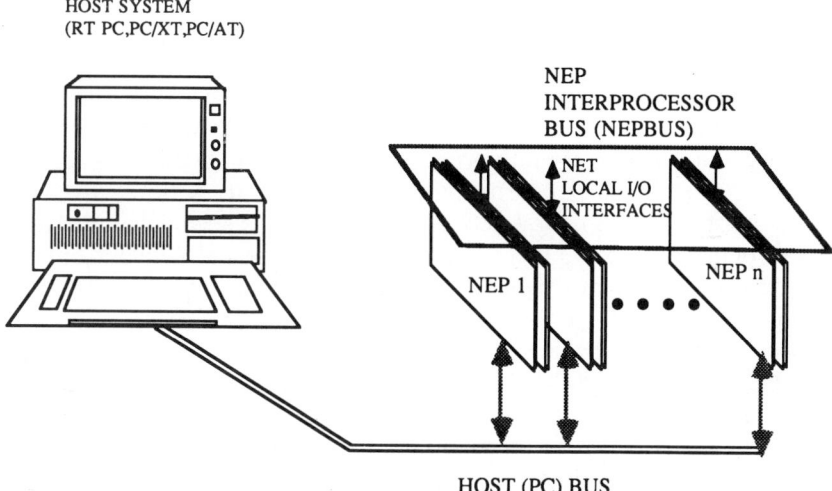

Figure 9.9 *System architecture of NEP [35].*

They developed a software environment together with a hardware implementation. The specification of FAN is programmed using a generalized network language (GNL). The GNL compiler uses the source statement, together with application-specific compiler extensions, to produce a generic intermediate network description called NETSPEC. The assembler takes a NETSPEC description of a network as an input and produces a data structure that describes the network in a format that is consistent with the item description routines coded for that emulator's engine(s). This approach isolates the process of high-level network compilation from its low-level implementation, permitting execution of a network on a variety of engines. This description can be verified using a software called interactive execution program (IXP). Taken together, the GNL compiler, the IXP network emulation program, and the (optional) NEP parallel emulation engine constitute the computational network environment (CONE).

The local processing characteristics of FANs allows a large network to be partitioned, with each subnetwork being implemented on a distinct hardware processor. Such partitioning allows the implementation of very large nets. Signal flow connections across the boundaries of partitions must be handled by an appropriate interprocessor communication mechanism. In the NEP system, a high-speed interprocessor communication bus called NETBUS is used to meet the requirement, and TMS320-based processor boards are used to accelerate the floating-point computations. It also has a high-speed "local I/O" (input/output) facility to preserve interprocessor communications bandwidth. Figure 9.9 presents the hardware configuration of NEP system.

Because the NEP's control store is programmable, the processing performed by nodes and links can easily be altered. Also, because the executable description of a network is contained in programmable memory, FAN nets of arbitrary topology may be emulated.

NNETS Since the transputer provides high-speed interprocessor communication facilities with on-chip four full-duplex links, it is well suited to construct distributed-memory multiprocessor simulating neural networks. Thus there have been a lot of neurocomputers based on it. A transputer-based processor array for the neural network simulation has been developed at the NASA Johnson Space Center in Houston, called neural network environment transputer system (NNETS) [37]. It consists of forty 32-bit/10-MIPS transputers.

The user can communicate with NNETS through an object description language. If the user finds this interface too constraining, a C language procedures library is available to design a unique neural network algorithm. The object description language utilizes a structure known as frame or script.

GRIFFIN In the GRIFFIN system [39], developed through a collaboration of Texas Instruments (UK) and Cambridge University, the neurons in a neural network are grouped into *nets* in which the neurons are fully interconnected. A net containing n inputs and m outputs contains $n \times m$ weights but needs only $n + m$ external connections. The basis of the system under consideration in the GRIFFIN is shown in Figure 9.10. It consists of a number of fully interconnected nets whose inputs and outputs may be joined to arbitrary places within the system or to the external inputs/outputs. The logical interconnection map is fully defined in software and may be easily modified to simulate different structures.

The physical structure of GRIFFIN is a distributed array of autonomous neural network simulators (NETSIM), as shown in Figure 9.11. Each NETSIM card consists of a microprocessor, a solution integrated circuit

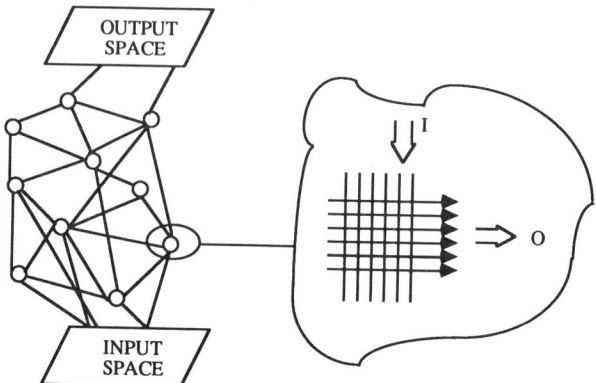

Figure 9.10 *The organization of nets in the GRIFFIN [39].*

9.4 IMPLEMENTATION OF NEURAL NETWORKS: NEUROCOMPUTERS

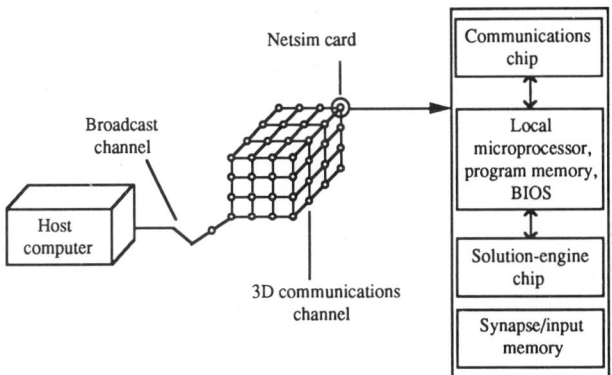

Figure 9.11 The system architecture of GRIFFIN [39].

(IC), which is a specialized vector processor to implement the neural network function at high speed, and a communication IC to allow large numbers of NETSIMs to be connected together to form the parallel processing GRIFFIN.

The solution engine has the instruction for data fetching and looping options as well as the multiply-and-sum. It is independent from the local microprocessor, so that the microprocessor can compute other elements of the simulation, including the nonlinear function and the simulation of interconnection independently. The communication IC is a three-dimensional serial message-passing device that passes the output of a neuron to which the neuron is logically connected.

Warp at CMU The Warp, developed at CMU, is a programmable linear array of processors, called *cells*, as shown in Figure 9.12a. Each cell is a 32-bit processor capable of very high bandwidth communication (80 Mbyte/s) with its left and right neighbors in the linear array. The two data streams (X and Y) can flow in the same or in opposing directions. Two cluster processors, an interface unit, and a support processor situated between the Warp array and the host act in concert to provide efficient I/O between the processor array and the large cluster memory.

The neural network models implemented on the Warp are the multilayered back-propagation model and the Kohonen feature map model, in which the training parallelism (or data partitioning algorithm) is used. Figure 9.12b shows the information flows of Warp when the training parallelism is used. The first nine cells perform the forward and backward pass of a number of I/O pattern pairs, and the tenth cell is reserved for computing the weight updates. Weights are no longer partitioned among the cell's local memories; instead they are pumped through the array from the large cluster memory. Since cells store only the activation levels of the units, it is possible to simulate much larger networks.

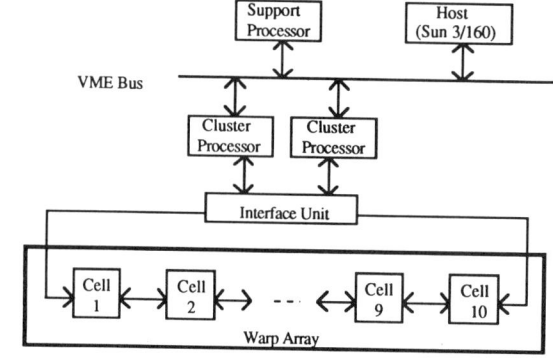

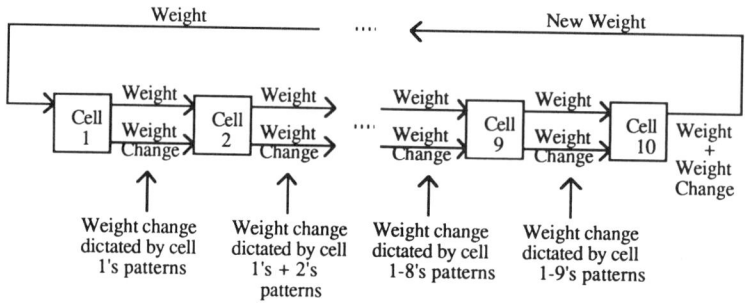

Figure 9.12 The back propagation algorithm on Warp: (a) the Warp architecture; (b) information flow in the data partitioning algorithm [6].

Multiple-Bus Architecture at USC A multiple-bus architecture [8] to implement the back-propagation neural network model has been proposed at the University of Southern California. In this system, one synaptic neuron element (SNE) is assigned to a neuron, and the connection weights are duplicated among these SNEs to exploit full parallelism in both passes. The group of SNEs in adjacent layers is connected to a common bus controlled by an LBC. While the backward pass operation between layer j and layer $j - 1$ SNEs is carried out, LBC(j) copies the weights from the row memories of layer j SNEs to the column memories of layer $j + 1$ SNEs. This ensures that the modified weights are used during the forward pass of the next iteration. Figure 9.13 shows the mapping of a multilayered neural network onto a multiple-bus network.

TLA at Sharp In the toroidal lattice architecture (TLA) neurocomputer [40], there is a virtual processor for each connection as well as each neuron.

9.4 IMPLEMENTATION OF NEURAL NETWORKS: NEUROCOMPUTERS

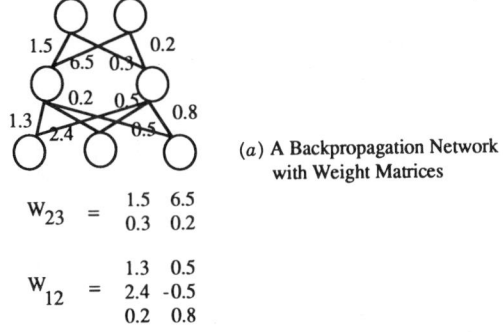

(a) A Backpropagation Network with Weight Matrices

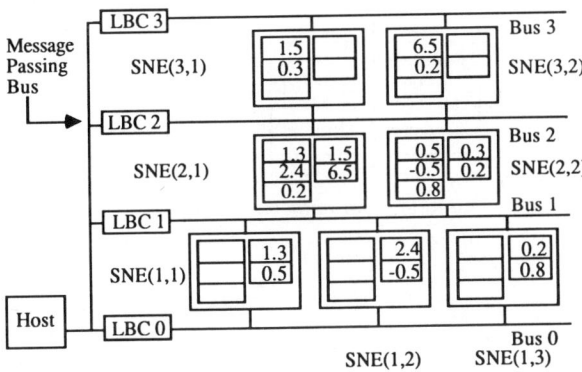

(b) Multiple-bus Network

Figure 9.13 Multiple-bus architecture and mapping scheme: (a) back-propagation network with weight matrices; (b) multiple-bus network [8].

They are partitioned and mapped onto the physical processor. To balance the load of each processor, a permutation of virtual processor matrix is performed before partitioning of neural network. Each submatrix is then assigned to the PE of a transputer-based neurocomputer, which itself has a toroidal lattice architecture, as shown in Figure 9.14.

NERV at University of Heidelberg The NERV neurocomputer [22] is a multiprocessor system based on a VME bus, in which the VME bus has been extended by a *broadcast feature* and a *global max finder logic*. Up to 320 MC68020 processors are used in a single VME crate together with a Macintosh II as a host computer, as shown in Figure 9.15. The host computer offers a friendly graphical user interface.

Each multiprocessor board contains 16 processing elements, and the local memory of each processing element stores the state of all neurons and the

258 NEURAL NETWORKS ON PARALLEL COMPUTERS

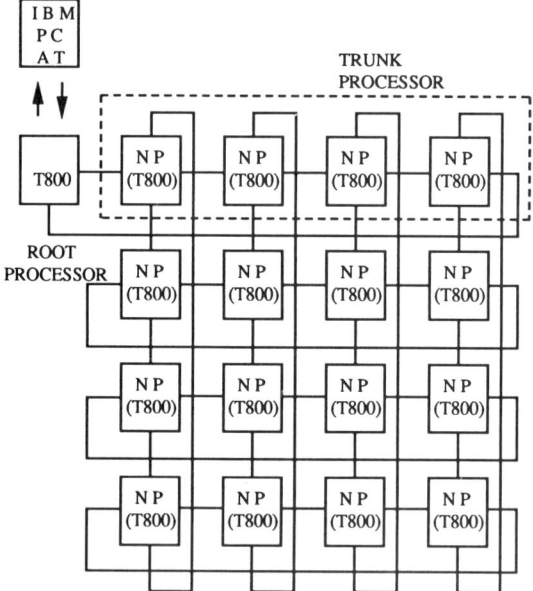

Figure 9.14 *TLA neurocomputer implementation using 17 transputers [40].*

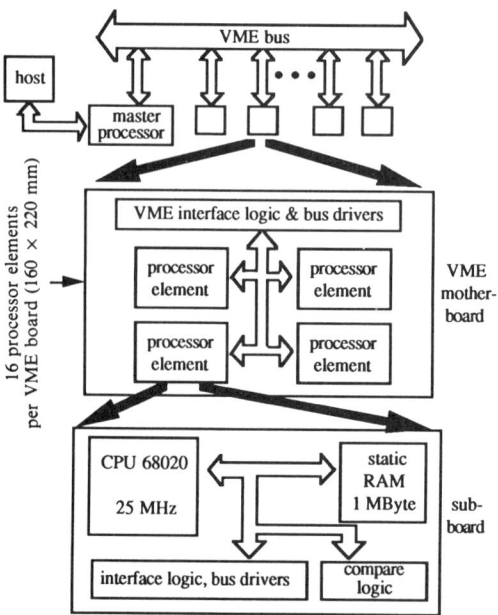

Figure 9.15 *Block diagram of the NERV system architecture [32].*

related connection weights for the mapped neurons. Because each processing element has a copy of the state of all neurons, state changes have to be synchronized in all local memories. This is done via broadcast transfers using specialized VME bus control lines, in which a write transfer into a certain part of the address space causes a simultaneous write operation to all local memories. The global max finder logic that is required for the simulation of asynchronous update models is also implemented on the VME bus. Its operation is similar to the priority arbitration logic of the FutureBus system or the OR network of the connection machine.

The software system used in NERV consists of an operating system kernel, standard application software, and user software. All of them are programmed using Modula-2 and are run under Hypercard on the Macintosh.

Neural RISC at UCL   The neural RISC [29], under development at University College, London (UCL), is a building block for a MIMD general-purpose neurocomputer. It is a 16-bit RISC processor with a communication unit and local memory on one chip, and employs the von Neumann structure because it provides the most efficient and general way to implement algorithms. To attain scalable clustering of processing elements without increasing the pin number of the VLSI chip, the chip integrates a cluster of complete processing elements connected in the form of a linear array, as shown in Figure 9.16. The integrated communication unit supplies two bidirectional, point-to-point links and simple protocol to support a *logical bus* for broadcasting and routing packet. Since a dual channel is employed, the processing element can send and receive messages at the same time.

Since the main design consideration in the neural RISC is how to integrate the necessary components into a single VLSI chip, the processor adopts a very reduced instruction set (only 16 instructions). Although a smaller set of

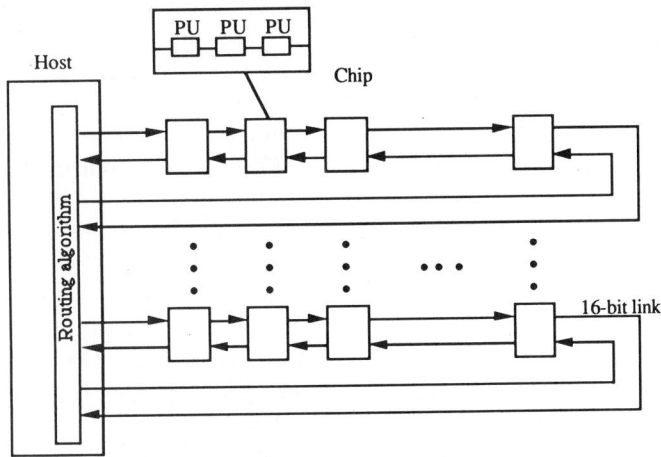

Figure 9.16 *Neural RISC network architecture [29].*

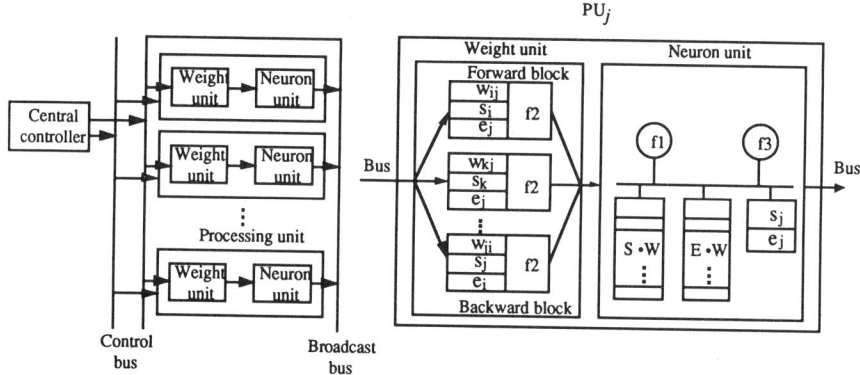

Figure 9.17 Generic neuron architectural framework [29].

instructions (four or eight instructions) may be enough for the neural network computations, it would significantly increase the required memory size.

Another noticeable development at UCL is a generalized neuron processor (GNP) [29], which combines the high performance of special-purpose hardware with the flexibility offered by a general-purpose neurocomputer. In this work, a silicon compiler takes the user program that specifies the characteristics of a neural network model and generates the target architecture using a regular bus structure to interconnect identical GNPs. There is one GNP that mimics a neuron, and each GNP gains access to the broadcast bus sequentially to simulate the interconnection of neurons.

The structure of GNP is divided into two logic units as shown in Figure 9.17: the *weight unit*, which accomplishes the synaptic functions, and the *neuron unit*, which carries out the neuron function. The main characteristics of the GNP framework are (1) to avoid transmitting the updated weight between PEs (both interconnected PEs have a copy of the weight block); (2) to use the silicon compiler (using the designer's specification) which digitizes the threshold function and implements it in a look-up table.

The use of programmable logic arrays (PLAs) in the control structure for both the weight and neuron units provides the required generality to support a variety of neural models; that is, the changes in PLA microcode can allow different neural models.

Sandy/8 at Fujitsu Sandy/8 [15] is a 256-PE neurocomputer under development at Fujitsu. Each PE consists of a floating-point DSP TMS320C30 that can multiply and add two 32-bit floating-point numbers in its one machine cycle, and a local memory for both program and data storage. They are connected in a ring topology with one tray per PE as shown in Figure 9.18a. Each tray functions as a container and a router that is connected to its two neighbors, and a ring is formed that functions as a cyclic shift register.

9.4 IMPLEMENTATION OF NEURAL NETWORKS: NEUROCOMPUTERS

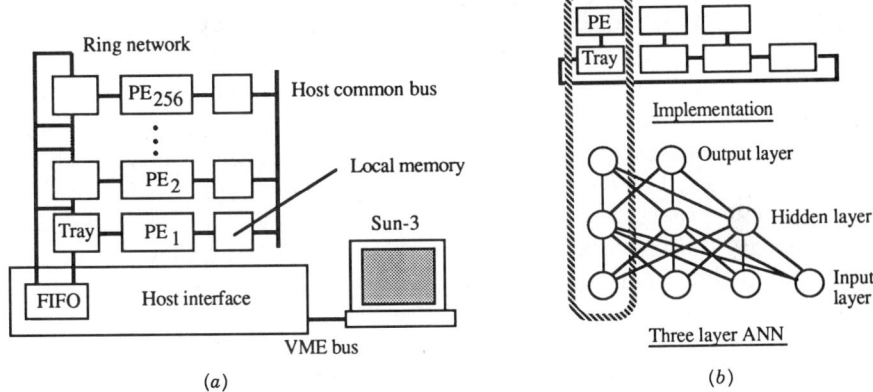

Figure 9.18 The Sandy/8 system [15]: (a) System architecture; (b) mapping.

The way to divide and map a multilayered neural network is shown in Figure 9.18b. The neurons at the same column in all layers are mapped onto the same tray and simulated successively by the associated PE. In [15], a distributed back propagation algorithm on Sandy/8 is presented, in which a partial-summing method in the forward and backward passes is used.

GCN at Sony The GCN (Giga CoNnection) [41] is a mesh-interconnected processor array, in which a 80860 microprocessor-based PE is used. A high-speed (160 Mbytes/s) first-in–first-out (FIFO) buffer is used for communication between adjacent PEs, and 80860 uses LOAD/STORE instructions to access it.

As shown in Figure 9.19, the training parallelism (data partitioning) is used in the horizontal ring, and the spatial parallelism (network partitioning) is used in the vertical ring to speed up the back propagation algorithm.

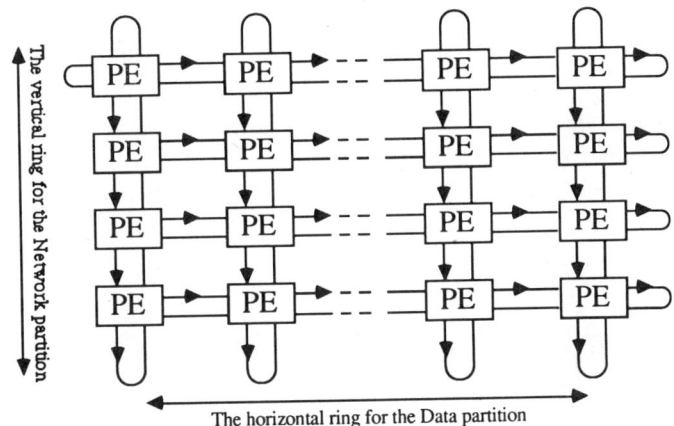

Figure 9.19 The GCN neurocomputer developed at Sony [41].

9.5 MULTILAYERED NEURAL NETWORK ON DISTRIBUTED-MEMORY MULTIPROCESSOR: A CASE STUDY

As a case study on the parallelization and implementation method, we present a way to parallelize the back-propagation model on a distributed-memory multiprocessor. It is based on the spatial parallelism in which the neurons on each layer are partitioned into p disjoint sets, and each set is mapped on a processor of a p-processor system. Let us present the basic parallel algorithm, necessary communication pattern among the processors, and their analytic and experimental time/space complexities.

9.5.1 Multilayered Neural Network and Back Propagation Algorithm

The neural network we are focusing on is a fully connected multilayered neural network using the back-propagation learning algorithm, which is the most popular one, and has been widely used in vision, speech, sonar, radar, signal processing, and robotics applications [1]. A fully connected multilayered network consists of $L + 1$ layers, as shown in Figure 9.2. The first layer at $l = 0$ is the input layer and contains n_0 neurons. Subsequent layers are labeled $1 \leq l \leq L$; the lth layer contains n_l neurons. Each neuron in a layer is connected to all neurons in the next layer. Associated with each neuron i on layer l is an activation value $a_i(l)$, and attached to each connection, connecting neuron i on layer l to neuron j on layer $l + 1$, is a weight $w_{ji}(l, l+1)$.

The back-propagation learning algorithm, a kind of supervised learning one, consists of three passes: forward execution, back propagation of the error, and weight update. In the forward execution pass, the activation value of a neuron a_i at layer l, given an input pattern presented at the input layer, is obtained by the feedforward equations

$$a_i(l) = f\left(\sum_{j=1}^{n_{l-1}} w_{ij}(l-1,l) a_j(l-1)\right), \quad i = 1, \ldots, n_l, l = 1, \ldots, L$$

(9.1)

where f is a nonlinear sigmoid function of the form $f(x) = (1 + e^{-x})^{-1}$. The second pass involves the comparison between the actual output pattern and the desired one, and the propagation of the error, which is governed by the equations

$$\delta_i(l) = \begin{cases} [t_i(l) - a_i(l)][a_i(l)(1 - a_i(l))], & l = L \\ \left[\sum_{k=1}^{n_{l+1}} \delta_k(l+1) w_{ki}(l, l+1)\right][a_i(l)(1 - a_i(l))], & l = L-1, \ldots, 1 \end{cases}$$

(9.2)

where $\delta_i(l)$ is the error value of neuron i on layer l and $t_i(L)$ is the desired value of neuron i in the output layer. Weight update is performed in the third pass according to the equations

$$\Delta w_{ij}(l-1,l) = \eta \delta_i(l) a_j(l-1) \tag{9.3}$$

when η is a learning rate. The second and third passes may be combined into one pass.

9.5.2 Mapping Multilayer Neural Network on a Distributed-Memory Multiprocessor

The parallelism used in our model is a kind of spatial parallelism, in which a multilayered network is vertically partitioned into p subnetworks and each subnetwork is mapped on a processor of the p-processor distributed-memory multiprocessor (DMM). This partitioning and mapping scheme are presented in Figure 9.3.

If a processor maintains only the output or the input weight values, an excessive interprocessor communication is required in either the forward or backward execution phase of the back propagation algorithm. Therefore, to simulate the subnetwork as independently as possible, each processor maintains in its local memory the activation values, the error values, and the input and output weight vectors of the assigned neurons. Since an input weight value of layer l is the output weight value of layer $l-1$, the same value is stored in two processors. Though this partitioning scheme results in the duplication of weight values, it avoids the complex communication requirement during the execution of the distributed back propagation algorithm. On the other hand, all the values of activations and errors are completely partitioned into p disjoint sets. Figure 9.20a shows our partitioning scheme on a four-processor DMM when a fully connected network consists of three layers and each layer has four neurons, and Figure 9.20b presents the distributions of activation values, error values, and weights when a DMM is connected using the ring topology.

9.5.3 A Fully Distributed Back Propagation Algorithm

Each processor in a DMM basically executes the three passes expressed in Eqs (9.1) to (9.3), but some communications are necessary because data in layer l, such as activation values $[a_i(l)]$ and error values $[\delta_i(l)]$, are fully distributed over p processors.

In the following, it is shown the necessary communication patterns and the distributed version of the back propagation algorithm for each pass, where it is assumed that $N_t(l)$ is a set of indices of neurons in layer l assigned to a processor p_t. Note that $|N_t(l)| = n_l/p$.

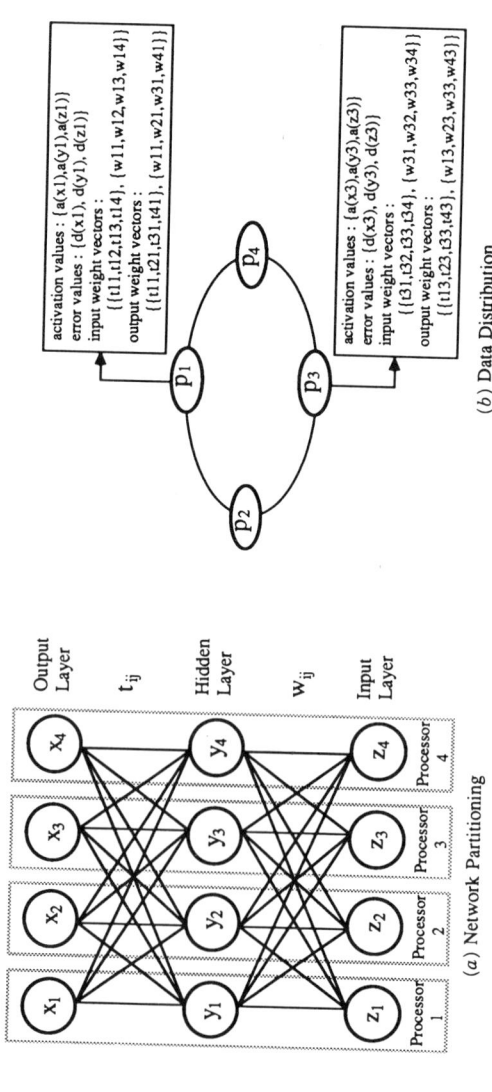

Figure 9.20 Partitioning and data distribution example: (a) network partitioning; (b) data distribution.

First Pass: Forward Execution To compute the $a_i(l)$ using Eq. (9.1) the p_t should know all $a_j(l-1)$ stored in different processors. This requirement could be satisfied by the way in which every processor p_t broadcasts the $a_j(l-1)$ for each neuron $j \in N_t(l-1)$, and receives the $a_{j'}(l-1)$ for each neuron $j' \in N_t(l-1)$. It is a process, called *all-to-all broadcasting* [43], whereby a set of messages, with a distinct message initially residing at each processor, is disseminated so that eventually a copy of each message comes to reside at each processor.

Once all-to-all broadcasting is completed, every processor is informed of all the necessary activation values of all other neurons in layer $l-1$, and can execute Eq. (9.1) for the assigned neurons independently. The forward pass of the distributed back propagation algorithm for p_t is presented in Algorithm 9.1, in which all-to-all broadcast is a procedure to broadcast and receive $a_j(l-1)$. (This procedure will be described in more detail in Section 9.5.4.)

ALGORITHM 9.1: A Distributed Forward Execution Algorithm

```
for l = 1 to L do
      /* Broadcasts and Receives aⱼ(l – 1) */
      for each neuron j ∈ Nₜ(l – 1) do
           all-to-all-broadcast(aⱼ(l – 1));
      end_for
      /* Computes aᵢ(l) */
      for each neuron i ∈ Nₜ(l) do
```
$$a_i(l) = f\left(\sum_{j=1}^{n_{l-1}} w_{ij}(l-1,l)a_j(l-1)\right);$$
```
      end_for
end_for
```

Second and Third Passes: Back Propagation of the Error and Weight Update The back propagation of the error pass is similar to the first pass except for broadcasting $\delta_k(l+1)$ rather than $a_j(l-1)$. To compute the error value $\delta_i(l)$ where $i \in N_t(l)$, all $\delta_k(l+1)$ stored in different processors are required. This requirement could be satisfied by the way in which every processor p_t broadcasts the $\delta_k(l+1)$ for each neuron $k \in N_t(l+1)$ and receives $\delta_{k'}(l+1)$ for each neuron $k' \in N_t$.

Now, let us explain how to update the weights stored in two different processors. A naive method to maintain the consistency of weight values is by communication; that is, one processor computes the new weight value and sends it to the other processor as presented in [9]. This kind of information exchange can be accomplished by *all-to-all personalized communication* [43], because computing and sending the new weights are necessary for all connections between the neurons mapped onto different processors. The all-to-all personalized communication is a process by which each processor sends a unique message to every other processor. However, the time complexity of this method depends on the processor interconnection topologies [9]. Furthermore, because it is proportional to the number of connections in the neural networks, it is a very time-consuming process when the size of a neural network is very large.

Another method to maintain the consistency of weight values, used in our model, is a *recomputation method* with an additional all-to-all broadcasting of activation or error values. Assume that neuron k in layer $l+1$ and neuron i in layer l are mapped onto p_x and p_t, respectively. The difference of weight $\Delta w_{ki}(l, l+1)$ can be computed in p_t if it knows the $\delta_k(l+1)$ stored in p_x, since the $w_{ki}(l, l+1)$ is kept locally in p_t as an output weight. Similarly, the difference of weight $\Delta' w_{ki}(l, l+1)$ can also be computed in p_x independently if it knows the $a_i(l)$ stored in p_t, since the $w_{ki}(l, l+1)$ is kept locally in p_x as an input weight. Furthermore, $\Delta w_{ki}(l, l+1)$ and $\Delta' w_{ki}(l, l+1)$ are identical because p_t and p_x use the same values for computation, and thus the weight value consistency is guaranteed. When p_t sends $a_i(l)$ and p_x sends $\delta_k(l+1)$, they should be broadcast because each neuron in layer l is connected to all neurons in layer $l+1$.

A distributed error back propagation and weight updating scheme for p_t is presented in Algorithm 9.2.

ALGORITHM 9.2: A Distributed Error Back Propagation and Weight Update

```
/* Computes Error Values for Output Layer */
for each neuron i ∈ N_t(L) do
```
$$\delta_i(L) = [t_i(L) - a_i(L)][a_i(L)(1 - a_i(L))]$$
```
end_for
```

```
for l = L - 1 to 1 do
    /* Broadcasts and Receives δ_k(l + 1) */
    for each neuron k ∈ N_t(l + 1) do
        all-to-all-broadcast(δ_k(l + 1));
    end_for
```

```
    /* Computes δ_i(l) */
    for each neuron i ∈ N_t(l) do
```
$$\delta_i(l) = \left[\sum_{k=1}^{n_{l+1}} \delta_k(l+1) w_{ki}(l, l+1)\right][a_i(l)(1 - a_i(l))]$$
```
    end_for
```

```
/* Updates Output Weight Vector w_ki(l,l + 1) */
for each neuron i ∈ N_t(l) do
    for k = 1 to n_{l+1} do
        Δw_ki(l,l + 1) = ηδ_k(l + 1)a_i(l);
        w_ki(l,l + 1) = w_ki(l,l + 1) + Δw_ki(l,l + 1);
    end_for
end_for

/* Updates Input Weight Vector w_ij(l − 1,l) */
/* Broadcasts and Receives a_j(l − 1) */
for each neuron j ∈ N_t(l − 1) do
    all-to-all-broadcast(a_j(l − 1));
end_for
/* Updates w_ij(l − 1,l) */
for each neuron i ∈ N_t(l) do
    for j = 1 to n_{l-1} do
        Δw_ij(l − 1,l) = ηδ_i(l)a_j(l − 1);
        w_ij(l − 1,l) = w_ij(l − 1,l) + Δw_ij(l − 1,l);
    end_for
end_for
end_for
```

9.5.4 Time Complexity and Speedup Ratio

The time required for a single processor to execute the back propagation algorithm given in Eqs. (9.1) to (9.3) for a layer of n neurons is given by $T_1 = t_1 + t_2 + t_3$, where t_i is the time to execute the ith pass. They could be expressed approximately as follows:

$$t_1 = n(nM_a + F)$$

$$t_2 = n(nM_a)$$

$$t_3 = n(nM_a)$$

$$T_1 = t_1 + t_2 + t_3$$

$$= n(3nM_a + F) \tag{9.4}$$

where M_a is a multiply-and-add time of two floating-point numbers, and F is the time to evaluate the sigmoid function. It is assumed in evaluating the time complexity that $n_l = n$ for $l = 0, \ldots, L$ for the sake of simplicity.

The time complexities of our distributed back propagation algorithm for a layer of n neurons running on a p-processor DMM, T_p, are shown in Eq. (9.5). In Eq. (9.5), t_1' is the time to execute Algorithm 9.1, t_2' is the time to execute Algorithm 9.2, and $AAB(p)$ is the time to complete all-to-all broadcasting on the p-processor DMM. The factor n/p is due to the fact that every processor is assigned n/p neurons per layer.

$$t'_1 = \frac{n}{p} \text{AAB}(p) + \frac{n}{p}(nM_a + F)$$

$$t'_2 = \frac{n}{p}(2\text{AAB}(p) + 3nM_a)$$

$$T_p = t'_1 + T'_2$$

$$= \frac{n}{p}(3\text{AAB}(P) + 4nM_a + F) \tag{9.5}$$

Now, let us consider the all-to-all broadcasting algorithm and its time complexity on a p-processor DMM. To evaluate this, one must first assume the communication capability of each processor. In our work, we adopt the *one-port* communication [43] in which a processor can send and receive a unit of message on one of its ports during each period of unit time. The port on which a processor sends and receives can be different. The unit time of communication for a processor sends and/or receives a unit of message is defined as C, which includes the overhead of setting up the necessary communication mechanism and the actual message transfer time. The message unit is a word representing the floating-point number of the activation or error value.

In one-port communication, the lower bound for all-to-all broadcasting on a p-processor DMM is $(p-1)C$, since each processor needs to receive from every other processor, that is, $p-1$ processors. This lower bound can be achieved in the ring topology. Note that as far as one-port communication is assumed, other topologies such as the complete connection, hypercube, and mesh, which have more connections than the ring, do not perform all-to-all broadcasting better than the ring. If a processor can send and receive on its d-ports, $d > 1$, concurrently during each time unit, more richly connected topologies would result in less all-to-all broadcasting time. However, the d-port communication is not realistic in most commercial message-passing multiprocessors, which is the reason why we assume one-port communication.

An AAB(p) algorithm for p_t when processors are connected by a ring topology is presented in Algorithm 9.3. In this algorithm, we assume that p processors in the ring-connected DMM are numbered from 0 to $p-1$ successively such that processor p_t is directly connected to $p_{(t+1) \bmod p}$ and $P_{(t-1) \bmod p}$.

9.5 NEURAL NETWORK ON DISTRIBUTED-MEMORY MULTIPROCESSOR

ALGORITHM 9.3: An AAB(p) Algorithm on the Ring Topology

```
/* Sends an element (A[0]) and
   Receives p − 1 elements (A[1]-A[p − 1]) */
   for i = 0 to p − 2 do
         parbegin
               send A[i] to p_{t+1};
               receive A[i + 1] from p_{t−1};
         parend
   end_for
```

Using Eqs. (9.4) and (9.5), the p-processor speedup ratio, $S(p) = T_1/T_p$, can be obtained as follows:

$$S(p) = \frac{T_1}{T_p} \quad \text{(by } C = \Delta M_a, F = \theta M_a\text{)}$$

$$= \frac{p(3nM_a + \theta M_a)}{3\Delta M_a(p - 1) + 4nM_a + \theta M_a}$$

$$= \frac{p(3n + \theta)}{3\Delta(p - 1) + 4n + \theta} \quad (9.6)$$

Once the learning is complete for a specific application, the application can be hard-wired, and the network may execute only the forward execution pass. The p-processor speedup for forward execution, $S'(p)$, can be obtained as follows:

$$S'(p) = \frac{t_1}{t'_1} = \frac{p(n + \theta)}{\Delta(p - 1) + n + \theta} \quad (9.7)$$

If the size of a neural network is much larger than p, the $S(p)$ is $\frac{3}{4}p$ and the $S'(p)$ is p in a p-processor DMM;

$$\lim_{n \to \infty} S(p) = \tfrac{3}{4}p \qquad \lim_{n \to \infty} S'(p) = p$$

The reason that $S(p)$ is not p but $\frac{3}{4}p$ when p processors are used is that the same computations for weight updates are performed twice in the weight update pass.

In Eqs. (9.6) and (9.7), the most important parameter is the communication/computation ratio Δ, whose value lies between 0.5 and 256 [44]. The theoretical learning speedup aspects for various Δ values are shown in Figure 9.21a graphically, where $n = 2048$ neurons/layer and $\theta = 40$. The theoretical forward execution speedup is also shown in Figure 9.21b under the

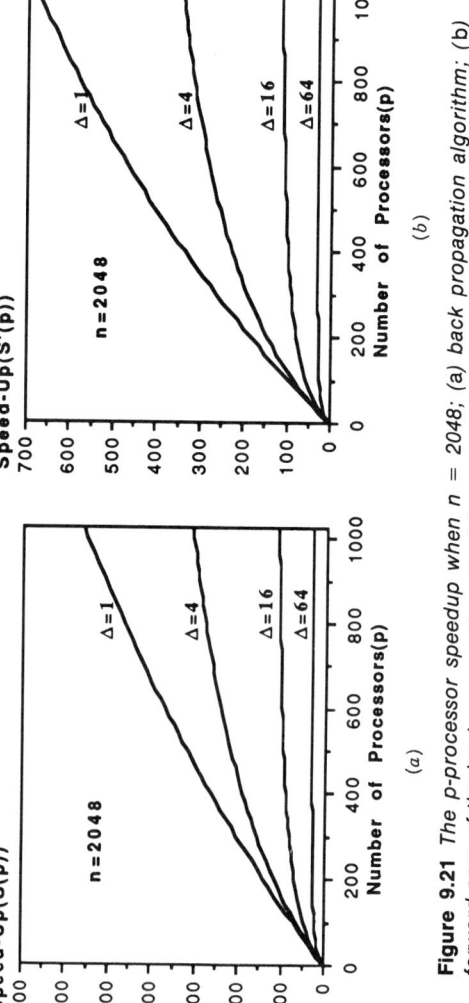

Figure 9.21 The p-processor speedup when n = 2048; (a) back propagation algorithm; (b) forward pass of the back propagation algorithm.

same assumptions. For the distributed simulation of a multilayered neural network, it is seen from Figure 9.21 that there is a cost-effective number of processors depending on the Δ values, such that even if more processors are added to the simulation the speedup ratio is not increased as much.

9.5.5 Space Complexity

The total space requirement to execute the back propagation algorithm for a fully connected neural network consisting of L layers of n neurons on a single processor M_1 could be computed as follows:

$$M_1 = nL + nL + n^2(L - 1)$$

The total space requirement of the distributed back propagation algorithm for the same neural network on a p-processor DMM, M_p, could be computed as follows:

$$M_p \approx nL + nL + 2n^2(L - 1) + pn$$
$$= n(2L + p) + 2n^2(L - 1)$$

Since the M_p could be equally partitioned into p spaces in a p-processor DMM, the space reduction ratio, $M(p)$, could be computed as follows:

$$M(p) = \frac{M_1}{M_p/p} = \frac{p(2nL + n^2(L - 1))}{n(2L + p) + 2n^2(L - 1)}$$

If the size of a neural network becomes large, the space reduction ratio approaches $p/2$.

$$\lim_{n \to \infty} M(p) = p/2$$

9.5.6 Experimental Speedup on a Network of Transputers

Our distributed back propagation algorithm has been implemented on a network of transputers. The transputer is a 32-bit microprocessor with four 10-Mbit/s serial bidirectional communication links, and T414 is a kind of transputer without floating-point units.

The Δ Value of Transputer The transputer provides a relatively small Δ value because its capabilities of communicating with other transputers are embedded in its architecture, which motivated us to adopt the network of transputers as our test bed system.

TABLE 9.3 The Δ Value of Transputer

Transputer All-to-All Broadcasting	T414	T800
One-by-one AAB (p) (Algorithm 9.3)	0.55	13.10
Grouped AAB (p)	0.05	1.26

Let us consider the Δ value of the transputer, which is the ratio of the time to send and receive one 32-bit floating point number over the time to multiply-and-add two 32-bit floating-point numbers. The only communication pattern required in our distributed back propagation algorithm is the all-to-all broadcasting of the activation/error values for the n/p neurons in a layer. If we directly use Algorithm 9.3 to broadcast the activation/error values of n/p neurons, an AAB(p) is required for every n/p neurons. However, it can be optimized because the number of neurons in a layer mapped onto a processor is larger than 1 (i.e., $n/p \gg 1$).

If all data for n/p neurons are grouped and broadcasted together with one AAB(p), called the *grouped* AAB(p), which is an optimized AAB(p), the additional overhead to setting up the necessary communication links can be eliminated. Table 9.3 shows experimental Δ values for T414 and T800. Since T800 is an enhanced version of T414 with a 64-bit floating-point unit, it can perform 32-bit floating-point operations more than 20 times faster than T414. Thus, T800 provides a relatively larger Δ value than T414.

An Experimental Speed-up An experimental speedup of our distributed backpropagation algorithm with the grouped AAB(p) on a ring of 32 T414 transputers (17 MHz) were obtained. All tested applications consist of three layers, in which the neurons in adjacent layers are fully connected. Table 9.4 shows the characteristics of bench-mark applications, in which binary images are used as input, and *structure* represents the number of neurons in the input layer, hidden layer, and output layer, respectively:

Figure 9.22a shows the speedup of the distributed back propagation algorithm, and Figure 9.22b shows the speedup of the forward pass of it.

TABLE 9.4 The Characteristics of Bench-mark Applications

	Structure (input × hidden × output)	Total Connections	Input Patterns
Number	80 × 9 × 4	756	8-by-10 Numbers
Hangul	1600 × 50 × 256	92800	40-by-40 Korean characters
Num 18 × 12	216 × 216 × 216	93312	18-by-12 Numbers

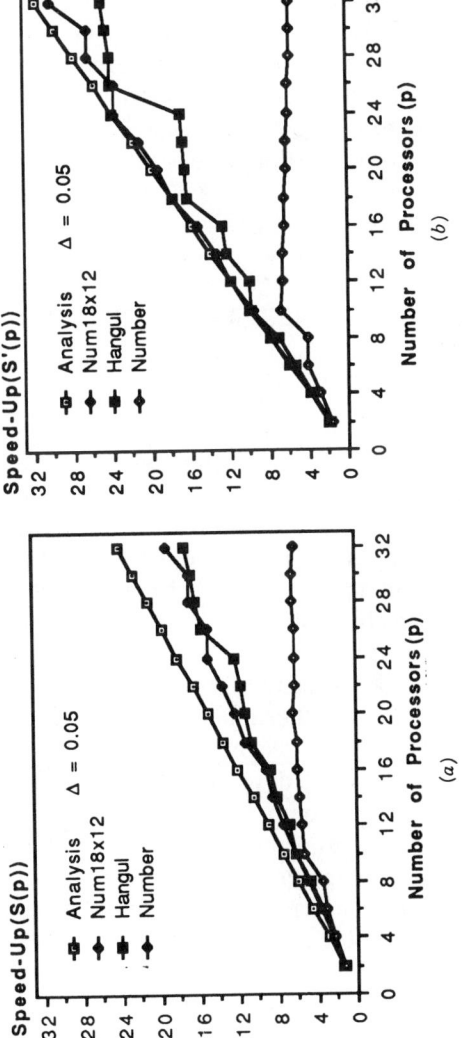

Figure 9.22 Experimental speedup with ring of 32 transputers: (a) distributed back propagation algorithm; (b) forward pass of the distributed back propagation algorithm.

The speedup of Hangul and Num18×12 applications are generally the same as the analyzed speedup, but Number is not. This aspect comes from the fact that the Number application does not have enough neurons (or connections) to utilize the full power of the multiprocessor system.

Another interesting feature in Figure 9.22 is that the speedup of the forward pass of the Hangul application is not linear, but is almost a step function. The reason is that the number of neurons in a layer, n_l, is not always divisible by p in the p-processor DMM. Therefore, though a processor in the p-processor DMM has finished the computation for $\lfloor n_l/p \rfloor$ neurons, it should wait for other processors to take charge of $\lceil n_l/p \rceil$ neurons. Because of this synchronization overhead, the speedup ratio is increased in a stepwise fashion. For example, the speedup ratios of the Hangul application with 18 and 24 transputers are almost the same (see Fig. 9.22b).

9.6 SUMMARY AND FUTURE RESEARCH AREAS

9.6.1 Summary

Since neural network simulation requires many computations, there have been extensive research efforts to exploit the inherent parallelism in neural network models. They can be classified as model-specific, which can be applied only to a specific neural network model, and general, which can partition and map any neural network. However, almost all research efforts on parallel algorithms and implementation methods are focused on the model-specific type, especially on back-propagation model. Although some algorithms or implementations assume a general model, there is no well-defined algorithm of good performance.

These parallel algorithms, whether model specific or not, have been implemented on the coprocessor, systolic array, SIMD, and MIMD hardware. Table 9.5 summarizes the currently developed parallel algorithms and implementation hardware (i.e., neurocomputers).

The speed and capacity of neural network simulation is not the only factor to use to evaluate the neurocomputer. Since it is quite tedious to specify and find the neural network parameters, a neural network simulation system should also provide a developing environment including a specification language and a graphic interface for portraying the structure and performance of the network. Several researchers have expressed the need for a standard simulation language to simplify comparisons of different types of networks, and suggested their own language and tool. However, there is no one powerful enough to specify various neural network models.

9.6.2 Research Areas

To develop a fast and easy-to-use neurocomputer system, the simulation speed should be high and it should be easy to specify and modify the various

TABLE 9.5 Parallel Algorithms and Neurocomputers

Parallelism H/W Architecture	Spatial Parallelism	Training Parallelism	Others
Systolic (VLSI)	BP [13]		
SIMD			BP [16–18]
MIMD (point-to point)	BP [6,10,14,15,19] KM [23] BM [27] GE [3,28,29,35,37,39,40]	BP [6,12,20] KM [22]	BP [41]
MIMD (bus)	BP [8] GE [34] GE [22]	BP [11]	

BP = back propagation; KM = Kohonen map; BM = Boltzmann machine; GE = general model.

neural network parameters, such as learning rate, network topology, activation function of the neurons, and learning rule.

A general system architecture* to simulate arbitrary neural network model can be represented as Figure 9.23. Let us discuss the research topics on general-purpose parallel neurocomputers.

Specification Language A language to specify the parameters of a neural network model, such as neuron characteristics, network topologies, and learning algorithms, is required. It should be powerful enough to specify various neural network models, and should be easy to use. Currently there are several neural network specification languages developed: for example, P3 [45], running on the Symbolics workstation; ANSpec [33], developed at SAIC; Axon [30], developed at HNC; and GNL [36], used in NEP.

User Interface Another research area is the interactive graphic user interface. It may help the user to modify the various neural network parameters, and to display both the structure and state of neural network such as the state of neurons/connections. However, it is a very difficult and complex task because of the multidimensional property of neural network.

Network Compilation One of the central issues in an efficient parallel simulation of neural networks is partitioning the problem to decrease communication and to balance the load among the processors. We must therefore use general-purpose partitioning techniques instead of the regular decompositions that are often possible with linear systems.

The physical structure of a neural network is given by its underlying network graph. Each vertex of this graph represents a neuron, and a directed

---

* Actually it is a neurocomputer system, called PANSi, developed at KAIST.

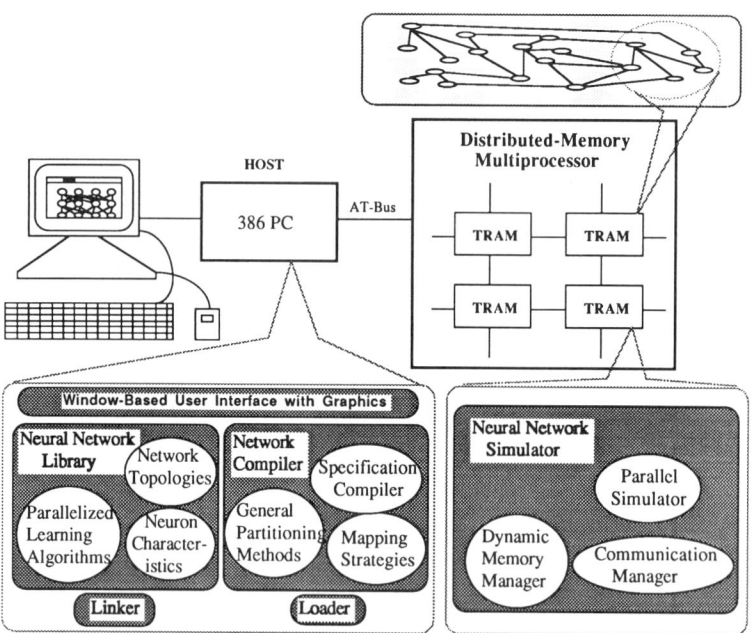

Figure 9.23 The system architecture of general-purpose neurocomputer.

edge exists between two vertices if and only if there is a nonzero weight associated with the corresponding pair of neurons. The algorithm to partition the graph, known as the min-cut problem, is an NP-complete problem. An efficient heuristic graph-partitioning algorithm [46] can be used in neural network partitioning with some neural network specific heuristics.

Another possible partitioning method is to use the neural network itself, such as the Boltzmann machine, to optimize the partitioning of the network. It can be performed on a parallel neurocomputer.

Parallel Computing Engine Although neural network computations consist mainly of floating-point sum-of-product operations, many specialist DSP processors are not well suited to them because of their small address spaces [38]. In addition, the cache memory cannot improve memory access time because memory accesses in them do not have locality of references. A more appropriate processor architecture for neural network simulations would be one with a very large address space, and with a small number of very specific instructions that are dominant in any neural network simulation. In addition, the repetitive nature of the operations permits considerable use of pipelining. Therefore, it is necessary to investigate a processor architecture as a building block of the parallel neurocomputer to meet these requirements, and a parallel system architecture to guarantee high performance.

REFERENCES

[1] *DARPA Neural Network Study,* AFCEA International press, 1988.//
[2] P. Treleaven, "Neurocomputers," Research Note 89/8, Univ. College London, Jan. 1989.
[3] J. Ghosh and K. Hwang, "Critical issues in mapping neural networks on message-passing multicomputers," *Proc. 15th Internat. Symp. on Computer Architecture,* IEEE Computer Society Press, Washington D.C., pp. 3–11, 1988.
[4] R. P. Lippmann, An introduction to computing with neural networks, *IEEE ASSP Magazine,* 4, 4–22 (April 1987).
[5] G. E. Hinton, Connectionist learning procedures, *Artificial Intelligence,* 40, 185–234 (1989).
[6] D. A. Pomerleau et al., Neural network simulation at warp speed: How we got 17 million connections per second, *Proc. IEEE Internat. Conf. on Neural Networks,* 2, 143–150 (1988).
[7] D. E. Rumelhart, G. E. Hinton, and R. J. Williams, "Learning Internal Representations by Error Propagation," in D. E. Rumelhart and J. L. McClelland, Eds., *Parallel Distributed Processing: Explorations in the Microstructure of Cognition, Vol. 1: Foundation,* MIT Press, Cambridge, Mass., 1986, pp. 318–362.
[8] D. K. Panda and K. Hwang, A multiple-bus network for implementing very-large neural networks with back-propagation learning, *Proc. Internat. Joint Conference on Neural Networks,* Lawrence Erlbaum Associates, Hillsdale, N.J., Vol. 2, 175–178 (January 1990).
[9] H. Yoon and J. H. Nang, "Multilayer neural networks on distributed-memory multiprocesors," *Proc. of Internat. Neural Network Conference,* Vol. 2, 669–672, Paris, Kluwer Academic Publishers, Dordrecht, (July 1990).
[10] H. Yoon, J. H. Nang, and S. R. Maeng, "Parallel simulation of multilayered neural networks on distributed-memory multiprocessors," *EUROMICRO Journal: Microprogramming and Microprocessing,* 29(3), 185–195, October 1990.
[11] K. L. Parker and A. L. Thornburgh, Parallelized back-propagation training and its effectiveness, *Proc. Internat. Joint Conference on Neural Networks,* Lawrence Erlbaum Associates, Hillsdale, N.J., Vol. 2, 179–182 (January 1990).
[12] J. Bourrely, Parallelization of a neural learning algorithm on a hypercube, *Proc. 1st European Workshop on Hypercube and Distributed Computers,* North-Holland, Amsterdam, pp. 219–229, October 1989.
[13] S. Y. Kung and J. N. Hwang, Parallel architectures for artificial neural nets, *Proc. IEEE Internat. Conf. on Neural Networks,* 2, 165–172 (1988).
[14] J. R. Millan and P. Bofill, Learning by back-propagation: A systolic algorithm and its transputer implementation, *Neural Networks,* 1(3) (July 1989).
[15] H. Kato, H. Yoshizawa, H. Iciki, and K. Asakawa, A parallel neurocomputer architecture towards billion connection updates per second, *Proc. Internat. Joint Conference on Neural Networks,* Lawrence Erlbaum Associates, Hillsdale, N.J., Vol. 2, 47–50 (January 1990).

[16] G. Blelloch and C. R. Rosenberg, Network learning on the connection machine, *Proc. of Tenth Internat. Joint Conf. on Artificial Intelligence*, Morgan Kaufmann, Los Altos, Calif., Vol. 1, 323–326 (1987).

[17] E. Deprit, Implementation recurrent back-propagation on the connection machine, *Neural Network*, 2, 295–314 (1989).

[18] T. Watanabe, Y. Sugiyama, T. Kondo, and Y. Kitamura, Neural network simulation on a massively parallel cellular array processor: AAP-2, *Proc. Internat. Joint Conference on Neural Networks*, Lawrence Erlbaum, Hillsdale, N.J., Vol. 2, 155–161 (1989).

[19] J. Cook and J. Gilbert, Parallel neural network simulation using sparse matrix technique, *EUROMICRO Journal: Microprocessing and Microprogramming*, 24, 621–626 (1988).

[20] A Petrowski, L. Personnaz, G. Dreyfus, and C. Girault, Parallel implementations of neural network simulations, *Proc. 1st European Workshop on Hypercube and Distributed Computers*, North-Holland, Amsterdam, pp. 205–218, October 1989.

[21] D. E. Rumelhart and D. Zipser, "Feature Discovery by Competitive Learning," in D. E. Rumelhart and J. L. McClelland, Eds., *Parallel Distributed Processing: Explorations in the Microstructure of Cognition, Vol. 1: Foundation*, MIT Press, Cambridge, Mass., 1986, pp. 151–193.

[22] R. Mann and S. Haykin, A parallel implementation of Kohonen feature maps on the warp systolic computer, *Proc. Internat. Joint Conference*, Lawrence Erlbaum Associates, Hillsdale, N.J., Vol. 2, 84–87 (January 1990).

[23] R. E. Hodges, Parallelizing the self-organizing feature map on multi-processor systems, *Proc. Internat. Joint Conference*, Vol. 2, 141–144 (January 1990).

H. K. Kwan and P. C. Tsang, Systolic implementation of multi-layer feedforward neural network with back-propagation learning scheme, *Proc. Internat. Joint Conference on Neural Networks*, Lawrence Erlbaum Associates, Hillsdale, N.J., Vol. 2, 155–158 (January 1990).

[24] G. E. Hinton and T. J. Sejnowski, "Learning and Relearning in Boltzmann Machines," in D. E. Rumelhart and J. L. McClelland, Eds., *Parallel Distributed Processing: Explorations in the Microstructure of Cognition, Vol. 1: Foundation*, MIT Press, Cambridge, Mass., 1985.

[25] E. H. L. Aarts and J. H. M. Korst, Computations in massively parallel networks based on the Boltzmann machine: a review," *Parallel Computing*, 9, 129–145 (1989).

[26] D. R. Greening, "A Taxonomy of Parallel Simulated Annealing Techniques," Tech. Report CSD-890050, Computer Science Dept., Univ. of California at Los Angeles, August 1989.

[27] A. Iazzetta, R. Vaccaro, and U. Villano, A transputer implementation of Boltzmann machines," *Proc. of 1st Italian Workshop on Parallel Architecture and Neural Networks*, World Scientific Pub., Singapore, pp. 128–145 (1988).

[28] J. A. Feldman, M. A. Fanty, N. H. Goddard, and K. J. Lynne, Computing with structured connectionist networks, *Communications of the ACM*, 31(2), 170–187 (February 1988).

[29] P. Treleaven, M. Pacheco, and M. Vellasco, VLSI architectures for neural networks, *IEEE MICRO*, 8–27 (December 1989).

[30] R. Hecht-Nileson, Neurocomputing: picking the human brain, *IEEE SPECTRUM*, 25(3), 36–41 (1988).

[31] B. Souček and M. Souček, "Artificial Neural Systems or Neurocomputers," in *Neural and Massively Parallel Computers: The Sixth Generation*, Wiley, New York, 1988, pp. 245–276.

[32] G. A. Works, The creation of DELTA: A new concept in ANS processing, *Proc. ICNN88*, IEEE, San Diego, Vol. 2, pp. 159–164 (1989).

[33] *Delta II Floating Point Processor*, SAIC, San Diego, 1990.

[34] R. M. Kuczewski, M. H. Myers, and W. J. Crawford, Neurocomputer workstations and processors: approaches and applications, *Proc. IEEE First Internat. Conf. on Neural Networks*, Vol. 3, 487–500 (June 1987).

[35] C. A. Cruz, W. A. Hanson, and J. Y. Tam, "Neural network emulation hardware design considerations," *Proc. IEEE First Internat. Conf. on Neural Networks*, 3, 427–434 (June 1987).

[36] W. A. Hanson et al., CONE—Computational network environment, *Proc. IEEE First Internat. Conf. on Neural Networks*, Vol. 3, 531–538 (June 1987).

[37] R. T. Savely, The implementation of neural network technology, *Proc. IEEE First Internat. Conf. on Neural Networks*, Vol. 4, 477–484 (June 1987).

[38] S. C. J. Garth, A chipset for high speed simulation of neural network systems, *Proc. IEEE First Internat. Conf. on Neural Networks*, 3, 443–452 (June 1987).

[39] S. Garth and D. Pike, An integrated system for neural network simulations, *ACM SIGARCH Computer Architecture News*, 16, No(1) (March 1988).

[40] N. Fukuda, Y. Fujimoto, and T. Akabane, A transputer implementation of toroidal lattice architecture for parallel neurocomputing, *Proc. Internat. Joint Conference on Neural Networks*, Lawrence Erlbaum Associates, Hillsdale, N.J., Vol. 2, 43–46 (January 1990).

[41] A. Hira et al., A two level pipeline RISC processor array for ANN, *Proc. Internat. Joint Confernece on Neural Networks*, Vol. 2, 137–140 (January 1990).

[42] G. Frazier, Ariel: A scalable multiprocessor for the simulation of neural networks, *ACM SIGARCH Computer Architecture News*, 18(1) (Mar 1990).

[43] S. L. Johnsson and C. T. Ho, Optimum broadcasting and personalized communication in hypercubes, *IEEE Trans. Computers*, 38(9), 1249–1268 (1989).

[44] M. Annaratone, C. Pommerell, and R. Ruhl, Interprocessor communication speed and performance in distributed-memory parallel processors, *Proc. of the 16th Annual Internat. Symp. on Computer Architecture*, IEEE Computer Society, Washington, D.C., pp. 315–324 (1989).

[45] D. Zipser and D. E. Rabin, "P3: A Parallel Network simulation System," in D. E. Rumelhart and J. L. McClelland, Eds., *Parallel Distributed Processing: Explorations in the Microstructure of Cognition, Vol. 1: Foundation*, MIT Press, Cambridge, Mass., 1986, pp 488–506.

[46] B. W. Kernighan and S. Lin, An efficient heuristic procedure for partitioning graphs, *Bell System Tech. J.* 49, 291–307 (February 1970).

CHAPTER 10

Neurocomputing on Transputers

MARTIN J. DUDZIAK

10.1 INTRODUCTION

Neural network models have been demonstrated as effective tools for solving a variety of computationally difficult tasks, particularly in areas of pattern recognition and classification. The essence of most neurocomputing models rests in the importance given to connectivity among a vast number of computationally simple units (neurons), rather than to large calculations performed by a comparatively small number of modules. This emphasis upon the connections rather than the nodes that are connected leads to such features as the ability to process incomplete or noisy data, match fuzzy data with known templates, and maintain fault tolerance. The connectionist principle of distributing information across a large number of point-to-point relations, usually through an implementation of weighted links from neuron to neuron, also provides an opportunity for a system to adapt (i.e., learn) new patterns. This amounts to being able to modify the ways in which the network will respond to particular patterns of data over time.

One of the major roadblocks to effective and widespread use of neurocomputing architectures in general has been the issue of performance, due primarily to the need for representing thousands or millions of interconnections between the neural units. The implementation of neural networks on serial computers has been limited due to the von Neumann machine's inability to perform more than one task at a time or to efficiently communicate

---

Neural and Intelligent Systems Integration, By Branko Souček and the IRIS Group.
ISBN 0-471-53676-8 ©1991 John Wiley & Sons, Inc.

with other processors that may be employed to distribute some of the tasks. Briefly put, neural networks operate in parallel fashion in the biological environment and in the abstract, mathematical models; thus the problem of implementing neural networks in hardware and software begs for a parallel solution also.

10.2 THE TRANSPUTER AS A NEURAL PLATFORM

While that ideal parallel solution might at first glance seem to consist in a vast array of fully or heavily connected simple processors, each emulating a single neuron, this is neither practical nor necessary. Having a mechanism that can be highly connectable and which can emulate the activity of a large number of neurons can provide a means of building systems that will have real-world engineering value. It is argued here that the INMOS transputer provides the basic building block component for constructing such a mechanism. The key points of this argument rest upon three foundations:

- The dual special features of the transputer (full on-chip computation facilities and multiple point-to-point communications);
- The dual nature of what is parallel processing (algorithmic performance gain and multiple-source, nondeterministic input processing); and
- The fundamental architecture of transputer modules (TRAMs), which allow for integration of special-purpose coprocessor silicon with transputers.

The features of the transputer can be found in associated documents and manual available directly from any of the worldwide offices of SGS-THOMSON Microelectronics (Parent company that acquired INMOS in 1989). It remains to be pointed out that there are at least two major values to employing multiple processors to handle a given computational problem. Obviously one advantage is in performance speed. By dividing up a task among several central processing units (CPUs), it is often possible to radically accelerate the completion of that task. This is the most commonly seen and understood value of parallel processing, perhaps because it has the most visible, easily verifiable results.

A second attribute of the parallel approach (and particularly of the CSP-based transputer model) is that one can have a large number of input/output (I/O) devices communicating in parallel, asynchronously, and indeed nondeterministically, with a network of processors. These devices could be disks, analog-to-digital converters (ADCs), digital signal processing (DSP) engines for data compression or decompression, and a variety of other mechanisms. By so doing one may, depending upon the application, gain more than the performance increase in either processing volumes of input or in responding to various ready devices. In processing segment A of some data set in parallel with processing segment B, there may be intermediate results that can be used as either shortcuts or heuristics for processing the other segment or some third other data segment.

For example, the system may be charged with processing image data collected from a satellite. It may be able to input and process in parallel the data from four quadrants or segments of the complete picture. Segment A data may indicate that there are some interesting features that overlap into segments B and D. This information may be obtainable through a first-pass, high-level analysis of the image data in segment A, such as through examining intensities of edges and line features. This information could then be used by the system to drop any processing of segment C for the moment and to allocate processor resources for examining, in parallel to segments A, B, and D, a new segment region that includes the areas of special interest that overlap portions of A, B and D. The net result (may be) that the features of special interest get processed sooner and that unnecessary processing is avoided (discarded) more often than in a sequentially ordered machine.

It may be argued that instead of a CSP-type model used by the transputer, one should build systems completely with special-purpose hardware. Certainly a number of neural processing devices emerging into production promise exciting performance. However, there is always the issue of communication with the outside world and among subsystems that demands a solution that will not be cumbersome or expensive. It is quite reasonable to expect that a transputer-only neural system will and should evolve into one that uses specialized silicon (e.g., digital analog neural chips embodying three-layer networks of 1024 or 2048 neurons) and that the transputer will end up serving mainly as a communication device for those devices. But how convenient that one can build the prototype and evolve it into the full-blown custom hardware implementation all from essentially the same platform (i.e., a B008 or B014 transputer motherboard).

In summary, the main advantages in using transputers for neural applications include:

True parallel processing with independent memory/processor (no bottleneck by relying exclusively upon shared memory);
True asynchronous performance;
High-speed links for interprocessor communication;
Dynamic reconfigurability and code-loading/distribution;
On-chip parallelism with multipriority processes;
On-chip RAM and FPU and ASIC potential;
Modularity through TRAM architecture;
Real-time sensitivity capabilities;
Multiple I/O via links;

Compatibility of multiple transputer types (16/32, IPU/FPU);
Ease of expandability with little or no change in program code;
Ease of building physical fault tolerance.

Furthermore, the advent of the next generation (T9000) transputer in mid-1991 brings to the fore a 150 MIPS, 20+ MFLOPS processor combined with 80 Mbytes/sec. link bandwidth and a novel virtual-channel architecture for interprocessor communications. Neural system builders will have an even wider range of computing options for prototyping and building deliverable neural networks. The T9000 will be fully compatible with existing transputer architectures (via a link converter module and a custom device, and C100) and it will be well-suited for handling high-speed communications among a cluster of specialized neural processors provided that the latter devices have sufficient I/O capabilities themselves.

These aspects will be examined with respect to a generic model of neural network that shares many of the characteristics found in more specific classes such as back propagation, generalized Delta rule, simulated annealing, adaptive resonance theory, and unsupervised clustering.

10.3 NEURAL COMPUTATIONS

Figure 10.1 shows a simple depiction of the level of connectivity among multiple layers of neurons in a common network structure (feedforward). Typically, the input layer receives input stimulation from a number of different receptors or sensors, and the data flow is upward through n hidden layers to a final output layer. The actual processing that occurs within each neuron is simple and not the type of calculation that would itself benefit from being distributed across several processes in parallel. However, the vast number of connections between neurons of any one layer and the next ($2n$ for fully connected layers) makes some type of parallel solution imperative for satisfactory performance. Collectively, the computation is typically a vast multiply-accumulate operation on the outputs of one layer of neurons and the weight values that are assigned to the individual layer-to-layer neural interconnects.

Given an individual neuron j in any layer, one then finds its input v[j] to be (typically the value of the summation

$$v[j] = \sum_{i=0}^{v} u[i] * w[i][j]$$

where k is the number of neurons in the previous or underlying layer connected to neuron

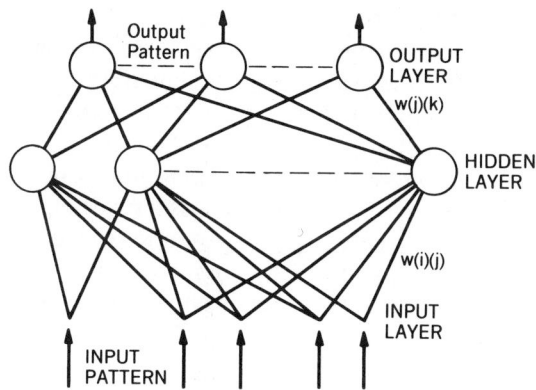

Figure 10.1 *Typical multilayer network.*

Note that in any given network, the neurons from one layer or set are not necessarily connected to each neuron in a subsequent layer; the connections may be stochastically distributed or they may evolve in the course of the network learning phase. The fact that connectivity between layers of neurons may change during the course of the network learning phase is a reason for building flexibility into the computational architecture, in order that the vector-product load, for instance, may be redistributed among processing resources. The transputer design is particularly well-suited for implementing such processes. This stems from both the manner in which parallel processes are managed on-chip through the workspace and the use of multiple queues for different types of processes (e.g., active, waiting on a channel, waiting for a timer, waiting for a link).

Suppose the purpose of a given network is to learn a set of frequency-response patterns and to recognize a member of that set when it is presented as input to the system. Either in a supervised mode (involving the operator as presenter and teacher of template patterns to the system) or in an unsupervised mode (using built-in algorithms, such as computing Euclidean or Hamming distances between possible match patterns), the system learns and stores n patterns. Next an unknown pattern is input, and (typically) based upon the weights that have been established and adjusted for the many interneuron connections, that unknown pattern is classed as being enough like one of the learned patterns to be treated as such by the rest of the program, or else so unlike the learned patterns that it is rejected or added as a new learned pattern. The reason underlying the use of weighted interconnects is that a given pattern can be represented as having n features or attributes. A feature could be as simple as a pixel value or as complex as something like a grammatical construct or a configuration of symbolic values (frames, schemata, rules). The sum of these features, when present, will activate a definite collection of neurons and such activation patterns indicate a match for the given pattern. That collection can be represented as a vector, and the basic operation is one of comparing one vector to a class of other vectors, searching for that which is most similar.

Clearly any architecture that is designed to implement such a system will have more than fast computation on its list of concerns. There must also be a method for efficient communication between processors, so that the workload can be divided up among many processors without creating high overhead for sharing information. There are often huge arrays of numbers (mostly floating point) that must be frequently accessed by any process working out the neural activation and weighting values. One often finds code that resembles the following:

```
...  float pattern [max_pattern][num_attributes]
     float weights [max_pattern][num_attributes]
     int table [max_pattern][num_attributes]
     float distance [max_pattern] ...
```

For working with as few as 100 different patterns, each with upward of 100 attribute (property) elements, something like the foregoing code alone will require 120,400 bytes—not much in terms of memory, but a lot for transferring even over transputer links if the transfers had to be performed repeatedly for each pattern that was being processed. This implies that the division of any workload among processors, and the communications between processes on those devices, must make use of the fact that on transputers one can be processing (e.g., doing multiply-accumulates) and performing I/O over all of the links simultaneously.

10.4 TRANSPUTER IMPLEMENTATION EXAMPLE

An illustration using a simple neural model will help to convey both the ways in which transputers may be employed for neural designs as well as the power of the device as a dual function unit (calculation and communication). Figure 10.2 shows a design for an unsupervised learning model based upon clustering of pattern data. The algorithm determines whether a given pattern falls into any existing cluster of similar patterns or is in a unique category of its own. It employs the calculation of a straightforward $a^2 = b^2 + c^2$ Euclidean distance measure and uses that value to determine how to class the unknown pattern. If the pattern falls into an established cluster, the weights for that cluster are modified to reflect the addition of the new member.

There are two general solutions for this problem. One is to establish parallelism for the actual computation, as is done in the model in Figure 10.1. Once a block of input patterns have been read in through the user interface and preprocessed by the pattern handler, each pattern is fed into a pipeline consisting of the processes for computing a Euclidean distance

10.4 TRANSPUTER IMPLEMENTATION EXAMPLE

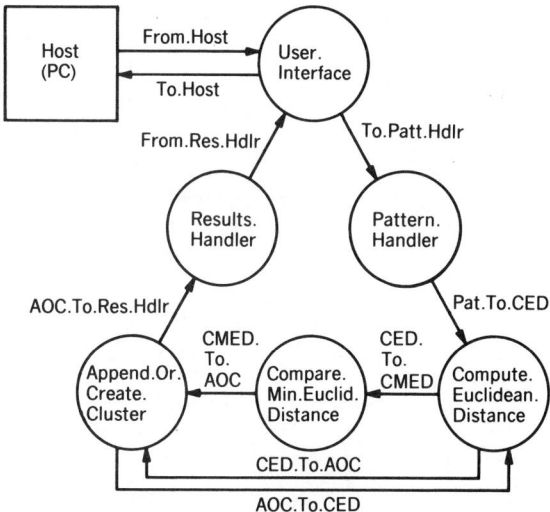

Figure 10.2 *Elementary unsupervised clustered learning.*

(CED), comparing the minimum Euclidean distance (CMED), and evaluating the result by either adding the pattern to an existing cluster or generating a new cluster (AOC). The true parallelism takes effect within the operation of this pipeline—as soon as a given pattern is processed by CED, it is off-loaded to CMED, freeing up CED for the next pattern.

For a five-processor network of T800s, the division of processes among transputers can be done as shown in Figure 10.3. Two processes share one processor, but the effects of time slicing between the two processes are minimal because each process is highly dependent upon host communica-

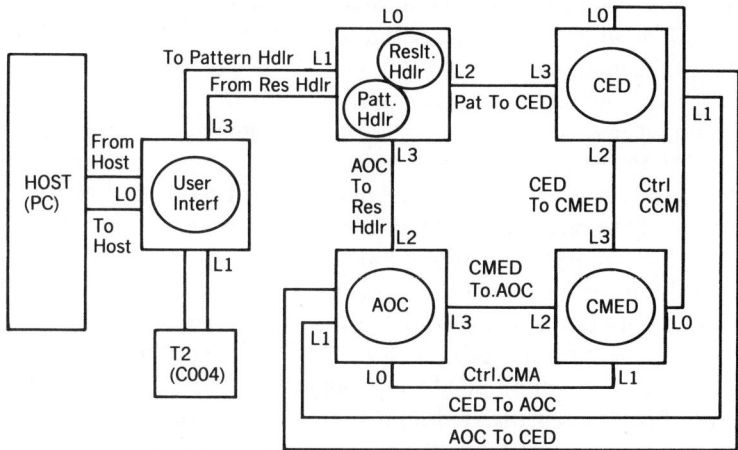

Figure 10.3 *UCL model on five transputers.*

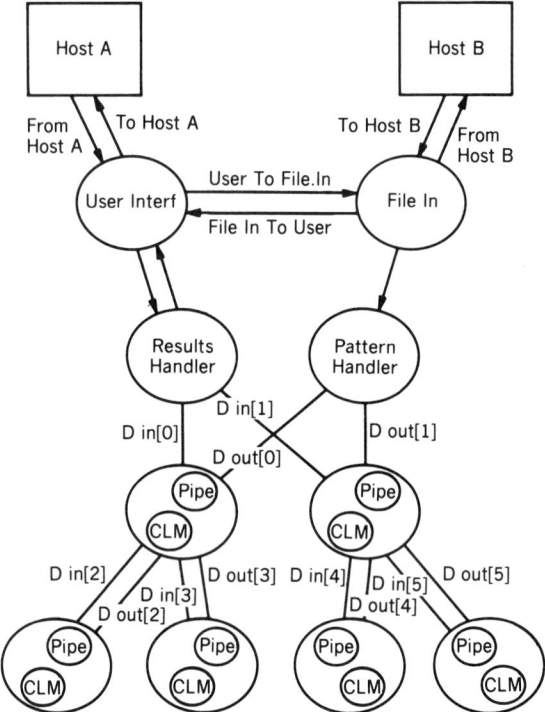

Figure 10.4 *Multiple units of the UCL network.*

tions and the operations are essentially sequential tasks that must be performed at the outset of the overall job.

An alternative solution is to put parallelism into the data flow by dividing up the set of input patterns among several processors (or subsets of processors), each of which is computing the complete learning algorithm, now on a portion of the input set rather than on the full set. This model is illustrated by Figure 10.4. Note that each instance of the unit UCL process could itself be implemented on a group of processors that maintain communication only with the I/O management process and that are otherwise completely independent. The results from this type of model will be several sets of clusters that are ready to be used independently or postprocessed in order to collate the clusters from each set into one set.

The worker farm concept is a common mechanism employed in conjunction with a data flow pipeline for maximizing the benefits of parallelism. Its use for a connectionist model such as the UCL algorithm is most effective when the communication bandwidth can be minimized; thus the bifurcation of data subsets rather than portions of a single network.

The division of tasks among processes and the partition of processes among transputers has as its goal maximizing the compute activity of each

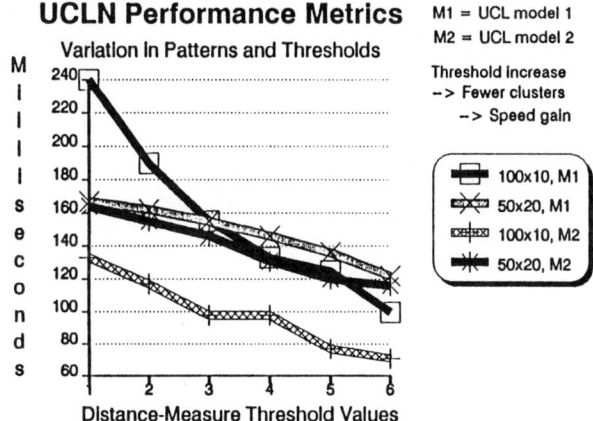

Figure 10.5 *UCL performances. M1 = UCL model 1, M2 = UCL model 2.*

processor. Performance of the model varies significantly, depending upon the number of different multipattern clusters that are established, since the more patterns are classified together, the fewer independent clusters there will be for a given input set. In addition to the type of data input, the selection of a threshold value for determining the cutoff level on cluster-pattern similarities affects the number of clusters recognized. On a single transputer, operations for a group of 80 patterns with 40 attributes each ranged from approximately 1.5 to slightly over 2 s, whereas on a 12-MHz 80286 performance for the same data was in the 55–75 s range. While the algorithm was relatively simple, the fact that most calculations had to be iterated over all patterns and for each attribute within a pattern meant that increases in the number of either patterns or attributes, but especially the latter, would cause significant lags in performance. Such lags could easily make the model unworthy for real-time tasks (image processing, for instance, with input patterns that even after DSP preprocessing might range in the hundreds or thousands of attributes). Some specific test results are shown in Figure 10.5.

10.5 UNSUPERVISED CLUSTERED LEARNING PROGRAM OVERVIEW

The following code provides an opportunity to view the overall program structure from a high level. It is written in OCCAM for a single transputer and within the constructs of the Toolset it will be processed first by the compiler and linker and then by the IBOOT utility that generates the .Bxx run file.

This code is written for the OCCAM D705 and related Toolset which uses a network configuration language that has been standard for some years.

Revisions to the configuration language have been made for the new INMOS D7214 Parallel C and D7205 OCCAM toolsets, which are being released in product form in 1991.

```
--UCL2001.OCC--Main bootable program for an OCCAM
version of the EDCL Unsupervised
--Learning Model based upon Clustering by Euclidean
Distance Measure
--
--Model 2: Use a pipeline architecture to distribute
--the sequential
--tasks for processing each pattern across several
--processes but
--maintain the fundamental algorithm in a single se-
--quential process.
--This model (UCLMAIN) is set up for a single trans-
--puter.
--Version AA
--
-- --Basic Design:
#USE "clustmod.t8h"
#USE "reshdlr.t8h"
CHAN OF PATTERN.PACKET to.pattern.handler, pat.to.
--cluster:
-- PATTERN.PACKET is a protocol for moving around pat-
--tern data sets
CHAN OF SYSTEM.PACKET cluster.to.res.hdlr, from.re-
--sults.handler:
-- a generic protocol for moving variable types of
--data
-- over a given channel
VAL model.id IS "Model 2 ":
VAL version.id IS "Version AA ":
VAL config.id IS "Config: 1 T800":
PAR
  -- edcl.user.interf handles all i/o with host
  edcl.user.interf (fs, ts, to.pattern.handler,
                    from.results.handler,
                    model.id, version.id, con-
                    fig.id)
  -- pattern.handler does rudimentary pre-processing
  of
  -- each pattern
  pattern.handler (to.pattern.handler, pat.to.cluster)
  -- cluster-learning handles the actual euclidean-
  --distance
```

10.5 UNSUPERVISED CLUSTERED LEARNING PROGRAM OVERVIEW

```
-- measurements and classifications into pattern
clusters
cluster.learning (pat.to.cluster, cluster.
to.res.hdlr)
-- results-handler does rudimentary post-processing
results.handler (cluster.to.res.hdlr, from.results.
handler)
```

As an example of how this code, the single transputer version of a configuration file (.pgm), would look when the program is transformed for operations on multiple transputers, we can look at the following example, UCL2003.PGM. This is the configuration file for a version that uses three transputers hard-wired in a pipeline and using the C004 for additional link connections.

```
-- UCL2003.PGM
-- Configuration file for OCCAM version of the Euclidean-Distance
--      Cluster-Learning Network
-- Model 3(00): Use a pipeline architecture to distribute the sequential
--      tasks for processing patterns across n processors, with the
--      main
--      algorithmic process on one processor
--
-- Version 03: Two processors, T800s on a B008
-- Basic Design:
--
--      Processor 0 : eucl.user.intf
--              Get user cmds, read in all patterns at once
--      Processor 0 : pattern.handler
--              Pre-process, transfer pattern to cluster learning
--              module
--      Processor 1 : results.handler
--              Handle all intermed. or final output to user
--      Processor 2 : cluster.learning
--              Computer euclidean distance betw pattern and cluster
--              centers
--              Find satisfactory minimal distances threshold
--              Either update weights or form new active node
INCLUDE "hostio.inc"
#INCLUDE "edclprot.inc"
#INCLUDE "confprot.inc"   -- relating to configuring of C004 #USE
                          -- "edcluser.c8h"
#USE "patthdlr.c8h"
#USE "clustmod.c8h"
#USE "reshdlr.c8h"
#INCLUDE "linkaddr.inc"   -- standard link addresses
CHAN OF SP fs, ts:
CHAN OF PATTERN.PACKET to.pattern.handler, pat.to.clmod:
CHAN OF SYSTEM.PACKET  clmod.to.results, from.results.handler:
VAL model.id IS "Model 2 ":
VAL version.id IS "Version AA ":
VAL config.id IS "Config: 3 T8s":
```

```
PLACED PAR
  PROCESSOR 0 T800
    PLACE fs AT link0.in:
    PLACE ts AT link0.out:
    PLACE from.results.handler AT link2.in:
    PLACE to.pattern.handler AT link2.out:
    edcl.user.interf (fs, ts, to.pattern.handler,
                from.results.handler, model.id, version.id, config.id)
  PROCESSOR 1 T800
    PLACE to.pattern.handler AT link1.in:
    PLACE from.results.handler AT link1.out:
    PLACE clmod.to.results AT link2.in:
    PLACE pat.to.clmod AT link2.out:
    PAR
      -- two procs on one processes because they are lightweight tasks
      pattern.handler (to.pattern.handler, pat.to.clmod)
      results.handler (clmod.to.results, from.results.handler)
  PROCESSOR 2 T800
    PLACE pat.to.clmod AT link1.in:
    PLACE clmod.to.results AT link1.out:
    cluster.learning (pat.to.clmod, clmod.to.results)
```

Extensive use can be made of variant protocols to handle communication of both pattern data, network results and control information, as seen in some of the following protocol definitions employed:

```
PROTOCOL PATTERN.PACKET
  CASE
    -- a single pattern of n attributes, as input to sys in integer
    -- format
    pat; INT::[]INT
    -- ditto but in real32 format, and an int for other data like
    -- active.nodes
    realpat; INT::[]REAL32
    -- parameters
    params; REAL32; INT; INT
    -- quit running
    terminate
PROTOCOL CTRL.PACKET
  CASE
    -- general parameters
    parameters; REAL32; INT; INT
    -- debugging data
    debugdata; INT; INT
    -- quit running
    cease

PROTOCOL WORK.PACKET
  CASE
    -- ditto but in real32 format, and an int for other data like
    -- active.nodes
    patblock; INT::[]REAL32
    -- cluster i.d. number
    clustnum; INT
    -- pattern i.d. number
```

10.5 UNSUPERVISED CLUSTERED LEARNING PROGRAM OVERVIEW

```
    patnum; INT
    -- cluster (node) i.d. number
    wts.request; INT
    actnodes; INT
    -- benchmarking timer value
    timedata; INT

PROTOCOL SYSTEM.PACKET
  CASE
    --     descriptor/ code|count/ no. of ints in msg/ the msg array
    results; INT; INT; INT::[]INT
    -- an error code
    errset; INT
    debugout; INT; INT
    time; INT
    acknowl
    taskdone
    quit
```

Notice that only single-dimensioned arrays are moved over the links, whereas in the actual clustering algorithm, two-dimensional arrays are common. This can be seen by examining the main clustering test processes in the procedure CLUSTER.LEARNING:

```
PROC cluster.learning (CHAN OF PATTERN.PACKET pat.to.cluster,
                       CHAN OF SYSTEM.PACKET cluster.to.res.hdlr)
  #INCLUDE "edclcnst.inc"
  -- all the patterns that are being processed
  [max.patterns][max.attribs]REAL32 patt.table:
  -- set of weights, one per attribute, per pattern
  [max.patterns][max.attribs]REAL32 wts:
  -- results are stored in the cluster table indicating which pattern(s)
  -- fit into which clusters
  [max.patterns][max.clust.members]INT cluster.table:
  [max.patterns]REAL32 distance:
  BOOL going, get.input:
  REAL32 min, threshold, x:
  INT ip, j, n, ninput, ninattr:
  INT active.nodes, cl.num.result, cluster.num:
  INT total.time:
  [max.patterns+1]INT time.check:
  TIMER tclock:
  SEQ
    -- Init vars
    ...   WHILE going
      SEQ
        ALT
          -- Intput on pat.to.cluster (all except terminate only at
          -- beginning!)
          get.input & pat.to.cluster ? CASE
            --{{{ Get special parameters
            params; threshold; ninput; ninattr
              SKIP
```

```
-- Read input pattern into array (must be done before
-- continuing)
realpat; n::patt.table[ip]
  IF
    (ip = 0)
      --Set up the first node
      SEQ
        SEQ i = 0 FOR ninattr
          wts[0][i] := patt.table[0][i]
        active.nodes := 1
        cluster.table[0][0] := 1
        cluster.table[0][1] := 0
        ip := 1
    (ip  (ninput-1))
      ip := ip+1
    (ip = (ninput-1))
      -- All patterns from input set read in (future action
      -- here)
      SEQ
        get.input := FALSE
        -- all done until current set processing completed
        tclock ? time.check[0]    -- Start timing here
    TRUE
      SKIP
  terminate
    SEQ
      cluster.to.res.hdlr ! quit
      going := FALSE
IF
  (NOT get.input)
    SEQ
      tclock ? time.check[1]    -- Get start time for first
                                -- pattern
      --{{{ Cycle thru all patterns in set just read in
      SEQ i = 1 FOR (ninput-1)
        SEQ
          -- Cycle thru all active nodes (clusters)
          j := 0
          WHILE (j  active.nodes)
            SEQ
              -- Compute eucl. dist. measure betw pattern &
              -- actnodes
              distance[j] := 0.0 (REAL32)
              SEQ k = 0 FOR ninattr
                SEQ
                  x := (wts[j][k] - patt.table[i][k])
                  distance[j] := distance[j] + (x * x)
              j := j + 1
          --Analyse the distances, seeking a mininum threshold
          min := minbase
          SEQ k = 0 FOR active.nodes
            IF
              (distance[k]  min)
                SEQ
                  min := distance[k]
                  cluster.num := k
```

10.5 UNSUPERVISED CLUSTERED LEARNING PROGRAM OVERVIEW

```
          TRUE
            SKIP
        x := SQRT(distance[cluster.num])
        IF
          (x >>= threshold)
            cl.num.result := cluster.num
          TRUE
            cl.num.result := out.of.bounds
      -- Process a pattern record
      IF
        (cl.num.result = out.of.bounds)
          -- {{{  No fit w any existing cluster; form a
          -- new node
          SEQ
            SEQ j = 0 FOR ninattr
              wts[active.nodes][j] := patt.table[i][j]
            cluster.table[active.nodes][0] := 1
            cluster.table[active.nodes][1] := i
            active.nodes := active.nodes + 1
            cluster.to.res.hdlr ! debugout; 0; 1
        (cl.num.result = 0) AND (cl.num.result  max.patterns)
          --{{{   Update weights and cluster table
          REAL32 x, y, a, b:
          INT cx, num.member:
          VAL REAL32 one IS 1.0 (REAL32):
          SEQ
            x := REAL32 ROUND cluster.table[cl.num.result][0]
            y := x + one
            SEQ k = 0 FOR ninattr
              SEQ
                a := (x/y) * wts[cl.num.result][k]
                b := (one/y) * patt.table[i][k]
                wts[cl.num.result][k] := a + b
            cx := cluster.table[cl.num.result][0]
            cluster.table[cl.num.result][0] := cx + 1
            num.member := cx + 1
            IF
              (num.member  max.clust.members)
                cluster.table[cl.num.result][num.member]
                     := i
              TRUE
                cluster.table[cl.num.result]
                     [max.clust.members-1]
                     := MOSTPOS INT
            cluster.to.res.hdlr ! debugout; 0; 2
        TRUE
          SKIP
      tclock ? time.check[i+1]
-- Final time processing and output of cluster table
-- Note: time.check[0] has initial start, tc[1] at end
-- of processing 1st pattern, etc. and
-- tc[max.patterns] has final time
-- calculate total time spent (in ticks)
total.time := 0
PAR
  SEQ x = 0 FOR ninput
```

```
              cluster.to.res.hdlr ! results; cldata; x;
                  max.clust.members::cluster.table[x]
        SEQ x = 1 FOR ninput
          total.time := total.time +
                       (time.check[x] MINUS time.check[x-1])
          cluster.to.res.hdlr ! time; total.time
          --{{{ Reset vars for next cycle
          get.input := TRUE
          ip := 0
    TRUE
      SKIP
```

10.6 CONNECTIVITY ISSUES

When building neural applications, the size and number of layers in a single network may dictate some kind of division of neurons within a layer or among layers between multiple processors. Consider a back-propagation engine that has three layers: A (input), B (hidden), and C (output) with 1000 nodes in the A layer, 200 in the B layer, and 50 in the C layer. It is desirable to split the interconnect computations among multiple transputers but with a minimization of communications over the linkds. There is no point in using multiple processors if the data traffic between them negates the value of having additional CPUs at work! So one wants to avoid moving large arrays or segments of arrays from one processor to another.

A solution lies in mapping the connections rather than the nodes among a set of worker processors. Figure 10.6 draws the distinction graphically. If nodes are mapped to select processors, then weight values must be transported from one set of processors to another. The communication load can become counterproductive. However, if connections are mapped to processors, then a specific processor p is the only one that requires certain weights, and data traffic is minimized to the outputs of a group of nodes being sent to a select few other processors.

Consider that processors $p0$ through $p9$ each are responsible for managing the connections between 100 nodes from layer A and the 200 nodes in layer B, where A and B are fully connected as in a typical back-prop implementation. For instance, $p0$ computes $u[i]w[i][j]$ where i varies from 0 to 99 for every jth node in layer B. The result is that $p0$ has produced a partial result for the input values for the 200 nodes in layer B. Likewise, $p1$ through $p9$ have computed partial results for the same 200 nodes in layer B.

What happens next? The partial results must be pooled together and summed so that there are complete input values for each of the 200 nodes. Thereafter the appropriate activation functions for those nodes will determine if they "fire" and have outputs that will be carried into the layer C above. Note that the accumulation of partial results for various layer B nodes can be performed by a set of processors (e.g., $p10$ through $p14$) other than $p0$ through $p9$, meaning that there can be a pipeline effect in operation;

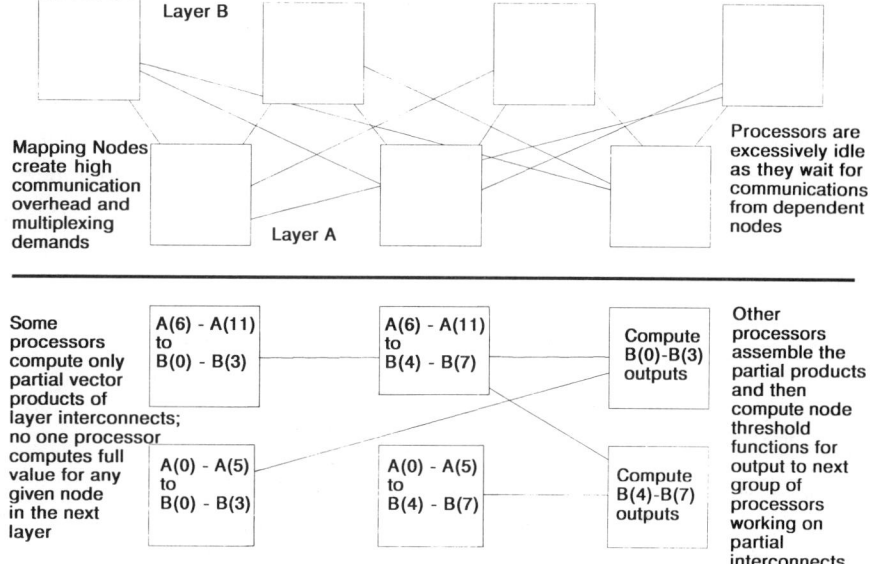

Figure 10.6 *Mapping modes versus connections.*

$p0$ through $p9$ are free to handle the next set of outputs from layer A coming into layer B.

Some added communication requirements imposed by this strategy is that now the partial results must be moved around the transputer network to the appropriate handlers. Also there will be some software overhead in the form of tables and code responsible for doing this routing traffic. However, the overhead is worth the savings, since what is being communicated is a very small set of numbers compared to the larger amount required for passing all the requisite input and weight values for a particular node to a given processor.

This particular kind of approach, dividing the network into regions of connectivity, has been successfully implemented by a number of researchers, including the developers of the N-NETS simulator at NASA Johnson Space Flight Center [see Villareal and McIntyre].

10.7 ANOTHER PROGRAMMING EXAMPLE, USING C

It is important to note that parallel processing on the transputer is not at all limited to the "native" OCCAM language. Several languages are available for the transputer, notably C, PASCAL, FORTRAN, MODULA-2 and ADA. Understandably, C is the more popular of these, particularly in the United States. A special convenience is that implementations can incorporate mixed code (i.e., an OCCAM program can have C processes on some transputers and vice versa). As an example of a C-based neural implementation, the following illustrates a portion of code from a program that implements a simple instar–outstar model of stimulus-response associative learning. This C version is in 3L Parallel C, the original C implementation for transputers. A substantially more powerful C compiler, the D7214 Parallel C Toolset, is now available through SGS-THOMSON and still other compilers are available through third-party providers in the United States, Europe, and Asia.

```
/**************************************************************
STMRSP.C
Main program for STIMULUS-RESPONSE Neural Network Simulator
        Based upon Grossberg instar-outstar neural models

This program demonstrates the instar/outstar model of
neural network for representing stimulus-response learning
and recall. Inputs consist of patterns (e.g., symbols, images)
and 'associated events' (which correspond to the various
'outstars'. Patterns are input and 'fed' to 1 or more activation
grids (arrays of instar neurodes) for training, with 1 or more
associated stimuli (outstar neurodes) simultaneously activated.
Forgetting or decay can be modelled for various grids and outstars.
Recall is modelled by stimulation of outstar nodes with no action
in the instar grids.

*********************************************************************

#include <stdio.h>
#include <chan.h>
#include <string.h>
#include <thread.h>
#include <sema.h>
#include "stmconst.inc"

/* Global declarations */
int errflag,statusflag;
SEMA wbuf_free;
CHAN **in_p, **out_p;

/**************************************************************
main()
   Main program - responsible for communications with host and for
driving the work process(es) that perform all network inits
and computations.
   Communication with host is via PAR C filter.
   Communic. with STMNET (network master) is via in_ports[2] &
out_ports[2] - minimal for now and mostly just in the out direction.
   Communic. with STMDATA (data i/o controller) is via in[3] & out[3]
and is very minimal at present.
**************************************************************/
main(argc,argv,envp,in_ports,ins,out_ports,outs)
int argc, ins, outs;
char *argv[], *envp[];
CHAN *in_ports[], *out_ports[];
```

10.7 ANOTHER PROGRAMMING EXAMPLE, USING C

```c
{
   ...

   int prun, choice, automode;
   int msglen, glsize, slsize, i, s;
   char *msgblock;
   FILE *datfile;      /* with intro of stmdata process, not clear why this
is needed */
   int dispflag, duration, op, pid, psource;
   int pat_info[2];
   int grouplist[MAXGRIDS];
   int stimlist[MAXOUTSTARS];

   in_p = in_ports;
   out_p = out_ports;

   errflag = NULL;
   sema_init (&wbuf_free,1);
   for (i=2; i<ins; i++)
      thread_create (wbuf_read,2*sizeof(int),1,i);

   glsize = sizeof(grouplist) * 4;      /* size in bytes */
   slsize = sizeof(stimlist) * 4;       /* size in bytes */
   datfile = fopen(DATFILENAME,"w");
   choice = NULL;
   prun = ON;
   while (prun != OFF)
      {
      if (choice != RESUME)
         /* references main arrays AND does user i/o - needs major mods */
         initialize(automode,dispflag,out_ports[2]);
         /* init/reset (vars,arrays, etc.) */
      choice = PROCESS;
      while (choice != QUIT)
         {
         choice = init_choice(PROCESS);      /* process, analysis or quit */
         if (choice == PROCESS)
            {
            pid = select_pattern(psource);   /* choice of pattern to work */
            /* send pattern id to data(pattern) i/o proc (stmdata) */
            chan_out_word (PATTERN_SPEC, out_ports[3]);
            /* note temporary sizing */
            pat_info[0] = pid;
            pat_info[1] = psource;
            chan_out_message (sizeof(pat_info),pat_info,out_ports[3]);
            select_grouping(pid, grouplist);
            /* pattern will apply to selected grids */
            select_assoc_stims(pid, stimlist);
            /* choice of outstars to work with */
            op = select_operation(pid);
            /* train, decay, recall, quit */
            duration = select_duration(op);
            /* duration of this operation */
            }
         switch (choice)
            {
```

```
      case PROCESS:
        switch (op)
          {
          case TRAIN || FORGET || RECALL:
            /* build msg block for executing the operation */
            msglen = glsize + slsize + 16;  /* 4 ints + 2 int arrays */
            chan_out_word (msglen, out_ports[2]);
            memcpy (&msgblock[0], &op, 4);
            memcpy (&msgblock[4], &pid, 4);   /* why needed now? */
            memcpy (&msgblock[8], &duration, 4);
            memcpy (&msgblock[12], &dispflag, 4);
            memcpy (&msgblock[16], grouplist, glsize);
            memcpy (&msgblock[glsize+16], stimlist, slsize);

            /* send params to waiting STMNET network master process */
            chan_out_message (msglen, msgblock, out_ports[2]);
            /* formerly in single, serial mode...
            run_cycles(op,pid,duration,dispflag,grouplist,stimlist); */
            /* train/forget/recall with pattern p in group(s) g for
               outstar(s) n */
            break;
              case QUIT:
                choice = QUIT;
                break;
              default:
                disp_string
                    (USER,stdout,"\n Invalid op code - program error!");
                break;
          }
        break;
      case QUIT:
        break;          /* fall out of while loop... */
      default:
        disp_string(USER,stdout,"\n Invalid response - try again!");
        break;
      }
    /* test for error msgs from stmnet or stmdata processes */
    if (errflag != NULL)
      {
      /* later do some branching on diferent values */
      printf ("\nError number %d occurred. Stopping.",errflag);
      choice = QUIT;
        /* no need at this point to do sema_signal(&wbuf_free); */
      }
    /* test for non-error msgs from stmnet or stmdata */
    if (statusflag != NULL)
      {
      printf ("\nStatus msg %d received.",statusflag);
      sema_signal(&wbuf_free);
      }
    }
  /* want to start all over (allows modification of constants)? */
  if (errflag == NULL)
    choice = init_choice(RESET);
  switch (choice)
    {
```

```
      case RESET:
        chan_out_word (RESET,out_ports[2]);
        chan_out_word (RESET,out_ports[3]);
        break;
      case RESUME:
        break;            /* no change to other processes */
      case QUIT:
        prun = OFF;
        if (errflag == NULL)
           {
           disp_string
             (USER,stdout,"\n Program terminated at user request.");
           chan_out_word (TERMINATE,out_ports[2]);
           chan_out_word (TERMINATE,out_ports[3]);
           }
        else
           disp_string(USER,stdout,"\nTermination due to error.");
        break;
      default:
        disp_string(USER,stdout,"\n Invalid response - one more try!");
        choice = init_choice(RESET);
        break;
      }
   }
   disp_string(USER,stdout,"\n\n STMRSP end.");
   fclose(datfile);
   return;
}

/************************************
void wbuf_read(i)
************************************/
void wbuf_read (i)
int i;
{
   int msg;
   for (;;)
     {
     chan_in_word (&msg, in_p[i]);
     sema_wait(&wbuf_free);
     if (msg >= ERROR_RANGE)       /* is it an error or a status msg? */
        errflag = msg;
     else
        statusflag = msg;
     /* the main task is responsible for clearing the semaphore
     once it has analysed the 1-word msg */
     }
}
```

The principle network computations are weight modifications and they occur in the following process, which communicates via channels with the main ( ) process as shown above. Note that the preceding process does not call a function—it sends data back and forth over channels to another main( ) process. The two processes may be on the same transputer (in which case time-slicing occurs) or on separate devices. A configuration map file is used to specify which processes are connected via channels and links.

```
/****************************************************
main (argc,argv,envp,in_ports,ins,out_ports,outs)
****************************************************/
main (argc,argv,envp,in_ports,ins,out_ports,outs)
int argc,inc,outs;
char *argv[], *envp[];
CHAN *in_ports[], *out_ports[];
{
 /* For the moment all the network computation will be performed
    within this single process in order to test the parallelism that
    has been established thus far. Later as each load of parameter
    data comes in over the channel from the root program (stmrsp.c)
    it will be routed to the next available processor who will be
    given those sections of the gridwts, activity, outstar and other
    arrays as are needed */
  int automode,control,prun,i,j;
  char *msgblock;
  int op,pid,dispflag,duration;
  int stimlist[MAXOUTSTARS];
  int grouplist[MAXGRIDS];
  int glsize, slsize;

  sema_init (&inbuf_free,1);
  glsize = sizeof(grouplist)*4;
  slsize = sizeof(stimlist)*4;
  prun = ON;
  while (prun != OFF)
    {
    chan_in_word (&control, in_ports[0]);
    switch (control)
      {
      case RESET:
        break;  /* for now */
      case TERMINATE:
        prun = OFF;
        break;

      case INITIALISE_AUTO || INITIALISE_MANUAL:
        automode = control;
        for (i=0; i<MAXGRIDS; i++)
          randomize_activity(i);
            /* randomize activity of grid neurodes */

        for (i=0; i<MAXOUTSTARS; i++)
          for (j=0; j<MAXGRIDS; j++)
            {
            randomize_wts(i,j,automode);
              /* randomize weights to grid neurodes from outstar */
            if (dispflag == FULL)
              show_wts(i,j);
                /* display resulting grid neurode weights */
            }
        curtime = 0;  /* reset current time to 0 */
```

10.7 ANOTHER PROGRAMMING EXAMPLE, USING C

```
        /* move the next line into stminit... */
        /* set_constants(automode); */
        /* set the constants to user specified values */
        break;

      case PATTERN_DATA:
        chan_in_message ((ROWSIZE*COLSIZE), pattern, in_ports[0]);
        break;

      default:
        if (control > 0)  /* if > 0, then no. of bytes in next msg */
          {
          chan_in_message (control, msgblock, in_ports[0]);
          /* move data from buffer into local vars */
          memcpy (&op, msgblock[0], 4);
          memcpy (&pid, msgblock[4], 4);
          memcpy (&duration, msgblock[8], 4);
          memcpy (&dispflag, msgblock[12], 4);
          memcpy (grouplist, msgblock[16], glsize);
          memcpy (stimlist, msgblock[glsize+16], slsize);
          /* run the appropriate operational fn */
          run_cycles (op, pid, duration, dispflag,
                      grouplist, stimlist);
          break;
          }
        else
          break;        /* some error handling s/b here */
      }
    }
}
```

Additional functions that can be used to implement a worker farm of processes that can be used to further parallelize the tasks of updating weights and activation levels are shown below. Relevant array types and constant values are shown for reference; they are given as "externs" only because these particular functions were declared in a separately compiled unit.

```
extern char pattern[MAXROWSIZE][MAXCOLSIZE];
   /* this contains the pattern to be impressed on the grid of rim
neurodes */
extern float gridwts[MAXOUTSTARS][MAXGRIDS][MAXROWSIZE][MAXCOLSIZE];
extern float activity[MAXGRIDS][MAXROWSIZE][MAXCOLSIZE];
   /* this contains the current activation levels of each grid neurode */
extern float outstar[MAXOUTSTARS]; /* activation level of outstar */
extern unsigned int outstar_save[MAXOUTSTARS]; /*saves history of
outstar's output */
extern int curtime; /* current time (in integral units)   */
/**********************************************************************
****/

/******* Grossberg Activation Constants (set by user or default values)
*****/
extern float A;/* activation decay constant */
extern float T;/* threshold */
```

```c
extern int t0 ;/* transmission time between neurodes */
extern float F ;/* forgetting constant for long term memory */
extern float G ;/* learning constant for Hebbian learning term */

extern SEMA inbuf_free;
extern char inbuf[5000], outbuf[5000];
extern CHAN **in_p, **out_p;

#define NUM_TOP_WORKERS 4
/****************************************************************
void net_worker_thread
  Note! The STMNET task uses one port to communic. with STMRSP on
the root, port [0]. That leaves 1, 2 & 3 free. But the number of
threads may be n. So any <counter> > the number of actual free ports
(1 or 2) must mean: pass this packet on to someone further on in the net.
*****************************************************************
/
void net_worker_thread (cnt, op, as, pg, stim_grid, stim_outstar, groups)
int cnt, op, as, pg, stim_grid, stim_outstar, groups[];
{
  int msglen;
  /* package argument data into packet for channel transmission */
  if (cnt <= NUM_TOP_WORKERS)
    {
    /* send to a directly-wired worker for actual processing */
    chan_out_message (msglen, &outbuf[0], out_p[cnt]);
    sema_wait (&inbuf_free);
    chan_in_message (msglen, &inbuf[0], in_p[cnt]);
    sema_signal (&inbuf_free);
    return;
    }
  else
    {
    chan_out_message (msglen, &outbuf[0], out_p[3]);
    chan_in_message (msglen, &inbuf[0], in_p[3]);
    }
}
/* This will constitute the worker task - either do the work
 or try to pass the job on to a connected transputer. */
/*      process_activation(as,pg,stim_grid,stim_outstar);
     if (op == TRAIN)
       process_weights(as,pg,stim_outstar);   */
/***********************************************/
void process_activation (as,pg,stgr,stout)
int as,pg,stgr,stout;
{
/*  work_cmd = ACTIVATION; */

  compute_activation(as,pg,stgr,stout);

}

/***********************************************/
void process_weights (as,pg,stout)
int as,pg,stout;
{
```

10.7 ANOTHER PROGRAMMING EXAMPLE, USING C

```c
/* int work_cmd;
   work_cmd = WEIGHTS; */
   change_weights(as,pg,stout);
}


/****************************************************************
run_cycles(op,pat,groups,stims,duration,dispflag,chan_info)
trains/forgets/recalls the outstar and grid for "duration" timeunits
weights are modified during this training period
After each synchronous update of the grid, the
activations are displayed.

If cycle duration time is negative, only the grid status after
all "duration" time units will be displayed.
****************************************************************/
void run_cycles (op,pat,duration,dispflag,groups,stims)
int op,pat,duration,dispflag;
int groups[],stims[];
{
  int displayflag, stoptime;
  int a, as, b, pg, counter;
  int stim_grid, stim_outstar;

  displayflag = dispflag;
  if (duration<0)
    {
    duration = -duration;
    displayflag = OFF;
    }
  stoptime = curtime+duration;
  switch (op)
    {
    case TRAIN:
      stim_grid = ON;
      stim_outstar = ON;
      break;
    case FORGET:
      stim_grid = OFF;
      stim_outstar = OFF;
      break;
    case RECALL:
      stim_grid = OFF;
      stim_outstar = ON;
      break;
    }
  for ( ; curtime<stoptime; curtime++)
    {
    /* USER OUTPUT printf("\n Time = %d",curtime); */
    counter = 0;
    for (a=0; a<sizeof(stims); a++)
      {
      as = stims[a];
      if (as >=0)
        {
        save_outstim(as,stim_outstar);
```

```
    for (b=0; b<sizeof(groups); b++)
    {
      pg = groups[b];
      if (pg >= 0)
        {
        counter++;
        /* send over linkout1 or linkout2:
             counter,op,as,pg,stimgrid,stimoutstar,(groups)
             AND the appropriate parts of gridwts and activity */
        /* There must be 2 threads to handle the return input */
        }
    }
}
...
```

The preceding code samples give a flavor of the type of functional constructs that are used to implement parallelism among C processes. As can be expected, the fundamental difference lies in the use of channels, implemented by special functions. The critical factor is that transmission of array segments should be kept to a minimum, lest an excessive amount of time be spent in communication. Maintaining a master-worker network adds some additional cost in software overhead but overall it is one of the most effective ways of addressing this communication bandwidth problem, since a number of workers can be handling smaller chunks or bites of a larger whole simultaneously.

10.8 EVOLUTION AND SPECIALIZATION

With the advent of specialized neural silicon devices (Intel, AT&T, Neural Semiconductor, Micro Devices, Adaptive Solutions are among a few of the institutions developing such devices as products), there is an express need to develop robust methods of interfacing such specialized hardware with the conventional digital processing world. Whether the neural chip in question is of an analog or a digital design, high-speed interfaces are necessary to make sufficient use of the neural chip's computational speeds. It does not suffice to have a high-speed analog engine for implementing neural interconnect and threshold computations in analog hardware if those devices must sit idle and wait for a particular digital processor to send or receive data. Moreover, such interfaces must be flexible, to allow for both different configurations of the same neural device and also different types of neural hardware that may be very specialized for certain tasks. An architecture should for practical purposes be robust enough that a new generation of neural devices will not require an extensive modification of the whole system.

In the same way that 16-bit T2s and 32-bit T800s may coexist and interact within the same network, or that 17-MHz and 25-MHz transputers can be part of the same communicative subsystem, so there should be a capability for building complex neural networks that make use of diverse types of

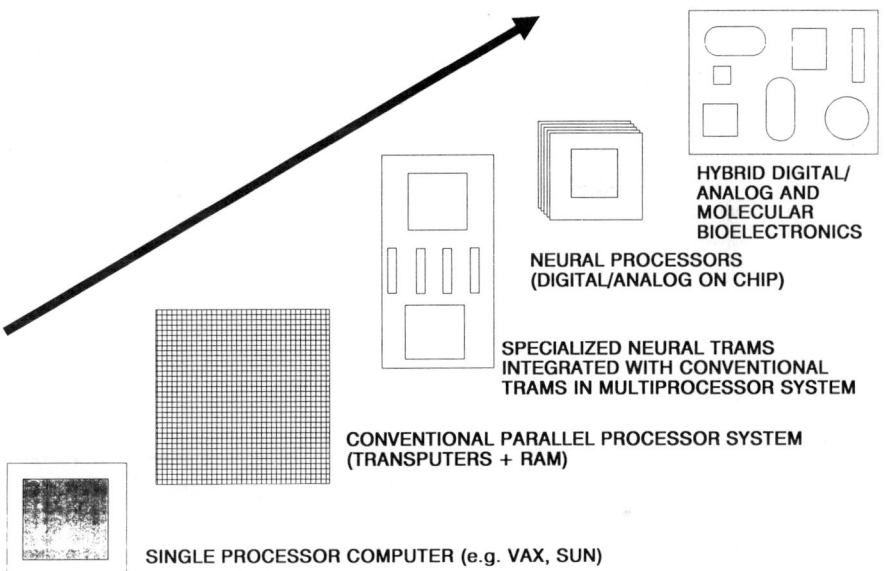

Figure 10.7 *Intelligent evolution of neural systems.*

neural processing "specialists" but without making extraordinary demands upon the types of standard processing that must provide source and destination for data in the network. Figure 10.7 illustrates the concept of neural hardware evolution from the perspective of communicating sequential processors that change over time to include variants that make use of the latest microprocessor and memory technologies but without sacrificing continuity with the present and recent past (i.e., with fielded and tested architectures and systems). This is a value that is often missed but it may be as significant in the long run for the neural system developer as the specific connections-per-second (CPS) or Mips/Mflops ratings of a particular device.

10.9 CLOSING REMARKS

While the emphasis here in this paper has been upon the value of distributed multiple processors for the standard "connectionist" algorithms found in neural networks, there are other important values to be considered. Not the least of these is the ability to easily integrate the neural system with a wide variety of conventional or expert-system modules that can share the same compute-resources. Effective use of dynamic resource schedulers and allocation mechanisms can manage the transputer network in such a way as to share processors among the tasks that most need resources for a given job, be it the neural network in its learning/relearning phase or the expert system in its evaluation of rules and constraint-satisfactions. Additionally there are the advantages of integrating the neural application with real-time I/O devices through a variety of transputer modules (TRAMs). Often the neural network has been severed from the real world into a black box, whereas in real life the multisensory input and output is a continuous and integral part of neural activity—the brain is never severed from the world beyond. Many of these advantages may be viewed as implementation issues but in the domain of neurocomputing it is an increasingly false distinction to separate the implementation and real-time performance of the network from the algorithmic foundations.

BIBLIOGRAPHY

The following articles and books provides a reasonably broad selection of scientific writings on the implementation of neural networks on transputer environments.

Abbreviations:

IEE_89 = Proceedings of the First International IEE Conference on Neural Networks, IEE Press, London, 1989.

NATUG1_89 = Proceedings of the First Conference of the North American Transputer Users' Groups, Salt Lake City UT, April 5-6, 1989.

NATUG2_89 = Proceedings of the Second Conference of the North American Transputer Users' Groups, Durham NC, October 18-19, 1989.

NATUG_90 = Proceedings of the Third conference of the North American Transputer Users' Groups, Santa Clara CA, April 26-27, 1990.

OUG_12 = Proceedings of the 12th OCCAM Users' Group Technical Meeting, Exeter UK, April 2-4, 1990.

Allinson, N. M., Brown, M. T. and Johnson, M. J., "{0,1}n Space Self-Organizing Feature Maps - Extensions and Hardware Implementation," IEE_89.

Abbruzzese, Francesco, "A Transputer Implementation of a McCulloch & Pitts Network," in Chiricozzi, E. & D'Amico (eds), *Parallel Processing and Applications,* Elsevier, North-Holland, 1988.

Beynon, Tricia and Dodd, Nigel, "The Implementation of Multi-Layer Perceptrons on Transputer Networks," OUG-9.

Board, John A. Jr. and Shue-Jen Lu, Jeffrey, "Performance of Parallel Neural Network Simulations," NATUG2_89.

Cavallo, J. M., Baran, R. H. and Restorff, J. B., "Neural Network Simulation with Glauber Dynamics on a Generic Multiprocessor," privately distributed by the author (contact: Naval Surface Warfare Center, Silver Spring, Maryland).

Chong, Michael W. H. and Fallside, F., "Implementation of Neural Networks for Speec Recognition on a Transputer Array," CUED/F-INFENG/TR.8, Cambridge University Engineering Dept., Cambridge, UK, 1988.

Dudziak, M. J., "Partitioned, replicative neural networks for cooperative robot systems," NATUG_90.

Duranton, M. and Mauduit, N., "A General Purpose Digital Architecture for Neural Network Simulations," IEE_89.

Feild, William B. Jr. and Navlakha, Jainendra K., Transputer Implementation of Hopfield Neural Network (in press 1989).

Forrest, B. M., Roweth, D., Stroud, N., Wallace, D. J. & Wilson, G. V., "Implementing Neural Network Models on Parallel Computers," *The Computer Journal,* **30** (5) 1987, 414–419.

Gaudiot, J. L. and Lee, L. T., "Occamflow: a methodology for programming multiprocessor systems," *Journal of Parallel and Distributed Computing,* (in press, 1989).

Gaudiot, J. L., von der Malsburg, C. and Shams, S., "A Data-Flow Implementation of a Neurocomputer For Pattern Recognition Applications," Proceedings of the 1988 Applications of Artificial Intelligence Conference, Dayton, Ohio, October, 1988.

Hameroff, S., Rasmussen, S., and Mansson, B., "Molecular automata in microtubules: basic computational logic for the living state?", in Artificial Life, the Sante Fe Institute Studies in the Science of Complexity, Vol. VI, C. Langton (ed.) Addison-Wesley, Reading, MA 1989, pp. 521–553.

Higuchi, T., Furuya, T., Kusumoto, H., Handa, K. and Kokubu, A., "The Prototype of a Semantic Network Machine IXM," Proceedings of the International Conference on Parallel Processing, 1989.

Jha, S. K., Soraghan, J. J., and Durani, T. S., "Equalization Using Neural Networks," IEE_89.

Johannet, A., Loheac, G., Personnaz, L., Guyon, I. and Dreyfus, G., "A Transputer-Based Neurocomputer," OUG-9.

Kamangar et al., "Implementing the back-propagation algorithm on the Meiko parallel computing surface," Proceedings of the First International Conference on Applications of Transputers, Liverpool, UK, August 23–25, 1989.

Kindermann, J., Muhlenbein, H. and Wolf, K., "Implementierung von Neuronalen Netzen," Technical Report, Gesellschaft fur Mathematik und Datenverarbeitung, 1988.

Lee, F. H. and Stiles, G. S., "Parallel Simulated Annealing: Several Approaches," NATUG2_89

Lee, Y. N., "A Parallel Semantic Net Engine and its Application to Data Modelling," OUG-12.

Leman, Marc and Van Renterghem, Patrick, "Transputer Implementation of the Kohonen Feature Map for a Music Recognition Task," Second International Transputer Conference, Belgian Institute for Automatic Control, Antwerp, October 23–24, 1989.

McFarlane, Donald and East, Ian, "An Investigation of Several Parallel Genetic Algorithms," OUG-12.

Muhlenbein, H., "The dynamics of evolution and learning-towards genetic neural networks," International Conference on Connectionism in Perspective, Zurich, (in press 1989).

Obermayer, K., Heller, H., Ritter, H. and Schulten, K., "Simulation of self-organizing neural nets: a comparison between a transputer ring and a connection machine CM-2," NATUG_90.

Oglesby, J. and Mason, J. S., "Dynamic Scheduling For Feed-Forward Neural Nets Using Transputers," IEE_89.

Patry, Pierre, Salome, J.-C. and Kuchler, P., "Optical Character Recognition of a Network of Transputers," OUG-9.

Prager, R. W., Clarke, T. J. W. and Fallside, F., "The Modified Kanerva Model: Results for Real-Time Word Recognition," IEE_89.

Rasmussen, S., Karampurwala, H., Vaidyanath, R., Jensen, K., and Hameroff, S., "Computational Connectionism Within Neurons: A model of Cytoskeletal Automata Subserving Neural Networks," *Physica D* **42** (1990), 428–449.

Richards, Gareth D., "Implementation of Back-Propagation on a Transputer Array," OUG-9.

Smith, S., Watt, R., and Hameroff, S., "Cellular automata in cytoskeletal lattices," *Physica D* **10** (1984), 168–174.

Schulten, Klaus & Ritter, Helge, "Adaptive Map Algorithms for Visuo-Motor Control of Robots," unpublished paper, Beckman Institute and Dept. of Physics at University of Illinois at Urbana/Champaign.

Soucek, Branko, *Neural and Concurrent Real-Time Systems: The Sixth Generation*, Wiley, New York, 1989.

Sutherland, John, "A Holographic Model of Neurological Function," privately distributed (contact: ANBD Corporation, Hamilton, ONT).

Takada, Hiroshi, "Implementation of Neural Networks on Parallel Systems," Second International Transputer Conference, Belgian Institute for Automatic Control, Antwerp, October 23–24, 1989.

Vanhala, J. and Kaski, K., "Simulating neural networks in distributed environments," Proc. of the 11th technical meeting of the Occam User Group, Edinburgh, Scotland, September 25–26, 1989.

Villareal, James and McIntire, Gary, "Simulation of Neural Networks Using a Transputer Array," Proceedings of the 1987 Space Operations Automation and Robotics Conference, Houston, Texas, 1987.

CHAPTER 11

Neural Bit-Slice Computing Element

JOSEPH YESTREBSKY
PAUL BASEHORE
GERALD REED

11.1 INTRODUCTION

In this chapter we describe a neural bit-slice (NBS) building block integrated circuit (IC) constructed in digital logic. The NBS is intended to be a hardware solution for the construction of synchronous neural networks and for general parallel processing applications. The slice architecture of the NBS allows devices to be interconnected efficiently, allowing many different neural network architectures to be constructed. Networks may be expanded along either rows or columns efficiently by simply adding NBS ICs to the network. Each slice contains eight digital neurons, with 15 "hard" parallel synaptic inputs per neuron. Expansion to beyond 256 synapses per neuron is possible with modest additional hardware (for a minimum of 2048 synapses per IC). The NBS may be trained under microprocessor control, with 16-bit integer synaptic weights stored in external memory. Processing delay through a single NBS with eight synapses per neuron is 7.2 μs, providing performance equivalent to that of a 55-MIPS processor.

11.2 LOW-COST DIGITAL NEURAL DEVICES

People have always been fascinated with the workings of the human mind. Its ability to rationalize, synthesize ideas, and perform immediate recall of a face or name just from partial information, ranks it as the most powerful and versatile computer known to man. For nearly a century, psychologists and neurobiologists have researched the subtleties of the human mind: its microstructure, internal information transfer, and the information format. Researchers have found the fundamental elements of the brain, neurons, to be densely interconnected and highly parallel, with interactions that are complex analog functions of time [1,2]. Still issues of intense research, however, are how these neurons are globally connected, and how they are taught to perform different functions.

Neural networks derive their power from massive parallelism of neurons, requiring large numbers of neurons, and many interconnections between neurons. A neuron by itself functions basically as an integrator and is not believed to be particularly intelligent. Neural networks "learn" to perform certain tasks by systematically regulating the kind and amount of information that flows between neurons. This regulation is accomplished through synaptic weights.

Most artificially created neural networks model neurons and their interconnections mathematically. A neuron is modeled by an equation that expresses the internal state of the neuron as a number that is a linear function of the outputs of all neurons connected to it. This index of the neuron's internal state is often called the *dot product*. The output of a neuron is customarily some nonlinear function or thresholding of the neuron's internal state.

The network as a whole is modeled as a system of equations and is simulated using matrix manipulation techniques. Most software simulation approaches remove the parallelism of biological neural networks by sequentially calculating the state of each neuron in turn and then iterating the calculations over the entire network to capture the interactions among neurons. Such calculations are complex and lengthy, and often require specialized floating-point acceleration hardware.

The NBS approach is also a simulation of a biological neural network, but one which, unlike most software simulations, retains much of the parallelism of its organic counterpart. By providing interconnectability of multiple digital very large scale integration (VLSI) slices, all operating in parallel, the NBS approach is able to solve for the internal state of many interconnected neurons simultaneously.

In recent years, engineers and computer scientists have begun studying both analog and digital neural network models and architectures [3]. These models are approximations to the mathematical models derived by neurobiologists. The investigative work to date has been limited primarily to computer simulations and rather simple hardware implementations. More sophisticated hardware implementations are limited by the practical problems of packaging large numbers of neurons with the required interconnectivity. However, this research has established that some interesting practical problems can be solved using these neural network models, particularly in regards to classification, pattern recognition, and image processing.

The advent of inexpensive VLSI/ULSI provides the opportunity to construct low-cost digital neural devices with significant numbers of neurons and synapses. In this chapter we describe such a device. The NBS is a digital neural network element containing eight neurons, whose design allows multiple devices to be interconnected to form arbitrarily large networks. The NBS design is independent of network architecture, and device architecture easily accommodates synapse expansion. The NBS contains no training algorithms directly, but may be trained with any appropriate algorithm implemented on an external central processor unit (CPU) connected to the NBS. Sixteen-bit integer synapse weights are stored in external memory, which may be expanded as the network and number of synapses expands. Consistent with a slice philosophy, the NBS is not intended to be a single-chip solution for network construction, but when used in conjunction with some inexpensive "glue" logic may be used to construct arbitrarily large networks. The device is constructed in complementary metal-oxide semiconductor (CMOS) technology, contains 22,000 transistors, and is packaged in a 68-pin PLCC.

11.3 DEVICE ARCHITECTURE

Figure 11.1 illustrates the NBS architecture. The NBS is composed of eight individual neurons, plus internal registers and control. Each of eight individual synapse inputs is connected in parallel to each neuron. Additionally, the output of any neuron within an NBS IC can be coupled into any other neuron in the same device through a programmable synaptic interconnect. This effectively provides 15 "hard" parallel synaptic inputs into each neuron. An external weight memory stores the synapse weights and the neuron bias values.

Registers and control within the NBS allow an NBS based network to operate autonomously or under external CPU control. Under CPU control the NBS can be "taught" to perform specific functions. In such real-time learning applications, the CPU is responsible for implementing the learning algorithm. For the purposes of teaching an NBS network, the CPU can access either the dot product or the thresholded outputs of the neurons through an external CPU bus. New synaptic weights computed by the CPU can then be written into the weight memory (random-access memory; RAM) using the CPU and register address buses. Weight storage external to the NBS allows for more efficient network expansion, and reduces the NBS cost and complexity. For applications where the network is trained once and never modified, the synapse weights (perhaps generated on a separate com-

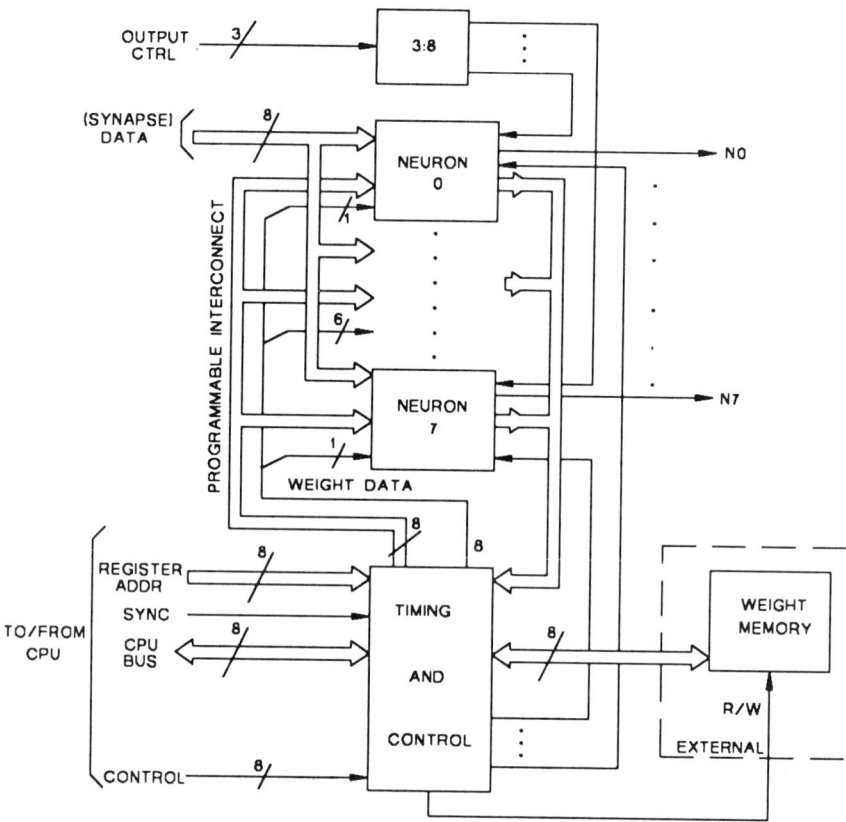

Figure 11.1 NBS architecture.

puter) may be loaded into a read-only memory (ROM), which would become the weight memory.

Additional control functions are included for the purposes of constructing and controlling large networks. These include output parallel-to-serial conversion for expansion purposes, and external synchronization, pause, and reset control. Accumulator overflows are also flagged to the CPU.

11.4 NEURAL NETWORK ARCHITECTURE

Figure 11.2 illustrates the NBS neural network architecture, with neuron 1 shown in the dashed block. The neurons are digital, requiring logic inputs. Each neuron is composed of an accumulator which sums the (weighted) inputs of a number of synapses [4]. The device has parallel and serial operating modes. Irrespective of mode, however, data is processed in the neuron in bit-serial fashion, since serial adders and multipliers are easy and efficient

11.4 NEURAL NETWORK ARCHITECTURE

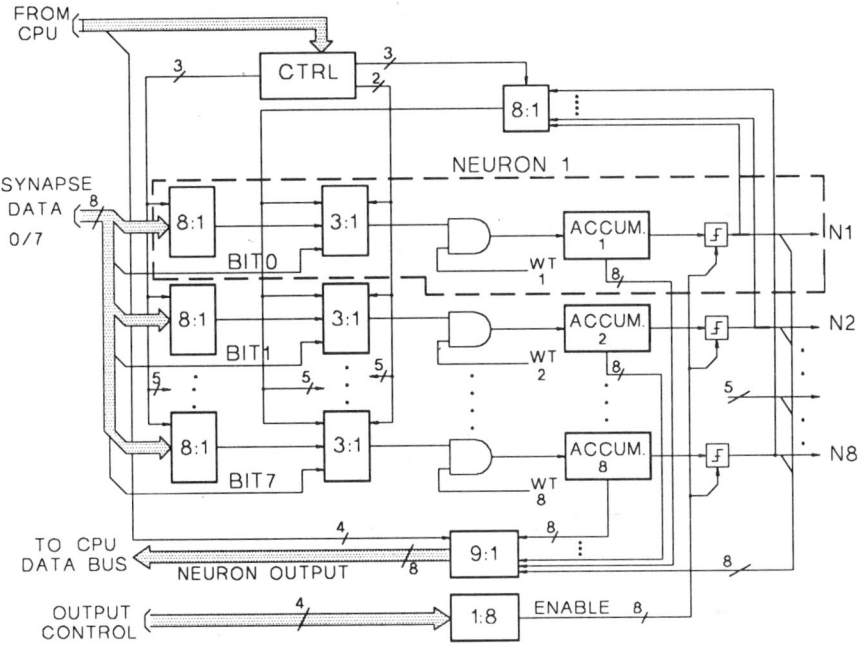

Figure 11.2 Neural network architecture.

to construct. This approach is powerful in that it is efficient, without increasing processing time in pipelined data processing systems, which are in common usage today.

With the NBS in parallel input mode, each neuron has eight external synapses (8:1 selector), and seven feedback synapses (3:1 selector).

When in serial mode, each neuron is connected to only one hard synaptic input. Serial mode may be used to provide synaptic expansion, or to "groom" certain inputs (i.e., connect certain inputs only to certain neurons), typically through some external "glue" logic. Neurons 1 through 8 operate in parallel and synchronously. For each neuron, the output of the steering functions (i.e., the selected input source) is bitwise multiplied with a 16-bit weight value (AND gate). The results of successive multiplies are summed in an accumulator. The resultant accumulator value is then applied to a nonlinear threshold function, whose output is the neuron output. By using a 16-bit weight (15 bits plus sign), weight values can range from $+32767$ to -32768. If the weight values are constrained to the range $[+127, -128]$, then using 16-bit weight lengths minimizes the chance of accumulator overflow. Utilizing such an approach thus guarantees synapse expansion to a minimum of 256 per neuron (2048 per IC). (This underscores an important point regarding digital neurons which sum a set of positive and negative values sequentially, that the summation results are a function of the summa-

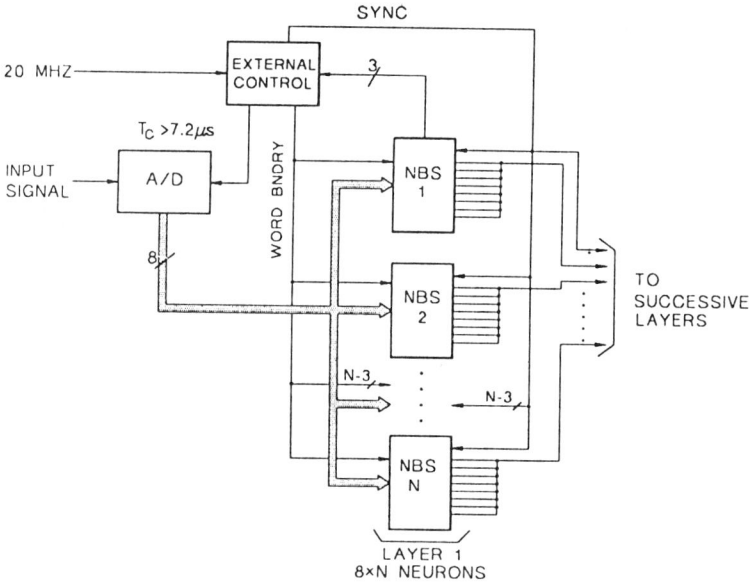

Figure 11.3 *Real-time pipelined neural network (parallel mode).*

tion order. Some analog neuron approaches may also exhibit this phenomena.)

We term one processing cycle of a neuron a frame. The following discussion pertains to parallel mode operation. Over one frame, each of the eight external inputs is sampled under NBS control, and multiplied bitwise by the appropriate synaptic weight in the neuron. Thus, to multiply and sum over eight bits requires 8 × 16 or 128 clock cycles. If feedback is not selected, then the frame is complete; with all seven feedback paths enabled into neuron 1 (and assuming that only thresholded outputs are fed back), the maximum frame time is (8 + 7) × 16 or 240 clock cycles.

Normally, the neuron outputs a new value every frame. Through external control, however, the neuron will update its output only on user-defined input word boundaries. By this technique, synapse expansion is effectively accomplished (up to a minimum of 256 for [+127,−128] weight values), albeit at the expense of processing speed. (Note that this speed penalty does not apply when an NBS network is employed in an 8-bit pipelined application, such as is illustrated in Fig. 11.3). The bias for each neuron is loaded once per frame or input word boundary, as appropriate, which adds 16 clock cycles to the neuron processing time.

The user may select from three possible nonlinear threshold functions. All threshold functions are hard-limiting, and are depicted in Figure 11.4. The double threshold transfer function (#2) is known to minimize the number of neurons and layers necessary to solve some linear inseparability problems (e.g., XOR), and is thus included as an option [5]. Alternatively, the user may choose to pass the accumulator dot product directly to the neuron

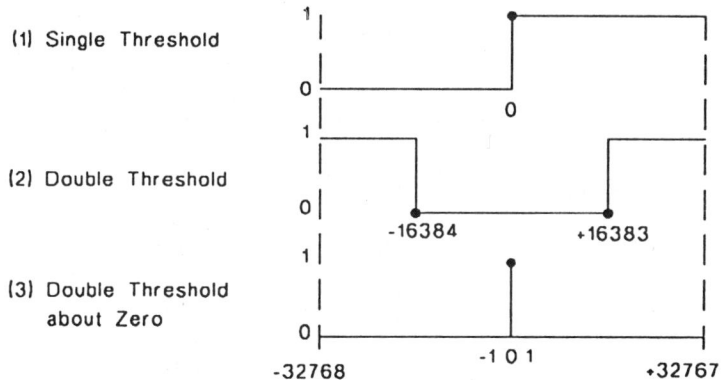

Figure 11.4 *Threshold function options.*

output. The dot product of any neuron may also be fed back as inputs to other neurons in the same device using the feedback mode. This operating mode adds 8 × 16 clock cycles to the neuron processing time for each frame in which feedback is selected.

11.5 NETWORK EXPANSION/ARCHITECTURE

Network expansion takes three basic forms: (1) synapse expansion, (2) expansion of neurons on a layer, and (3) layer expansion. For the purposes of the following discussion, global interconnectivity between neurons in successive layers is assumed. As discussed earlier, synapse expansion may be accomplished readily by external control of input word boundaries. (Alternative parallel approaches for synapse expansion exist that reduce processing time, but require many more NBS devices.) The parallel, globally connected input structure of the NBS guarantees that each input is connected (by a weight) to each neuron within that device. To expand neurons on a layer, then, simply requires adding NBS devices to that layer, and connecting together the input buses of all devices on that layer. If synapse expansion were being performed on a layer, then all devices on that layer would be controlled by the same word boundary control signal. Note that expanding a network layer does not increase the processing time through that layer.

Assuming a significant number of neurons on a given layer, one technique for expanding to successive layers is indicated in Figure 11.5. In this example, the neuron 1 outputs of all NBS devices on a layer may be wire-Ored, and so forth, for all neurons up to neuron 8, creating an 8-bit bus that can be connected to the parallel inputs of all NBS devices in the next layer. The NBS devices in the previous layer would then be enabled sequentially to communicate to the next layer using some simple control logic. (Disabling an NBS disables only its outputs; the device may continue to process inputs.) This bus-oriented approach to layer expansion is efficient in NBS utilization,

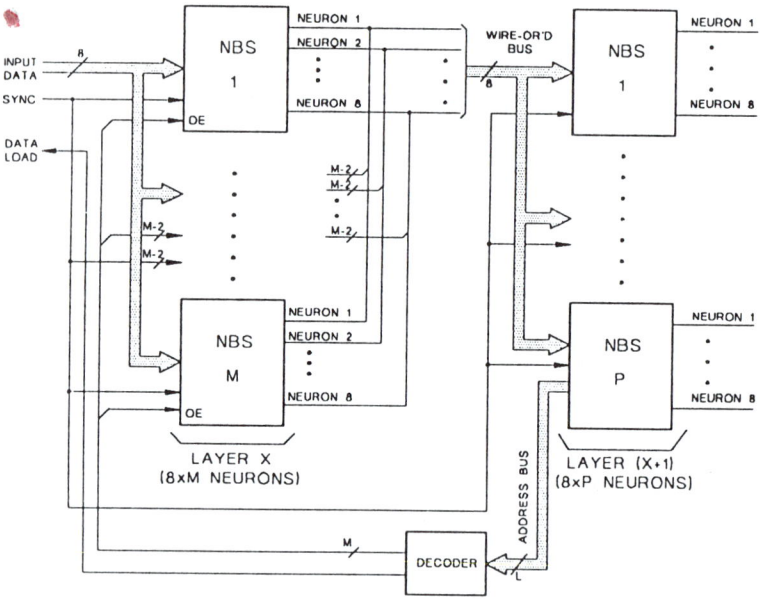

Figure 11.5 *Network expansion of layer X to layer X + 1.*

but increases processing time. In Figure 11.5, the processing time through layer X + 1 is given as $(N + 1) \times 16$ clock cycles, where N represents the number of neurons on the previous layer. As well, data cannot be applied to layer X at a rate faster than this. Where processing time is a factor, other, more parallel, expansion approaches are possible, but require more NBS devices.

A large number of different network architectures have been studied, including cascades of two or more differing architectures [3]. Even with this diversity, however, networks possess only two basic underlying characteristics, feedforward and feedback. The NBS is inherently a feedforward computing element, but can also provide feedback efficiently on the same layer, as is necessary when constructing Hopfield networks. Sparsely interconnected networks are also easily constructed with the NBS simply by setting the appropriate synaptic weights to zero. Thus the NBS is network architecture independent and may be used to construct almost any conceivable network.

11.6 NBS/NETWORK PERFORMANCE

Computing performance of the NBS can be viewed in both absolute and relative terms. In absolute terms, a three-layer feedforward network with eight neurons on each layer, fully interconnected, can process an 8-bit input in 21.6 μs using a 20-MHz clock. Each layer exhibits a processing delay of 7.2 μs. Delay through a layer is independent of the number of neurons on that layer. However, processing delay across a layer does increase as the number of neurons N in the previous layer increases beyond 8. The processing time increases approximately linearly with increasing N.

Comparing the NBS against a conventional CPU takes into account the parallel processing nature of the device. To perform the same logical tasks as a single NBS would require a general-purpose CPU, inherently a sequential function, to perform conservatively 396 instructions, equal to a 55-MIPS (million instructions per second) processing rate. Thus we expect an NBS-based network to outperform many currently available software neural network simulators, and to be more cost-effective than current hardware simulators or hardware-accelerated software simulators.

In terms of neural network benchmarks, the NBS can provide a *minimum* of 256 $\times$ 8, or 2048 interconnects (dependent on maximum weight values and limited only by accumulator overflow), and 10 million interconnects per second.

11.7 CONCLUSIONS

We have presented a neural slice computing element whose architecture allows convenient and efficient network expansion. The element implements digital neurons and provides features necessary to incorporate learning. Device design is independent of network architecture, allowing it to be useful in constructing varying types of architectures. Network processing time increases approximately linearly with network size. The element is inexpensive and possesses a small processing time, so that a network constructed of such elements should be very competitive with other currently available hardware and software approaches.

REFERENCES

[1] W. S. McCulloch, and W. Pitts, A logical calculus of the ideas immanent in nervous activity, *Bull. Mathematical Biophysics* 5, 115–133 (1943).

[2] F. Rosenblatt, *Principles of Neurodynamics,* Spartan Press, New York, 1962.

[3] R. P. Lippman, An introduction to computing with neural nets, *IEEE ASSP Magazine* (April 1987).

[4] M. Minsky, and S. Papert, *Perceptrons,* MIT Press, Cambridge, Mass., 1969.

[5] California Scientific Software, *"Brainmaker" software manual,* 1988, Ch. 2, p. 56.

PART II

Integrated Neural-Knowledge-Fuzzy Hybrids

Data preprocessing for AI systems
Fuzzy data comparator with neural postprocessor
Neurally inspired rule-based reasoning
Symbolic-connectionist processing
Adaptive algorithms and multilevel processing

CHAPTER 12

Data Preprocessing for AI Systems: A Maximum Information Viewpoint

JOSEPH YESTREBSKY
MICHAEL ZIEMACKI

12.1 INTRODUCTION

Artificial intelligence (AI), including neural network and fuzzy logic systems, shows promise in its abilities to process and classify noisy or incomplete data and disjoint data sets. To date, however, classification of such inputs is still not to the levels of success necessary to warrant application of such systems in manufacturing or control environments. In many situations, a significant amount of the problem lies in the variance of the input data, and the problem is compounded when input signals are corrupted by additive Gaussian white noise, which itself exhibits a significant variance. As a result, of the input data AI systems must process, some are either not related to or necessary for the classification process, or hide the classification related information.

The classification success of AI systems may be improved either by (1) constructing larger and more complex systems with greater discrimination capability, or (2) preprocessing the source data to remove redundant or noisy information that is not required as part of the classification process. The first approach, while viable in some cases, quickly leads to large networks with all the attendant problems of delayed network response and increased training times when the input data are extremely noisy. With data preprocessing, however, one might expect that for equivalent networks, the

---

Neural and Intelligent Systems Integration, By Branko Souček and the IRIS Group.
ISBN 0-471-53676-8 ©1991 John Wiley & Sons, Inc.

system that employs preprocessing will be more robust (i.e., greater classification success rate) and exhibit a reduced training time.

In this chapter, we consider the notion of data preprocessing as a means of improving AI system classification performance. Several data transforms commonly used in communication and information theory are considered as the preprocessing operators. The concept of differential entropy is used to measure the reduction of noise and redundancy of data applied to an AI system. As data transformations invariably distort or destroy some aspects of the original data, the consequences of such features are explored and discussed, along with uniqueness of the transform.

12.2 DIFFERENTIAL ENTROPY MEASURES INFORMATION CONTENT

Figure 12.1 illustrates a typical neural-network-based classification system. Such systems are under study as replacements or augments for expert systems or human classifiers, especially in situations where such systems can outperform either of the latter two classifiers. Unfortunately, these situations usually exist where the input data are either very noisy, contain significant information not essential to the classification process, or where the information content that distinguishes one class from another is very small.

Figure 12.2 illustrates the general form of the proposed preprocessing neural network classification system. In this system, to achieve the goal of reducing neural network complexity and improving classification success, the preprocessing transform function should remove as much of the unnecessary or corrupting information from the incoming data as possible, without otherwise impairing the classification process. Thus, choosing a transform requires an understanding of the characteristics of the transform, and how they may affect the classification process.

To measure the effectiveness of the transform in achieving the stated goals, we must measure how the transform function changes the information content of the data applied to its input. One approach is to measure the entropy change between the data at the input and output of the transform function. Entropy is used commonly to measure the information content of a data source. Source data entropy H is defined as follows for a data source with memory [1]. Consider a random (data) vector X composed of n successive discrete-valued events generated by a random stationary process. Let

Figure 12.1 Neural-network-based classification system.

12.2 DIFFERENTIAL ENTROPY MEASURES INFORMATION CONTENT

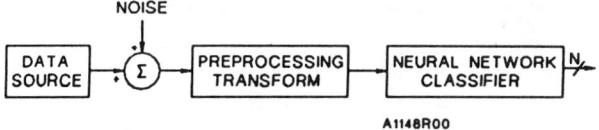

Figure 12.2 *Preprocessing neural network system.*

$P(x)$ be the probability that the vector X takes the value x (a specific n-event sequence). The entropy on a vector basis is

$$H(X) = - \sum_{\text{all } x} P(x) \log_2 P(x)$$

and the entropy per event is given as

$$H = \lim_{n \to \infty} \{n^{-1} H(X)\}$$

For a memoryless source, the entropy of its output data reduces to the (common) form:

$$H = - \sum_{i=1}^{M} P_i \log_2 P_i$$

(For sampled data the units of entropy are bits.) While the probabilities involved in these relationships are almost never known a priori, they may be approximated from samples of the source data.

The definition of entropy is based on the concept that information is defined in terms of a measure of uncertainty: the less likely a message, the greater its information; the more likely, the lesser its information. Thus, random noise has a large entropy, since its probability distribution function (pdf) is uniform. However, not all information is *useful*.

As an example, we compute the entropy of zero-mean white Gaussian noise digitized into 8-bit data words. Over a sufficiently long sample of (digitized) noise values, the probabilities of values are uniformly distributed, so that the probability of any word occurring is 1/256, or .0039. So the entropy

$$H = -256 \left(\frac{1}{256} \log_2 \frac{1}{256} \right)$$
$$= 8 \text{ bits}$$

implying that to precisely express the noise data, all 8 bits of information are important. Note, however, that while the entropy of noise is quite large (implying a large information content), the useful information content is zero.

We argue that if the (classical) entropy of a data source is reduced without degrading the classification success, then the information that has been removed is at least nonuseful information. Since the maximum entropy of a data source is $\log_2(M)$, where M is the number of quantization levels, then for a noiseless source

$$\text{Redundancy} = \log_2(M) - H_1$$

serves as a definition of redundancy, where H_1 is the useful information content of the data source.

12.3 PREPROCESSING TRANSFORMATIONS

Transformations that exhibit appealing characteristics for the problem at hand fall into two basic categories: (1) redundancy reduction and (2) noise reduction. (Some transforms have characteristics that fall into both categories.) We examine two transformations from category 1 (Fourier and Gabor) and one from category 2 (correlation). The histogram transformation, commonly used in image processing is also examined and is found to be related to the magnitude of the Fourier transform.

12.3.1 Redundancy Suppression Using the Fourier Transform

Removal of redundant or nonessential source information is closely related to data compression in communication systems, a field that has seen significant advancement [1,2]. Among those techniques that reduce data redundancy is transform coding, of which a subset is Fourier transform (FT) coding.

Generally in such systems where redundancy reduction is being attempted, sampled data are Fourier-transformed and then quantized. Redundancy reduction is not primarily attributable to the transform, however. The function of the transformation is to increase the independence between the data samples and to compact the signal energy into a small number of components, making the quantization process more efficient [2].

The power of the Fourier transform in such applications is attributable to two primary characteristics. (1) FT representation of a signal is composed of separable magnitude and phase components. (2) Only the phase function of a Δt time-shifted signal FT is modified, and then only by an additive linear phase ($\omega \, \Delta t$). Property 1 has been successfully employed in voice compression systems where only the magnitude information is transmitted, with some compromise in voice quality. However, while the phase function of the FT is unique, the magnitude function is not [3].

Considering property 2, it should be clear that for successive data that are merely time-shifted, only the subsequent differential phase information need be transmitted. This reduces the source data rate by requiring only data that are useful (particularly important in image processing) to be transmitted or processed.

From Lynch [1], the data compression ratio of a communication system employing transform coding is

$$R = \frac{S_u k_u}{C_s \bar{k}_c + \text{Sync}}$$

where S_u = original number of samples in a block,
k_u = number of bits per original sample,
C_s = number of selected coefficients in transformed block,
$\bar{k}_c$ = average number of bits per selected coefficient, and
Sync = number of bits in the block sync word.

According to Lynch, data compression ratios of between 2:1 and 10:1 are achievable. Assuming a 10:1 compression ratio, then for a signal quantized into 8 bits/sample, the resultant data would be 0.8 bit/sample, a significant reduction in information that must be applied to an AI system. Finally, by the uniqueness property of the Fourier transform, the resultant compressed data must be unique, eliminating the possibility of AI classification errors associated with nonunique input data.

The histogram transformation used commonly in image processing is closely related to the magnitude component of the Fourier transform. For example, to compute the histogram function of a (digital) image data sample, one merely counts the occurrences of binary 1's in the data sample, of 2's in the data sample, and so forth, and then plots the number of occurrences versus the ordered binary values. This corresponds to the FT magnitude function, which is a plot of the strengths of frequency components versus the ordered frequency values. This transform does not preserve positional information of the original data sample, however, and so the transformation results are not unique.

12.3.2 Redundancy Suppression Using the Gabor Transform

Daugman [4] has proposed a method of image compression using a set of generally noncomplete and nonorthogonal elementary functions attributable to Gabor. In this approach, Daugman develops a (finite) series approximation $A(x,y)$ to a set of image data $I(x,y)$ according to

$$A(x,y) = \sum_{i=1}^{n} a_i G_i(x,y)$$

where the a_i's are the unknown coefficients (similar to the coefficients of a Fourier series expansion). We take for granted that a set of a_i's can be found that sufficiently minimize the summed squared difference between $I(x,y)$ and $A(x,y)$. The $G_i(x,y)$ are the Gabor elementary functions, given as

$$G_{mnrs}(x,y) = \exp(-\pi[\alpha^2(x - mx_0)^2\beta^2 + (y - ny_0)^2]) \cdot \exp(-2\pi i[rx + sy])$$

These elementary functions have the interesting property that they contain positional (x_0, y_0), spatial scaling (α, β), preferred orientation and spatial frequency, and tuning bandwidths for orientation and spatial frequency. (Note that the second term on the RHS is comparable in form to the frequency translation property of the Fourier transform.) On the other hand, while it is unclear at this time whether the a_i's are unique, it is suspected that since the elementary functions are generally neither orthogonal nor complete, uniqueness problems may result.

Through this transformation, data compression is accomplished in that rather than processing image data, all that need be processed are the coefficients a_i. Note that while many more coefficients a_i exist than pixels of the original image, these a_i can be efficiently encoded (e.g., Huffman coding), which leaves only the minimum set $\{a_i\}$ necessary for application to the AI system.

In an experiment conducted by Daugman, a 256 × 256 pixel image quantized to 8 bits, with an initial entropy of 7.57 bits (nearly the entropy of white noise) is compressed to a set of Gabor coefficients. The entropy of the coefficients was found to be 2.55 bits, a reduction in information of about 3:1. The image was then reconstructed with minimal distortion from the coefficients alone, suggesting that the coefficients contained nearly all of the necessary image information, and that redundant information had been removed.

12.3.3 Noise Suppression Using Correlation

As regards data processing, data correlation exhibits some interesting and useful properties. While these properties are easily derived and well documented [5], our focus will be on the noise reduction and translation-invariance properties of the correlation operator.

The general cross-correlation function (CCF) relating data signals $x(t)$ and $y(t)$ may be expressed as

$$R_{xy}(\tau) = \int_{-\infty}^{\infty} x(t)y(t + \tau)\,dt$$

where t is the variable of integration and τ represents a time delay. By replacing y with x in the expression, the autocorrelation function (ACF) results. This expression evaluated at some delay value τ represents the degree to which the two data signals x and y are similar when y is delayed by time τ.

In some applications, additive white noise may be a problem (e.g., classification of signals in biomedical applications). Basically, the desired signal is buried in noise, such that it is indiscernible when viewed on an oscilloscope. In the frequency domain, the uniform power spectral density (PSD) of white noise is added to the PSD of the desired signal. Fortunately, the noise and signal components are uncorrelated, and even the noise samples are uncorrelated among themselves. Since a signal is always correlated with itself, then performing an autocorrelation results in a set of correlation scores based on the input data but with the noise component eliminated. This correlation data, rather than the original signal, may then be applied to the AI classification system.

To further illustrate this point, assume that signals x and y are at least partially correlated and corrupted by noise, and that the correlation value is determined over a (statistically) sufficiently large time interval. Then any noise averages to zero, and only a correlation score relating the (noiseless) sequences x and y remains for the given delay value τ. Repeating this procedure over a range of τ results in a set of correlation scores that relates the similarity of x and y in the absence of noise.

For a given signal, since the signal and corrupting white noise are uncorrelated, then the ACF of the signal plus noise is independent and unaltered by the noise component, no matter what its amplitude. This can be shown as follows. Consider a signal $x(t) = y(t) + n(t)$, where $y(t)$ is the desired signal, and $n(t)$ is uncorrelated noise. Then from the correlation integral,

$$R_x(\tau) = \int_{-\infty}^{\infty} x(t)x(t + \tau)\, dt = \int_{-\infty}^{\infty} [y(t) + n(t)][y(t + \tau) + n(t + \tau)]\, dt$$

$$= \int_{-\infty}^{\infty} y(t)y(t + \tau)\, dt + \int_{-\infty}^{\infty} n(t)y(t + \tau)\, dt$$

$$+ \int_{-\infty}^{\infty} y(t)n(t + \tau)\, dt + \int_{-\infty}^{\infty} n(t)n(t + \tau)\, dt$$

$$= \int_{-\infty}^{\infty} y(t)y(t + \tau)\, dt$$

since the signal and noise are unrelated. So the resulting autocorrelation function is noise-invariant and based only on the desired signal. In image and pattern recognition applications where signals are corrupted by noise, one may consider using the correlation scores rather than the original image data.

Due to the integrating nature of the correlation operator, correlation functions are at least continuous and piecewise smooth. As noted earlier, for a signal with additive white noise, if the signal-to-noise ratio is small, the

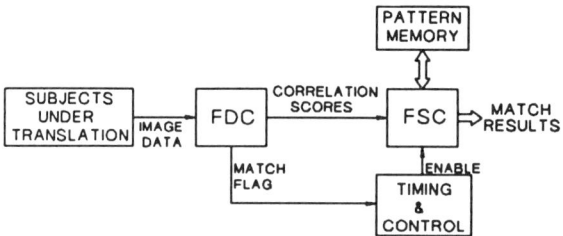

Figure 12.3 *Autocorrelator and fuzzy set comparator.*

entropy of the signal plus noise is approximately $\log_2(M)$. However, after correlation, the noise is removed *and* the original signal is smoothed. The entropy of the resultant ACF will thus be smaller than that of the original signal plus noise.

One should be careful to note, however, that, strictly speaking, the ACF is not unique, although rarely do differing input signals generate identical ACFs. On the other hand, cross-correlation operations, which indicate the relationships between two unique but different signals, further increase the possibility of uniqueness of the resulting function.

12.3.4 A Translation-Insensitive Image Recognition System

As with the Fourier transform, the magnitude and shape characteristics of the ACF are unaltered by a translation or shift of the input signal. However, in such a situation, the entire function is shifted by the amount of translation. This makes the correlation operator useful in data synchronization and time measurement problems.

As an initial experiment in AI data preprocessing, we used the translation invariance of the correlation operator to construct a high-speed system to

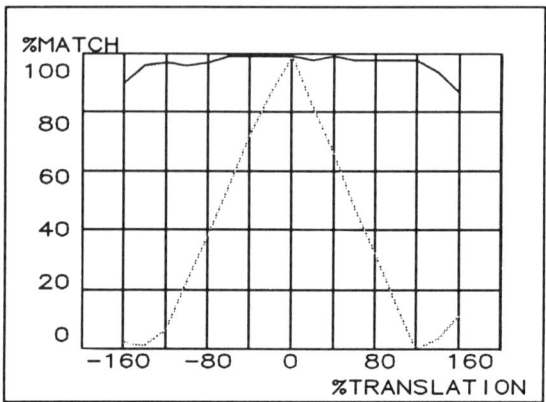

Figure 12.4 *Translation sensitivity of a square image. (The solid line is image correlation data and the dotted line is unprocessed image data.)*

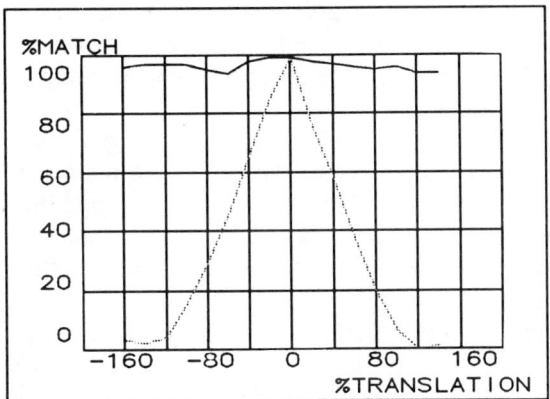

Figure 12.5 *Translation sensitivity of a circular image. (The solid line is image correlation data and the dotted line is unprocessed image data.)*

recognize images irrespective of the position of the subjects in the field of view. The autocorrelation operation was performed by a software simulation of a recently developed 128 × 1-bit correlator IC, the MD1212, while pattern matching was accomplished by the MD1210 fuzzy set comparator (FSC) IC. Figure 12.3 illustrates the system.

In this system, the video signal is digitized using a four-gray-level frame grabber, and cross-correlated with a data sequence related to the gray level common to all of the subjects. (The gray level of all the subjects was intentionally made identical, although the shapes vary.) Initially, the cross-correlation functions for each of the test images were computed and stored in the FSC pattern memory. When storing subject images, only correlation scores subsequent to the first match found by the correlation operation (i.e., maximum correlation of the input with the reference pattern) are stored in pattern

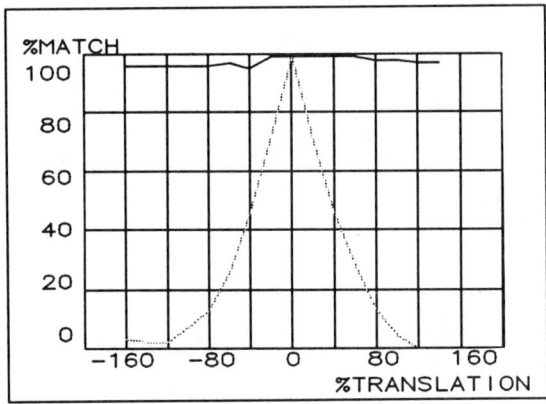

Figure 12.6 *Translation sensitivity of a triangular image. (The solid line is image correlation data and the dotted line is unprocessed image data.)*

memory. During the pattern recognition segment of the experiment, all correlation scores were discarded except those subsequent to the initial correlation match. In this way, the position of the subject in the camera's field of view is compensated for.

The experiment was performed in two steps. First, unprocessed image data for each subject was stored in the FSC pattern memory. The subjects were then translated relative to their initial position, and the match between the subject and stored pattern was computed. In the second phase, the procedure was repeated, but with the image data passed through a CCF and all matching operations performed on CCF scores.

Experimental results are displayed in Figures 12.4 to 12.6. On the vertical axis is plotted %Match versus %Translation on the horizontal axis. The subjects of the experiment were simple shapes; square (Fig. 12.4), circle (Fig. 12.5), and triangle (Fig. 12.6). Clearly, irrespective of the translation of the subjects, the correlation scores of each of the subjects matched very closely to those that were stored prior to the start of the experiment. This was not the case when matching was attempted on the unprocessed image data. While this experiment is admittedly simple, it suggests uses for correlation in such applications as translation-invariant edge detection and subfield pattern searching.

REFERENCES

[1] T. J. Lynch, *Data Compression—Techniques and Applications,* Van Nostrand Reinhold, New York, 1985.

[2] R. H. Stafford, *Digital Television—Bandwidth Reduction and Communication Aspects,* Wiley, New York, 1980.

[3] A. V. Oppenheim, and J. S. Lim, The Importance of Phase in Signals, *Proc IEEE,* 69(5) 529–541 (May 1988).

[4] J. G., Daugman, Relaxation Neural Network for Non-Orthogonal Image Transforms, *IEEE Internat. Conference on Neural Networks,* 1547–1560 (July 1988).

[5] R. E. Ziemer, and W. H. Tranter, *Principles of Communication,* Houghton Mifflin, Boston, Mass., 1985.

CHAPTER 13

Fuzzy Data Comparator with Neural Network Postprocessor: A Hardware Implementation

PAUL BASEHORE
JOSEPH YESTREBSKY
GERALD REED

13.1 INTRODUCTION

In this chapter we describe a fuzzy set comparator (FSC) designed for adaptive ranking, and ranking fuzzy data in groups by certain predetermined characteristics (i.e., modified single index ranking). The FSC is intended to simplify the implementation of systems where decisions must be made rapidly from inaccurate, noisy, or real-time variable data. Either Hamming or linear distance may be selected as the comparison metric. A neural network postprocessor ranks the comparisons, providing a simple hardware implementation and superior rank calculation speed. The FSC is implemented in 1.5-micron complementary metal-oxide semiconductor (CMOS) technology. As it is bus-oriented, the FSC may be incorporated into a microprocessor environment, or operated autonomously.

13.2 FUZZY SET COMPARATOR

In practice, many data sources, such as those from optical, speech, telemetry, and judgmental sources, tend to be inaccurate, noisy, or otherwise variable. Such data are therefore difficult to use directly in decision-making or pattern recognition applications. The action of evaluating (ranking) fuzzy

---

Copyright (C) 1989, Micro Devices, Special Products Division of Chip Supply.

Neural and Intelligent Systems Integration, By Branko Souček and the IRIS Group.
ISBN 0-471-53676-8 ©1991 John Wiley & Sons, Inc.

data defuzzifies those data. Defuzzification must take place to make the data useful in any decision-making process. However, the defuzzification of data by single index ranking [1] may result in reduced resolution and loss of potentially vital information. Techniques for handling this problem are being investigated (e.g., multiple indexed ranking [2]), but these techniques are in their early stages of development. It is expected that a modified single index ranking will be adequate for most practical problems.

In this chapter we describe the theory and performance of a recently developed FSC integrated circuit (IC). The processing element of the FSC is based upon the mathematics of fuzzy set theory. A postprocessor comprised of a tandem pair of neural networks computes the fuzzy set ranking. The FSC is designed for adaptive ranking and ranking fuzzy sets in groups by certain predetermined characteristics (i.e., modified single index ranking). The neural network features a simpler hardware implementation and a superior rank calculation speed than more conventional hardware implementations. The device is implemented in 1.5-micron CMOS technology, using about 20,000 transistors.

The FSC accomplishes limited "learning" by allowing input data streams to be stored in memory for subsequent use as comparison data.

13.3 TYPICAL APPLICATIONS AND DEVICE OBJECTIVES

Since many practical data are fuzzy, there are numerous potential applications for fuzzy data comparison. Most notable among these are text and voice recognition, robotics control, security, surveillance, and computer-aided manufacturing (CAM) systems. A primary design goal of the FSC is compatibility with a broad range of applications. This includes applications where high data throughput, high-speed processing, or both, are necessary. To meet these objectives, the FSC architecture provides expansion to allow simultaneous comparison of up to 256 fuzzy sets (Fig. 13.1). In addition, we employ a high-speed neural network postprocessor whose processing speed

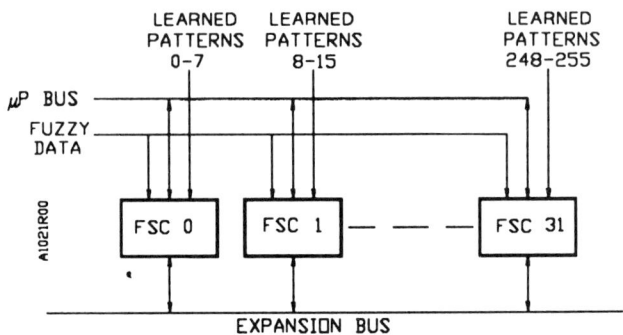

Figure 13.1 *FSC system expansion.*

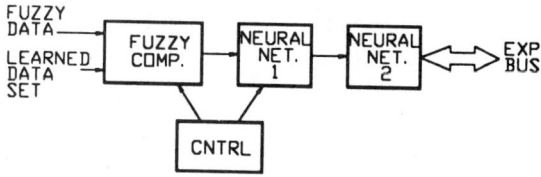

Figure 13.2 *FSC block diagram.*

is independent of the number of fuzzy sets being analyzed. Fuzzy set ranking analysis and control are accomplished through a microprocessor interface.

The FSC does not preprocess or transform the input data in any way before performing the fuzzy ranking, though this may be required for certain applications. This elimination of preprocessing circuitry lowers device cost and helps maintain our compatibility goals.

13.4 DEVICE ARCHITECTURE

The FSC is composed of eight data comparison circuits, two neural networks, and circuitry to perform overhead functions (Fig. 13.2). Each comparison circuit compares a "fuzzy" data set with a "learned" data set. Each comparison circuit generates an error value using one of two externally selectable comparison metrics. Once obtained, the error values are passed on to the neural networks.

The ranking postprocessor is configured as two tandem neural networks. The first network receives inputs from each of the eight data comparison circuits (Fig. 13.3) and from the threshold register (ranking value). This network settles on the lowest value of the nine 16-bit inputs in two clock periods. (During the first clock time, the high bytes of the nine inputs are compared. The low bytes are compared during the second period. While slightly decreasing throughput, this approach was chosen to reduce circuitry.) The second neural network provides for expansion to other FSC

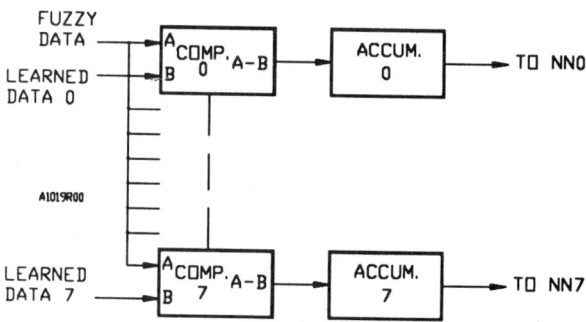

Figure 13.3 *Comparison circuit.*

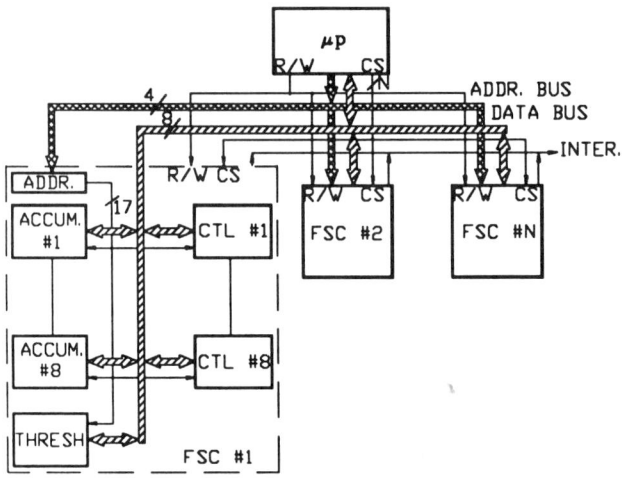

Figure 13.4 *FSC bus structure.*

devices. Comparison errors less than the ranking values contained in the corresponding FSC devices are compared on the expansion bus, which forms a part of the second neural network. The output, from as many as 32 FSC devices, settles on the lowest of the up to 256 total inputs within two additional clock periods. Arbitration circuitry is provided to handle tie conditions. This arbitration consumes one additional clock period. A solution is thus reached within five clocks from the time data are initially presented to the first neural network.

The FSC is bus-oriented and may be operated autonomously or under microprocessor control (Fig. 13.4). Any accumulator or control register from any FSC is addressable through a (4-bit) address bus and chip select control signal. Data may be moved to and from the accumulator and control registers via an 8-bit data bus and the read/write control signal. Certain device operating conditions and alarms are signaled via an open-collector output interrupt signal.

13.5 FUZZY COMPARATOR PERFORMANCE

Comparison of a fuzzy data set with a learned data set quantifies the fuzzy data. This defuzzification process is controlled by the ranking threshold (known as the α-*cut*) of the membership characteristic function μ associated with the fuzzy set [1]. For those applications requiring dynamic adjustment of this parameter the ranking threshold may be externally controlled in the FSC. A threshold value of zero would be used in ranking a nonfuzzy data set; hence, only exact matches will pass the comparison. The largest value of the ranking threshold in the FSC is 65535.

A number of metrics exist for ranking fuzzy digital data. In general, metric choice is application dependent, but two choices prevail (linear and Hamming distance). The FSC comparator segments both the fuzzy and learned data into identical word lengths of n-bits ($1 < n < 8$). It then computes and sums the linear distance (L_1 norm) associated with m n-bit words,

$$E = \sum_{i=1}^{m} | \text{FDW} - \text{LDW} |$$

where both m and n are externally selectable. FDW and LDW are fuzzy and learned data word, respectively, and E is the resultant error. The value for m may be any integer greater than or equal to 1. For example, if pixel data were being compared, m would be equal to the number of pixels in the image. Colors or gray levels, represented by bits/pixel, would determine the value of n. For applications where the Hamming distance is the more appropriate metric, simply set n equal to 1.

13.6 NEURAL NETWORK DESIGN AND PERFORMANCE

Starting from the position that a logic element is a limiting form of a neuron, [3,4] we constructed a three-layer asynchronous competitive neural network from simple logic elements. The network contained in one FSC device consists of 128 neurons and solves the problem of identifying the lowest 8-bit value from up to eight error accumulators. (See Fig. 13.5).

Information from the accumulators feeds forward through the network, while feedback from layer 3 to layer 1 provides reinforcement or inhibition of the input data, as appropriate. Some feedforward of the input data directly to layer 3 also occurs. The lowest 16-bit word from among all accumulators in a particular FSC eventually settles at the layer 2 output.

To support applications involving multiple FSC devices (i.e., more than eight input data patterns), a second three-layer neural network is attached to the output of the first network (Fig. 13.6). The structure of these two networks is similar, and the second network identifies the lowest 16-bit accumulator result from a group of FSC devices. Connectivity between neurons across FSC devices in the second network is provided through an (open-collector) expansion bus.

The interconnection weights of the networks are fixed at 0 or +1, so the neural networks can perform only the tasks for which they were created, and cannot learn other tasks. The advantage of fixed interconnection weights for this application is the high-speed parallel data processing is achievable. Though employing positive feedback, simulation has shown these networks to be stable. As any of the 2^8 possible inputs may occur, the stable solution

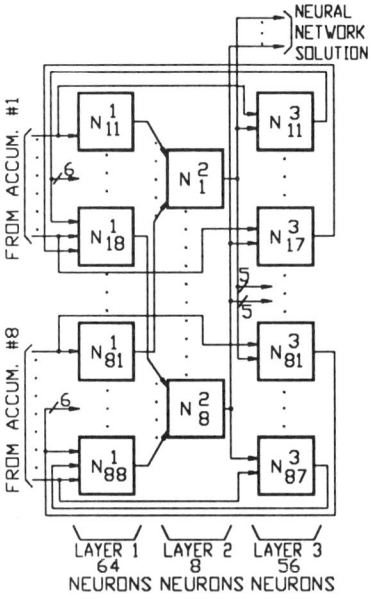

Figure 13.5 *Neural network structure.*

space for each neural network is an eight-dimensional hypercube, with stable solutions appearing at the corners of the cube.

The global neural network selects the minimum of up to 256 error values in five cycles of a 20-MHz clock, illustrating the parallel processing nature of neural networks. The network also exhibits reduced hardware implementa-

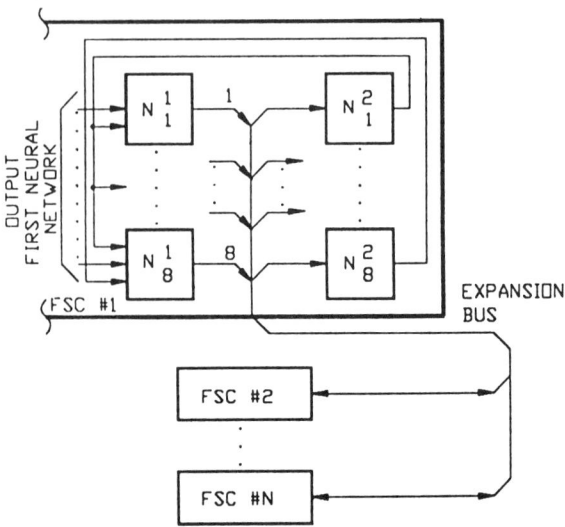

Figure 13.6 *Neural/expansion bus control.*

tion relative to more conventional approaches. Each neuron of the network may be individually deselected, thus allowing fuzzy sets with a particular range of ranking values to be grouped.

13.7 CONCLUSION

A monolithic IC has been developed to aid in the ranking of fuzzy data sets, providing

1. Dynamic control of the ranking parameters,
2. Flexibility in choice of the comparison metric,
3. Simple expansion capabilities.

A postprocessing neural network computes ranking comparisons rapidly, allowing the device to operate in high-speed applications. An IBM PC-compatible evaluation PC board with video interface and supporting software have been developed, aiding prototype evaluation and continuing research into fuzzy information processing.

REFERENCES

[1] A. Kandel, *Fuzzy Mathematical Techniques with Applications*, Addison-Wesley, Reading, Mass. 1986.
[2] S. Mabuchi, An approach to the comparison of fuzzy subjects with an α-cut dependent index, *IEEE Transactions on Systems, Man and Cybernetics,* **18,** (2) (1988).
[3] J. Hopfield, Natural networks and physical systems with emergent collective computational abilities, *Proc. Nat. Acad. Science,* **79** (April 1982).
[4] B. Widrow, R. G. Winter, and R. A. Baxter, Learning Phenomena in Layered Neural Networks," *Proc. IEEE First Internat. Conference on Neural Networks,* II-411–II-429 (June 1987).

CHAPTER 14

A Neurally Inspired Massively Parallel Model of Rule-Based Reasoning

RON SUN
DAVID WALTZ

14.1 INTRODUCTION

This chapter presents a localist connectionist model of approximate rule-based reasoning, which utilizes a high degree of parallelism. The model is inspired by neurobiological data regarding firings of neurons, signal transmissions between neurons, and other aspects of real neural networks. The model accounts for the approximate, evidential, and adaptive nature of commonsense reasoning. This introductory section is intended to motivate the work in this chapter. The next section presents and analyzes two classes of neural models—the discrete neuronal models and the probabilistic discrete neuronal models, which include internal states and state transitions as a main mechanism, as the substrate for our model of rule-based reasoning. Then in Section 14.3 we develop a unified scheme for approximate reasoning that is inspired by fuzzy logic and possesses some characteristics that are not present in other approaches. Based on the neural model and the scheme for approximate reasoning, a localist rule-based reasoning system CRBR (Connectionist Rule-Based Reasoner), is presented in Section 14.4, and its advantages and characteristics as a forward-chaining commonsense reasoner are analyzed and compared with other existing models. Section 14.5 discusses issues related to learning rules. Section 14.6 discusses some more general

---

This work was supported in part by the Defense Advanced Research Projects Agency, administered by the U.S. Air Force Office of Scientific Research under contract F49620-88-C-0058.

Neural and Intelligent Systems Integration, By Branko Souček and the IRIS Group.
ISBN 0-471-53676-8 ©1991 John Wiley & Sons, Inc.

issues and future research directions. Section 14.7 concludes the chapter. Appendixes contain some more details: Appendix 14A contains the proof of a theorem, Appendix 14B describes a complete example of rule representations, and Appendix 14C presents a compilation procedure from rules to network representation.

It is clear that commonsense reasoning is approximate and evidential in nature, and the reasoning processes are usually fast and spontaneous. The reasoning follows some common patterns. There are many recurring, domain-independent reasoning patterns used in commonsense reasoning, which are well identified and studied (for example, by Collins and Michalski [1]). Most of these patterns are fuzzy, uncertain, and evidential, in accordance with the nature of commonsense reasoning.

One instance of a common pattern is as follows, which illustrates the approximate reasoning based on the partial information:

Q: Do you think they might grow rice in Florida?
A: Yeah. I guess they could, if there is enough fresh water supply. Certainly a nice, flat, warm area.

Here a conclusion is reached even though some crucial evidence is missing. Another instance is as follows:

Q: Is Florida moist?
A: The temperature is high there, so the water-holding capacity of the air is high too. I think Florida is moist.

Those concepts involved are not all-or-none, but graded somehow. Yet another instance illustrates approximate decision making:

Q: What do you do for exercises?
A: Well, let me see. If it is sunny and warm, I go swimming. If it is cool but not raining and not too windy, I play tennis. If it is cool and raining, then I will try to stay indoors.

In this case the conditions are not exhaustive and mutually exclusive, so in some circumstances an exact match may not be found. If there is no exact match, the best partially matched rule will be used to make a decision.

The reasoning patterns exemplified by the foregoing instances cannot be accounted for by traditional models, such as traditional logic. The approximate, evidential, and adaptive nature of commonsense reasoning and its spontaneity and speed require us to look into different formalisms and frameworks. It is our contention that this type of reasoning can be better accounted for by massively parallel connectionist models of approximate rule-based reasoning.

In terms of the relationship between connectionist models and rule-based paradigms, connectionism has made some claims regarding rule-based behavior and systems. One point of view is that rule like behavior is the result of complicated interactions of network components, in deterministic or statistical ways, and therefore there is no fundamental difference between rules and nonrules. Another approach attempts to implement rules directly in connectionist networks. Significant early research of this type includes Touretzky and Hinton [2], Barnden [3], and Shastri and Feldman [4]. In each of these schemes, parallelism is lost in some way, either because of a serial matching process of hard-wired rules or a centralized working memory. For example, in Touretzky and Hinton [2], an elaborate "pull-out network" is designed to pick up a rule from a rule network and to match the data (triples) in the working memory. The mechanism performs only one match at a time. There is no room for approximate reasoning. In Barnden'scheme, the rules are wired in symbolic form into a network that has a grid form. Symbolic manipulation necessary to match the rule against data is a complicated process, and as a result the inherent parallelism is not fully utilized. In Shastri and Feldman [4], a mathematical formalism is developed, and a network architecture is designed to implement the formalism. Many different types of neurons have been devised, and each has a special activation function specifically designed for it. This scheme can handle property inheritance in a conceptual hierarchy but not the full capability of rule-based reasoning. It is important to point out that although they have limitations as indicated above, they are all important early work. The most recent work, concurrent with ours [37], includes Ajjanagadde and Shastri [5] and Lange and Dyer [6], which will be discussed and compared with our work later on. Besides these, there are other schemes that employ different techniques for high-level data or knowledge representation that are related to rule-based reasoning, for example, Feldman and Ballard [42], Hendler [7], Blelloch [8], Kosko [9], Fanty [10], Ackley [11], Dolan and Smolensky [41], and Derthick [12]. A full treatment of these schemes is beyond the scope of this work.

14.2 DISCRETE NEURAL MODELS

In this section, we will look into a more general neural network formalism, and demonstrate some of its capabilities that are important to the development of connectionist models of rule-based reasoning in general and to variable binding mechanisms in particular. Readers are referred to Chapter 6 for more detailed information on the formalism itself and associated learning algorithms.

Our aim is to devise a model more general than conventional parallel distribution processing (PDP) models and capable of incorporating many intricate phenomena found in real neural networks. Our generalization goes along several dimensions: internal states, different synaptic outputs, and temporal responses. The resulting model can solve several important problems in developing a reasoning scheme (i.e., rule matching, variable binding, and certainty factor propagation). Thus the model forms a reasonable basis for an inference system for conceptual level processing.

14.2.1 DN Models

A automaton-theoretic description of a discrete neuronal model (DN) is a 2-tuple

$$W = \langle N, M \rangle$$

where $N = \{\langle S, O, A, I, B, T, C \rangle\}$ (a set of neurons)
    S = set of all possible states of a neuron,
    O = set of all outputs,
    A = set of all actions (output symbols),
    I = set of inputs,
    B = set of all input symbols,
    T = state transition function: $S \times I \rightarrow S$,
    C = action function: $S \times I \rightarrow O$,
    M = connectivity among neurons in the set N: $M = \{(i,j,k) \mid i$ and j are indices of neurons and k is a label unique for a particular synapse$\}$.

In this model a set of discrete states is explicitly specified instead of a continuous activation function. Hopefully, this can capture more accurately the biological information processing mechanisms built into a real neuron. The idea came from the modeling study of lobster stomatogastric ganglion neural networks (see Sun et al. [13]). Evidence from physiological data observed by biologists overwhelmingly points to a more powerful neural network model, which is capable of accounting for more phenomena than conventional models. By introducing the state variables, we hope to be able to model more aspects of the working of real neurons.

In the rest of this subsection, some arguments based on biological data are presented to justify our models.

An issue is the importance of the membrane properties and, therefore, the endogenous firing of individual cells. According to our study (Sun et al. [13]), the dynamics and emergent properties of a neural network can mostly be attributed to two factors: the endogenous firing (determined by membrane properties of the cell, which could be affected by current inputs and the input history) and the synaptic connectivity. Because of the physiological properties of the cell membrane, each cell is capable of firing endogenously even when it is insulated from any external influence. The endogenous firings are important as a source of influences that help to shape the behavior of a network. This fact is indicated in many biological papers (e.g., Selverston and Moulin [14]). However, the importance of membrane properties and endogenous firings is overlooked in conventional connectionist models, because of the highly approximate nature of these models. In the discrete neural model, this feature can be captured by a state variable that represents a particular moment of internal changes. The mechanism works this way: from the formula specified earlier, $s(t)$ is the internal state that determines a particular endogenous firing curve; for example, suppose the weighted sum model is used (see Sun et al. [13]):

$$C(t) = f(s(t), I_1, I_2, \ldots, I_k) = w_0 E(s(t)) + w_1 I_1(t)$$
$$+ w_2 I_2(t) + \cdots + w_k I_k(t),$$

where $E(s(t)) = \sin(s(t))$ and C is the action function.

In a real neuron, unlike in conventional connectionist models, there is no continuous input or output through synapses. Instead, an all-or-none action potential is generated if the cell is depolarized to a certain degree, which in turn causes the release of neurotransmitters. The input to the postsynaptic cell is dependent upon two factors: the type and the amount of neurotransmitters released (Kandel and Schwartz [15] and Edelman [16]). This powerful mechanism cannot be captured by conventional neural network models. In a conventional neural network model, the continuous output is meant to represent the frequency in which the action potentials are generated, it is doubtful that the firing frequency is a primitive feature (instead of an emergent feature that is caused by other more primitive activities) in the neuronal information processing mechanism. On the contrary, we have shown in a simulation study (Sun et al. [13]) that the firing frequency, as well as phase relationships, is an emergent property of the network created by the complex interaction of the components of the network, at least in lobster stomatogastric ganglions. The proposed discrete model can easily capture the neural information processing mechanism through action functions and state transition functions, by specifying a sequence of states to go through, and specifying actions associated with each state. For example,

$$s(t) = s(t-1) + 1 \bmod n,$$

and

$$C(t) = f(s(t), I_1, I_2, \ldots, I_k),$$

where f is a predetermined function such as weighted sum or Goldman equation (see Kandel and Schwartz [15]).

This formalism can explain the firing frequency, its initial formation and later modulation by the change of firing frequencies in other neurons, as well as

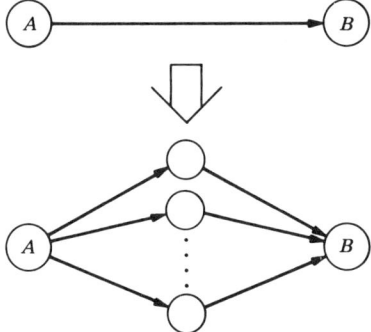

Figure 14.1 *Replacing multiple output values with multiple intermediate nodes.*

many other intricate phenomena found in real neural networks such as phase responses, neuronal modulation, and phasic relationship. These properties are important in terms of the functional capability and versatility of a network, as seen in many different domains (e.g., Richmond and Optican [17]).

Yet another issue concerns the different presynaptic actions performed by the same cell at different sites of the axon. Different sites on the same axon can release different types of neurotransmitters (thus cause different types of reactions in postsynaptic cells) or different amounts of transmitters of the same type. Some types of neurotransmitters may have long lasting effect, while others may act instantaneously. Each can cause a different reaction in a postsynaptic cells. The "action" taken by an individual postsynaptic cell is determined mainly, but not exclusively, by the following factors: the endogenous properties of the cell, the type(s) and amount of transmitters it received, and the current that is injected into it in case of electric synapses. The issue of different postsynaptic actions is not dealt with in conventional connectionist models either. In our model, the variety in presynaptic actions can be modeled by A (the set of actions) and C (the action functions).*

14.2.2 Comparisons with Weighted-Sum Model

The equivalence property of this model to the more conventional models has been studied. It is at least capable of the same computational power as well as expressive power as weighted-sum models. Beyond that, it has the advan-

---

* There is one aspect of the model that may raise some questions. Usually in a real neuron, one synapse can only release a certain type of transmitters (or a certain group of transmitters). But in our model each synaptic site can release different transmitters (i.e., different messages or synaptic actions). This contradiction can be easily resolved by realizing the fact that we can implement this discrete neural model using only the type of neurons in which each synaptic site can only release one type of transmitters, by adding a group of intermediate neurons each of which represents a particular message and hook them together as in Figure 14.1.

State	Weighted sum	Output
Don't care	≤0	0
Don't care	>0	1

Figure 14.2 Action function C simulating a thresholding function.

tage of greater generality and versatility. It is more general because it can accommodate the conventional connectionist models as special cases, as discussed below. It is also more versatile because, by introducing state variables and a set of synaptic actions, the model can handle more elaborate processing at neuronal levels, not covered in other models at all. The properties are important, because later on when we develop a representational scheme for rules and predicates, we will rely heavily on these properties.

To see how our model simulates other connectionist models, we have to look at various neural network models. In general, neural network models can be divided into four classes:

- Continuous input/discrete activation models (e.g., the linear threshold unit model [34]),
- Discrete input/discrete activation models (e.g., Feldman and Ballard [42] model),
- Continuous input/continuous activation models (e.g., McClelland and Rumelhart's interactive activation and competition model [34,44]),
- Discrete input/continuous activation models (as another possibility).

All of them can be easily handled by our general formalism. To simulate a continuous input/discrete activation neural model (suppose using a weighted-sum activation function with the threshold equal to 0), let a discrete neuron be

$$\langle S,A,B,I,O,T,C \rangle$$

where $A = \{0,1\}$
$B = \{0,1\}$,
$C = T =$ a table specifying which action in A to perform (see Fig. 14.2).

This model can simulate the original model and produce the same output: 0 if input < threshold and 1 if input > threshold. The model will carry out the computation exactly as its conventional counterparts.

State	Weighted sum	Output
0	0	0
0	0.1	0.1
0	0.2	0.2
.	.	.
.	.	.
.	.	.
0.1	0	0.1
0.2	0	0.2
.	.	.
.	.	.

Figure 14.3 Action function C in table form simulating a continuous function.

For the ease of computation, sometimes we may want to represent action or state transition functions in a table form. Then we may need to discretize a continuous function; action function C will be a table specifying a sequence of points sampled from the response curve of the neuron in the original model. If we sample enough points on the continuous output curve, we can approximate the behavior of the original model closely enough for any practical purpose. For example, Figure 14.3 shows the approximation of the model: output = potential + $\Sigma(w_i i_i)$.

Discrete neurons can be implemented with the simplest conventional neural network type: the thresholding weighted-sum model. The question is how to implement the state variable and the state transition function of a DN unit. It has been shown that a multilayer network can do the job: The hidden layer represents the current state and each input/output value is explicitly represented by an individual node (cf. Fang and Wilson [18], Servan-Schreiber [19]). The output from the hidden layer is fed back into the input layer to help decide, together with current inputs, which state to enter next. See Figure 14.4.

14.2.3 PDN Models

The probabilistic discrete neuronal model (PDN) is a simple extension of DN. It assumes that each entry (of the table specifying the state transition function and the action function) is a probabilistic distribution instead of being deterministic. If a state transition or action is not specified by the table, then it is completely random, based on a uniform distribution. If it is specified with a probability less than 1, then the other choices are deter-

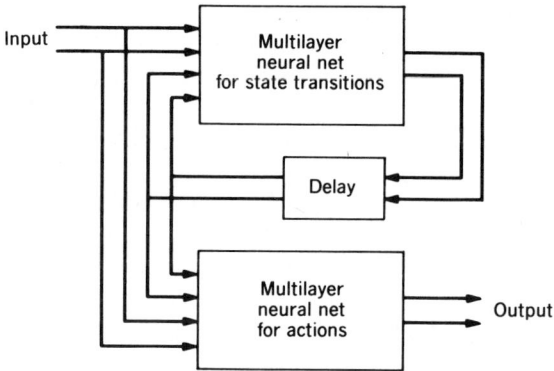

Figure 14.4 Implementing a finite state automaton in neural nets.

mined by a uniform distribution over all other choices with the remaining probability.

Formally, it can be defined as

$$W = \langle N, M \rangle$$

where
$N = \{\langle S, A, B, I, O, T, C, P_1, P_2 \rangle\}$
S = set of all the possible states of a neuron,
A = set of all actions,
B = set of all input symbols,
I = inputs,
O = outputs,
T = state transition function: $S \times I \to S$; P_1,
C = action function: $S \times I \to O$; P_2,
M = connectivity among neurons in N.
P_1 and P_2 = sets of probabilities associated with each state transition or each action.

The state transition function is a mapping to a statistical distribution of states. The action function is a mapping to a statistical distribution of actions.

The probabilistic nature of the model can be attributed to the unreliability and noise in cells, the statistical nature of membrane operations (channel openings and closings), and random environmental influences, and so on.

The purpose of using this formalism is that (1) DN models can be viewed as a special case of PDN models, in which all probability distributions are all one or zero, and (2) most importantly, some useful learning algorithms can be readily applied to this formalism (see Sun [20]). These algorithms are used to learn appropriate variable bindings in a concept assembly as will be discussed later, though it is beyond the scope of this chapter to explain these learning algorithms (see Chapter 6.).*

14.3 APPROXIMATE RULE-BASED REASONING

In this section, we will present a theory of approximate reasoning, which is inspired by and similar to fuzzy logic (Zadeh [21]) and implemented in a network model CRBR. This approximate reasoning scheme provides a solid foundation for carrying out rule-based reasoning in connectionist networks, for it treats the weighted-sum computation as combining evidence to reach a conclusion based on a rule. Alternatively, we can view this theory as providing a semantics for the weighted-sum connectionist models. A methodical way of implementing the theory with the DN/PDN formalism will be discussed in the next section. We will call this approximate reasoning scheme *fuzzy evidential logic* (FEL).†

Although this reasoning theory is presented in a symbolic framework, it will become evident in this section and the next section that it differs from symbolic approaches in that it does not require indexing, retrieving, reorganizing, and updating of large databases. Rather, the computation is done in place, without the hassle of selecting data and moving data around, by directly connecting related facts (related by rules) together so that only local computation is necessary. This way we can achieve massive parallelism and cognitive plansibility.

14.3.1 A Propositional Theory

We will present a propositional theory first for the sake of clarity: Here we will deal with fuzzy propositions only (i.e., mappings from propositions to values ranging continuously from 0 to 1, without any argument).

Propositional inference is the computing of values (i.e., certainty) of some propositions from the values of a given set of propositions. A *propositional inference theory* is a well-defined way of doing propositional inference. A *rule* is the specification of the value of a proposition, given a set of propositions and their values, computed according to the propositional inference theory. A *fact* is a proposition and its value, computed according to the propositional inference theory.

---

* Link weights representing rule strengths can be learned by another set of algorithms presented in later sections.
† This scheme is presented in the context of combining ideas from fuzzy logic and neural networks, for the purpose of a better model of commonsense reasoning. Many ideas presented here can be traced back to early literatures, such as contained in Dubois and Prade [22]. Particularly, the idea of a grade of membership originates from Zadeh [23], and that of linguistic variables from Zadeh [24].

14.3 APPROXIMATE RULE-BASED REASONING

This is an extension of traditional mathematical logic (cf., Chang and Lee [25]), in which instead of using a Boolean variable we use a variable $\in [0,1]$. So instead of being clear-cut true or false, we have a degree of truth. We will call this degree of truth (the value of a proposition) certainty or certainty factor, to follow what is commonly used in literature (e.g., Buchanan and Shortliffe [26], or Hayes-Roth et al. [27]). (As we will see, this is not an accurate name for it.)

In commonsense reasoning, we tend to "add up" evidence when evaluating the plausibility of something (accumulating evidence). And different pieces of evidence have different "weights" in the decision making because of their relative relevance to the conclusion. A reasoning system has to simulate those effects.

A formal definition follows.

Scheme 14.1. *A propositional theory $T = (F,R)$, where F is a set of facts and R is a set of rules.*

$$F = \{(P,C)\}$$
$$R = \{(I,C_R)\}$$
$$I = (LHS \rightarrow RHS)$$

where $LHS = F_1 \cap F_2 \ldots \cap F_n$ and $RHS = F$.

Associated with LHS is a weight distribution $(W_1, W_2, \ldots, W_n)$ over all facts in it. The total weights sum to 1. C denotes the certainty of a fact or a rule.

The question is: given $P_1, P_2, \ldots, P_n$, Is P_1 true?

T only gives an answer with some degree of certainty (i.e., the value of the proposition). If there is a sequence of R's, $R_1, R_2, \ldots, R_m, R_i \in R$, LHS in R_i is a subset of the union of RHS in previous R's and F, and $P \in RHS$ of R_m, then we may conclude that P is true with certainty C.

C is calculated this way: Given a rule (LHS $\rightarrow$ RHS) with a certainty C_R and weight distribution $(W_1, W_2, \ldots, W_n)$, and a set of facts with certainty: F_1 with C_1, F_2 with C_2, $\ldots$, and F_n with C_n, assuming these facts constitute LHS, then C (the certainty of RHS) is $C_R * (C_1*W_1 + C_2 * W_2 + \cdots + C_n * W_n)$. When implementing it in a computational way, we have

$$C = C_1 * W_1' + C_2 * W_2' + \cdots + C_n * W_n'$$

where $W_i' = C_R * W_i$, the combined weight distribution.*

---

* A problem here is that uncertainty and vagueness are mixed together (see Dubois and Prade [28] and Sun [29]). But, on the other hand, this provides a uniform treatment of various forms of inexactness, and therefore provides a computationally efficient way of dealing with it. Vague-
Continued

When all the certainty factors are 1's, then we have the following theorems (the proofs are in the appendix):

THEOREM 14.1. *The above formalism is complete in terms of Horn clause logic (in propositional forms)*

THEOREM 14.2. *The above formalism is sound in terms of Horn clause logic (in propositional forms)*

We can extend the FEL formalism beyond Horn clause logic by allowing a limited form of negation. In the condition or conclusion part of a rule, a fact can be a "negative fact," for example, not raining (vs. raining), or not success (vs. success). When there are no facts appearing in both negated forms and positive forms, we can simply treat a negated fact as a different fact. To get a positive fact out of a negated fact, or vice versa, we simply subtract the certainty of it from 1.

A simple example follows. From the cases analyzed in the first section, we have the following rules:

if warm (not raining), then swimming

if (not warm) (not raining) (not windy), then tennis

if raining (not warm), then indoor

For each rule, we assign a weight distribution to facts in the conditional part, where the total weights sum to 1. And we attach a certainty to each rule, representing its plausibility. And we calculate weighted sum of the certainties of the conditions, and determine the certainty of the conclusion by multiplying the certainty of the rule with the weighted sum. It is easy to see that by multiplying the certainty of the rule with the weight distribution beforehand, we only need to calculate weighted sum to reach the certainty of the conclusion.

The system should be able to handle partial matching situations. For example, when warm, not raining and tired, there is no rule fully matching the situation, but there are rules partially matching it (when a fact in a condition of a rule is unknown, assume its certainty to be zero). Then according to the weight assignment, the system should fire one rule or the other.

---

ness (or fuzziness) is the ill-definedness of concepts. It is handled usually with a linguistic variable (Zadeh [24]). For example, in the statement, "John is tall," it is very difficult to know what exactly "tall" means: 6 ft, 6.5 ft, or 7 ft? A linguistic variable can be defined that draws its value from a fuzzy subset that defines "tall" with a grade of membership (Zadeh [23]). On the other hand, uncertainty refers to the nondeterminism of an event, which has to be dealt with statistically. For example, "John might be home by eight o'clock," which is a probabilistic statement. The foregoing method handles both the same way.

14.3 APPROXIMATE RULE-BASED REASONING

The system should also be able to handle uncertain or fuzzy information. For example, "warm" is a fuzzy set. It has no clear-cut boundary. So this variable will have to take a value between 0 and 1, to be used in the calculation of the overall confidence of the conclusion reached from the rule.*

The system allows evidence to accumulate, or in other words, it "adds up" various pieces of evidence to reach a conclusion with a certainty that is calculated from the certainties of the different pieces of evidence. This system can do that because of its additive nature of computation. Instead of max/min operations used in fuzzy logic, by using addition (in the weighted-sum computation), we have a simple way of combining evidence from different sources cumulatively, without incurring too much computational overhead (such as in probabilistic reasoning [30], or Dempster-Shafer calculus [33]).

The scheme above can be viewed as an extension of traditional mathematical logic in propositional calculus form with two additional operators for computing certainty factors along the way when doing inferencing. But why do we choose the two operators: + and *, in extending traditional logic? There are a number of other obvious choices for operator 1 (+): max, mean, median, sum-minus-overlap, Bayesian model, Dempster-Shafer model, or adjustment and anchoring [51], and others, for the effect of accumulating evidence, and a number of other choices for operator 2 (*): min, thresholding, fuzzy logic approach, possibility theory approach, Shastri approach [4], and so on, for the effect of modulating the C with the certainty of the rule used. There are three reasons for the choice we have made:

- It allows partial match—even without knowing all the facts in the LHS of a rule, as long as we get enough evidence, we still can deduce something with partial certainty.
- It accumulates evidence—we want the resulting C to be cumulative; that is, knowing two facts in LHS results in a bigger C than knowing only one of them.
- it reduces to Horn clause logic—when everything is 100 percent certain, we want the system perform exactly like Horn clause logic

---

* However, it is evident enough that this model is not perfect regarding various human decision making situations. It cannot handle many intricate cases. For example,

if A, then B is possible ($CR = 0.50$).
if A and C, then B is certain ($CR = 1.00$).
if (not A) and C, then B is impossible ($CR = 0$).

This situation cannot be modeled by one rule with the above method, but neither can the other methods (actually if we allow the use of two rules for this case, our method works just fine after weights are properly tuned). To fully account for the various human reasoning and decision-making scenarios, we need far more complicated mechanisms and mathematical tools.

The two operators we have chosen have all of these three characteristics, and furthermore they seem to be the only ones that work according to these criteria.

14.3.2 A Predicate Theory

A predicate theory follows now. The difference from the propositional theory is that in this case we allow arguments to be associated with each proposition to represent different individuals. In other words, the truth of a proposition not only depends on the proposition itself but also depends on the set of individuals associated with it. This way the theory can handle more complicated cases.

SCHEME 14.2 *A predicate theory $T = (F,R)$, where F is a set of facts and R is a set of rules.*

$F = \{(P(A),C)\}$, where A is a set of arguments for predicate P

$R = \{(I,C_R)\}$

$I = (LHS \rightarrow RHS)$, where $LHS = F_1(X_1) \cap F_2(X_1) \cap \cdots \cap F_2(X_n)$, and $RHS = C(X)$. A weight distribution is associated with facts in LHS.

Given $P_1(X_1), P_2(X_2), \ldots, P_n(X_n)$, is $P(X)$ true?
If there is a sequence of R's, $R_1, R_2, \ldots, R_m, R_i \in R$, LHS in R_i is a subset of the union of RHS in previous R's and F, and $P \in RHS$ of R_m, then we conclude that $P(X)$ is true with certainty C.
The certainty is calculated in the same fashion as before.

The same theorems as before can be proven easily. But for completeness there is a restriction: to implement the predicate calculus formalism, there must not be the same predicate appearing more than once with different arguments in RHS (in other words, one predicate can only appear once with one particular set of argument variables, for keeping chaining separate). This should not be a serious problem, because one can always duplicate a part of the network and rename predicates.

THEOREM 14.3 *The preceding formalism is complete in terms of Horn clause logic (predicate calculus).*

THEOREM 14.4 *The preceding formalism is sound in terms of Horn clause logic (predicate calculus).*

Appendix 14D gives a set of procedures used to compile a set of predicate rules into a DN network automatically. The procedures also take care of details such as variable bindings, constants, consistency checking among different arguments and combining evidence.

The reason that we develop this theory is that there is no existing computational model fully accounting for the type of reasoning we deal with in this system; therefore we modify the fuzzy logic formalisms into this theory to describe reasoning processes at an abstract level. It differs from Zadeh's fuzzy logic (Zadeh [21]) in the following ways:

- It uses addition and multiplication, which are natural for neural networks, instead of max and min.
- It allows the cumulation of evidence and partial matching.
- It accounts for the phenomenon that, in a chain of reasoning, the certainty of conclusions weakens along the way.
- It distributes weights to different pieces of evidence, and thus can emphasize certain things and disregard others as appropriate.

14.4 A LOCALIST NETWORK OF RULE-BASED REASONING

We will now look at an implementation of this theory (FEL) in a localist network CRBR. The representation and reasoning scheme is developed in the DN network model, which maximizes the inherent parallelism in a neural network model and performs all possible inference steps in parallel.

14.4.1 Representing Predicates

It is easy to see from the previous discussion that FEL evidence combination functions are the same as the commonly used neuronal activation function: the weighted-sum computation. Besides the advantage of allowing partial matching and cumulative evidential combination, it has an extra advantage of being directly implementable with neural networks; The structure in a FEL reasoning system can be directly mapped onto a neural network topology (using the DN/PDN formalism).

Details of this mapping is as follows: A concept (predicate) is represented by an assembly of DN nodes and is encoded in a network by the connections it has to the other nodes. Those unidirectional connections help to shape the concept as well as to guide the reasoning. For example, Figure 14.5 shows how a concept is wired into a network. In each assembly, there are $k + 1$ nodes of DN type: $C, X_1, X_2, \ldots, X_k$. The C node contains the certainty factors used in reasoning and connects to all other concepts related to it in two directions, either as LHS or RHS. The other k nodes take care of variable bindings for a maximum of k variables. The set of states in the variable nodes represents all possible bindings, fixed a priori. The signal from the C node tells the variable nodes what to do. The formulas used in

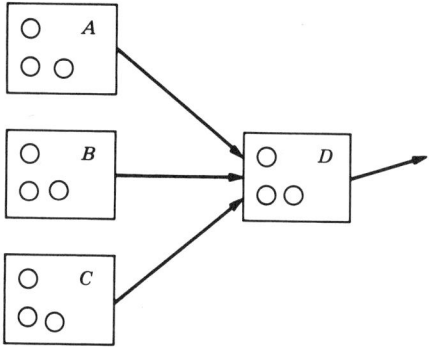

Figure 14.5 Wiring of a network for representing the rule: if A B C, then D.

these DN type nodes are summarized below (just one simple case as an example):

For C: $\quad O_C(t) = f(I_1(t), I_2(t), \ldots, I_k(t)) \in A_C$.

Action function f here is a mapping to a value represents the activation level. The I_i's are outputs of other C nodes in other assemblies. A_C is the set of output symbols in node C. Node C can also receive inputs from other nodes in the same assembly, useful in certain situations which will be explicated in Appendix 14B. This function calculates the certainty of the conclusion with the weighted sum computation.

For X_i: $\quad i = 1, 2, \ldots, k, \quad O_{X_i}(t) = f_i(I_1(t), I_2(t), \ldots, I_n(t)) \in A_{X_i}$.

Action function f_i here is a mapping specifying the output of X_i at time t based on inputs at that moment from X_i nodes in other assemblies. It can also receive inputs from C nodes, as will be explained in Appendix 14B. A_{X_i} is the set of output symbols in the node X_i.

For example, if we want to code two concepts: "x is human" and "x is mortal," we will have two separate assemblies, one for each, as in Figure

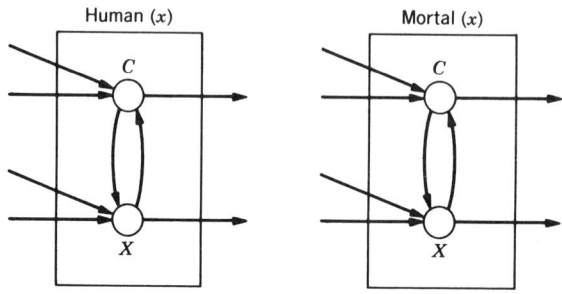

Figure 14.6 Two assemblies.

14.6. There is an argument node for x, and a C node in each assembly. In this case, with some simplification, we have for C, $O_C(t) = \Sigma W_i I_i$, and for X, $O_X(t) = I$. See Appendix 14B for a detailed derivation of this example.

It is important to point out that argument nodes do not necessarily contain values that represent a particular domain object, in other words, they can be free variables. For example, if we have a rule $P(x) \rightarrow Q(x)$, given $P(z)$, we can derive $Q(z)$, by representing z with a particular value in argument nodes of the assemblies for $P(x)$ and $Q(x)$.

14.4.2 The Architecture

The basic architecture of CRBR consists of three layers of DN nodes: the input layer, the processing layer and the output layer, with additional modules attachable to them, each of which can handle learning and pre- and postprocessing correspondingly. The input pattern is stored in the input layer, clamped until next presentation of new input patterns. Each node in the input layer is connected to the corresponding nodes in the processing layer, which is a signal propagation network (not clamped). The latch opens briefly to allow signals to pass. See Figure 14.7.

The processing layer can have complicated internal structures. Nodes are connected to each other according to the rules that the system implements. Basically, each assembly representing a fact in the conditional part of a rule is connected to the assembly representing the conclusion of the rule, and nodes in each assembly are hooked up with corresponding nodes in other assemblies. The information passed to each node is processed and propagated to all the post synaptic nodes from each presynaptic node. This way, a node will fire as soon as it can, that is, as soon as it receives some activation. Correspondingly, describing it at a higher level of abstraction, a rule will fire as soon as it receives some evidence for its conditional part. This scheme guarantees a high degree of parallelism.

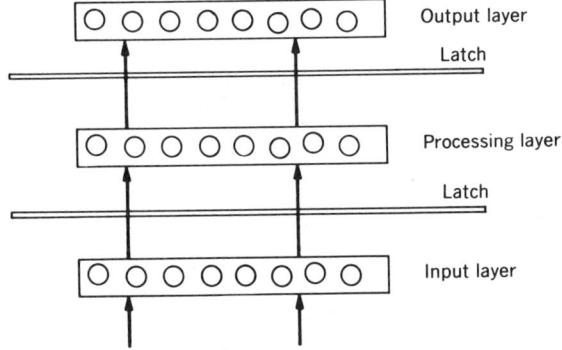

Figure 14.7 *The overall architecture.*

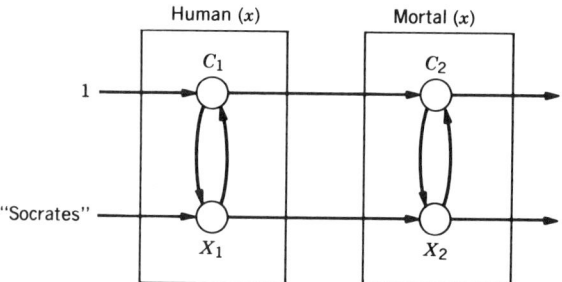

Figure 14.8 *A network for the Socrates example: Output(X_1) = Input(X_1) = "Socrates", Output(C_1) = Input(C_1) = 1, Output(X_2) = Input(X_2) = Output(X_1) if Output(C_2) ≠ 0, Output(C_2) = Input(C_2) × (certainty factor) = O(C_1).*

Now we can have a complete example using one single rule: If x is human, x is mortal (i.e., Human (x) → Mortal (x). The reasoning goes like this:

If one is human, one is mortal.
Socrates is human.
So Socrates is mortal.

Figure 14.8 shows the wiring of the network for the rule, and the reasoning process when the input data for x is Socrates. Equations for the links can be easily derived from the diagram and the foregoing discussion and specification, and are shown in Appendix 14B. The input layer and the output layer are not shown for this example, because they are not needed in this case.

The other example presented before regarding exercises options can be implemented the same way.

The way in which this system performs reasoning is what we call data-driven reasoning. It is mainly a forward-chaining process: From available data derive all conclusions possible simultaneously by network propagation based on links that represent rules.* Another way to look at it is that the

---

* The additive operation in FEL weighted sum computations replaces logical conjunction in the condition part of a rule. Logical conjunction is only "emulated" in a loose sense. When strict logical operations (AND, OR, and NOT) are required in the reasoning, we can devise a simple method for handling them with DN nodes (though this is not needed for the foregoing theory). The reason for this is that a number of researchers have cited logical operations as an important factor in determining the adequacy of a neural network model as a universal computational model (Abu-Mostafa [33], Rumelhart and McClelland [34], etc.). The discrete model can handle all logical operations very efficiently because of the nature of the state transition function and the action function. An AND operation of two inputs can be modeled by the state transition/ action function shown in Figure 14.9. The other logical operations such as OR and NOT can be modeled exactly the same way.

14.4 A LOCALIST NETWORK OF RULE-BASED REASONING

State	Input 1	Input 2	Output
Don't care	0	0	0
Don't care	0	1	0
Don't care	1	0	0
Don't care	1	1	1

Figure 14.9 *Implementing AND gates with action function C.*

system is performing parallel breadth-first search where all nodes constitute states and links are search operators. However, the system does not have to be purely data driven: We can add some goal-orientedness to the system by adding a goal component to each rule, such as

If warm (not raining) goal(exercise), then swimming.
If (not warm) (not raining) (not windy) goal(exercise), then tennis.
If (not warm) raining goal(exercise), then indoor.

The model is capable of dealing with reasoning with partial and uncertain information, based on the characteristics of FEL. For example, if $A(X)$ and $B(X)$ and $C(X)$ then $D(X)$, where X is the vector of variable bindings, can be coded as shown in Figure 14.5. Node D then combines the evidence by doing $\Sigma_i w_i I_i$. In case C is unknown, based on partial information available, the system can still deduce D, but with activation weaker than it would be if C were known. This case can be easily mapped onto the exercise option example mentioned in the previous sections, so that when *warm, not raining* and *tired*, even though there is no rule fully matching the situation, we can still deduce a conclusion based on the partial match. Suppose the weight assignments are as follows:

> The first rule: (0.6 0.4);
> The second rule: (0.2 0.5 0.3);
> The third rule: (0.3 0.7).

Suppose *warm* = 0.9, (*not raining*) = 0.7, but nothing is known about windiness. Assuming *windy* = 0, an easy calculation according to the approximate reasoning theory (FEL) shows

> Swimming: 0.82;
> Tennis: 0.37;
> Indoor: 0.24.

So the best choice is going swimming.

The model can also deal with some forms of default reasoning. For example,

$A \rightarrow C$, weight distribution = (0.8),
$AB \rightarrow D$, weight distribution = (0.5,0.5).

assume the weight distributions already take into account the certainty of rules. In this case, when only A is known, C is concluded. When A and B are known, D is concluded. This results from the parallelism of this neural network model. In typical production systems, one rule and only one rule will be picked, but in CRBR all reachable conclusions can be reached, and the final outcome is the combination of these partial conclusions. For contradictory propositions, there can be inhibitory connections between each pair of them, with strengths corresponding to the degrees of contradiction.

This discussion shows the capability of the system as a commonsense reasoner. It provides a framework in which commonsense knowledge can be stored and inferencing about the knowledge can take place.

14.5 LEARNING CONNECTIONS

Some natural questions to ask are: Is there a way that in this network weights that represent rules can be learned through observing the regularity of the environment, instead of being given? What is the type of learning that is best suited to this task? How should learning algorithms deal with noise, interference from similar but different rules, and complex conditions or definitions? Learning is needed to account for the adaptive nature of commonsense reasoning.

In the following paragraphs, we will develop step by step a learning algorithm that is suitable for networks and try to answer these above questions.

The purpose of an internal model of the outside environment is to allow for the prediction of events in the environment and therefore the planning of the best course of action to achieve internal goals. What is important here is the good correspondence between the model and the environment. Based on this understanding, it is obvious that supervised learning and reinforcement learning are better suited for the task. Both types of learning can be looked at this way: mapping input to an output, and then receiving or computing an error signal and adjusting internal parameters so that error will be reduced. The error signal could be binary (right or wrong) or continuous, either actual distance between target output and actual output or merely a degree of correctness. We will develop a learning algorithm using only a generic error signal.

The basic idea follows: There are a set of input nodes and a set of output nodes, and each output node has a fan-in connection from all input nodes. The learning occurs at the fan-in junction. Assume the nodes are all binary

14.5 LEARNING CONNECTIONS   361

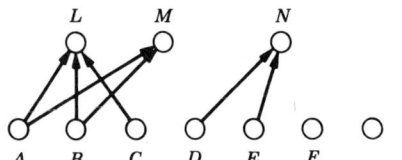

Input nodes: A B C D E F
Output nodes: L M N
Rules:
   if A B C then L
   if A B then M
   if D E then N

Figure 14.10 *Learning simple rules. (Input nodes: A–F, output nodes: L–N. Rules: If ABC, then L; if AB, then N; if DE, then N.)*

weighted-average (similar to weighted-sum except the sum is divided by the total weight) threshold units, and input/output is binary. If the node fires and error is zero, then increase by a certain small amount the weights of the links whose corresponding inputs are 1's. If the node fires and error is not zero, then decrease by a certain small amount the weights of the links whose corresponding inputs are 1's. If the node does not fire, and error is zero, then do nothing. If the node does not fire, and error is not zero, then increase by a certain small amount the weights of the links whose corresponding inputs are 1's. The threshold is between 0 and 1. We call this algorithm LA1. It is easy to notice that this is very similar to the idea of learning matrices, except that it starts with random initial weights. It is also easy to see that this will eventually correctly set up a set of simple rule mapping (of nondisjunctive rules).

For example, suppose we have a set of training data according to the following set of rules, adopted from the example analyzed in the first section, in which A = (not warm), B = (not raining), and so on:*

$$ABC \rightarrow L$$
$$AB \rightarrow M$$
$$DE \rightarrow N$$

See Figure 14.10. After training, the network will look like the one shown in Figure 14.10. Because of the fact that we adopt weighted-average activation functions, the absolute value of the weight on an individual link is of no significance. This way learning never has to end, or in other words, the learning stage and the performance stage are the one and the same. Decay can be used throughout the working of the system, but it is not an essential feature because of the weighted-average activation function, that is, instead of decaying some weights, we can increase some other weights to get the same net effect. For example, if the initial weight from F to N is not zero, then instead of decaying it to zero, we can learn to acquire large weights

---

* We use letters here to represent concepts, not only for the ease of explaining ideas, but also for indicating the generality, or domain independence, of the algorithms. The letters can be easily mapped onto the examples stated before, or any other examples that might come to mind.

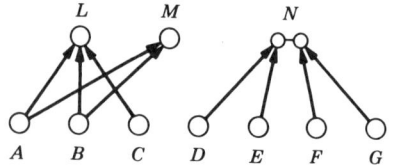

```
Input nodes: A B C D E F
Output nodes: L M N
Rules:
    if A B C then L
    if A B then M
    if D E then N
    if F G then N
```

Figure 14.11 Learning multiple rules, with sites competing to activate a node. (Input nodes: A–F, output nodes: L–N. Rules: if ABC, then L; if AB, then M; if DE, then N; if FG, then N.)

from D to N and from E to N, so that the weight from F to N will not have much effect. The activation function for N is $(w_{D-N}i_D + w_{E-N}i_E + w_{F-N}i_F)/(w_{D-N} + w_{E-N} + w_{F-N})$. It can be converted to the weighted-sum form easily.

Now the problem is if there are complex conditions (i.e., disjunctive conditions), how can we learn the rule? Obviously we cannot use the previous algorithm, because with that algorithm after we learned one disjunct, the presentation of the training data for another disjunct will destroy the learned connections. Suppose we have this set of rules:

$$ABC \rightarrow L$$
$$AB \rightarrow M$$
$$DE \rightarrow N$$
$$FG \rightarrow N$$

After learning the rule $DE \rightarrow N$, if we keep presenting data for the rule $FG \rightarrow N$, the previously learned connection will be destroyed due to much higher weights from F and G to N.

The solution to this problem is adopting neurons with multiple sites (as another special case of the DN model), each of which is a simple rule (one disjunct of a complex rule). Those sites (or groups of weights) compete to win the right to send activation to the neuron. So within each site the computation is the same as simple neurons, and overall the neuron will compute the maximum of all sites; that maximum becomes the current activation of the neuron. The learned connections are shown in Figure 14.11. With sufficient numbers of sites and different random initial weights, then the chances are that for different disjuncts, different sites will win and thus learn to represent the corresponding disjunct. We will call this learning algorithm LA2.

The foregoing cases are for binary input/output. When input/output is analog, a simple extension will suffice. While the previous algorithms (LA1 and LA2) increment weights by a fixed amount, now we increment each weight by an amount proportional to the corresponding input signal strength (call this algorithm LA3). This algorithm has the nice property of being noise tolerant (as do LA1 and LA2): Regarding the set of rules in the previous example, if occasionally we have data that amounts to

$$ABF \rightarrow M \text{ or}$$
$$BDF \rightarrow M$$

assume those data are of low frequency, then the presence of those noisy data will not affect the learning of correct rules, because only a small amount of weight increment between F and M or between D and M. This property is actually derived from a more general property that each weight is a frequency/strength sampling of the corresponding input signals.

After the learning of individual rules, we can expect the chaining of multiple rules, because during the learning of individual rules the set of input nodes can overlap with the set of output nodes and this overlapping enables the chaining of rules whenever needed. Learning, when chaining of rules is considered, can be done with the help of the bucket-brigade algorithm (Holland et al. [35]). The signal sent from the next rule to the previous serves as an error signal in our algorithm, and the rest is the same as LA3.*

These algorithms indicate a way in which the weights in the CRBR system can be formed, or in other words, they provide an operational definition of weights in the CRBR architecture.

14.6 DISCUSSION

14.6.1 Levels of Modeling

An interesting issue is the appropriate level (or levels) for cognitive modeling. Several obvious choices are LISP code, artificial neural network models (connectionist models), or biological neural network models.

First we want to discuss the biological plausibility of connectionist models. An artificial neural network model ought to be a biologically plausible model of neural computations aimed at providing a vehicle for building applications in various fields of computation. The key word here is *plausible,* meaning they should reflect our main understanding of the biological neural networks but should not either be confined by the requirement that

---

* The discussion assumes that there is no variable and mechanism for variable binding. If we include variable binding in the rule learning, then we have to use a more general neuronal model: DN/PDN. This general formalism allows variable binding and other mechanisms, and thus has its advantages. But we cannot use these learning algorithms any more, because they only deal with weighted-average models. Fortunately we have another set of algorithms that can be used for this more general case, as discussed in Sun [20]. These algorithms are very general, even allowing hidden units in a model. They are not specifically designed for rule learning, and thus it is expected that convergence of learning would take much longer time.

we model only what we know nor be overwhelmed by the detailed "hardware" operations found in the real neuron.

Biologists' concern of reality might be different from cognitive scientists' or computer scientists' concern of reality. The difference can be characterized as that of perspectives: We who are concerned only with the computation carried out by the brain look at things at a different level of abstraction, concerning mainly the information processing mechanism, not the underlying "hardware" implementation. By the same token, the information flow of the network is of interest, but not necessarily the details of how the information transfer is actually carried out.

We also want to argue for connectionist models against traditional symbolic artificial intelligence (AI) paradigms, although this has been done elsewhere (e.g., Waltz and Pollack [36] and Sun [37,38,48]). To further illustrate the point, we look at it from a different perspective: Are connectionist models symbolic or nonsymbolic? We think that a proper way to put it is that connectionist models start from nonsymbolic levels (but are not limited to that level). From neuropsychological perspective, neural networks are the ultimate substrate for all cognitive processes. Each and every high-level cognitive activity can be traced back to the activities at the neuronal network level. In other words, symbolic processing builds up from nonsymbolic processing. So no matter how complicated a symbolic processing task is, there is a way to implement it in a neural network fashion because nature already did it. Although we only have simple implementations of scaled-down versions of complex symbolic processing systems, it is not too far from full-fledged systems (i.e., comparable with symbolic AI systems in functional complexity). It is a just matter of engineering.

By looking at things at a lower level (looking at substrates rather than phenomena), we might be able to gain some unique insight into the matter and generate some useful constraints that will help us in finding right algorithms in building AI systems.

It is necessary for connectionist architectures to implement rules, *explicitly,* both for engineering reasons in developing applications and for theoretical reasons in understanding human intelligence. The engineering reasons include massive parallelism, fast relaxation time, tractability and ease of construction, and others (as in CRBR). The theoretical reasons follow from the need to model the conceptual processor (cf. Smolensky [39]). Without doubts, explicit symbolic rules *do* exist in reasoning (cf. Fodor and Pylyshyn [43] Pinker and Prince [45]).

14.6.2 Comparisons with Other Models

We wish to compare our work with two pieces of recent work on connectionist models of rule-based reasoning (both appeared first in Cognitive Science Conference Proceedings 1989, the same as this model [37]). First we want to compare our model with Ajjanagadde and Shastri [5]. Both are connectionist models that can do logical reasoning and handle variable bindings, and both use local representation. Ajjanagadde and Shastri explored the temporal dimension of activation patterns for representing variables and bindings, while our model extended the alphabet of signals between nodes in a network to handle variable bindings. It is easy to see that different frequencies in node outputs can be readily represented by a single output value (symbol) as in our model. Their system performs backward-chaining reasoning, different from our system, which performs forward chaining. Because of the nature of forward-chaining reasoning, the only constraints on the representation for our system are noncontradiction and acyclicity in case of the binary logic. So the scheme can handle Horn clause logic (without recursion), and more general cases. In addition to binary logic, by introducing the approximate reasoning scheme proposed earlier, the system has the power of approximate reasoning and it can accommodate contradictions.

Lange and Dyer [6] present an interesting rule-based reasoning system built out of connectionist models. This system is very close to ours, in both variable bindings and rule activation. However, their system does not have a unifying underlying neural network model that can explain all kinds of mechanisms appeared in the system: Signature (used for variable bindings) and numeric activation form two separate components. Our work is based on the foundation of a neural network formalism that is well studied in various aspects and is the force integrating the whole system. Our system also provides a logically sound scheme for doing approximate reasoning in connectionist models, without resorting to probabilistic theory, which is very difficult to be correctly implemented by weighted-sum connectionist models. Fuzzy evidential logic provides a semantics, a way to interpret the weighted-sum connectionist models.

Blelloch [8] and Kosko [9] also developed connectionist rule-based reasoning systems, except that they do not have mechanisms for representing variables and for variable bindings. Instead they adopt the multiple instantiation approach.

14.6.3 Reasoning at Both Conceptual and Subconceptual Levels

In our ongoing research [56] we are trying to augment rule-based reasoning by adding similarity-based components. There are two parts in the system: CD and CL. The CL representation is rule-based (exactly the same as CRBR), and the CD representation is similarity-based. The resulting system will be capable of generalization: adding continuity, filling in knowledge gaps, extrapolation, and interpolation.

This system combining reasoning at conceptual and subconceptual levels can help deal with a major problem in rule-based reasoning—the *brittleness* problem resulting from discretization of continuous thought processes. It is manifested in the inability in dealing with the following aspects:

- Partial information,
- Fuzzy or uncertain information,
- No match,
- Fragmented rule bases,
- Bottom-up inheritance and cancellation,
- Top-down inheritance and cancellation,

This ongoing research is concerned with the following question: in the rule-following sort of reasoning, how the interaction between explicit rule application and emergent rule following tendency helps to determine the rigidity, tractability, treatment of novel data, and correctness of the reasoning process. It stipulates that there are two separate processors (maybe more) for conceptual and subconceptual reasoning. The conceptual processor is embodied in a localist connectionist network, and the subconceptual processor is wired up in a more or less distributed network. The goal of the research is to see how these two components interact, compete, and cooperate in problem solving, and to develop a system capable of various reasoning capacities by combining the two components in a connectionist network.

14.7 CONCLUDING REMARKS

Using a new neural network formalism (the DN/PDN model), which is very general and powerful, we developed a massively parallel model of approximate rule-based reasoning called CRBR. The reasoning is carried out in a data-driven fashion, modulated by system goals. A scheme for approximate reasoning (FEL), which is well-suited for connectionist models, is incorporated into the system. Thus the system can deal with partial and uncertain information and cumulative evidence. The system is aimed for commonsense reasoning tasks. It illustrates that a combination of symbolic AI and connectionist models can result in efficient and practical systems.

ACKNOWLEDGMENT

We wish to thank James Pustejovsky and Tim Hickey for their comments and suggestions.

APPENDIX 14A: THE SOUNDNESS AND COMPLETENESS OF FUZZY EVIDENTIAL LOGIC

Fuzzy evidential logic is an inference system* which, given a set S of FEL propositional Horn clauses

$$(p_1, w_1), \ldots, (p_r, w_r) \to q \quad \text{(rules)}$$
$$\ldots$$
$$\to d \quad \text{(facts)}$$
$$\ldots$$

where p_i's are propositions, $w_i's$ are real, $w_i \geq 0$, and $w_1 + \cdots + w_r \leq 1$, and given a propositional symbol g, the system determines if g is a FEL logical consequence of S with certainty $\geq c$.

We will show that g is a FEL logical consequence of S with certainty 1 iff g is a logical consequence of the underlying propositional theory.

The inference rule for FEL is a variant of forward chaining, defined as follows: Let K be a set of FEL propositions, that is, pairs (d,c) where d is a proposition and c is a real number between 0 and 1 representing the certainty that d is true. We assume that all propositions are uniquely represented in K (though their certainty values may be zero).

The *inference rule* will simply add FEL propositions to K until no new propositions can be added.

Given the FEL rule $(p_1, w_1), \ldots, (p_r, w_r) \to q$
 If $(p_1, d_1), \ldots, (p_r, d_r)$ are in K, then
  let $d' = w_1 * d_1 + \cdots + w_r * d_r$ and let (q, d) be in K.
 If $d' > d$, then replace (q, d) by (q, d').
If there is no condition, that is, $\to q$
 then we simply replace (q, d) in K by $(q, 1)$.

Given a Horn clause theory T, one can produce a corresponding FEL theory F as follows: Each Horn clause $p_1, \ldots, p_r \to q$ is transformed into a FEL clause by associating a weight w_i with each atom p_i, in such a way that $w_i > 0$ and they sum to 1.

THEOREM. *Suppose that we are given a Horn clause theory* T. *Let* F *be a FEL theory corresponding to* T *(as before), and assume that all propositions*

---

* For the sake of clarity, we only present the propositional case here. It is straightforward to extend it to the predicate case. We disallow any functional symbol in our system, similar to DATALOG.

in K initially have certainty values 0, then $(g,1)$ is a FEL logical consequence of F iff g is a logical consequence of T.

Proof. Observe that a FEL proposition $(g,1)$ is added to K if and only if all of the propositions in the body of the clause have certainty values 1. Thus, $(g,1)$ is only introduced to K if there exist propositions $(p_1,1), \ldots, (p_r,1)$ in K. Thus, the FEL inference rule behaves exactly like the forward-chaining operator for Horn clause logic when we restrict our attention to propositions of weight 1. Since forward chaining is a sound and complete theorem prover for atomic queries and Horn clause theories, g is a logical consequence of P' if and only if g can be inferred by forward chaining iff $(g,1)$ can be inferred by FEL Inference. Q.E.D.

Because of the fact that Horn clause forward-chaining is sound and complete, it is obvious that FEL is sound and complete in its degenerate form as indicated by the theorem.

APPENDIX 14B: FUZZY EVIDENTIAL LOGIC RULES IN DN NETWORKS

Here we will derive a complete representation of the first-order FEL rules in DN networks for the following example: We prove that Socrates is mortal, based on the general rule: if one is human, one is mortal, and the knowledge that Socrates is human. The reasoning goes like this:

If one is human, one is mortal.
Socrates is human.
So Socrates is mortal.

The rule to be represented is

If x is human, x is mortal.

First we have to set up a network for representing the rule. Applying the DN formalism, we build up two assemblies (see Fig. 14.12). Each assembly contains a C node and a X node (for one argument). The input to j is the word "Socrates" or its representation in whatever form, and the input to i is the value of the fact "Socrates is human," which is one in this case.

For C_1,
$$I = \{i,l\}$$
$$O = \{k,q\}$$
$$S = \text{real}$$

where B_i = real, B_l = string, A_k = real, and A_q = real.

APPENDIX 14B: FUZZY EVIDENTIAL LOGIC RULES IN DN NETWORKS

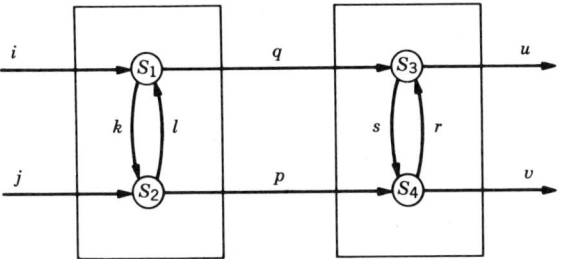

Figure 14.12 An example network.

The inputs from argument nodes to C nodes are to inform them if there is right binding present at each argument node. If the bindings are inconsistent, or do not satisfy what is required of the arguments for a particular predicate, the C node represents the predicate should not be activated. For example, for $A(x,x,y)$, the first two arguments must be the same for the predicate to be activated. The input from C nodes to argument nodes is to tell them if they should pass on bindings. In case the C node is not activated, the argument nodes in the same assembly should not, or need not, work on the variable bindings. Although these issues are not important for this simple example, we will still keep these two types of signals there for the sake of generality.

Now the state transition function and action function: Let R stand for real, and Σ stand for string.

$$U: R \times (R \times \Sigma) \to R$$

or

$$S \times (i,l) \vdash i$$
$$V: R \times (R \times \Sigma) \to A_k \times A_q = R^2$$

or

$$S \times (i,l) \vdash (k,q), \quad \text{where } k = i, q = i$$

Expressing them in functional forms, we have

$$U(S,(i,l)) = i$$

and

$$V(S,(i,l)) = (k,q) \vdash (i,l)$$

For X_1,

$$I = \{j,k\}$$
$$O = \{l,p\}$$
$$\Sigma = \text{string}$$

where B_j = string, B_k = real, A_l = string, and A_p = string.

Now the state transition function and action function: Again let R stand for real, and Σ stand for string.

$$U: \Sigma \times (\Sigma \times R) \to \Sigma$$

or

$$S \times (j,k) \vdash j$$
$$V: \Sigma \times (\Sigma \times R) \to \Sigma^2$$

or

$$S \times (j,k) \vdash (l,p), \quad \text{where } l = j \text{ if } k \neq 0, \quad l = \bot \text{ if otherwise;} \quad p = j$$

Expressing them in functional forms, we have

$$U(S,(j,k)) = j$$

and

$$V(S,(j,k)) = (l,p) \vdash (i \text{ if } l \neq 0, \quad \bot \text{ if } l = 0))$$

It is the same for C_2 and X_2. For C_2:

$$U(S,(q,r)) = q * c$$

where c is a constant.

$$V(S,(q,r)) = (u,s) \vdash (q * c, q)$$

For X_2:

$$U(S,(p,s)) = p$$
$$V(S,(p,s)) = (v,r) \vdash ((p \text{ if } s \neq 0, \quad \bot \text{ if otherwise}), p)$$

Simplifying the notations, we come up with simple assignment statements (so that they will be easily programmable) as follows: Assume prime (')

represents the next time period. For C_1,

$$S' = i$$
$$q' = l$$
$$k' = i$$

For X_1,

$$S' = j$$
$$l' = j$$
$$p' = j \text{ if } k \neq 0, \quad \perp \text{ if otherwise}$$

For C_2,

$$S' = q * c$$
$$u' = q * c$$
$$s' = q$$

For X_2,

$$S' = p$$
$$r' = p$$
$$v' = p \text{ if } s \neq 0, \quad \perp \text{ if otherwise}$$

Based on this, the whole working process can be described as follows:

$$p' = j \text{ if } k \neq 0, \quad \perp \text{ if otherwise}$$
$$q' = i$$

So

$$u'' = q' * c = i * c$$

and

$$v'' = p' \text{ if } s' \neq 0, \quad \perp \text{ otherwise}$$

According to the equations regarding p' and s', we have

$$v'' = j \text{ if } i \neq 0, \quad \perp \text{ if otherwise}$$

APPENDIX 14C: TIMING

There are some subtle timing considerations that need to be addressed. Outputs of a node depend on not only inputs from other assemblies but also inputs from other nodes in the same assembly. Thus it is important to check that there is no mutual waiting situation. This, in a way, resembles the deadlock problem in communicating concurrent processes: An order has to be imposed on the input/output relations, so that certain communication is always facilitated to keep the flow going. In this case, intra-assembly communication always proceeds first. In other words, it does not depend on input from other nodes in the same assembly, only on input from other assemblies. This way intra-assembly communication can be finished first, and then outputs to other assemblies can be done without any impediments.

So we will have to delineate the model as follows:

$$I = I_1 \cup I_2$$
$$O = O_1 \cup O_2$$

where subscript 1 denotes interassembly input/output, and subscript 2 denotes intra-assembly input/output. And also for state transition and action functions:

$$V = V_1 \cup V_2$$

where

$$V_1: S \times B^{I_1} \to A^{O_2}$$
$$V_2: S \times B^I \to A^{O_1}$$
$$U: S \times B^I \to S$$

The function V_1 is done during the first part of a cycle, communicating within an assembly, and V_2 is done during the second part of a cycle, sending outputs to other assemblies downstream.

APPENDIX 14D: COMPILATION PROCEDURES

The CFRDN (Compilation of FEL Rules into DN Networks) is a set of procedures for automatically transforming FEL propositions, predicates and rules into a forward-chaining reasoning network. The main procedures for compiling a set of rules into a DN network are as follows. Assume for each node there is a set of intra-assembly input and a set of interassembly input; Likewise, each node has a set of intra-assembly output and a set of interassembly output; There is no recursion in the rule set. Here is a bottom-up procedure for setting up rules.

```
                        SET-UP-RULE-NET
Initialize the rule set as all rules to be implemented.
Repeat until the set is empty:
  Choose a rule ( P1(X1) P2(X2) ... Pn(Xn) → Q(X) | w ) to work on,
      if there is no rule in the set that has Pi as conclusion.
  Identify all existing assemblies for Pi, i=1,2,...,n,
      if there is none make one.
  Identify all existing assemblies for Q,
      if there is none make one.
  If there is only one assembly for each Pi and there is no in-link to Q,
      link each Pi to Q, assign weights to each Pi.
  If there is in-link to Q,
      link Pi's to an intermediate assembly, then or-link this assembly
      with existing Q to a new Q assembly which assumes all out-links of Q,
      and assign weights as appropriate.
  If there are multiple assemblies for Pi,
      OR-link all assemblies for that Pi to an intermediate assembly,
      and then link that intermediate assembly to Q,
      and assign weights as appropriate.
  Delete the rule implemented from the set.
```

Notice that we are concerned only with hierarchical FEL. So the relations between LHS and RHS of various rules impose a partial ordering among the rule set. What the algorithm does is to choose a full ordering out of the partial ordering.

The predicates in the rule set can be divided into three classes:

$$P = P_I \cup P_P \cup P_O$$

where P_I = set of predicates that do not appear as conclusions in any rules; it constitutes the input layer.

P_P = set of predicates appearing in as both conditions and conclusions; it constitutes the processing layer.

P_O = set of predicates appearing only as conclusions; it constitutes the output layer.

The procedures for setting up each individual assembly are as follows:

SET-UP-ASSEMBLY

Set-Up-C
Set-Up-X

SET-UP-C

(I= (I1, I2, ...In) is the input from other assemblies)
(I'= (I1', I2', ...) are the inputs from Xs in the same assembly)
output to other Cs: W * I,   if I' are all 1's
                      0   otherwise
output to Xs: 1 if I <> 0
               0 otherwise
state: W * I

The output to Xs are to inform them if they should pass on bindings. I' is used to determine if there is right binding present at argument nodes.

SET-UP-X:

(I is the input from Xs in other assemblies)
(I' is the input from C)
output to C: 1 if I is not empty
              0 otherwise
output to Xs in other assemblies: I if I' <>0
                                   nothing, otherwise
state: I

OR-LINK-ASSEMBLY

For C node:
  (I is the input from other assemblies. k denotes the group)
  (each group is a vector by itself)
  output to other Cs: W * I(k) if W * I(k) = max(W * I(j))
  output to Xs: k if I(k) <>0
                0 otherwise
  state: same as output to other Cs.

For X nodes:
  (I is the input from other assemblies)
  (Each element of I is a scalor)
  (I' is the input from C)
  output to C: I=(I1, I2, ..., In)
  output to Xs in other assemblies: Ik if I'=k
  state: same as output to Xs

Sometimes we may need to generate bindings for arguments. For example, $A(X_1) \rightarrow B(X_1, X_2, C)$. Two numbers have to be generated to cover X_2 and C. The number generated for C is a genuine binding, but the number generated for X_2 is not really bound to anything, just denoting a possible binding, or pseudobinding.

GENERATE–ARGUMENT

For C node:
  output to Xs: m (where m > M is a number specific to the node)
  (the same as above otherwise)

For X:
  output to Xs: m if I' = m > M
  state: m if I' = m > M
  (the same as above otherwise)

Here, I' is the intra-assembly input to X nodes, and M is a fixed number, greater than n (the number of inputs to a X node) such that if $I' > M$, I' denotes a new binding. Each C node has some fixed m's unique to it for this purpose.

We may also need to check the consistency of binding.* For example, $A(X_1, X_2)\ B(X_2, X_3) \rightarrow C(X_1)$.

CONSTRAIN–ARGUMENT

For C node:
  if constraints are satisfied, then outputs and states are
    the same as above
  otherwise, output 0.

We may even perform unification of terms, based on the action function. This action function is implemented as a table look-up. This is relatively an easy task, since there is no function for arguments or high-order predicates.

UNIFY–ARGUMENT

For X node:
               (I is the input from Xs in other assemblies)
               (I' is the input from C)
output to C: 1 if all inputs can be unified and I is not empty
             0 otherwise
output to Xs in other assemblies:
             the result of the unification if I' $\diamondsuit$ 0
             nothing, otherwise
state: I

---

* This is implicit in the semantics of logic, but it has to be dealt with explicitly in implementation.

376 A NEURALLY INSPIRED MASSIVELY PARALLEL MODEL OF RULE-BASED REASONING

An example is in order here. Suppose we want to implement the following rule set:

```
P(x1, a) Q(a, x3) → R(x1, x3)
S(x4, x4) → P(x4, a)
P(x5, x6) → Q(x6, x5)
P(x7, x8) → R(x8, x8)
```

This example involves many issues already discussed, such as multiple rules having the same conclusion, constraint checking, generating arguments, and so forth. The solution obtained according to the algorithms is shown in Figure 14.13. Part (*a*) shows the connections of the network, and part (*b*) shows a diagram of one assembly. The weight assignment is not shown here, which could be anything as long as they add up to be equal to or less than one.

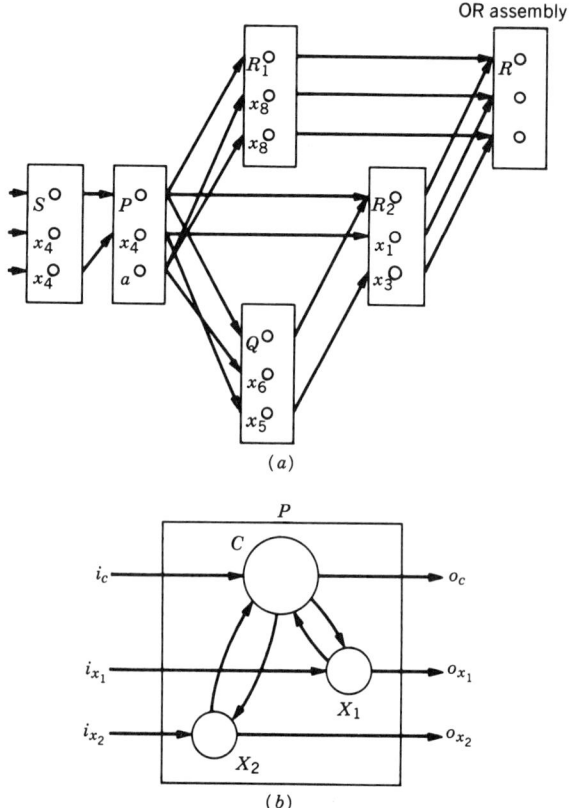

Figure 14.13 An example network: (a) the network; (b) an assembly in the network.

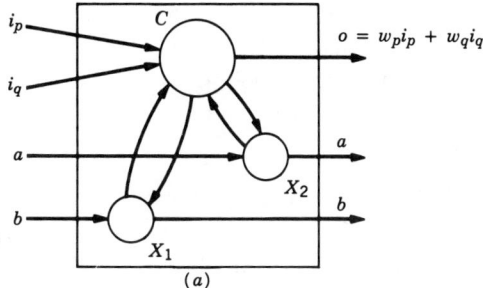

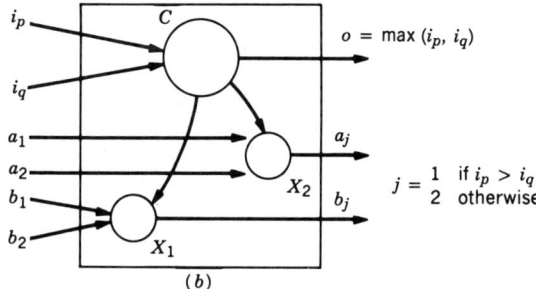

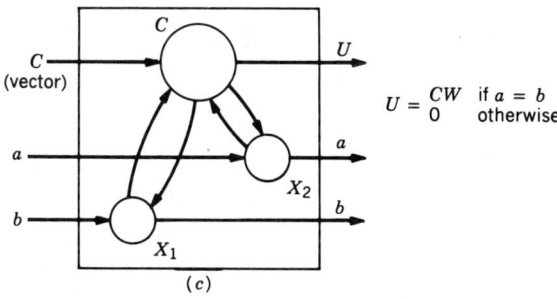

Figure 14.14 *The three types of assemblies.*

We will summarize what is needed for a node below. Usual operations for a node will include:

- Conditionals (e.g., if input > 0 then output $= 1$),
- Addition,
- Multiplication,
- Weighted sum,
- Weighted-average,

- Max,
- Selection (e.g., if $I(1) = 5$ then output $I(5)$),
- Equality (e.g., $I(1) = I(2)$) and inequality,

and so on, and their combinations.

There are basically three different types of assemblies (see Fig. 14.14):

- Ordinary assemblies—computing weighted sums of inputs and passing along bindings for the arguments.
- OR assemblies—computing the maximum (the MAX operation) of inputs and choosing one binding out of many based on who is the winner (the maximum), using the selection operation.
- Complex assemblies—the same as ordinary assemblies, except performing constraint checking and/or unification, and also argument generation when necessary, using the equality and inequality operations.

APPENDIX 14E: CORRECTNESS OF THE IMPLEMENTATION

In the previous algorithms, we assume that there is no recursion whatsoever. Under this condition, the implementation scheme is complete and sound with regard to FEL.*

With recursive rules, either direct or indirect, the system will lose its completeness, if implemented with cyclic connections. However we are mainly interested in hierarchical FEL, so this is not a problem. Besides, there are also a couple of ways we can use in dealing with recursions. One way is to replicate the predicate a number of times in the network, up to the maximum depth of the recursion, if we know that a priori. Another possible way is to use an external stack, which stores multiple bindings and pops one up when necessary.

For the propositional theory, the above algorithm for implementing rules works just fine, except there is no need for argument nodes. It is very interesting that in case of propositions (i.e., no arguments), no matter whether there is recursion, the implementation is always sound and complete. This is because there is no multiple binding (unification) problem there.

---

* It is also sound and complete for first-order Horn clause logic, when using binary FEL. The completeness comes under the assumption that you replicate and rename a predicate (and all rules involving that predicate) throughout the entire rule set if it appears more than once with different arguments in RHS of rules. Otherwise, the outcomes from these different rules will be MAXed, and therefore the implemented system loses its completeness.

REFERENCES

[1] A. Collins and R. Michalski, The logic of plausible reasoning: A core theory, *Cognitive Science,* 13(1) (1989).

[2] D. Touretzky and G. Hinton, "Symbols among neurons," *Proc. of 9th IJCAI,* Moran Kaufman, San Mateo, Calif., 1987.

[3] J. Barnden, "The Right of Free Association: Relative-Position Encoding for Connectionist Data Structures," *Proc. of 10th Conf. of Cognitive Science Society,* Lawrence Erlbaum, Hillsdale, N.J., 1988.

[4] L. Shastri and J. Feldman, A. connectionist approach to knowledge representation and limited inference, *Cognitive Science,* 12 (1988).

[5] V. Ajjanagadde and L. Shastri, "Efficient Inference with Multi-place Predicates and Variables in a Connectionist System," 11th Cognitive Science Society Conference, 1989.

[6] T. Lange and M. Dyer, "Frame selection in connectionist model of high level inferencing," *Proc. of 11th Cognitive Science Conference,* 1989.

[7] J. Hendler, "Marker Passing and Microfeature," *Proc. of 10th IJCAI,* 1987.

[8] G. Blelloch, "AFL-1: A Programming Language for Massively Parallel Computers," Master's thesis, MIT AI Lab, Cambridge, Mass., 1986.

[9] B. Kosko, Hidden patterns in combined and adaptive knowledge networks, *Internat. J. Approximate Reasoning,* 2 (1988).

[10] M. Fanty, Learning in structured connectionist networks, TR 252, University of Rochester, Rochester, N.Y., 1988.

[11] D. Ackley, Stochastic iterated genetic hillclimbing, TR CMU-CS-87-107, Carnegie-Mellon University, Pittsburgh, Pa., 1987.

[12] M. Derthick, Mundane reasoning by parallel constraint satisfaction, TR CMU-CS-88-182, Carnegie-Mellon University, Pittsburgh, Pa., 1988.

[13] R. Sun, E. Marder, and D. Waltz, "Model Local Neural Networks in the Lobster Stomatogastric Ganglion," *Proc. of IJCNN,* 1989.

[14] A Selverston and M. Moulin, Eds., *The Crustacean Stomatogastric System,* Springer-Verlag, Berlin, 1987.

[15] E. Kandel and J. Schwartz, *Principles of Neural Science,* 2d ed., Elsevier, New York, 1984.

[16] G. Edelman et al. Eds., *Synaptic Function,* Wiley, New York, 1987.

[17] B. Richmond and L. Optican, Temporal encoding of two-dimensional patterns by single units in primate inferior temporal cortex, *J. Neurophysiology,* 57(1) (January 1987).

[18] L. Fang and W. Wilson, "A Study of Sequence Processing on Neural Networks," *IJCNN,* 1989.

[19] D. Servan-Schreiber et al., Encoding sequential structure, CMU-CS-88-183, Carnegie Mellon University, Pittsburgh, Pa., 1988.

[20] R. Sun, The DN and PDN models, *Proc. of Internat. Neural Network Conference,* Paris, Kluwer 1990.

[21] L. Zadeh, Commonsense knowledge representation based on fuzzy logic, *Computer,* 16(10) (1983).

[22] D. Dubois and H. Prade, Eds., *Fuzzy Sets and Systems,* Academic Press, New York, 1980.
[23] L. Zadeh, Fuzzy Sets, *Information and Control,* 1965.
[24] L. Zadeh, The Concept of a Linguistic Variable and Its Application to Approximate Reasoning, *Information Science,* 1975.
[25] C. Chang and R. Lee, *Symbolic Logic and Mechanical Theorem Proving,* Academic Press, New York, 1973.
[26] B. Buchanan and E. Shortliffe, Eds., *Rule Based Reasoning: The Mycin Experiment,* Addison-Wesley, Reading, Mass., 1984.
[27] F. Hayes-Roth et al. Eds., *Building Expert Systems,* Addison-Wesley, Reading, Mass., 1983.
[28] D. Dubois and H. Prade, "An Introduction to Possibilistic and Fuzzy Logics," in P. Smets et al., Eds., *Non-Standard Logics for Automated Reasoning,* Academic Press, New York, 1988.
[29] R. Sun, "The Logic for Approximate Reasoning Combining Probability and Fuzziness," 1st Congress of International Fuzzy System Association, 1985.
[30] J. Pearl, *Probabilistic Reasoning in Intelligent Systems,* Morgan Kaufman, San Mateo, Calif., 1988.
[31] G. Shafer, *A Mathematical Theory of Evidence,* Princeton University Press, Princeton, N.J., 1974.
[32] A. Tversky and D. Kahneman, Extensional Versus Intuitive Reasoning: The conjuction Fallacy in Probabilty Judgement, *Psychological Review,* 1983.
[33] Y. Abu-Mostafa, "Neural Networks for Computing?" *Neural Networks for Computing,* American Physics Association, 1986.
J. Anderson and G. Murphy, "Concepts in Connectionist Models," *Neural Networks for Computing,* APA, 1986.
[34] D. Rumelhart and J. McClelland, *Parallel Distributed Processing,* MIT Press, Cambridge, Mass., 1986.
[35] J. Holland et al., *Induction,* MIT Press, Cambridge, Mass., 1986.
[36] D. Waltz and J. Pollack, Massive parallel parsing, *Cognitive Science,* 9 (1985).
[37] R. Sun, "A Discrete Neural Network Model for Conceptual Representation and Reasoning," *Proc. of 11th Cognitive Science Society Conference,* 1989.
[38] R. Sun, Rules and connectionism, *Proc. of Internat. Neural Network Conference,* Paris, 1990.
[39] P. Smolensky, On the proper treatment of connectionism, *Behavioral and Brain Sciences,* 11, 1988.
[40] T. Shultz et al., "Managing Uncertainty in Rule-Based Reasoning," 11th Cognitive Science Conference, 1989.
[41] C. Dolan and P. Smolensky, "Implementing a connectionist production system using tensor products," *Proc. 1988 Connectionist Model Summer School,* 1988.
[42] J. Feldman and D. Ballard, Connectionist models and their properties, *Cognitive Science* (July 1982).
[43] J. Fodor and Z. Pylyshyn, "Connectionism and Cognitive Architecture: A

Critical Analysis," in Pinker and Mehler Eds., *Connections and Symbols*, MIT Press, Cambridge, Mass., 1988.

[44] S. Grossberg, *The Adaptive Brain,* North-Holland, Amsterdam 1987.

[45] S. Pinker and A. Prince, On Language and Connectionism, in S. Pinker and J. Mehler, Eds., *Connections and Symbols,* MIT Press, Cambridge, Mass., 1988.

[46] R. Sun, Designing inference engines based on a discrete neural network model, *Proc. of IEA/AIE,* 1989.

[47] R. Sun, Reasoning at conceptual and subconceptual levels, TR-CS-90-136, Brandeis University, Waltham, Mass., 1990.

[48] R. Sun and D. Waltz, Neural networks and human intelligence, *J. Mathematical Psychology,* 1990.

CHAPTER 15

Injecting Symbol Processing Into a Connectionist Model

STEVE G. ROMANIUK
LAWRENCE O. HALL

15.1 INTRODUCTION

This chapter describes a hybrid connectionist, symbolic approach to integrating learning capabilities into an expert system. Specifically, rule-based systems are the ones that are concentrated upon. The system does its learning from examples that are encoded in much the same way that examples to connectionist systems would be presented. The exceptions are due to the injection of symbol processing for variable representation. The system can learn concepts where imprecision is involved. The network representation allows for variables in the form of attribute, value pairs to be used. Both numeric and scalar variables can be represented. They are provided to the system upon setting up for the domain. Relational comparators are supported. Rules from standard rule bases may be directly encoded in our representation. The system has been used to learn knowledge bases from some small examples originally done in EMYCIN [1]. It has also been used in learning defects in complementary metal-oxide semiconductor (CMOS) chips [2], the Iris data set, and the Soybean data set [3]. One data set with a lot of uncertainty, on the creditworthiness of a credit applicant, has also been processed.

The learning algorithm uses a network structure, which is configured based on the distinct examples presented to the system. For examples that resemble others previously seen, bias values of cells in the network are adjusted. The system can learn incrementally.

15.2 SC-NET—THE REPRESENTATION

A connectionist model is a network, which in its simplest format has no feedback loops. It consists of three types of cells (input, output, and hidden cells). Every cell has a bias associated with it, which lies on the real number scale. Cells are connected through links that have weights associated with them. In the SC net model of a connectionist network, each cell can take on an activation value within the range [0,1]. This corresponds to the fuzzy membership values of fuzzy sets.

The system is intended to be able to deal with imprecision and uncertainty. The uncertainty handling constructs come from fuzzy set theory [4].

In fuzzy logic one may define disjunction (fOR) as the maximum operation, conjunction (fAND) as the minimum operation, and complement (fNOT) as strong negation. Since fOR and fAND are defined as maximum and minimum operations, we let certain cells act as max and min functions in order to provide for the operators. In order to be able to distinguish cells as modeling the min (fAND) or the max (fOR) function, we use the sign of the bias of a cell to determine which of the two functions is to be modeled. Furthermore, we denote a bias value of zero to indicate when a cell should operate as an inverter (fNOT).

15.2.1 The Network Structure

We can think of every cell in a network accommodating n inputs I_n with associated weights CW_n. Every cell contains a bias value, which indicates what type of fuzzy function a cell models, and its absolute value represents the rule range. Every cell C_i with a cell activation of CA_i (except for input cells) computes its new cell activation according to the formula in Figure 15.1. If cell C_i (with CA_i) and cell C_j (with CA_j) are connected, then the weight of the connecting link is given as $CW_{i,j}$; otherwise $CW_{i,j} = 0$. Note, an activation value outside the given range is truncated. An activation of 0 indicates no presence, 0.5 indicates unknown, and 1 indicates true. In the initial topology, an extra layer of two cells (denoted as the positive and the

CA_i = cell activation for cell C_i, CA_i in [0,1].

$CW_{i,j}$ = weight for connection between cell C_i and C_j, $CW_{i,j}$ in R.

CB_i = cell bias for cell C_i, CB_i in [−1,+1].

$$CA'_i = \begin{cases} \min_{j=0,\ldots,i-1,i+1,\ldots,n} (CA_j * CW_{i,j}) * |CB_i|, & CB_i < 0 \\ \max_{j=0,\ldots,i-1,i+1,\ldots,n} (CA_j * CW_{i,j}) * |CB_i|, & CB_i > 0 \\ 1 - (CA_j * CW_{i,j}), & CB_i = 0 \text{ and } CW_{i,j} \neq 0 \end{cases}$$

Figure 15.1: *Cell activation formula.*

negative cell) is placed before every output cell. These two cells collect information for (positive cell) and against the presence of a conclusion (negative cell). These collecting cells are connected to every output cell, and every concluding intermediate cell (these are cells defined by the user in the SC net program specification). The final cell activation for the concluding cell is given as

$$CA_{output} = CA_{positive\ cell} + CA_{negative\ cell} - 0.5.$$

Note, the use of the cell labeled UK (unknown cell) in Figure 15.2. This cell always propagates a fixed activation of 0.5 and, therefore, acts on the positive and the negative cells as a threshold.

The positive cell will only propagate an activation ≥ 0.5, whereas the negative cell will propagate an activation of ≤ 0.5. Whenever there is a contradiction in the derivation of a conclusion, this fact will be represented in a final cell activation close to 0.5. For example, if $CA_{positive\ cell} = 0.9$ and $CA_{negative\ cell} = 0.1$, then $CA_{output} = 0.5$, which means it is unknown. If either $CA_{positive\ cell}$ or $CA_{negative\ cell}$ is equal to 0.5, then CA_{output} will be equal to the others' cell activation (indicating that no contradiction is present).

15.2.2 Translating Rules into the Network

An example of how a rule may be translated into the SC-net system follows. The rule is as given in Figure 15.3. Only if the premise evaluates to 1.0 (completely true) will the conclusion d_1 be derived with a belief of 0.8. In all other cases the final conclusion will be less than 0.8. The translation of the rule is shown in Figure 15.4.

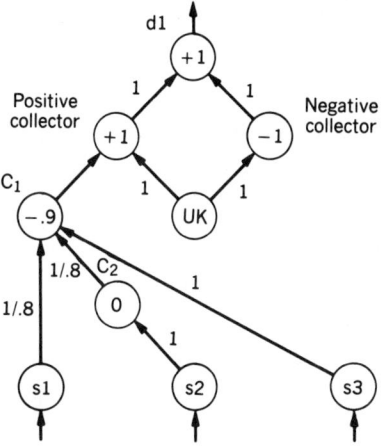

Figure 15.2 Network after learning V_1.

```
if and(or(s1,s2),s3) then d1 (0.8);
```
Figure 15.3: *Rule for translation into network.*

In doing translations, one can think of every cell in a network accommodating n inputs I_n with associated weights $CW_{i,j}$. Every cell contains a bias value, which indicates what type of fuzzy function a cell models, and its absolute value represents the previously mentioned rule range. By successively translating rules into subnetworks, and by combining these, we are capable of constructing complete networks.

15.2.3 Learning and Modifying Information

The system can be used (after learning) as an expert system. In steps 3.1–3.3 of the learning algorithm in this section, the algorithm is checking an error tolerance. If the example is within the tolerance, nothing is done. Otherwise, the biases of the appropriate cells connected to an information collector cell are modified to reduce the error, if the concept is close to an old one. The parameter α is used for tuning the algorithm. If the value is 1, the just-learned vector takes precedence over everything previous. If it was set to infinity, no change would occur as a result of the new learn vector.

Steps 3.4–3.6 of the algorithm are used when a new concept (or version of) is encountered. A totally unfamiliar concept is indicated by an activation value of 0.5. The algorithm essentially creates a new reasoning path to the desired output in a manner somewhat analogous to an induced decision tree [5]. We will next give a formal description of the learning algorithm used in the SC-net network model. This description is then followed by an example.

Recruitment of Cells Learning Algorithm Let V_c be given as a learn vector with the following format:

$$V_c = (\delta_{v_1}, \delta_{v_2}, \ldots, \delta_{v_h}, \delta_{v_{h+1}}, \delta_{v_{h+2}}, \ldots, \delta_{v_{h+i}})$$

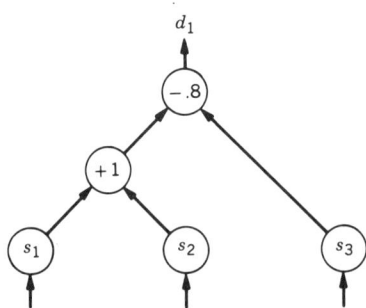

Figure 15.4 *The Figure 15.3 rule translated into an SC net representation.*

15.2 SC-NET—THE REPRESENTATION

with $v_k \in [1, N]$, where $N \geq h + 1$ is the number of input, intermediate, and output nodes. The number v_k represents a component of the learn vector. Each of the δ_m are thresholds for the individual components of the learn vector. Here the $\delta_{v_k}, k = 1, \ldots, h$, are either facts or intermediate results. The $\delta_{v_l}, l = h + 1, \ldots, h + i$, are either intermediate or final results. The v_k and the v_l components can both be intermediate results, since one can be an input to the other. For every learn vector V_c do

Step 1. Apply learn vector V_c to the current network by assigning the thresholds of all facts and intermediate results to the appropriate cells as their new activation. For all $k = 1, \ldots, h$ do
let $CA_{v_k} = \delta_{v_k}$; that is, assign the threshold δ_{v_k} of every component v_k to its corresponding cell C_{v_k} as an activation. This is called the initialization phase.

Step 2. Simulate the network in one pass by starting at level 1 and ending at level n (n is the maximum level of the network). In this simulation phase the new cell activation of every cell is calculated using the earlier listed formula for cell activation calculation. The only cells that are excluded from the simulation are input cells or intermediate cells that correspond to any $C_{v_k}, k = 1, \ldots, h$, of the learn vector. Their cell activation is provided through the use of the δ_{v_k}.

Step 3. For all the $l = h + 1, \ldots, h + i$ component values of the learn vector V_c, compare these (expected outputs δ_{v_l}) with the actual outputs CA_{v_l}. These are given through the cell activation of cell C_{v_l} after the simulation phase.

Step 3.1. Let δ_{v_l} denote the expected (i.e., desired) output of cell C_{v_l}, and CA_{v_l} the actual (activation) output of this cell. Let ε be the error threshold (currently $\varepsilon = 0.01$). If $CA_{v_l} - \varepsilon \leq \delta_{v_l}$ and $CA_{v_l} + \varepsilon \geq \delta_{v_l}$, then consider the vector V_c learned for the output component C_{v_l}.
return to step 3 until done with the $h + i$ component.
else go to step 3.2

Step 3.2. In case that there is a contradiction in expected output and calculated activation, it will be noted and the next 1 value processed at step 3. Otherwise, if ($|\delta_{v_l} - CA_{v_l}| > 5 * \varepsilon$), then go to step 3.4.
if $CA_{v_l} > 0.5$, then
for all cells C_j connected to the positive collector cell of cell C_{v_l} and $\neq C_{v_l}$ do calculate the new cell bias CB_j' as follows:

Step 3.2.1. If $|CA_j - \delta_{v_l}| \leq 5 * \varepsilon$, then
if $CA_j > \delta_{v_l}$

$$CB'_j = \text{sign}(CB_j) * \left(|CB_j| - \frac{||CB_j| - \delta_{v_l}|}{\alpha}\right)$$

else

$$CB'_j = \text{sign}(CB_j) * \left(|CB_j| + \frac{||CB_j| - \delta_{v_l}|}{\alpha}\right)$$

end for
where α indicates the degree in change of the cell bias based on the old cell bias and the given threshold (currently α is set to 6). The more α is increased, the less impact the newly learned information has.
return to step 3 until done with the $h + i$ component.
else go to step 3.3

Step 3.3. If $CA_{v_l} < 0.5$, then
for all C_j connected to the negative collector cell of cell C_{v_l} and $\neq C_{v_l}$ do

Step 3.3.1. if $|CA_j - \delta_{v_l}| \leq 5 * \varepsilon$ then
if $CA_j > \delta_{v_l}$

$$CB'_j = \text{sign}(CB_j) * \left(|CB_j| - \frac{||CB_j| - \delta_{v_l}|}{\alpha}\right)$$

else

$$CB'_j = \text{sign}(CB_j) * \left(|CB_j| + \frac{||CB_j| - \delta_{v_l}|}{\alpha}\right)$$

end for
return to step 3 until done with the $h + i$ component.

Step 3.4. Consider the vector V_c as unknown to the current network. Therefore, recruit a new cell, call it C_m, from a conceptual pool of cells. Make appropriate connections to C_m as follows:
For all C_{v_k} in V_c do for $k = 1, \ldots, h$.

Step 3.4.1. If the threshold $\delta_{v_k} \geq 0.5$, then connect C_{v_k} to C_m with weight $CW_{v_k,m} = 1/\delta_{v_k}$.

15.2 SC-NET—THE REPRESENTATION

Step 3.4.2. If the threshold $\delta_{v_k} < 0.5$, then recruit a negation cell, call it C_n, from the conceptual pool of cells (a negation cell has a bias of zero). Let $CB_n = 0$. Next, connect cell C_{v_k} to cell C_n and assign the weight $CW_{v_k,n} = 1$. Then connect cell C_n to cell C_m and assign as a weight $CW_{n,m} = 1/(1 - \delta_{v_k})$.

Step 3.5. If the threshold $\delta_{v_l} \geq 0.5$, then connect C_m to the positive collector cell $C_{v_{l_{positive}}}$ and let $CW_{m,v_{l_{positive}}} = 1$. Additionally, let $CB_m = -\delta_{v_l}$. Go to step 3 until the $h + i$ component has been processed.

Step 3.6. If the threshold $\delta_{v_l} < 0.5$, then recruit a new cell C_n from the pool of cells. Assign bias of 0 to CB_n. Connect C_m to C_n and make $CW_{m,n} = 1$. Let $CB_m = -(1 - \delta_{v_l})$. Now connect cell C_n to the negative collector cell $C_{v_{l_{negative}}}$. Let $CW_{n,v_{l_{negative}}} = 1$.
Go to step 3 until the $h + i$ component has been processed.

We now show how the algorithm works on a small example. Assume our initial network contains no information. The only thing we initially know about are its input, intermediate, and output cells. Let these be given as follows: input cells $s\{3\}$; output cells $d\{1\}$;

The foregoing represents a fragment of the SC-net programming language for setting up the initial network configuration (Note, that the s_i represent inputs, the d_i outputs). In our example we will not consider intermediate results. Let us further assume we want the uncertainty values of the inputs to be defined over the range [0,1]. This means we use a value of 1 to represent complete presence of a fact or conclusion, a value of 0 complete absence, and finally let 0.5 denote the fact that nothing is known. We also allow uncertainty in any of the responses. Our initially empty network is given in Figure 15.2, if you subtract cells C_1, C_2, and all connections associated with them. Let us next attempt to learn the following vector: $V_1 = \{s_1 = .8, s_2 = .2, s_3 = 1, d_1 = 0.9\}$.

Simulation of Cell Recruitment Learning Algorithm Since we have three facts and one conclusion, $h = 3$ and $i = 1$.

Step 1. In the initialization phase we assign the thresholds of every fact to the corresponding input cell as its new activation. This results in the following assignments: $CA_{s_1} = .8$, $CA_{s_2} = .2$, $CA_{s_3} = 1$.

Step 2. Next, the network is simulated. Since it is initially empty, the resulting cell activation of d_1 will be 0.5.

Step 3. Here l runs from 4 to 4. In the chosen example, there is only one conclusion (d_1). Therefore, $CA_{d_1} = 0.5$ and $\delta_{d_1} = 0.9$.

Step 3.1. The condition is false. We can conclude that V_1 is an unknown concept to the network.

Step 3.2. Since $\delta_{d_1} = 0.9$ and $CA_{d_1} = 0.5$, the condition is also false.

Step 3.3. Having reached this point in the algorithm, we can conclude that the concept represented by V_1 has no representation within the network. This leads to recruiting a new cell C_1 from the conceptual pool. We then connect the cells of all facts to the newly recruited cell C_1.

$k = 1$. **Step 3.4.1.** Since $\delta_{s_1} = 0.8$ we connect cell C_{s_1} to cell C_1 and assign it a weight of $1/0.8 = 1.25$.

$k = 2$. **Step 3.4.2.** $\delta_{s_2} = 0.2$, therefore a new cell C_2 is recruited. $CB_2 = 0$. Connect C_{s_2} to cell C_2 and assign a weight of 1. Connect C_2 to C_1, and let the weight be $1/(1 - 0.2) = 1/0.8$.

$k = 3$. **Step 3.4.1.** Since $\delta_{s_3} = 1$, it is connected to C_1 and the connection assigned a weight of 1.

Step 3.5. Since we connect to the positive collector cell for d_1, a weight of 1 is assigned. Finally CB_1 is set to -0.9.

Figure 15.2 shows the network after learning the first learn vector. Let $V_2 = \{s_1 = 0.9, s_2 = 0.1, s_3 = 1, d_1 = 0.85\}$ be a second learn vector. Then step 3.2.1 of the algorithm will cause the bias for cell C_1 to change. The bias will become -0.8916 after learning V_2.

In the case of contradictory information (over time or whatever), we will actually let the network learn the complement of what we want to forget. Therefore, if any fact fires in the network, so will its complement, and the ANDing of the two will result in an unknown conclusion. It may surprise the reader that we are not deleting information from the network by changing the appropriate weights, which seems to have the same effect. This is done because some rules may be given to (encoded in) the system a priori. When the network is constructed solely through learning, changing connection weights serves the same purpose.

15.3 IMPLEMENTATION OF VARIABLES

15.3.1 Overview

Variables are of great value in implementing powerful constructs in conventional symbolic expert systems. MYCIN [6] is an example of such an expert system. It uses ⟨Object,Attribute,Value⟩ triplets for the implementation of variables. Connectionist-type expert systems hardly make use of variables at all, since it seems that their implementation is far from being simple, if even possible, to realize. As Samad [7] pointed out, variable bindings can be handled more elegantly using microfeatures, which will be associated with every cell (slots in the cell could be used to hold certain information, like variable values, binding information, separation of attributes and values, etc.), or one could think of every cell representing a microfeature by itself. It is obvious that one can associate microfeatures with every cell in a neural network. The amount of microfeatures have to be fixed in the design phase, and once they have been chosen they cannot be changed.

In another implementation one can think of every cell having just one or two microfeatures (actually indexes into some memory module) that point to some memory area, containing all the information for the implementation of variables and their values. Since one of the main purposes for using connectionist networks is speed (parallel processing), making use of the information in some central storage place seems to contradict the notion of parallel processing.

15.4 VARIABLES IN SC-NET

The approach here differs from the above in an attempt to avoid the problems discussed. The SC-net system distinguishes between two different types of variables. The first category consists of fuzzy variables. They allow the user to divide a numerical range of a variable into its fuzzy equivalent. Consider for example the age of a person. Let us refer to the variable as *age*. Age can now take on values in the closed interval [0,100] (this is accomplished by defining age as a variable in the declaration part of a SC-net program). Let us further assume that the following attributes are associated with this variable:

Child: 0–12(0,16)   *Middle-aged:* 30–60(30,60)
Teenager: 13–19(5,25)   *Old:* 60–80(55,90)
Young: 0–30(0,34)   *Very old:* 80–100(75,100)

If age is assigned the value 15, then we might like the associated attribute teenager to be completely true and the attribute child somewhat true, too. The reason for this is that the age of 15 is still reasonably close to the child interval, causing it to be activated with some belief. The numbers in parentheses after the main range for the declarations indicate how far on either side of the range *some* belief will persist. For example, with the teenager variable, there is total belief that the person is a teenager while the value is within the interval. Outside of the interval the belief is less than total, decreasing as it approaches the end points and goes to 0 at the age of 5 and 25. The representation is accomplished by the use of a group of cells to represent the variable and attribute. This is true of both the fuzzy variables discussed and the scalar variables, such as color. An example of age[teenager] is shown here. The age for our example has been compressed into [0,1], such that an age of 15 would map to 0.15.

In general we will have the following format (π-shaped function definition):

attribute value : lower bound, upper bound
(lower plateau, upper plateau)

In case of the value teenager, we guarantee that a value ≤ 5 and ≥ 25 is certainly no longer a teenager (membership value equal 0), and a value of ≥ 13 and ≤ 19 returns a membership value of 1. Figure 15.5 shows the resulting network. The weights have been calculated as follows:

a. Weight: $\dfrac{1}{\text{upper plateau}} = \dfrac{1}{0.25} = 4$.

b. Weight: 1.

c. Weight: $\dfrac{\text{upper plateau}}{\text{upper plateau} - \text{upper bound}} = \dfrac{0.25}{0.25 - 0.19} = 4.167$.

d. Weight: $\dfrac{1}{1 - \text{lower plateau}} = \dfrac{1}{(1 - 0.05)} = 1.053$.

e. Weight: $\dfrac{1 - \text{lower plateau}}{\text{lower bound} - \text{lower plateau}} = \dfrac{1 - 0.05}{0.13 - 0.05} = 11.875$.

Note that in the cases in which our lower bound and lower plateau and/or upper bound and upper plateau are the same, a weight of an arbitrarily large integer is assigned.

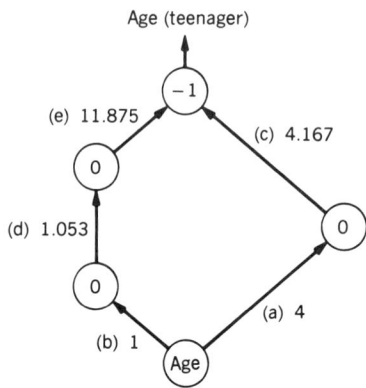

Figure 15.5 *Age (teenager) network.*

15.4 VARIABLES IN SC-NET

Next, we will introduce the second type of variable that is used in SC-net. This second category of variables is referred to as that of scalar type variables. In this scheme a variable can take on a symbolic value rather than a numeric one. Only one possible value will be selected together with its fuzzy belief value. The other possibilities will be considered false. Let us illustrate this through another example. In our next case we consider a variable *color* that can take on three possible values (red, blue, green) and a value (indicating the strength of belief in the selection of color) associated with the selection. The resulting network that will model this scalar variable is given as follows (Fig. 15.6).

In this figure a possible assignment to the cells is given for the case that red is selected as the value for the variable color together with a fuzzy belief factor of 0.9. By assigning a value of $\frac{1}{3}$ to the cell marked as color, the color red is selected. By the same means, a selection of $\frac{2}{3}$ would have selected blue as the color, and $\frac{3}{3}$ would have selected the color green. In the SC-net program, the user will be given a set of choices to select from. Each choice will have a number from 1 to total attributes associated with it. By selecting the appropriate number, the cell with label color will be given the cell activation 1/(selected choice). Additionally, a value of 0 is reserved for the case, that none of the listed choices is assumed to be true. Finally, users can associate a fuzzy belief factor with their choice, indicating the strength of belief they have in their choice.

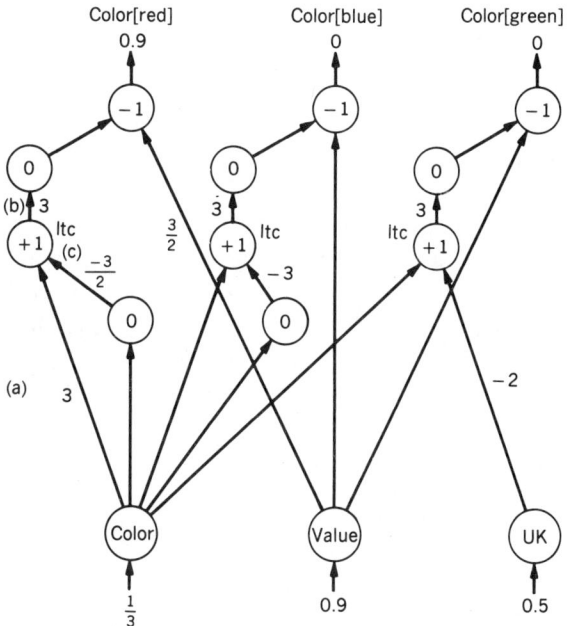

Figure 15.6 *Cell construct for scalar variable color with three attributes.*

We can summarize that the process of assigning the selector cell the value to select the attribute red and discard all the other possibilities is done by a specific program that keeps track of how many attributes are associated with a particular variable. An integer in the range [1,total attributes] is then divided by total attributes to give a number in the range [0,1], which will only select one of the attributes. In our example the user chose to select red as the color, which caused the variable color to receive the value $\frac{1}{3}$. By assigning this value, only the selector for color[red] will be turned on, whereas all the other choices will stay turned off. So, only color[red] will be able to propagate the associated value of 0.9 with this selection. The cells listed with the label ltc are linear threshold cells. In these cells the cell activation is calculated by multiplying every cell input with its associated weight and then adding each product to form the final result.

The variable network is set up and the weights determined as follows. For each of n possible attributes the network structure is the same with the exception of the nth attribute. Hence, for colors of red and blue the structure is the same. The structure for the nth attribute is also the same for any n. The biases of the cells are fixed. Let the numbers indicating the scalar values be denoted as follows, $m_1 = 1/n, \ldots, m_2 = 2/n, m_n = n/n = 1$. The weights on the links are then determined as follows.

a. $1/m_i$.
b. n.
c. $-1/(1 - m_i)$.

All the others may be determined by symmetry. Those link values with calculations not shown are fixed.

15.5 CONDITIONAL VARIABLE BINDING

Once variables have been implemented in the SC-net system, the next obvious step is to allow for a variable binding scheme. Allowing binding of variables in the network at any time, dependent only on the current state of the network, would give the system necessary versatility, which other connectionist expert systems lack. What is meant by conditional variable binding? We will answer this question through the use of an example.

Let us assume, we are given the following program:

```
define    color_type={red,blue,green};
inputcells  (q{2}):color_type, c{4};
outputcells (i1):color_type;

(1) if and(c1,c2) then bind_to(q1,i1) (0.9);
(2) if and(c3,c4) then bind_to(q2,i1) (1.0);
```

15.5 CONDITIONAL VARIABLE BINDING

The prefix operator bind to takes two arguments, which have to be variables, and assigns the value of the first operand to that of the second operand with a certainty given by the corresponding confidence factor (for rule (1) this factor is equal to 0.9) and the belief that is obtained by evaluating the premise. The strength of binding of variables q_1 and q_2 to variable i_1 is determined by the previously mentioned factors. If our network is to model this type of variable binding, a decision has to be made at any one time (state of the network) indicating what value i_1 will receive and how strong the connection actually is. The strongest of all incoming possible bindings will then be chosen as the actual binding of one variable to another variable based on the state of the network. Figure 15.7 illustrates this.

The cells labeled Cond. I and Cond. II accumulate the result of the premise of the two rules. The variable q_1 has a value of $\frac{1}{3}$ (color is red) and a certainty factor of $\frac{1}{3}$. Correspondingly, q_2 has a value of $\frac{2}{3}$ (color is blue) with a certainty factor of $\frac{2}{3}$. Let us assume $c_1 = c_2 = c_3 = 1$ (true) and $c_4 = 0.5$

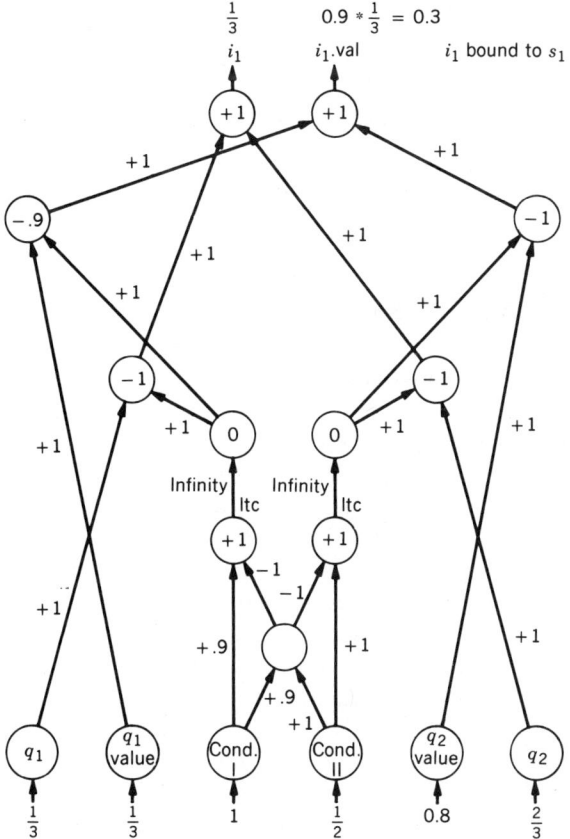

Figure 15.7 *Binding of variables.*

(unknown). In this case the premise of rule (1) evaluates to 1 and the premise of rule (2) to 0.5 (Cond. I = 1, Cond. II = 0.5). After evaluating the network variable i_1 will be bound to the value of cell q_1, since rule (1) was able to fire. Therefore, cell i_1 has a value of $\frac{1}{3}$ (color is red), and a binding strength of $0.9 * (\frac{1}{3}) = 0.3$.

If there were any other conditions that could bind variable i_1 to other variables values, the appropriate condition cells would be connected to the collector cell in Figure 15.7 (this cell has the two cells Cond. I and Cond. II as inputs). Otherwise, additional (symmetrical) connections of variable cells (like q_1 and q_2) and value cells (like q_1 value and q_2 value) would have to be made.

The network structure for binding variables is fixed with the following exceptions. The values on the links from *Cond I* and *Cond II* are the values of the respective rules certainties. The same connections, as shown in the example, will be made in a like fashion when there are more bindings that must be dealt with.

Variables can be used not only within rules but also within learn vectors. It is, for example, possible to assign age[old] as a component of a learn vector. In the GEMS knowledge base every possible input has been constructed using these attribute/value pairs, and the learn vectors incorporate variables as their basic components. This adds additional power to the system, since it allows complete distribution of knowledge acquisition using either rules, or learning from examples. Since a uniform format of representation has been chosen, variables can be incorporated from within rules, learning alone, or a combination of the two.

15.6 FUZZY AND RELATIONAL COMPARATORS

It is often important to make decisions on the relative size of two or more objects. Up to this time no connectionist expert system has provided for relational, and especially fuzzy comparators, as a means to implement comparisons. We believe, that they are as important as variables and their binding, and should be considered as an essential part of every expert system development tool.

We will start our explanation with the commonly found relational operators. Figures 15.8 and 15.9 depict the five commonly found comparators. Let us take a closer look at one of these comparators, to be more specific, let us look at the greater than or equal to operator (Fig. 15.8a). If we assign the cells x and y the values 0.8 and 0.6, then we will first determine the larger of the two values, using a max cell (this is, of course, x with 0.8). The cell labeled ltc (linear threshold cell) will essentially subtract x's value from the larger value due to the negative weight on the link from x. We hereby obtain a value that is either equal to zero or not (in our example it is zero). The so

15.6 FUZZY AND RELATIONAL COMPARATORS   397

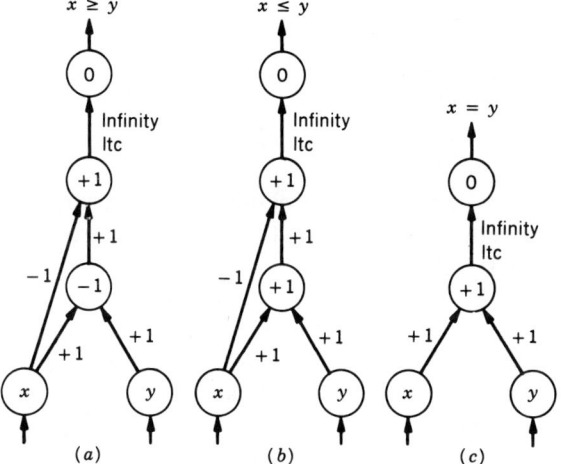

Figure 15.8 *Relational comparators:* (a) *greater equal;* (b) *less equal;* (c) *equal.*

obtained result is then multiplied with infinity (on a computer this would correspond to the largest representable positive number). If the two inputs to the ltc cell were different (indicating that x is less than y), then a result different from zero would be obtained. Multiplying it with infinity will result in a value larger than one, which sent through an inverter cell will result in a

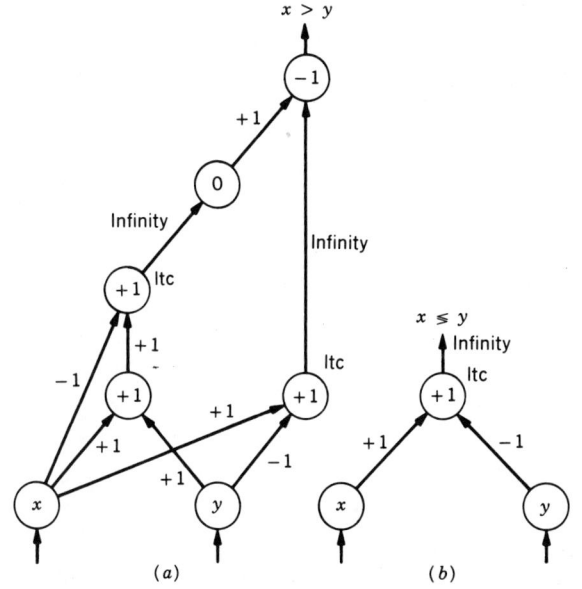

Figure 15.9 *Relational comparators:* (a) *greater;* (b) *not equal.*

final value of zero (x is not greater equal than y). In the case where the ltc cell returned a value of zero, multiplying it with infinity will cause no change. The value of the zero inverter will result in a final value of one (x is greater equal to y).

By taking a closer look at these comparator constructs, one will find a lot of similarities between them.

Constructing fuzzy comparators follows the same basic guidelines that were used for implementing their relational counterparts. Fuzzy comparators differ in that they will return a fuzzy value in the range [0,1], instead of a crisp value of {0,1}. Therefore, if we have the same assignments for cells x and y as before, the fuzzy comparator *fuzzy less than or equal* to (Fig. 15.10a) will return a value of 0.8 instead of 0. Figures 15.10 and 15.11 show five fuzzy comparators. Note, that in Figure 15.10 the weight associated to the link exiting the ltc cell is labeled scale factor. The weight can actually be assigned values larger than one, to achieve a faster decrease in the final fuzzy value, or by assigning values less than 1 a slower decrease can be obtained. For the example of Figure 15.10a, we could make the scale factor 10 and thereby allow for a much faster decrease in the output response of the fuzzy less than or equal comparator.

Fuzzy comparators can be useful in several situations. For example, suppose we wish to represent the concept *around 12 o'clock* within one of our rules. We can accomplish this by using the fuzzy comparator fuzzy equal.

$$\text{fequal(time\_is,12).}$$

This statement compares a variable time_is with the value 12. If the time is exactly 12, the statement will be completely true. If, on the other hand, the

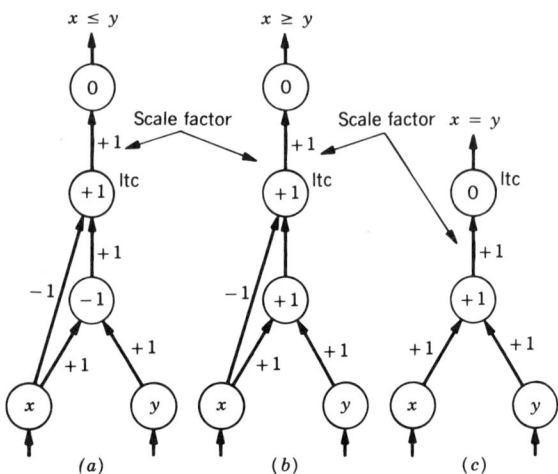

Figure 15.10 *Fuzzy comparators:* (a) *less equal;* (b) *greater equal* (c) *equal.*

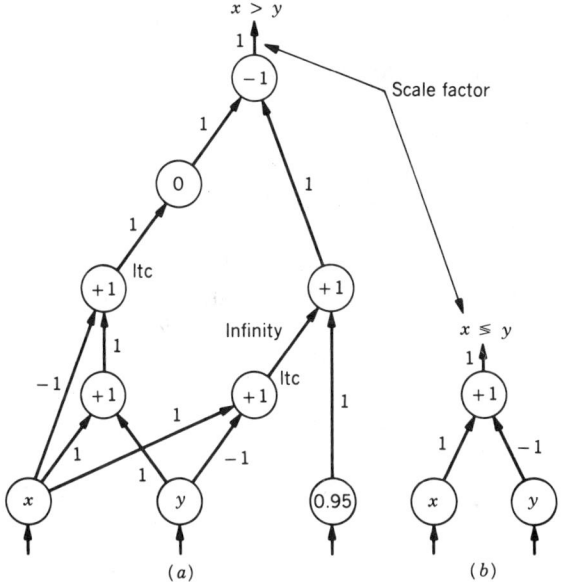

Figure 15.11 *Fuzzy comparators:* (a) *greater;* (b) *not equal.*

time is shortly before or after 12, a value close to 1 will be returned. The actual cutoff is determined by the setting of the scale factor.

15.7 PARTIAL PATTERN MATCHING

The SC-net system can do partial pattern matching on test cases using two different methods. This allows it to make *guesses* about examples that may belong to a previously seen class, but are not too close. One reason this is necessary is that parameters may be missing or wrong. The partial matching feature can be turned off, but would normally be used in learning for a domain.

In the first method a partial matching facility is implemented by dropping inputs. That is, in the case that no outputs are turned on with a value above the appropriate threshold and the partial pattern matching facility is in use, inputs to cells will be dropped. There will be a limit (set by the user) on the maximum number of inputs to be dropped. One input at a time will be dropped from each cell in the network that is not an input, output or negation cell. The lowest undropped input, below 0.5, is the one chosen for dropping. If this turns on an output, that will be reported and computation will halt. Otherwise, the input dropping process continues until either an output has turned on or the maximum number of inputs have been dropped.

The second method is for hardware implementation. In it, additional cells and links are added to the network for the partial pattern matching [2].

15.8 CURRENT RESULTS

Our current research has been dealing with medium-sized data sets. The results obtained have been quite promising. After allowing the system to learn the knowledge bases of GEMS and FEVERS, comparisons where made with an EMYCIN [1] version of these two knowledge bases running under MULTILISP [8]. All our tests were conducted by randomly sampling the rule bases, and determining possible assignments to the inputs (facts), that would fire the premise of some rule. In all cases the results where almost identical and only differed by a few percent in the certainty factor of the final conclusion. In the following, a couple of examples are given.

Figure 15.12 gives a set of selected rules that were taken from the FEVERS knowledge base for the purpose of this simulation. An approximate translation of rule 032 is the following.

Rule 032: *if* the patient has abdominal symptoms and the patient has right upper abdominal pain or left upper abdominal pain or lower abdominal pain *then* there is conclusive evidence (1000) that the patient's abdominal symptom is abdominal pain.

```
(putprops rule032
subject patientrules
action (conclude patient comb-ab abdom-pain tally 1000)
premise ($and (same patient sympl abdom-symp ($or (same
patient abdominal rt-upper-pain) (same patient abdominal
lft-upper-pain) (same patient abdominal lower-pain))))

(putprops rule033
subject patientrules
action (conclude patient disease fabry's tally 750)
premise ($and (same patient sex male)(same patient comb-ab
abdom-pain) ($or ($and (same patient comp-host dis) (same
patient dis-histl stroke)) ($and (same patient sympl
skin-symp) (same patient skin punctate-lesions)) (same
patient symp4 edema))))

(putprops rule 054
subject patientrules
action (conclude patient disease rheumatic-fever tally 800)
premise ($and (between* (vall patient age)4 16) ($or ($and
(same patient comp-host dis) ($or (same patient dis-hist3
strep-hist) (same patient dis-hist3 rheumatic-fever-hist)))
(same patient sympl cardiac-symp)(same patient symp5 chorea)
($and (same patient sympl skin-symp) (same patient skin
erythema-marginatum)))))
```

Figure 15.12: *Sample of randomly selected rules of the FEVERS knowledge base.*

TABLE 15.1 Example of Crisp and Uncertain Input

Rule	Facts	Goals
054	Dis (999) Strep hist (999) Cardiac symp (555) Age is 15 (999)	Rheumatic fever 799/799

Tables 15.1 and 15.2 demonstrate two examples of actual simulations after learning. Each table indicates what rule(s) were fired in the EMYCIN version, lists the facts that were supplied and their certainty values. The system provided that certainties are given under the goals derived, with the EMYCIN certainty given first and our systems' second. The accuracy of the certainty values of the results are in all cases, within several percent, almost identical. The range of certainty is given as [−1000, +1000]. A straightforward mapping was done.

We will next present the results obtained from the GEMS knowledge base. The GEMS knowledge base deals with the identification of gems and it consists of about a 100 rules. Figure 15.13 contains a listing of a subset of the rules that were used for simulation. Figure 15.14 shows one example from the GEMS domain.

In comparison with the Iris data of 150 examples, we followed the example of Weiss [3]. The data contain three classes of Iris plants, 50 examples in each class. There are four predictive numeric attributes. After learning all cases, there was no apparent error. In our system this means that the solution presented with the most certainty was correct. It was occasionally the case that one or more other conclusions would have some certainty. In some cases the certainties were close to the chosen value. When the system was tested by learning all the cases except one and then being presented the left-out case, the error rate was 0.04. These results compare favorably with the results shown by Weiss.

The system has also been run on the soybean disease domain [9]. The particular data used is that from Reinke [10]. In testing learning capabilities we followed Shavlik et al. [11]. This domain consists of 50 features that

TABLE 15.2 Example with Uncertainty and Chaining

Rule	Facts	Goals
032	Abdom symp (700) Rt upper pain (750)	Abdom pain 700/700
033	Male (800) Edema (750)	Cholecystitis 560/530 Fabry's 525/488

```
(putprops rule005
subject
mineralrules
action
(conclude cntxt crystal-system orthorhombic tally 960)
premise
($and (same cntxt crystal-form (oneof ||4-sided-prism||
pyramid pinacoid)
(same cntxt cross-section (oneof rhombus rectangle))))

(putprops rule015
subject
mineralrules
action
(conclude cntxt rock-environment basalt tally 960)
premise
($and (same cntxt rock-color (oneof dark-gray black brown
red))
(same cntxt fabric dense) (same cntxt rock-type hard-rock)
(same cntxt structure layered)))

(putprops rule067
subject
mineralrules
action
(conclude* cntxt mineral-species tally
'((olivine 920) (apatite 60) (garnet 20)))
premise
($and (same cntxt crystal-system orthorhombic)
(notsame cntxt fluorescence yes)
(same cntxt cleavage two)
(same cntxt color
(oneof yellowish-green yellowish-brown)))

(putprops rule069
subject
mineralrules
action
(conclude cntxt gem peridot tally 910)
premise
($and (same cntxt mineral-special olivine)
(same cntxt transmission translucent)
(same cntxt rock-environment (oneof gabbro peridotite
basalt))))
```

Figure 15.13: *Sample of randomly selected rules of the GEMS knowledge base.*

Rule	Facts	Certainty	EMYCIN goals	SC-net goals
	Color=yellowish-brown	766		
067	Fluorescence=no	999	Peridot (641)	Peridot (619)
005	Cross section=rectangle	870		
			Crystal-system=orthorombic (835)	Crystal-system=orthorombic (832)
069	Crystal form=pyramid	920		
015	Cleavage=two	888	Mineral-species=olivine (705)	Mineral-species=olivine (695)
	Transmission=translucent	940		
	Structure=layered	910	Rock-environment=basalt (729)	Rock-environment=basalt (725)
	Rock-color=black	870		
	Rock-type=hard-rock	760		
	Fabric=dense	900		

Figure 15-14: *GEMS example of crisp/uncertain input and multiple uncertain results with chaining.*

describe soybeans. The features are encoded in 208 binary bits in the following manner. If a feature can take on seven values, it will have seven inputs (bits), only one of which will be turned on for each example. There are 17 classes of disease and 17 examples of each class. Training was done by taking 10 of the examples for each class. Then the other seven were used to test. Five permutations of this were done, and 5 permutations with 11 train examples and 6 test examples. The results were averaged to provide the overall correct classification ratio. The system had the correct classification as its top diagnosis, 95.2 percent of the time.

15.9 SUMMARY

In this chapter we have presented a hybrid connectionist, symbolic system. It can represent variables, deal with imprecision, and learn from examples. The learning algorithm is based mainly on changing the network topology, then changing weights, as can be found in most conventional connectionist learning paradigms. For learning, only one pass must be made through the example data. This makes the learning system much faster than a back-propagation algorithm. Further, training can be done incrementally without the system forgetting what it knew before. The size of the network grows linearly with the number of distinct examples (examples that do not simply cause a weight change). The networks is not fully connected in general. Work on the growth rate of the network, and its complexity is still underway. Some large randomly generated example domains have been run through the system with many (thousands) distinct examples. The system has been able to successfully classify these.

This system has some important expert systems features, which normally have been neglected in other approaches to constructing connectionist expert systems. These include the representation of variables, variable binding, and relational and fuzzy comparators. Other features such as consultation and explanation facilities have been added and tested favorably in several different applications. The system is quite well suited to domains that have uncertainty and imprecision associated with them. Finally, tests with several domains show good performance. On the one widely tested data set used, the system compares well with other methods.

ACKNOWLEDGMENTS

This research was partially supported by a grant from the Florida High Technology and Industry Council, Computer Integrated Engineering and Manufacturing Division and by the Air Force Office of Scientific Research/AFSC, United States Air Force, under Contract F49620-88-C-0053. The United States government is authorized to reproduce and distribute reprints for governmental purposes notwithstanding any copyright notation hereon. We would like to thank Jude Shavlik and Geoff Towell for providing a copy of the data to us.

REFERENCES

[1] W. van Melle, E. H. Shortliffe, and B. G. Buchanan, EMYCIN: a knowledge engineer's tool for constructing rule-based expert systems, in B. Buchanan and E. Shortliffe, Eds., *Rule-based Expert Systems,* Addison-Wesley, Reading, Mass., 1984, pp. 302–328.

[2] S. G. Romaniuk and L. O. Hall, FUZZNET, A fuzzy connectionist expert system, Tech. Report CSE-89-07, Dept. of Computer Science and Engineering, Univ. of South Florida, Tampa, Fla., 1989.

S. G. Romaniuk and L. O. Hall, FUZZNET: Towards a fuzzy connectionist expert system development tool, in *Proc. IJCNN,* 483–486 (1990).

D. E. Rumelhart and J. L. McClelland, Eds., *Parallel Distributed Processing: Exploration in the Microstructure of Cognition,* Vol. 1, MIT Press, Cambridge, Mass., 1986.

[3] S. M. Weiss and I. Kapouleas, An empirical comparison of pattern recognition, neural nets, and machine learning classification methods, N. S. Sridharan *Proc. IJCAI '89,* 775–780 Morgan-Kaufman, San Mateo, CA. (1989).

[4] A. Kandel, *Fuzzy Mathematical Techniques with Applications*, Addison-Wesley, Reading, Mass., 1986.

[5] R. S. Michalski, J. G. Carbonell, and T. M. Mitchell, *Machine Learning: An Artificial Intelligence Approach*, Tioga Publishing, Palo Alto, Calif., 1983.

[6] D. A. Waterman, *A Guide to Expert Systems*, Addison-Wesley, Reading, Mass., 1986.

[7] T. Samad, Towards connectionist rule-based systems, *Proc. IEEE Internat. Conference on Neural Networks*, 525–532 (1988).

[8] R. H. Halstead, Multilisp: A Language for Concurrent Symbolic Computation, *ACM Trans. Programming Languages and Systems*, 7(4) 501–538, October 1985.

[9] R. S. Michalski and R. L. Chilausky, Learning by being told and learning from examples: an experimental comparison of the two methods of knowledge acquisition in the context of developing an expert system for soybean disease diagnosis, *J. Policy Analysis and Information Systems*, 4(2) 125–161 (June 1980).

[10] R. Reinke, "Knowledge Acquisition and Refinement Tools for the ADVISE Meta-Expert System," Master's thesis, University of Illinois at Urbana-Champaign, 1984.

[11] J. W. Shavlik, R. J. Mooney, and G. G. Towell, Symbolic and neural learning algorithms: an experimental comparison, Computer Sciences Tech. Report #857, University of Wisconsin-Madison, Madison, Wis., 1989.

CHAPTER 16

Adaptive Algorithms and Multilevel Processing Control in Intelligent Systems

JAMES S. J. LEE
DZIEM D. NGUYEN

16.1 INTRODUCTION

Many complex information processing applications such as industrial automation, automated recognition, object-of-interest tracking, automated medical screening, and diagnostic decision making involve common tasks such as data acquisition, noise filtering, feature (information) detection, extraction, and conversion, decision making, and process control. We call a system that devises appropriate processes to address these different task requirements an *intelligent system*.

In practice, the majority of intelligent systems are still in the research stage. Most of the practical systems are simple in processing and thus limited in capability. These systems are examplified by well-defined machine vision inspection or robot control tasks, human in the loop object-of-interest recognition, and specially prepared samples for medical screenings. They often require strictly controlled working conditions and often fail when experiencing variations in the data source or unexpected changes in the working conditions.

In order to develop high-quality intelligent systems, a methodology that integrates all available information and processing technologies is needed. This creates the ancillary requirement to provide for adaptive processing of the input data and for automatic capture of the features, followed by intelligent, automatic application of the application specific knowledge presently utilized by human experts in reaching a decision.

The major difficulty in adapting existing information processing technologies to practical high-quality intelligent system applications is the near-endless variety among data sources. Because the data in most of the applications are so disparate, no rudimentary rules or application-specific knowledge can be encoded into a simple processing system. Thus, it is desirable to have multiple processing layers with built-in learning, adaptive processing, and integration modules.

The key technology area in intelligent systems involves optimization of the processes required to perform information processing and decision making, and the strategies for their characterization and integration. In addition to development of on-line processing modules, it requires off-line learning and optimization schemes that are able to capture significant features and knowledge. These features and knowledge will then be used for synthesizing on-line processing modules and strategies for decision making. One of the key processing strategies used for this purpose combines probabilistic inference with artificial intelligence, symbolic processing, and artificial neural network learning. A general model for multilayer processing is the subject of the next section. As an application example, an intelligent object-of-interest tracking system integrating multiple layers of processings is described in Section 16.3. And in Section 16.4, we discuss a neural network model that plays a major role in a multilayer processing system as it enables the integration of neural networks with intelligent processing. The test results of the intelligent tracking system is provided in Section 16.5, followed by conclusions.

16.2 MULTILAYER PROCESSING MODELS

One promising approach to build such an intelligent system is an integration of rule-based reasoning with the adaptability and incremental learning characteristics of neural networks. Figure 16.1 shows the proposed scheme where rule-based reasoning components work together with a neural-based component. The recognition controller applies contextual information to interpret the results of a neural network adaptive preclassifier, then invokes the statistical pattern recognizer to further examine the potential detections. The statistical pattern recognizer incrementally accumulates statistics of the detected object and its contextual conditions and performs dynamic feature weighting and selection. The learning controller guides incremental training of a neural net classifier, including reinforced learning against classification errors. In addition, these processing modules can be locally integrated and coordinated by appropriate control modules. Multiple modules may perform similar processing in parallel with their outputs integrated by a combination module.

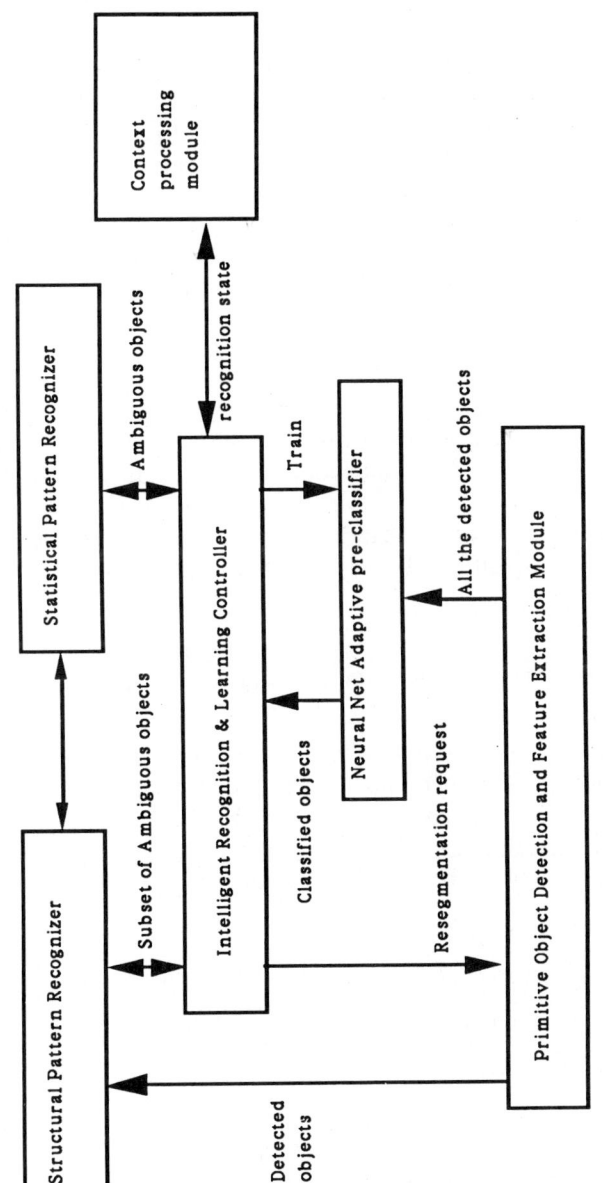

Figure 16.1 *An intelligent system block diagram.*

Finally, to facilitate system design and trade-off evaluation, information processing and intelligent decision making is posed as an optimization problem that minimizes error cost functions modeled in a Bayesian decision framework [1,2].

Given a measurement X and its associated contextual information and application knowledge CC, the intelligent decision maker assigns the measurement a decision label 1, where, for example, 1 can be "object of interest" or "clutter" in a tracking application, or 1 can be "normal," "abnormal" ("defective"), or any other meaningful indication in screening/inspection applications. According to the Bayesian framework for decision making, the intelligent decision-making problem can be formulated as one of determining the label 1 that minimizes the expected error cost (C):

$$C(1 \mid X, CC) = \sum_{1^*} c(1 \mid 1^*) P(1^* \mid X, CC)$$

where $c(1 \mid 1^*)$ is the cost imposed when the assigned label is 1 and the true label is 1^*, and $P(1^* \mid X, CC)$ is the conditional probability that the true label is 1^* given X and CC. To solve this problem and make design trade-offs tractable, global optimization criteria are decomposed into local optimization criteria involving different layers of processing.
Since

$$P(1^* \mid X, CC) = \frac{P(X, CC \mid 1^*) \times P(1^*)}{P(X, CC)}$$

to perform global optimization one has to determine the conditional probability $P(X, CC \mid 1^*)$. The decomposition of the global optimization problem into an optimization of subcomponents can be accomplished by decomposing $P(X, CC \mid 1^*)$ into products of simpler conditional probabilities. For example, the conditional probability for a six-layer processing system can be decomposed into

$$\begin{aligned} P(X, CC \mid 1^*) &= P(L_6, L_5, L_4, L_3, L_2, L_1 \mid 1^*) \\ &= P(L_6 \mid L_5, L_4, L_3, L_2, L_1, 1^*) \\ &\quad \times P(L_5 \mid L_4, L_3, L_2, L_1, 1^*) \times P(L_4 \mid L_3, L_2, L_1, 1^*) \\ &\quad \times P(L_3 \mid L_2, L_1, 1^*) \times P(L_2 \mid L_1, 1^*) \times P(L_1 \mid 1^*) \end{aligned}$$

where L_i designates the data and features derived from processing layer i. Determination of all the conditional probability components will yield the overall conditional probability $P(X, CC \mid 1^*)$. One can decorrelate certain components in the conditional probabilities based on application-specific knowledge. For instance, let L_6 be the context processing module, L_2 be the neural-based adaptive preclassifier, and L_1 be the primitive detection and feature extraction module in Figure 16.1. Since L_6 may not directly relate to L_1 and L_2, one can decorrelate them by setting

$$P(L_6 \mid L_5, L_4, L_3, L_2, L_1, 1^*) = P(L_6 \mid L_5, L_4, L_3, 1^*)$$

The function of the intelligent control module is to conditionally correlate and decorrelate the conditional probabilities and alter the ways the conditional probabilities are determined. As can be seen in the next section, in an object-of-interest tracking example different detection and decision-making strategies are adapted depending on the state of the tracking.

16.3 OBJECT-OF-INTEREST TRACKING APPLICATION

It is difficult to detect and track multiple objects in real time, particularly when image sequences change rapidly from frame to frame due to sensor motion or background variation. Problems with conventional algorithmic trackers include computational complexity, lack of higher-level functionality, such as prediction, learning, and adaptation, and dependence on prior knowledge or simplified models that inaccurately represent actual operating conditions. For example, trackers based on estimation theory (e.g., Kalman filtering [3,4]) are computationally expensive and rely on Gaussian or Poisson process models for noise and object maneuvers. Simple neural networks [5,6], can adaptively learn and classify (e.g., object of interest versus clutter), but perform poorly for sparse training data with varying distributions and uncertain reliability, as found in dynamic tracking situations. A final shortcoming in most tracking approaches is a failure to change process control in response to changing situations.

In response to these shortcomings in current trackers, a new tracker is developed based on multilayer processing control strategy. This intelligent processing and control scheme consists of the principal functional units shown in Figure 16.2. Spatiotemporal filtering [7–10] is used to detect and analyze motion, exploiting the speed and accuracy of multiresolution processing. A real-time neural net scheme adaptively prefilters out clutter to assist object-of-interest detection, using a Least Mean Square (LMS)–based scheme and a new push-pull "balancing" strategy to learn rapidly. A recognition and learning controller interprets these classifications as hypotheses to test by pattern recognition. A statistical recognizer accumulates object-of-interest statistics, performs dynamic feature weighting and selection, and, as needed, activates additional object-of-interest detection with changed processing parameters. A structural pattern recognizer verifies the final detection by structural matching and object grouping/splitting, and sends a tracking confidence value to a tracking controller. Good tracking is maintained by assigning to each tracking frame one of eight possible tracking states, each with associated strategies for detection, matching, and prediction. When the

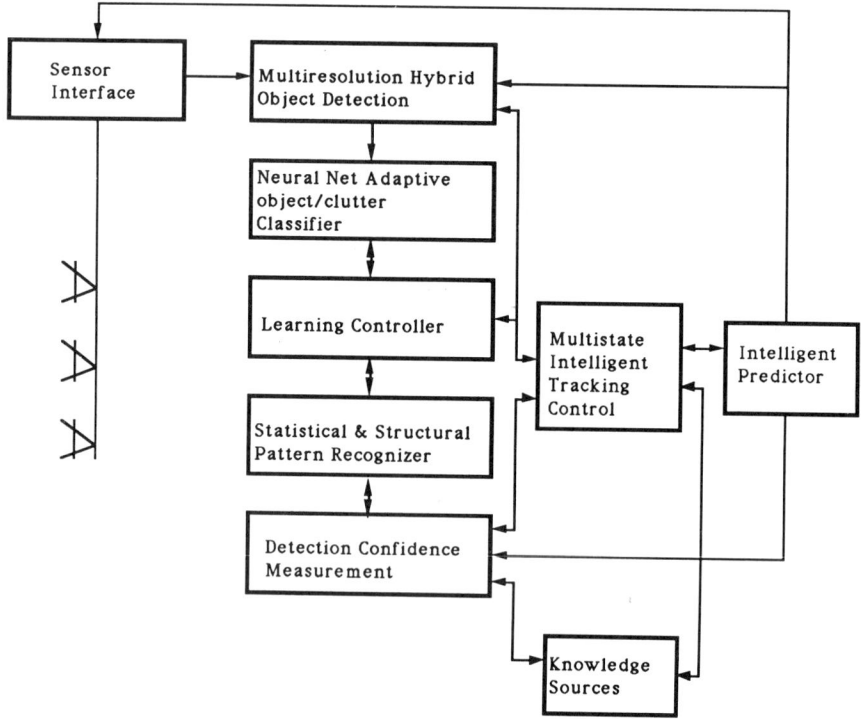

Figure 16.2 *Functional components of an adaptive real-time tracker.*

tracking state is in "stable tracking," the learning controller uses the data to incrementally train the neural net classifier. In the following, each functional component in this tracking system will be briefly described.

16.3.1 Object-of-Interest Detection

Figure 16.3 illustrates the multiresolution operation of the detection component, which cooperatively combines correlation [7] and motion-based and contrast-based segmentation. Detection based on contrast is a computationally efficient way to detect objects with definite contrast of gray-scale intensity characteristics. However, two drawbacks are possible preprocessing required to enhance contrast, and more false alarms or missed objects of interest, depending on contrast thresholds, than other methods. Real-time motion detection can be quite effective [11] but only for objects of interest undergoing significant motion; this is generally a poor way to segment distinct objects. Correlation is the most reliable method when the object's appearance is consistent from frame to frame, but can be the worst for dynamic imagery and is always computationally expensive.

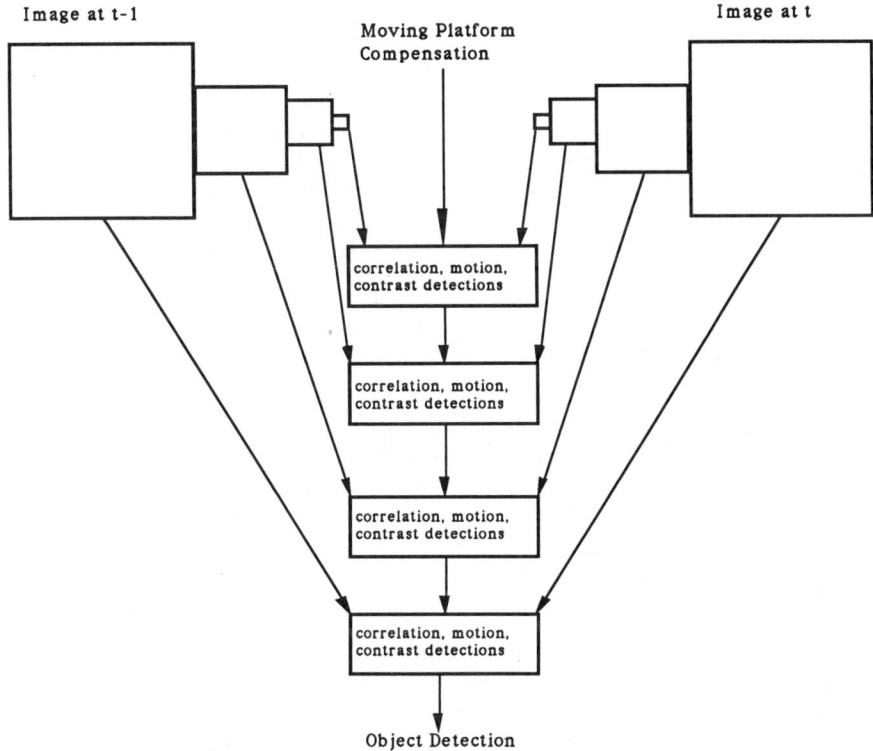

Figure 16.3 *The coarse-to-fine normal detection control of the multiresolution hybrid object detection process.*

The hybrid approach illustrated in Figure 16.3 combines all three modes adaptively. A low-pass image pyramid [12] is generated such that the image is low-pass filtered and downsampled between consecutive pyramid levels (image resolutions). Detection starts at the coarsest resolution (smallest image) by applying correlation to focus attention on a region-of-interest window. Preprocessing and contrast detection are used within the window in this downsampled image to segment the object of interest. If the object is moving significantly, motion detection replaces or assists the correlation-based focus of attention. The correlation coefficient and motion energy are used as performance measures of these two schemes. Selection of the detection scheme also depends on heuristics strategies based on the current tracking state; for example, correlation is usually better during stable tracking. After this initial detection, the process is repeated at successively finer resolutions, with a gain in overall speed by settling on one of the two schemes after accumulating sufficient detection confidence.

In addition to this coarse-to-fine detection control, the following fine-to-coarse scheme (Fig. 16.4) is applied during object-of-interest searching for

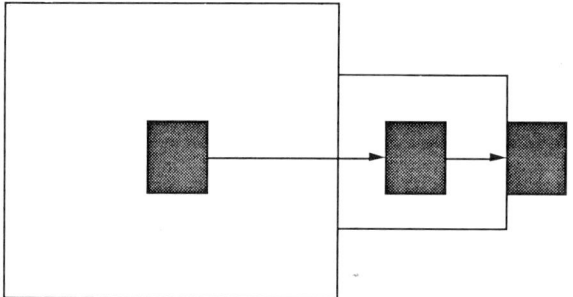

Figure 16.4 *The fine-to-coarse control for object-of-interest searching.*

"new detection" or "nonstable tracking" states (e.g., occlusion, radical change in appearance, departure from view). If no reasonable object-of-interest is detected within an image window, the search area is increased, while keeping window size fixed, by applying object detection to successively coarser (smaller) images.

16.3.2 Neural Net Classification

A neural network approach provides adaptive classification of potential detections as object of interest or clutter to avoid detailed processing of uninteresting objects. The Widrow-Hoff algorithm [5,6] was chosen as our starting point because its simplicity allows real-time implementation and its adaptiveness, demonstrated in application, allows incremental learning. However, several new strategies were developed to address the constraints imposed by dynamic tracking: real-time execution, training on sparse data, data of uncertain reliability, rapid change in data from frame to frame.

The main features of the algorithm are briefly reviewed in the following. The details of the implementation will be discussed in Appendix 16A. In operation, a feature vector for each detected object in a frame is extracted and used to train our adaptive classifier frame by frame. The training samples, or patterns, are of the form ⟨feature vector **X**, desired output y⟩. The feature vector **X** represents a point in a multidimensional vector space with each feature component defining an axis. This space is partitioned into two half-spaces, one corresponding to the class object of interest (**T**) and the other to clutter (**C**), by a hyperplane called the *discrimination surface S*, represented by a vector normal to it. This weight vector is denoted by **W**; which half-space contains a vector **X** is determined by the sign of the dot product **W · X**. At Frame(i) the current weight vector **W** (updated in Frame($i-1$)) positions the input patterns **X** to produce classification results to be verified by the statistical and structural pattern recognizer described in Section 16.3.3. The verified classification is used by the tracking controller to assign a tracking state to the current detection. When the assignment is stable tracking, the input patterns are put to further use as training values to update **W** for use in Frame($i+1$). This process is then repeated on Frame($i+1$), and so on. In addition, training on the patterns that are incorrectly classified based on data up to Frame($i-1$) constitutes a form of learning reinforcement using previously committed errors. This interaction means that the classifier serves as a preprocessor to narrow the search space for the pattern recognizers and tracking controller, and these components serve in turn as teachers to influence the training of the classifier.

The following critical new strategies, implemented to satisfy the tracking constraints, are discussed in detail in [13,14]. Real-time response is supported in several ways. A training time limit is set for any training step, emphasizing speed over thoroughness because the dynamic tracking application precludes "perfect" learning in any case. Tolerances are also set on each side of the discrimination surface S: **W** is updated (S moved) only when a training pattern lies closer to the surface than the tolerance for its half-space. Each tolerance expresses the precision and reliability of information available for its half-space and mitigates the effect of measurement noise. Finally, **W** is not normalized with each update, thus avoiding performing division. Instead, **W** is multiplied by a scale factor if the dot products grow too large in magnitude; this leaves classification unchanged since it depends only on the *sign* of the dot products.

Two *centroid* vectors **T** and **C** are introduced. They estimate the centers of the "clusters" in feature space of, respectively, all previous objects of interest and clutter patterns. These centroids **T** and **C** are updated for each frame and used to deal with sparse or unreliable training data, and rapid change from frame to frame. If recent frames contain sparse or dubious patterns, the learning controller (Section 16.3.3) suppresses their training impact. The centroids compactly summarize all lessons learned up to the current frame, and are used for classification in the current frame with dubious recent past. To adapt to frame to frame changes and trends, a centroid is updated by averaging its previous value with current data. This binomial weighting has the effect of gradually discounting training data as it recedes into the past, and is very computationally efficient. To avoid subsequent misclassification due to overtraining (Fig. 16.5), **T** and **C** are used in the following "balancing strategy." If **W** is changed (S is moved) in training with respect to a particular pattern (say on the clutter side **C**), it will be followed with a "balancing" training with respect to the "opposite" *centroid* (in this case, **T**). This will again be balanced with training using its opposite centroid (**C**), and so on. Iteration continues until S converges or the training time limit is exceeded. An extension of the single neuron classifier to a multiple-neuron-based classifier will be discussed in Section 16.4.

16.3.3 Recognition and Learning Control

The control scheme (Fig. 16.6) is organized to perform detailed, complex processes only on demand and utilizing the least possible data. After prelimi-

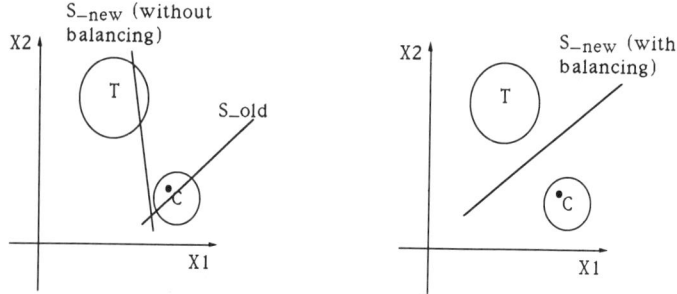

Figure 16.5 *The difference between training without balancing and with balancing.*

nary classifications by the neural net, control may go one of several paths. If there are no ambiguities, the recognition controller passes the detection result to the tracking controller for final tracking state assignment (see Section 16.3.6). If this assignment is stable, the learning controller trains the classifier on the new patterns. Ambiguous classifications are first attacked using statistical pattern recognition. If that fails the controller invokes structural pattern recognition, which may in turn require a controller request for resegmentation by the detection module. In stable tracking, resolved ambiguities are used by the learning controller to train the neural net against its errors for reinforced learning.

Intelligent use of context is the key to our strategies to control learning and recognition [15]. Intraframe context says that at most one object of interest is expected within a tracking window. Another intraframe cue is the prediction of current tracking state. Stable suggests that one may expect straightforward object-of-interest detection, and that it is appropriate to train the classifier; unstable or momentum tracking imply that the object of interest may be undetectable or alternative detection modes may be required.

Interframe context eliminates wasted processing time and misclassification due to poor object segmentation or ambiguous patterns in feature space. The interframe consistency of an object's image supplies a context that is applied differently to object of interest than to clutter. The object of interest is well characterized by its features, but must be tracked even during nonstable states when occlusion, noise, or viewing angle affect its appearance. The recognition controller may need to redeploy the segmentation process with different parameter settings or detection mode, taking into account the tracking state predicted for the current frame. In contrast, clutter is "everything else," so no individual clutter object is well characterized. However, false alarms can be avoided by eliminating detected objects similar to previous-frame clutter.

16.3.4 Weighted Nearest-Neighbor Statistical Pattern Recognition

The recognition controller invokes interframe context, as needed, by measuring both spatial distance and an adaptive, weighted feature space distance

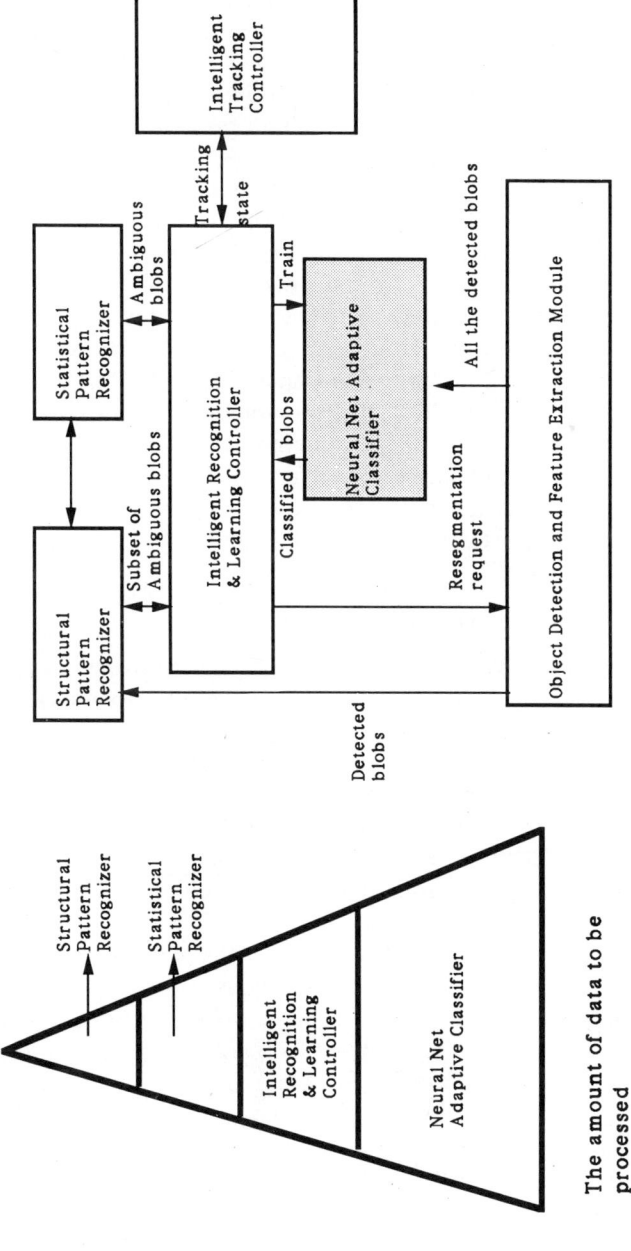

Figure 16.6 *Organization of the intelligent recognition and learning control.*

(from previous-frame object of interest and clutter to a current ambiguous **X**). The feature space distance measure gives more credence to more consistent features by weighting each component feature inversely proportional to its deviation, dynamically accumulated over its values for the object of interest in all preceding frames. The current **X** is disambiguated by comparing its feature space distance from the previous-frame object of interest, with its minimum feature space distance from all previous-frame clutter in close *spatial* proximity.

Figure 16.7 shows how weighted feature space distance resolves ambiguous detection, with two potential objects-of-interest O_1 and O_2. Intraframe context says at most one of the objects is the object-of-interest. O_2 lies a longer unweighted distance d_2 from the discrimination surface S, and would be selected as object of interest if unweighted object-to-surface distance were the selection criterion. However, interframe context "distorts" feature space distances around the previous-frame object of interest.

The correct selection O_1 lies a shorter weighted feature space distance D_1 from that object of interest. If the tracking state is stable, the neural net is first trained (S moved) using O_1 as object of interest and O_2 and all other detections as clutter. The neural net is then retrained on O_2 for learning reinforcement using previously committed errors.

16.3.5 Structural Object Splitting/Grouping

Occasionally, the tracked object of interest may be merged with clutter or broken up by bad segmentation or noisy tracking environments. To make up for the noisy tracking conditions, structural object splitting/grouping operations may be required. In such a case, it is necessary to group the "broken" components into one, and reperform object-of-interest classification. Object splitting/grouping requires structural and numerical rules for combining or separating object features. Since moment features can be easily combined, features derived from moments are used for such operations.

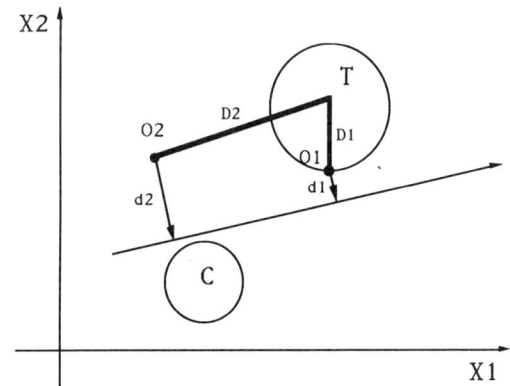

Figure 16.7 *Using weighted feature space distance to resolve ambiguous detection.*

16.3.6 Intelligent Tracking Control

The tracking controller [10] is our central control unit, overseeing system function by monitoring key processes. At each tracking stage, the controller mediates among several types of values for each parameter it evaluates: previous-frame, predicted current-frame, measured current-frame, current (combination of predicted and measured) frame, and predicted next-frame. Feedback/feedforward control determines how values are combined for one parameter using the results of combining values for another. For example, tracking state depends on detection confidence, which is based on object features, which depend on measurement technique, which is determined by detection status, which in turn depends on predicted tracking state and an intermediate detection confidence.

The controller maintains good tracking by assigning to each tracking stage one of the eight tracking states shown in the state transition diagram; Figure 16.8. Each tracking state has associated strategies for detection, matching and prediction, efficiently implemented by look-up tables. The current tracking state is based on the previous state, the predicted current one, and the current detection confidence. Detection confidence is determined by summing over differences between predicted and measured features, using the dynamic weights assigned by the learning controller after preprocessing by the neural net. Weights are also assigned to the various levels of our multire-

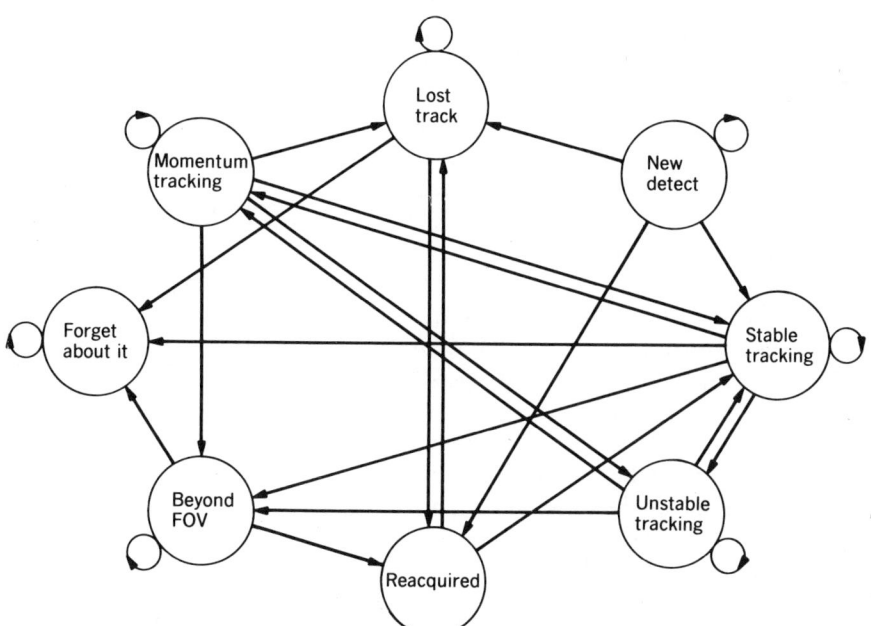

Figure 16.8 *The tracking state transition diagram.*

solution pyramid according to attributes such as motion energy and coherence. Previous and predicted tracking states and detection confidence also determine the detection status (e.g., "good detection at coarse resolution"), for goal-directed feedback to control processes such as the fine-to-coarse object-of-interest searching strategy discussed in Section 16.3.1.

16.4 THE IMPROVED ADAPTIVE CLASSIFIER ALGORITHMS

Many conventional neural network models emphasize massively parallel and distributed processing. Each neuron follows a simple transition function and does not necessarily have any physical meaning; it is the global representation and behavior of the whole network that is meaningful [16-18]. Our neural network approach, which is inspired by the works on Adaline/Madaline by Widrow et al. [4,5], assigns physical meaning to each neuron. In addition, it exploits the existing application domain-specific knowledge to organize the network compactly and to increase learning speed. Our neural learning network provides adaptive classifications and supervised incremental learning characteristics for applications where (1) real-time response is required, (2) the classifier is subjected to repeated train-then-apply cycles, such that very few training samples are available for each training period before the results must be used, and (3) the object-of-interest and clutter distributions may change dynamically.

Three new strategies are developed for single-neuron learning: (1) a cluster center representation, (2) a "balancing" learning strategy, and (3) guard bands for enforced learning [13,14]. These new schemes allow us to handle the dynamic tracking conditions and identify the potential misclassifications and prevent them from happening. It also speeds up the training process so that only a small number of training data are required before the neural network can be applied to perform classifications. Because of the innovative learning control scheme, the number of neurons required in our network for a real application environment is small. In fact, a single-neuron system, as experimental evidence demonstrated, can do a sufficiently good job.

The single-neuron classifier forms a discrimination surface to partition the multidimensional feature space into an object of interest and a clutter half-space. This structure may not work well when the clutter is distributed dynamically all over the multidimensional space rather than a half-space. A multineuron structure is designed to handle those more complicated application conditions. In the multineuron structure, each neuron forms a surface in the multidimensional feature space and multiple surfaces are used to enclose the object-of-interest distribution, each surface separates a subset of clutter from object-of-interest. The structure of each neuron is the same as that for the single-neuron classifier, and each neuron is trained and balanced by the global object-of-interest center and its associated clutter center in the same way as those for the single-neuron classifier.

In our multineuron structure, each neuron forms a surface in the multidimensional feature space and multiple surfaces are used to enclose the object-of-interest distribution, each surface takes care of a subset of clutter. Each neuron is trained and balanced by the global object-of-interest center and its own clutter center. To learn a new clutter sample, statistical distance (rather than the distance to the surface) is used to determine the nearest clutter center and only the neuron associated with it is trained. To learn a new object-of-interest sample, all neurons will be trained. In this way, the system can dynamically adjust itself to adapt to the dynamic changing conditions. A dynamic network construction scheme is developed that dynamically determines the network architecture by incrementally creating new neurons and merging old neurons in the network [19].

To learn a new clutter sample, statistical distance (rather than the distance to the surface) is used to determine the nearest clutter center and only the neuron associated with the nearest clutter center is trained (thus, it does not increase the complexity of clutter learning). To learn a new object-of-interest sample, all neurons are trained to adapt to the dynamic changing conditions. This is still a real-time architecture, since there is normally more clutter than object of interest in a tracking frame and the number of neurons in the learning network is small.

The learning network architecture is dynamically determined by incrementally creating new neurons and merging old neurons. The tolerable cluster size for each cluster set is estimated dynamically and is used to decide whether a new neuron (surface) should be created for a given clutter training sample. If two clutter centers are close enough and their discrimination surfaces are similar, they are merged.

As shown in Figure 16.9, the learning network includes two layers of processing. The first layer includes several single-neuron classifiers. Each takes the input from the input feature vector and outputs its classification result to an output neuron and a learning control neuron. The output and learning control neurons constitute the second layer of the learning network. The output neuron is dedicated for feature classification and the learning control neuron is used only in the learning phase.

Each first-layer neuron forms a surface in the multidimensional feature space which partitions the space into two half-spaces (a **T** side and a **C** side). The combination of the first-layer neurons partitions the multidimensional space into different volumes. The network is constructed in a way that the intersection of the **T** half-space of all the neurons is nonempty and the object-of-interest distribution is enclosed in the volume formed by this intersection (see Fig. 16.10). The guard band embedded in each neuron determines the volume size. In the classification phase, each first-layer neuron performs a classification of the input feature vector (into **T** or **C** classes) and the output neuron performs an AND on the **T** class; thus, its output is **T** if and only if all the inputs are **T**, and is **C** otherwise.

In the learning phase, the learning control neuron controls the learning of

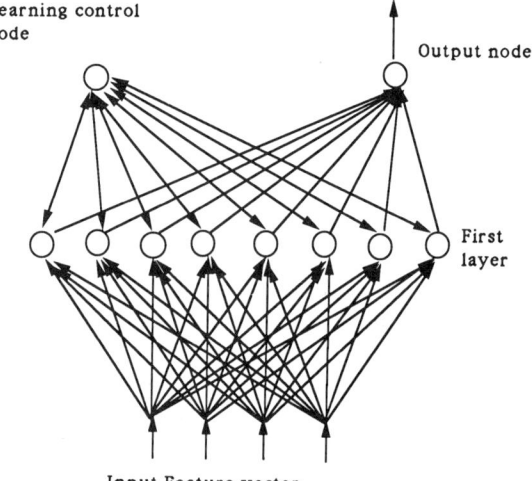

Figure 16.9 *The learning network includes two layers of processing. The first layer includes several single-neuron classifiers, and the output and learning control neurons constitute the second layer.*

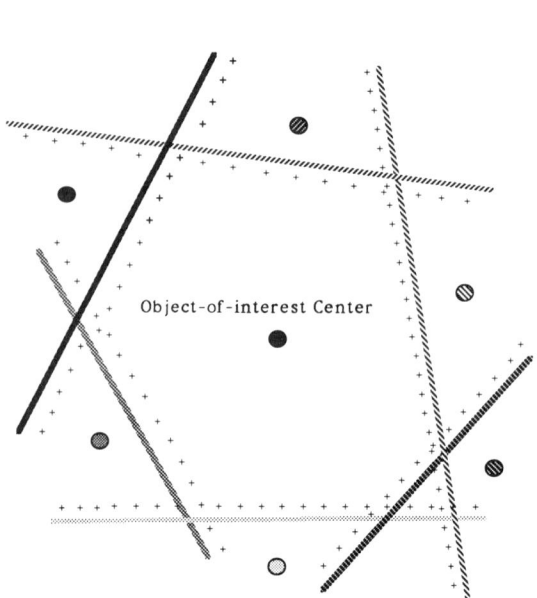

Figure 16.10 *The combination of the first-layer neurons partitions the multidimensional space into different volumes. The object-of-interest distribution is enclosed in the volume formed by their intersection.*

each first layer neuron given an input training vector **X** and its true class label **l***. If **l*** is **T**, the object-of-interest center is updated, and the single-neuron training and balancing process takes place at each first-layer neuron. Since each clutter is handled only by one neuron in the first layer, when **l*** is **C**, the learning control neuron determines a set of potential neurons **P** that may cover the training clutter sample **X**. The potential neuron set is defined as the set of all neurons in the input layer that identify **X** as clutter.

Then, from **P** the neuron whose cluster center is closest to the input training vector **X** is selected. To reflect the dynamically changing environments, the average distance between clutter centers is also incrementally maintained. If the distance between the training sample **X** and the discrimination surface of the selected neuron is less than certain threshold, the network is updated by training the selected neuron using **X**. Otherwise, since the new sample is too far away from the cluster centers of all the existing neurons, a new neuron is created to handle this clutter. For the newly created neuron, **X** is used as its cluster center and its weight **W** is set up in a way that its discrimination surface bisects the vector between the global object-of-interest cluster center and the new training sample **X**. A scheme is developed to dynamically delete redundant neurons. It checks the distance between neuron cluster centers. If the distance between a pair of neurons is too small (compare to average cluster distance) and their discrimination surface are similar (in terms of the orientation and distance to the object-of-interest center), the two neurons are combined together by averaging their cluster center and discrimination surface.

16.5 EXPERIMENTAL RESULTS

Our single-neuron classifier is applied as a tracked object-of-interest/clutter filter for three Forward Looking InfraRed (FLIR) image sequences. In the first image sequence consisting of 24 frames, a moving sensor follows four moving objects of interest occluded occasionally by trees or by each other. The 36 frames of the second image sequence was taken by a sensor circling around a rotating vehicle as it leaves the ocean, crosses the surf line, and parks on the beach. This sequence is challenging because there are two sources of nonuniform, nontrivial image rotation and significant changes in thermal background contrast from cold water to warm sand. A hand-guided sensor acquired the 24 image frames of the third sequence, following the noisy, fragmented appearances of a plane taxiing through intermittent snow, occluded by various airfield structures. The following feature vector components are used: area, centroid, gray-scale mean and standard deviation, elongateness, rotation angle, and circularity.

The performance of the single-neuron classifier is bench-marked against the Widrow-Hoff algorithm. Errors are partitioned to differentiate between false detections (misclassifying clutter as object of interest) and missing an

TABLE 16.1 Comparison of Classification Error Rates for Widrow-Hoff, Best and Worst Cases Using our Single-Neuron Classifier

Algorithm	Data Set	β	θ	% False Alarms	% Missed Obj. of Int.	# Errors
Widrow-Hoff	1	—	1.00	2.94	12.50	4
	2	—	1.00	1.25	29.20	8
	3	—	1.00	5.62	21.40	8
Basic Lee-Nguyen	1	0.8	1.00	1.47	0.00	1
classifier	1	1	1.00	4.41	0.00	3
	2	0.8	0.80	8.69	0.00	4
	2	1	1.00	2.17	0.00	1
	3	0.8	0.75	6.06	4.34	4
	3	1	1.00	3.03	0.00	2

object of interest by misclassifying it as clutter. Table 16.1 shows the Widrow-Hoff results on each data set, and the best and worst results for our classifier. The parameters β and θ tune our classifier's sensitivity, with β corresponding to the "guard band" factor and θ to the "learning constant." Widrow-Hoff's algorithm does not use β. Note that our classifier outperforms the Widrow-Hoff algorithm on all three sets, especially in the "missed object-of-interest" category, but shows particular improvement for the second data set, for which training data are sparse and stable training frames are few.

The data set extracted from the third sequence of image frames are object of interest and clutter within a tracking window. To show the effectiveness of the multineuron learning network, another image sequence is generated from the third data set referred to as data set 4. This time no tracking window is placed. Thus, there is clutter coming from all over the images with different sizes and shapes. The performances of the learning network against the single-neuron classifier and the Widrow-Hoff algorithm are compared. The results are plotted in Figure 16.11. It shows the results of false detection and missed object of interest for different numbers of allowable neurons. Figure 16.11a is the results for $\tau = 0.25$, Figure 16.11b is the results for $\tau = 0.5$, and Figure 16.11c is the results for $\tau = 0.75$. Note that the single-neuron classifier does not miss any object of interest, yet its false detection rate is very high (21.1 percent). The false detection rate decreases as the number of neurons increases, it reaches the local minimum at around four or five neurons and goes up as more neurons are used. The missed object-of-interest rate of the multiple-neuron network fluctuates but is always small. It is apparent that the introduction of more than one neuron does improve the classification results, since the more discrimination surfaces (neurons) the better chance of enclosing the whole object-of-interest distribution. It is also evident that there is limit on the number of neurons for a given data set beyond which the performance cannot be improved or may even become worse.

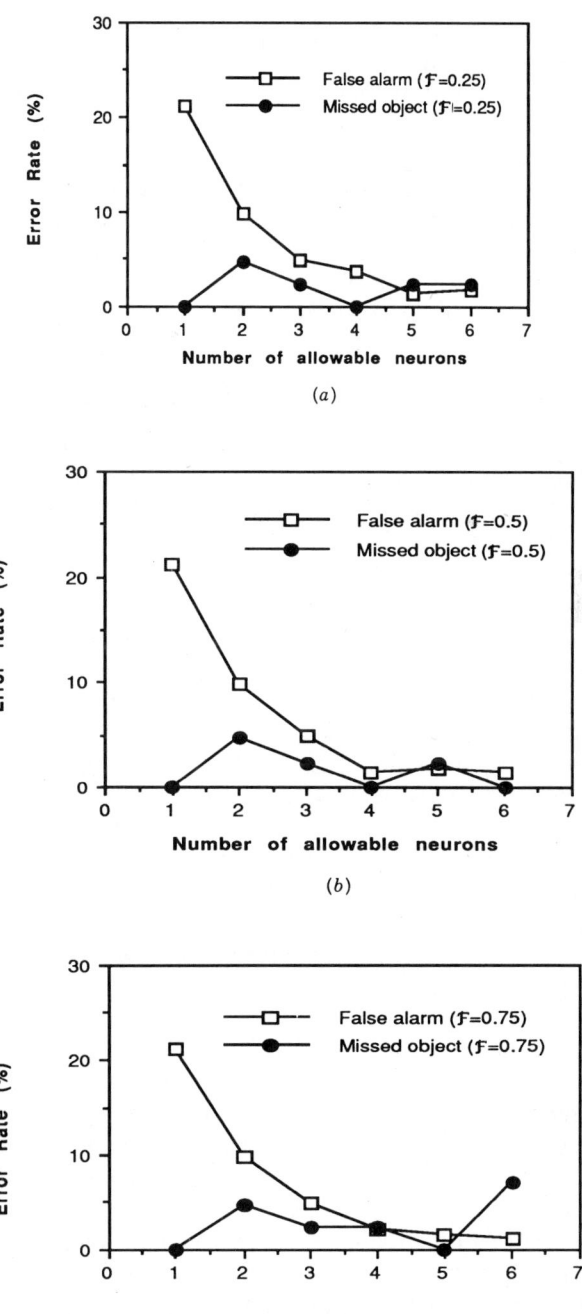

Figure 16.11 The results of false detection and missed object of interest for different number of allowable neurons: (a) $\tau = 0.25$; (b) $\tau = 0.5$; (c) $\tau = 0.75$.

The parameter τ is used in our algorithm to determine how large is the neighborhood of each individual clutter cluster. The results show that 0.5 is a good value. Since the neighborhood "radius" is defined as $\tau \times$ average distance between clutter centers, it is intuitively that $\tau = 0.5$ is a good value to use. The standard Widrow-Hoff algorithm performs poorly. Its false detection rate is 0.0 percent, yet its missed object-of-interest rate is 85.7 percent. It seems to have difficulty assigning any detection as object of interest.

16.6 CONCLUSIONS AND SUMMARY

We believe the development of a methodology that integrates all available information and processing technologies for information processing is a key to the development of high-performance intelligent systems. This creates the ancillary requirement to provide for adaptive processing of the input data and for automatic capture of the features, followed by intelligent, automatic application of the application-specific knowledge presently utilized by human experts in reaching a decision.

We pose information processing and intelligent decision making as an optimization problem that minimizes error cost functions modeled in a Bayesian decision framework [1,2]. In the context of multilevel processing, this optimization problem can often be decomposed into the optimization of subprocessing stages, and application-specific knowledge can be used to simplify the problem.

One configuration in an intelligent system is to integrate rule-based reasoning with the adaptability and incremental learning characteristics of neural networks. This is exemplified by a practical application of this strategy in an object-of-interest tracking system. In this system, a rule-based recognition controller works with a neural network adaptive classifier. The recognition controller applies contextual information to interpret the results of a neural network adaptive preclassifier, then invokes a statistical pattern recognizer to further verify the potential detections. The statistical pattern recognizer incrementally accumulates statistics of the detected object and its contextual conditions and performs dynamic feature weighting and selection. The learning controller guides incremental training of one neural net classifier, including reinforced learning against classification errors. In this way, the overall system performance can achieve a level that is not achievable by any of the processing components alone. More importantly, this system does not rely on strict control of working conditions and application environments. Thus, it is more robust and can better deal with the variations of data source and unforeseen conditions.

We describe a neural network model that plays a major role in a multilayer processing system as it enables the integration of neural networks with intelligent processing. The network model and learning algorithm are designed for a practical application, and many application specific insights are incorporated to enhance the network.

Performance is bench-marked against the Widrow-Hoff algorithm, for object-of-interest detection scenarios presented in diverse FLIR image sequences. Efficient algorithm design ensures that this recognition and control scheme, implemented in software and commercially available image processing hardware, meets the real-time requirements of tracking applications.

APPENDIX 16A

Figure 16.A1 shows the interactions between the adaptive classifier algorithm and the "teacher." The latter generates all training data (feature vectors of detected objects with known or already determined class labels) as well as feature vectors of detected objects to be classified. In addition, the teacher performs additional intelligent processing on results of the classifier algorithm for potential improvements as well as to provide guidance to the adaptation process of the classifier algorithm. In the object-of-interest tracking application, the modules object detection and feature extraction module and the intelligent recognition and learning controller together play the role of teacher.

Each cycle begins with the teacher passing training data of the current frame to the adaptive classifier, which will enter the training phase. This

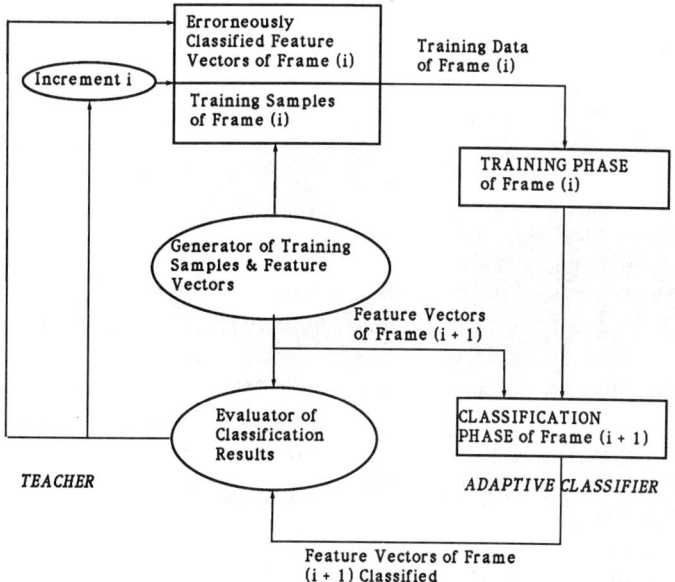

Figure 16.A.1 *Interactions between the adaptive classifier and its "teacher".*

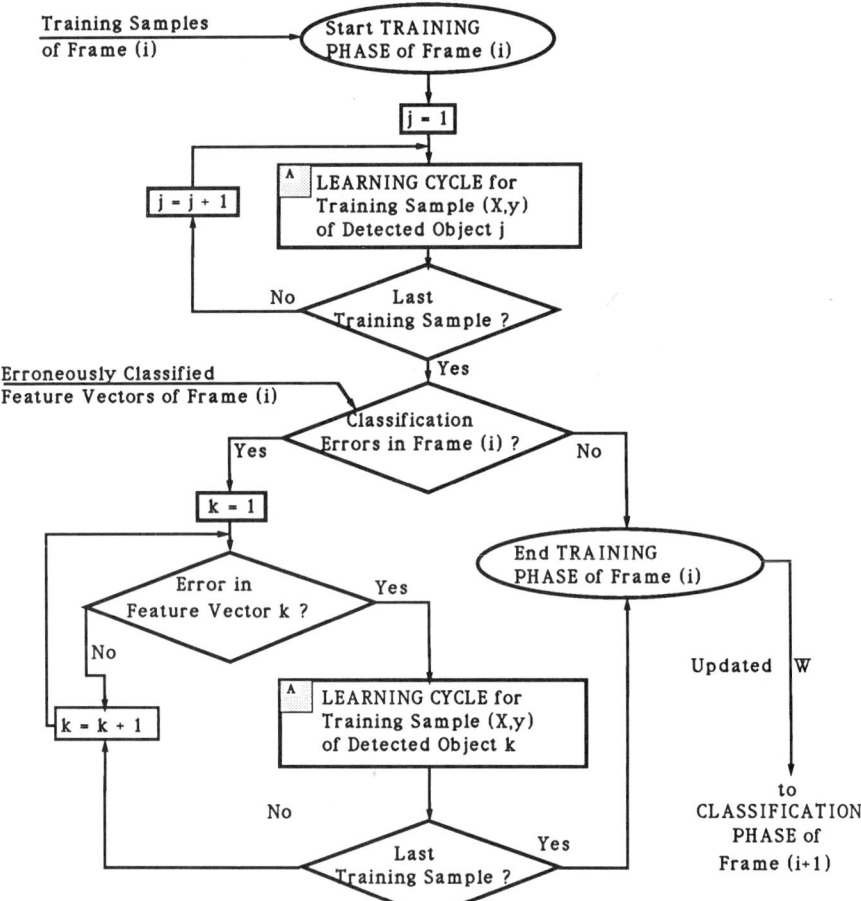

Figure 16.A.2 Flowchart of the training phase.

phase (Fig. 16.A2) consists of two parts: (1) training with each of the training samples—that is, feature vector and label [**X**,y] of each of the detected objects in the current frame, and (2) reinforcement training with those samples in the same frame that have been previously erroneously classified or labeled. Details of the algorithm of each learning cycle (boxes A of Fig. 16.A2) are provided in Figure 16.A4. After the learning phase, feature vectors of detected objects in the next frame will be furnished to the adaptive classifier for the classification phase.

In this phase (Fig. 16.A3) the adaptive classifier applies the classify algorithm (shown in shaded box) to the input feature vectors using the new weight vector **W** updated by preceding classification phase. At the end of this phase, all classified feature vectors are returned to the teacher for evaluation. It is assumed that the teacher, with additional supporting information,

APPENDIX 16A

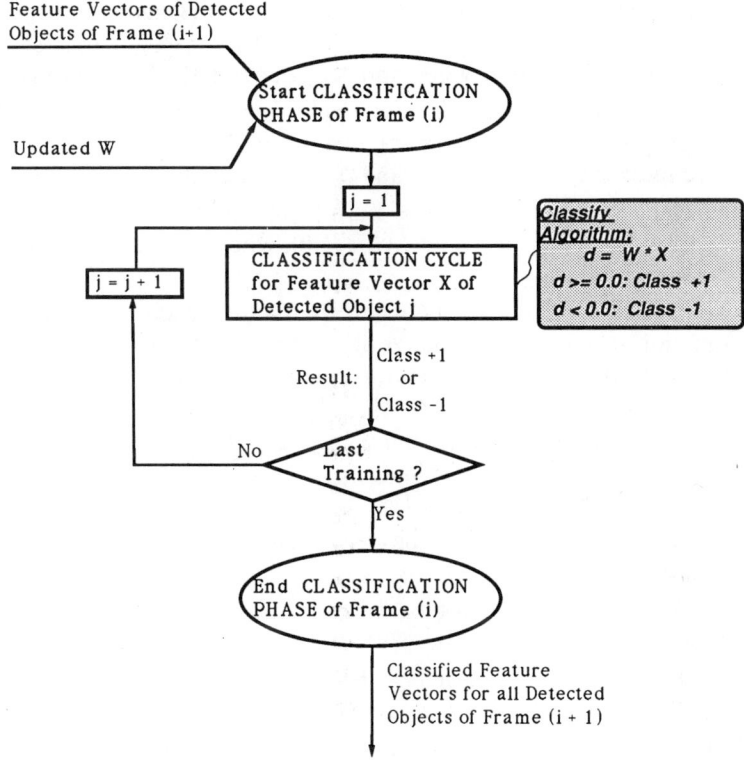

Figure 16.A.3 *Flowchart of the classification phase.*

may be able to confirm or reject these classification results. The evaluated classified feature vectors (including those that have been determined to be incorrectly labeled) form the training data for the next frame.

Before more details of the implementation are described, the terminology pertaining to the new algorithm is given:

- Class$_{+1}$: The first object class (corresponding to the object of interest in the tracker example).
- Class$_{-1}$: The second object class (corresponding to the clutter in the tracker example).
- [**X**,*y*]: A training sample.
- **X**: Feature vector of a detected object, $\mathbf{X} = (x_0, x_1, \ldots, x_{m-1}, 1.0)$.
- *m*: Dimension of the feature vector **X**.
- x_i: A feature, $x_i \in [0.0, 1.0]$, for all $i \in \{0, \ldots, m\}$.
- *y*: Class assignment or label of **X**; *y* is binary $+1.0$ (if corresponds to Class$_{+1}$) or -1.0 (if corresponds to Class$_{-1}$).

- $\mathbf{X}_{\text{avg}}^+$: The estimated cluster center associated with Class$_{+1}$; it is updated by binomially weighted averaging formula $\mathbf{X}_{\text{avg}}'^+ = (\mathbf{X}_{\text{avg}}^+ + \mathbf{X})/2$, where $\mathbf{X}_{\text{avg}}^+$ denotes the current value and $\mathbf{X}_{\text{avg}}'^+$ denotes the updated value.
- $\mathbf{X}_{\text{avg}}^-$: The estimated cluster center associated with Class$_-$; it is updated by binomially weighted averaging formula $\mathbf{X}_{\text{avg}}'^- = (\mathbf{X}_{\text{avg}}^- + \mathbf{X})/2$, where $\mathbf{X}_{\text{avg}}^-$ denotes the current value and $\mathbf{X}_{\text{avg}}'^-$ denotes the updated value.
- **W**: Weight vector (the surface normal of the current discrimination surface S); $\mathbf{W} = (w_0, w_1, \ldots, w_{m-1}, 1.0)$, where $w_i, i \in \{0, \ldots, m\}$ is a real scalar value.
- d: The normal "distance" between the feature vector **X** and the discrimination surface S along the **W** direction, $d = \mathbf{W} \cdot \mathbf{X}$; $d^+ = \mathbf{W} \cdot \mathbf{X}_{\text{avg}}^+$ is the normal "distance" between the cluster center $\mathbf{X}_{\text{avg}}^+$ and the updated discrimination surface S; $d^- = \mathbf{W} \cdot \mathbf{X}_{\text{avg}}^-$ is the normal "distance" between the cluster center $\mathbf{X}_{\text{avg}}^-$ and the updated discrimination surface S. Note that d (d^+ or d^-) usually is not the true distance between **X** (X^+ or X^-) and S unless **W** ($\mathbf{W}^+$ or $\mathbf{W}^-$) is normalized.
- **W'**: Updated weight vector based on current feature vector **X** and weight vector **W**.
- $\mathbf{W}^+$: Updated weight vector based on estimate cluster center $\mathbf{X}^+$ and weight vector **W'**.
- $\mathbf{W}^-$: Updated weight vector based on estimate cluster center $\mathbf{X}^-$ and weight vector **W'**.
- β: Guard band factor, a constant used to compute the guard bands δ^+ and δ^- shown below.
- δ^+: Guard band associated with the Class$_{+1}$ half-space; $\delta^+ = \beta * d^+$.
- δ^-: Guard band associated with the Class$_{-1}$ half-space; $\delta^- = \beta * d^-$.
- θ: Learning constant, a constant used in the weight adjustment formula shown in Figure 16.A4. It controls the rate of adaptation of the algorithm.
- f: Weight adjustment factor, a variable determined by the training sample [**X**, y] (or "balanced" training sample [$\mathbf{X}^+$, 1.0] or [$\mathbf{X}^-$, −1.0]) with respect to the weight vector **W** (or $\mathbf{W}^+$ or $\mathbf{W}^-$ respectively).
- ϕ: Goodness check factor, a constant for checking variations in d^+ and d^- after each learning cycle.
- N: Limit on number of iterations in the "balancing training" period.

Figure 16.A4 illustrates the algorithm flow in each learning cycle (boxes A of Fig. 16.A2). Each learning cycle consists of a variable updating period, a normal training period by the new training sample [**X**, y] followed by a balancing training period based on the two class centers $\mathbf{X}^+$ and $\mathbf{X}^-$, using the updated [$\mathbf{X}^+$, +1.0] and [$\mathbf{X}^-$, −1.0] as training samples. In the variable updating period, new guard bands δ^+, δ^- are determined and depending on the

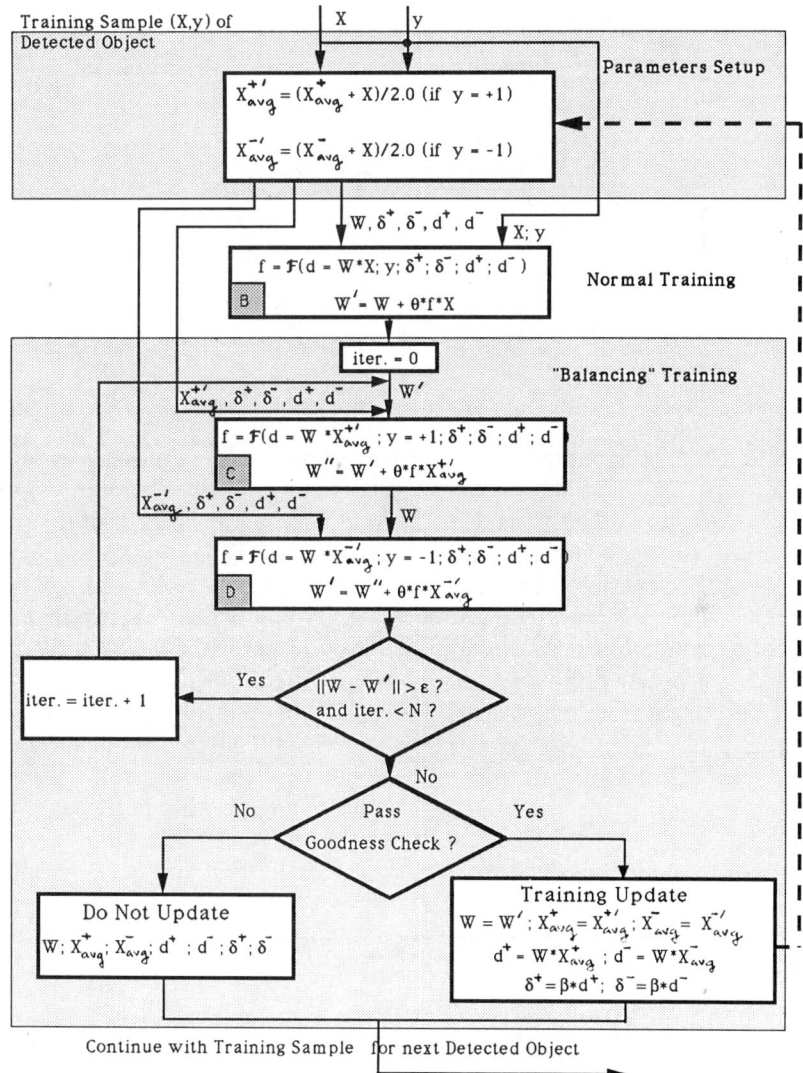

Figure 16.A.4 *Details of algorithm used in each learning cycle (boxes A of Fig. 16.A2).*

value of y, the new Class$_{+1}$ or Class$_{-1}$ cluster center is estimated using the binomially weighted averaging formula. In the normal training period, the weight adjustment factor f is computed based on the new values of $\mathbf{X}$, δ^+, δ^-, and the current value of the weight vector $\mathbf{W}$ following the rule given in Figure 16.A5. The new weight adjustment factor f is used to generate the new weight vector $\mathbf{W}'$, where $\mathbf{W}' = \mathbf{W} + \theta \cdot f \cdot \mathbf{X}$ (box B).

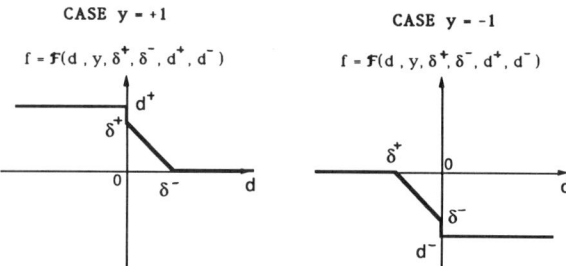

Figure 16.A.5 *Weight adjustment factor computation rule in boxes B, C, and D of Figure 16.A1.*

The balancing training is then applied to adjust **W**′. In this period, the algorithm will go through many weight vector adjustment iterations until either the weight vector stops changing or a limit on the number of iterations is reached. In each iteration, the weight vector **W**′ is first trained with respect to the updated Class$_{+1}$ cluster center, using **X**$^+$, **W**′, δ^+, and δ^- to produce a temporary weight vector **W**$^+$ (box C). **W**$^+$ is then trained with respect to the updated Class$_{-1}$ cluster based on the values of **X**$^-$, **W**$^+$, δ^+, and δ^- to produce a second temporary weight vector **W**$^-$ (box D). Except for the changes in variables, boxes B, C, and D have identical computational rules for **W**′, **W**$^+$, and **W**$^-$, respectively. If **W**′ still differs significantly from **W**$^-$, based on a limit ε in a norm measure of distance or if the iteration number does not exceed a limit N, this balancing training is repeated with **W**′ updated as **W**$^-$ and the iteration number incremented. After the iterations stop, a "goodness check" is performed. This goodness check allows the algorithm to recover from potentially "bad" data that may arise in applications using real data. Such bad data may throw the learning off-course or cause a pertubation that adversely affects the performance of the algorithm. Our technique to detect such events is to monitor the variations of important variables in the adaptation algorithm. After a training cycle, a drastic variation in the "distance" measures (such as $d'^+ < \phi * \delta^+$ or $d'^- < \phi * \delta^-$) indicates that the current training sample is risky. To avoid the potential damages mentioned, one can simply restore the state of the algorithm to one that exists before this training cycle. If it is satisfactory, **W** is updated as **W**$^-$, **X**$^+$ as **X**$^{+\prime}$, **X**$^-$ as **X**$^{-\prime}$, and new values for d^+ and d^- are computed. Otherwise, these variables keep their old values at the beginning of the learning cycle, which has the same effect as discarding the training sample [**X**,y] and the training associated with it.

Figure 16.A5 presents a graphical interpretation of the weight adjustment factor computation rule used in boxes B, C, and D in Figure 16.A4 to compute f given the values of d, y, δ^+, δ^-, d^+, and d^-.

REFERENCES

[1] T. Bayes, An essay toward solving a problem in the doctrine of chances, *Philosophical Trans. Royal Soc.*, 53, 370–418 (1763). E. S. Pearson and M. G. Kendall (Eds.), *Studies in the History of Statistics and Probability,* Hafner, New York, 1970, pp. 131–153.

[2] R. O. Duda, P. E. Hart, and N. J. Nilsson, Subjective bayesian methods for rule-based inference systems, *Proc. AFIPS National Computer Conference,* 45, 1075–1082 (1976).

[3] P. S. Maybeck, R. L. Jenson, and D. A. Harnly, An adaptive extended Kalman filter for target image tracking, *IEEE Trans. Aero. and Electr. Syst.,* AES-17(2), 173–180 (March 1981).

[4] T. R. Kronhamm, Adaptive target tracking with serial kalman filters, *Proc. 24th Conf. Decision and Control,* 1288–1293 (Dec. 1985).

[5] B. Widrow and M. E. Hoff, Jr., "Adaptive Switching Circuits," in *IRE WESCON Conv. Rec.* Pt. 4, IRE, New York, 1960, pp. 96–104.

[6] B. Widrow, R. G. Winter, and R. A. Baxter, Learning phenomena in layered neural networks, *Proc. IEEE Internat. Conf. on Neural Networks,* pp. 411–429 (1987).

[7] P. J. Burt, C. Yen, and X. Xu, "Local Correlation Measures for Motion Analysis, a Comparative Study," *Proc. Pattern Recognition and Image Processing,* 269–274 (1982).

[8] E. H. Adelson and J. R. Bergen, Spatio-temporal energy models for the perception of motion, *J. Optical Soc. America,* A2, 284–299 (February 1985).

[9] E. H. Anderson, P. J. Burt, and G. S. Van der Wal, Change detection and tracking using pyramid transform techniques, *Proc. SPIE Intelligent Robots and Computer Vision,* 579 (1985).

[10] J. S. J. Lee and C. Lin, An intelligent real-time multiple moving object tracker, *Proc. SPIE Appl. of A.I.,* 4, 328–335 (April 1988).

[11] J. S. J. Lee and C. Lin, A Novel Approach to Real-time Motion Detection, to appear in *Proc. IEEE Computer Vision and Pattern Recognition,* pp. 730–735, Ann Arbor (1988).

[12] P. J. Burt, Fast filter transforms for image processing, *Comp. Graphics and Image Processing,* 16, 20–51 (1981).

[13] D. Nguyen and J. S. J. Lee, A new LMS-based algorithm for rapid adaptive classification for dynamic environments: Theory and preliminary results, *Proc. IEEE Internat. Conference on Neural Networks,* Vol. 1, 455–463 (1988).

[14] D. Nguyen and J. S. J. Lee, A new LMS-based algorithm for rapid adaptive classification in dynamic environments, *Neural Networks* 2, 215–228 (1989).

[15] J. S. J. Lee, D. Nguyen, and C. Lin, "Intelligent Recognition and Learning: Integrating Intelligent Processing and Neural Networks for Adaptive Moving Object Tracking," Poster session, *IEEE Internat. Conference on Neural Networks,* 1988.

[16] T. Kohonen, *Self-Organization and Associative Memory,* Springer-Verlag, Berlin, 1984.

[17] R. Lippmann, "An Introduction to Computing with Neural Nets," *IEEE ASSP Magazine,* (April 1987).

[18] J. J. Hopfield, "Neural Networks and Physical Systems with Emergent Collective Computational Abilities," *Proc. National Academy of Science USA,* Vol. 79, pp. 2554–2558, (April 1982).

[19] J. S. J. Lee and D. Nguyen, A learning network for adaptive target tracking, to appear, *Proc. IEEE Internat. Conference on Systems, Man and Cybernetics,* pp. 173–177, Los Angeles, Nov 1990.

PART III

Integrated Reasoning, Informing, and Serving Systems

Distributed Knowledge Processing
Japanese Fifth-Generation Computer Systems
Software for Intelligent Systems Integration
Integrated Complex Automation and Control
Robotic and Flexible Automation Simulator/Integrator
Intelligent Database and Automatic Discovery
Artificial Intelligence, Symbols, Knowledge, Cognition

CHAPTER 17

Architectures for Distributed Knowledge Processing

YUEH-MIN HUANG
JERZY W. ROZENBLIT

17.1 INTRODUCTION

Artificial intelligence (AI) techniques have been widely applied to various problem domains, such as natural language understanding, computer vision, robotics, and manufacturing. With more AI applications outside of academia, it is desirable to improve the efficiency of AI computations. To achieve this task, we investigate the different characteristics of conventional and AI computations. Then, we can select or design a suitable computational architecture to make AI applications more practical.

The behaviors of conventional numeric algorithms on sequential von Neumann computers have been observed for a long time and are well understood [1]. In contrast to those behaviors, research [2] has indicated that AI computations have the following characteristics: (1) symbolic processing, (2) nondeterministic computations, (3) dynamic execution, (4) large potential for parallel and distributed processing, and (5) management of extensive knowledge. This research suggests that AI computations have a higher complexity than conventional ones.

Over the past decade, the development of parallel computer architectures and networked computer systems have created more research interest in distributed artificial intelligence (DAI). The idea of DAI [3] is to apply paral-

---

This research was supported by McDonnell Douglas External Relations Grant "Knowledge Based Support of Modular, Hierarchical Model Management."

Neural and Intelligent Systems Integration, By Branko Souček and the IRIS Group.
ISBN 0-471-53676-8 ©1991 John Wiley & Sons, Inc.

lel or distributed computational power to reduce the penalty caused by symbolic processing. The application has several advantages, such as speeding up computations, increasing efficiency, improving reliability via redundancy, and achieving increased modularity and reconfigurability [4]. The approaches currently being employed could be categorized into concurrency and distribution.

Knowledge engineering (KE), one of the AI research domains, is regarded as the fastest-growing area in AI. Many research results have been moved from laboratories to the real world. Examples include MYCIN, OPS5, and KEE. The ultimate goal of KE technology is to create a means for people to capture, store, distribute, and apply knowledge electronically, that is, apply knowledge stored on computers to solve the problems that ordinarily require human intelligence.

The advancement of very large scale integration (VLSI) and computer networking technologies has increased the variety and scope of knowledge engineering applications possible. Although a distributed approach cannot be used to increase the solvable problem size of AI problems, it is certainly useful in improving computational efficiency [2]. The idea of distributed knowledge processing systems is similar to that of conventional distributed computing systems: to take advantage of accessing other processing units' resources, or knowledge not existing in local processing units, as well as increasing the throughput by providing facilities for parallel processing.

From an organizational point of view, how to represent a chunk of knowledge within a distributed knowledge processing system is still a fundamental task. Besides the consideration of the problem domain, other factors must be examined when selecting a suitable representation scheme for a distributed knowledge processing system.

17.2 MOTIVATION AND PREVIOUS RESEARCH

17.2.1 Motivation

The trend of assigning a knowledge-based system to multiprocessing systems has become apparent with the demand for high-speed operations [5]. At the same time, the reduction of hardware costs also increases the feasibility of multiprocessing. The allocation of a knowledge base to processors is a difficult task. Although some studies indicate that the task could be done by analysis of dependencies [6], these studies are limited to production systems. In practice, the task is usually done by users due to the complexity of such allocation algorithms. It is obvious that a knowledge representation scheme with a high degree of distribution is preferable for multiprocessing systems.

Knowledge distribution is concerned with "parallelism" of inference in knowledge-based systems. In general, architecture for achieving this parallel operation can be categorized into two types: (1) to assign a knowledge base (KB) over a number of processing elements (PEs) and (2) to integrate several KBs located on an individual PE into a large-scale knowledge-based system (LSKBS). From a topological point of view, the former is a tightly coupled system while the latter is a loosely coupled system.

Our goal is to investigate architectures that facilitate the requirements of efficient knowledge sharing among a group of PEs. Primarily, we are interested in two aspects: (1) automatic knowledge distribution based on a treelike representation called FRASES and (2) object-oriented design approaches in distributed systems. FRASES representation will be introduced in next section.

The design methodology for KBS has been studied previously [7]. Currently, we are focusing on applying KBS techniques to improve the efficiency of design model generation, simulation, and planning environments [8]. Under those applications, solution procedures often involve search processes of exponential time complexity. As problem domains become more complicated, the degree of complexity increases tremendously and is intolerable [9]. Therefore, it is worthwhile to consider multiprocessing systems as test bed environments for further study. Current applications of multiprocessing systems to knowledge processing include tightly coupled and loosely coupled systems [2]. We now characterize these two types of systems.

17.2.2 Tightly Coupled Systems

The advantage of exploiting parallelism is the increased speed of matching operations in inferencing a knowledge base. As studies [10,11] indicate, pattern/object matching usually occupies most of the inferencing time on a uniprocessor system. Most of the issues discussed in research regarding parallelism have focused on the parallel inferencing of production rules using the RETE algorithm [10]. The RETE algorithm was originally proposed for speeding up pattern matching with the facts in the working memory and activating the appropriate rules in uniprocessor architecture. The algorithm was further improved and adopted for multiprocessing environments. Some architectures based on the RETE algorithm include PESA I [12] and MAPPS [13]. PESA I, a data flow architecture, exploits the parallel inference for a rule. On the other hand, MAPPS emphasizes executing rules simultaneously. In order to maximize the throughput by minimizing interprocessor communications, a common problem addressed by these studies is to find better algorithms for partitioning a knowledge base.

Some other applications of tightly coupled system include implementing knowledge-based systems into VLSI multiprocessor systems. The knowledge-based system using semantic nets representation is the first test bed to achieve this application [14]. Moldovan describes a model intended for semantic network processing and concludes the feasibility of implementation in VLSI.

17.2.3 Loosely Coupled Systems

A distributed problem-solving system is a network of problem solvers cooperating to solve a single problem. If the problem solvers are knowledge-based systems, then the network is called a distributed knowledge-based system (DKBS).

It is obvious that the formulation of strategies of cooperation among the nodes in such a system is the most difficult task in implementing a distributed problem-solving network. Although several approaches to the problem have been studied [15], only some of them have actually been applied in real-world situations. The approaches under study include multiagent planning, negotiation, and the functionally accurate, cooperative method. The multiagent planning approach relies on a *global view of the problem* and forming a primary planning node. Negotiation emphasizes the *decomposition of the problem* and distribution based on *bidding protocols*. For the latter approach, the solution of the problem is formed by the synthesis of partial solutions for each node [15]. No matter which approach is used, a common characteristic among them is that they seek to enable each node to predict its actions and, based on exchanged information, to predict the actions of other nodes. Therefore, how to exchange information effectively among these nodes is an important design issue for DKBSs. Typical implementations of distributed problem-solving networks include distributed sensor networks [16], distributed air traffic control [17], and distributed robot systems [18].

17.3 KNOWLEDGE REPRESENTATION FOR DISTRIBUTED SYSTEMS

There are several knowledge representation schemes that have been widely used, such as production rules [19], frames [20], and semantic networks [21]. For different problem domains, the adopted representation schemes are also different. For instance, frame representation is suitable for the classification domain, whereas production rules are preferred for diagnosis domains. However, there is a common characteristic for these representation schemes. That is, they were basically designed for the conventional sequential computations. At the time, multiprocessing systems were not widely used as they are today.

Since these representation schemes have been in use for a long time, the representation levels are already fixed. Research has focused on designing AI computers that are able to automatically detect the parallelism existing in a given representation [2]. Therefore, degree of independence becomes one of the criteria for evaluating particular representation schemes for a multiprocessing system. Wah and Li suggest that other criteria should include declarative power, degree of knowledge distribution, and structuralization [2]. Declarative representation has a higher potential for parallelism than procedural representation since it can represent tasks explicitly. In general, it is easy to control and partition tasks with the exact amount indicated. For instance, rules of production systems belong to this class. The second criterion refers to the recent evolution of physical devices for storing knowledge such as neural nets [22]. We shall not discuss neural nets in this chapter. An experiment conducted by Niwa et al. indicate that a structured knowledge representation usually needs less inferencing time than an unstructured representation. However, it requires more memory space [23]. For example, a frame system has a faster inferencing speed than other familiar systems representing the same amount of knowledge. Before developing our distributed models, it is necessary to illustrate the representation upon which they are based.

7.3.1 FRASES Representations

The frames and rules associated system entity structure (FRASES) is a scheme that combines an entity-based representation with production rules and frames. It is a superclass of the system entity structure (SES) [24], which encompasses the boundaries, decompositions, and taxonomic relationships of the system being represented.

An entity in FRASES signifies a conceptual part of the system that has been identified as a component. Each such decomposition is called an *aspect*. In addition to decompositions, there are relations that facilitate the representation of variants for an entity. Called *specialized entities,* such variants inherit properties of an entity to which they are related by the *specialization* relation. Each node of a FRASES tree is associated with a cluster of knowledge, termed the *entity information frame* (EIF). An EIF is a frame object [20] containing the following variable slots:

$$EIF = \langle M, ATTs, CRP, CH \rangle$$

where M = name of associated entity and represents the key to access EIF,
  $ATTs$ = attributes and parameters of M,
  CRP = constraint rules or procedures for describing EIF,
  CH = children entities of M.

The ATTs are attributes or parameters used to characterize the object being represented. Attributes of an entity are variables used to describe the object's general information. For example, typical attributes for a microprocessor are a source, chip size, technology type, and packaging style. Attributes are usually instantiated with values (i.e., quantitative or qualitative) while constructing the FRASES structure.

Constraint Rules and Procedures (CRP) slot contains knowledge for further describing EIF characteristics. The constraint rules are general production rules that provide relevant information for the object itself and the information for coupling relationships among its children. Procedures may be also associated with an EIF, such as a program to compute required ATT values from existing values.

Children (CH) indicates the children nodes of the entity M.

A FRASES structure for representing an abstract design level of a multiprocessing system for AI applications is shown in Figure 17.1. Each FRASES entity has an associated EIF. With EIF, both declarative and procedural knowledge about the entity are appropriately represented.

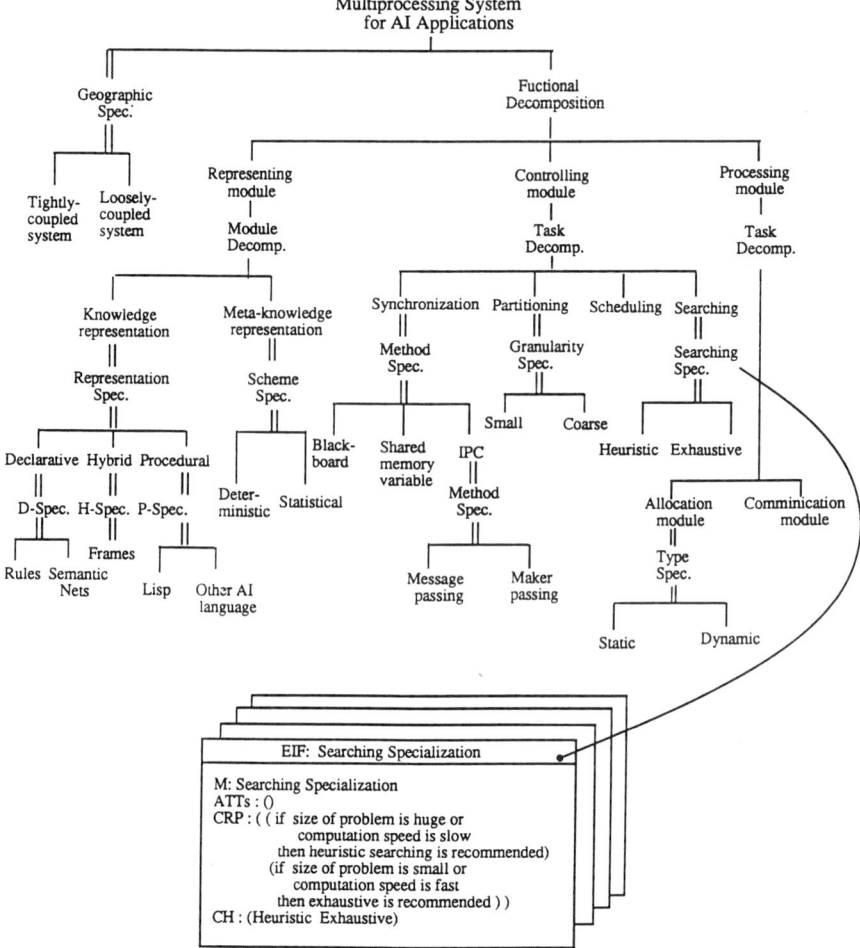

Figure 17.1 FRASES representation of multiprocessing system for AI applications.

Having described the FRASES structure, we illustrate its common characteristics with an object-oriented representation. The advantages of the object-oriented approach are described to illustrate the application of FRASES to the design of distributed knowledge processing systems.

17.3.2 Toward Object-Oriented Processing

Object-oriented programming is expected to be the major programming trend in software engineering. It is also regarded as having promise as a framework for concurrent programming that can be extended to knowledge bases [2]. Indeed, some studies have indicated the advantages of using the object-oriented representation approach to knowledge-based system design [25,26]. As in the frame-based system, objects themselves are entities combining attributes and procedures. However, the following differences exist between frame and object representations: (1) relationships are unique for objects (i.e., subclass link between two classes and is-a link between a class and an instance) rather than frames (i.e., subclass and member links existing among frames); (2) rules or procedures are only defined in class objects.

It is not difficult to convert a FRASES representation into object-oriented representation. Basically, the slots of ATTs and CRP of an EIF provide the same functions as the attributes and methods of an object. Additional work needed for the conversion is to have each object associate proper relationships related to other objects. Thus, the objects that represent FRASES have stronger semantic relationships than the objects with original definitions of object-oriented programming. Such research is currently being conducted in [27].

17.3.3 Object-Oriented Representation for Distributed Systems

The application of a distributed approach to knowledge processing systems is more complex than a nondistributed approach. The knowledge representation scheme will play an important role in this application. Object-oriented methodology is regarded as one of the successful schemes in developing a distributed knowledge processing system [28]. Several systems have been developed based on this methodology [29,30]. Indeed, object-oriented representation is a good paradigm for organizing knowledge and facilitating the communications of messages.

The collection of objects that describe knowledge in a certain problem domain is called *object space*. In the implementation, the entire object space is usually partitioned into a number of groups called *contexts* [31]. A single context is assigned to a processor if a free processor is available, otherwise several contexts may be assigned to a single processor. To maximize the throughput of processing contexts, it is desirable to minimize intercontext communications so that the preceding assignment strategy is used. Therefore, a good partitioning strategy for distributing these objects must minimize the communications among contexts.

The partitioning strategies on a production system have been studied in [5,6]. Dixit and Moldovan describe the method of dependence analysis in exploiting the parallelism of a set of production rules. Their study indicates that the problem of partitioning can be solved by a 0-1 linear programming technique. However, this method is infeasible for large systems. The study eventually concludes that a heuristic approach is necessary. A typical example of a distributed system called DEO using an object-oriented representation can be found in [28]. In a DEO environment, primitive operations of a distributed system such as remote, replicated, and migrating services are implemented efficiently using the object-oriented approach. From these studies, we summarize the advantages of object-oriented representation (OOR) over production rules representation (PRR) in distributed systems.

- *OOR Is Less Costly in the Analysis of Data Dependencies.* To analyze the dependencies among a set of production rules, we have to investigate the common attributes and their positions (premise or conclusion). The investigation is necessary in order to locate the relationships of parallelism and communications for all the rules. Analysis of dependencies is less costly for OOR since facilities such as class-class, class-instance, and inheritance are available. Those properties are useful in determining the partitions of objects.
- *OOR Exacts Less Penalty for the Change of Granularity of Parallelism.* The granularity refers to the size of tasks executed as an indivisible unit in a processor. To achieve maximum performance, the workload for every processor is preferred to be dynamic; that is, the original task is to be distributed evenly, and the objects of each context should be migratable. The distribution of production rules is relatively static since the coupling relationships among the rules are fixed. Prior to allowing the migration of some rules to idle processors, the allocation scheme must consider the ordering in a conflict set. The migration for OOR is relatively simple, since the only work needed is to determine which objects are more relevant to those objects residing on processors. Research has indicated that for a distributed production system, incorporating the object-based knowledge representation can improve the performance of production systems [32].

17.4 PARTITIONING AND ALLOCATION OF FRASES

Partitioning divides the original problem into several subproblems so that existing processing elements can execute those subproblems simultaneously. To exploit the parallelism from the partition, the parallelism level can be defined using the subproblem size ("granularity"). For instance, in a production system, parallelism could occur among rules or within a rule [13].

The former obviously has larger granularity than the latter. A tightly coupled system has smaller granularity than a loosely coupled system. However, common factors affecting the partition of both systems include processor performance, problem complexity, topology of network, and so forth. Regardless of those factors, the first step for partitioning is to investigate the inherent parallelism of the problem itself.

Thus, the representation scheme of domain knowledge will have a great impact on exploiting parallelism. As shown in Section 17.2, FRASES is a structured knowledge representation. The links among the nodes not only increase the speed of inferencing but also serve as a guide for partition. FRASES belongs to the class of hybrid knowledge representation; that is, it combines both declarative and procedural knowledge. The procedures associated with EIF CRP are the only procedural knowledge in existence at present. An EIF is treated as one object. A single object will reside on a PE, so that procedural knowledge will not create much overhead when executing objects simultaneously.

From a complexity point of view, decomposing a problem into subproblems and assigning them to a network of processing elements belongs to the NP-complete class. For example, we can assign k subproblems to n processing elements in n^k ways if k is larger than n (it is assumed that more than one problem can be assigned to one processor). Most DAI systems leave the task of allocation to system designers [15]. For simplicity, we ignore individual differences among those objects in order to reduce the complexity. In other words, we assume every object will take about the same time to process. The distributed models for both small and coarse granularities are discussed in the ensuing sections.

17.5 GRANULARITY OF PARALLELISM

17.5.1 Small Granularity Analysis

The FRASES representation is a scheme that allows users to describe information from abstract to more specific levels. Based on FRASES' properties, aspects correspond to *AND relationships* while specializations are *OR relationships*. If we consider an entity as a goal, then the aspect is used to decompose this goal into several subgoals so that the goal can be solved after these subgoals are all solved. On the other hand, the specialization node is employed to select alternatives for its parent entity or the goal node. Therefore, manipulation of a FRASES-based KB may be considered as an AND/OR tree search. In a previous study [7], a *pruning* tool for FRASES-based KB has been developed for supporting this manipulation. The tool, called model synthesizer, employs production rules to control searching path. Some undesired nodes are excluded from being processed by the control mechanism of the tool. In other words, some nodes are pruned off in searching paths.

Searching a FRASES tree is obviously different than searching other types of trees. The techniques for multiprocessing for combinatorial searches [9] may not be applied before observing these differences.

- *Solution Type.* In contrast to other searches that look for a best solution of a node or a path, the result of searching FRASES is a substructure of the FRASES tree.
- *Searching with Pruning.* The idea of searching with pruning is to restrict the search space. For instance, if one of the children of an aspect node cannot find its instantiation based on the constraints, then we can prune the remaining children under the node. This is called *AND pruning*. Similarly, in *OR pruning*, if one alternative is selected, then we can ignore the rest of the alternatives.
- *Searching with Unification.* Attached variables represent important information for a node. A value for a particular variable may be an input or an output so that a consistent binding of variables is required. Especially, for the entities under one aspect, variable binding will affect the parallel execution imposed on these entities. The reason for this is that every common variable for these entities must be bound to the same value.

AND and OR Parallelism of FRASES The computation models of AND and OR parallelism have been proposed for the parallel execution of logic programs [33]. These concepts can be accommodated and applied to multiprocessing of the FRASES representation.

Procedure 1. AND Parallelism For an aspect node, instantiating its parent is reduced to instantiating its children. These children can be instantiated simultaneously by the unification of common attached variables.

Procedure 2. OR Parallelism For a specialization node, instantiating its parent can always be done by instantiating its children simultaneously, if the number of children is more than one.

Procedure 1 indicates that each entity's variable inherited from an aspect node must be bound to the same value. In other words, the value of a common variable for the children should be identical during parallel execution since these children are the components of the aspect node. Thus, the message passing is required if the common variable is unbound originally. On the other hand, in procedure 2 we do not encounter this problem since the children of a specialization are independent entities. The common variables do not have to be the same for independent entities.

The above definitions provide us with the construction of a parallel processing model for FRASES. We regard the execution as goal-driven with a

17.5 GRANULARITY OF PARALLELISM

divide-and-conquer process. As mentioned previously, a context contains some related objects so that our contexts can be categorized into two types based on AND and OR parallelism as shown in Figure 17.2. Each context is basically a three-level tree which begins with an entity. Following the entity is an aspect or specialization relationship, and the leaves are their children. From an implementation point of view, only the root entity physically exists and the rest merely exist logically. The AND parallelism context corresponds to the synthesis of components for one entity, and OR parallelism stands for the selection of alternatives. The CRP slot of root entity provides the information necessary to achieve the operations of selection and synthesis. These constraints rules associated with CRP slot are further characterized as selection and synthesis rules [7].

Cluster PE Model for Context The dependencies and organization of data are main factors in determining the hardware structure. To minimize the communication overhead, the preferred topology of PEs is isomorphic to the data representation. For example, the Manchester ring [34], a data flow computer, has demonstrated its advantages in multiprocessing OR parallelism for logic programs. In other words, the architecture for executing a specific problem should be the inherent computational structure of the problem.

Therefore, a treelike topology is suitable for processing FRASES-based KB. From a practical point of view, the topology of a common bus efficiently fits the requirement. A cluster PE model can be constructed based on the computational structure of a context. A cluster PE must consist of the following models: (1) individual processing units (PUs) for processing production rules or attached procedures in parallel; (2) a shared memory for storing

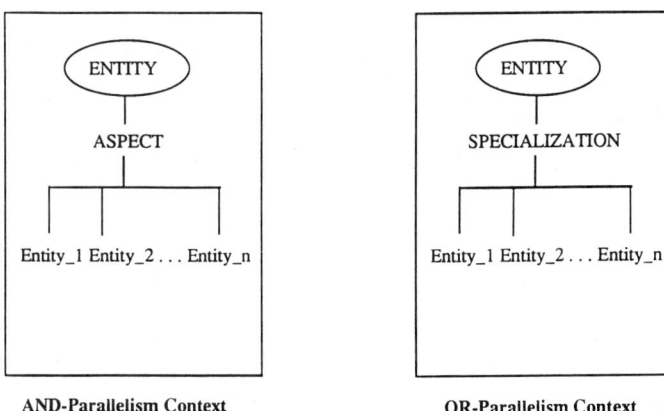

Figure 17.2 The contexts of AND and OR parallelism.

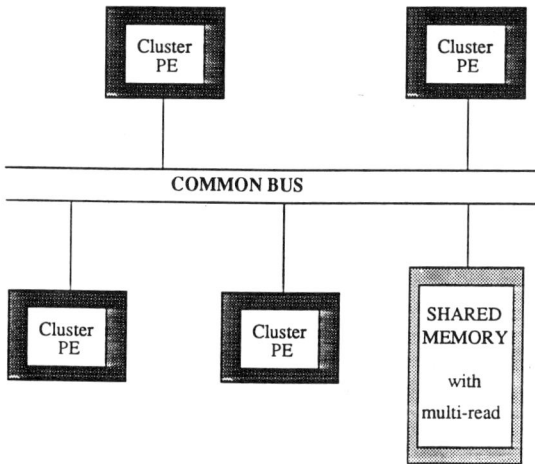

Global Cluster PE with a Multi-Read Shared Memory

Figure 17.3 *A common-bus tightly coupled system.*

the original structure of the EIF frame and for unification of common variables; (3) a cluster manager (CM) for arbitrating and distributing the EIF frame; and (4) a communication unit (CU) connecting to a global common bus. The architectures for a global cluster PE system and local cluster PE are shown in Figures 17.3 and 17.4, respectively. As Figure 17.4 indicates, a global shared memory is used to store the whole FRASES knowledge base and is allowed to be accessed by cluster PEs simultaneously. Furthermore,

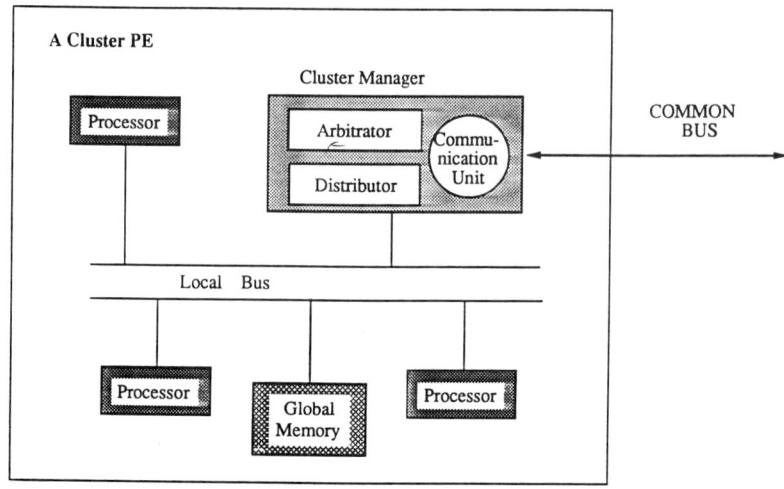

Figure 17.4 *A local cluster PE.*

the number of cluster PEs is determined by the average branching factor of FRASES and the required performance.

From an operational point of view, the following constraints must be observed: (1) the CM and each PU are associated with a small size of cache memory, serving as a working memory for inferencing; (2) the number of PUs is a function of the number of EIF slots and required throughput; and (3) shared memory is a read/write memory and is used for unifying common variables when executing AND parallelism.

Context-Flow Procedure Having described the functional modules and their constraints, we notice that the numbers of cluster PEs must meet the maximum branching factor of the FRASES tree in order to minimize the synchronization overhead. However, the size of shared memory must be able to store the EIF frame. The procedure of inferencing based on the foregoing assumptions is illustrated below. (The initial searching task is done by one host cluster PE.)

Algorithm of Backward Object-Oriented Parallel Inferencing

```
begin
    Allocate FRASES KB to Global Shared Memory
    Goal := external input;
    Address of Goal := searching (host PE, Goal);
    Entity := Value (Address of Goal);
    concurrent begin
        while (free Cluster PE) Post_Order_Allocate(Entity);
        while (Cluster PE allocated) Processing_Entity;
    Concurrent end;
end;
        Procedure Post_Order_Allocate(Entity);
        begin
            if Entity 〈 〉 leaf of FRASES
                then
                  begin
                    for i = 1 to number of children
                        Post_Order_Processing[Entity.Child(i)];
                    allocate Entity to free Cluster PE;
                  end
        end;

        Procedure Processing_Entity (Entity);
        begin
            case context parallelism
              :AND-parallelism
```

```
              begin
                distributing synthesis rules;
                concurrent_processing(PUs);
                unifying common variables;
                sending results to global memory;
              end
           :OR-parallelism
              begin
                distributing selection rules;
                concurrent_processing(PUs);
                sending results to global memory;
              end
end;
```

17.5.2 Coarse Granularity Analysis

The multiprocessing of a knowledge-based system of coarse granularity corresponds to the processing in a loosely coupled system. The system built on a loosely coupled architecture is called a DKBS. The node of a loosely coupled system is an individual KBS.

Typical Architecture for Distributed Knowledge-Based System A KBS consists of a knowledge base and an inferencing mechanism. The function of the inferencing mechanism is to manipulate the KB according to the input facts. The inferencing mechanism and KB are usually separated so that the KB can be replaced by another KB with a different problem domain.

A DKBS comprises a number of KBSs and a global coordinator called the distributed knowledge base management system (DKBMS). The KBSs are distributed geographically according to the topology of the network system. The individual KBS stores the knowledge that is often accessed by the local users. The tasks that a DKBMS executes include accepting a query (or plan) from the external user or one of the KBSs, generating subplans corresponding to the original plan, mapping these subplans to KBSs, and synthesizing responses generated by KBSs. In order to achieve these tasks, a DKBMS must be capable of "knowing" the following: (1) the organization of the whole system and (2) the domain knowledge managed by each KBS. According to the task described before, a DKBMS may be decomposed into these functional modules: a metalevel global KB (GKB), an arbitrator, and a distributor. The GKB is a passive module that stores the knowledge required to perform the given tasks. The arbitrator and distributor are responsible for performing these tasks according to the knowledge provided by the GKB. A typical architecture of a DKBS is shown in Figure 17.5.

Messages and Operations of a DKBS The messages in a DKBS can be categorized into external and internal messages. External messages are the

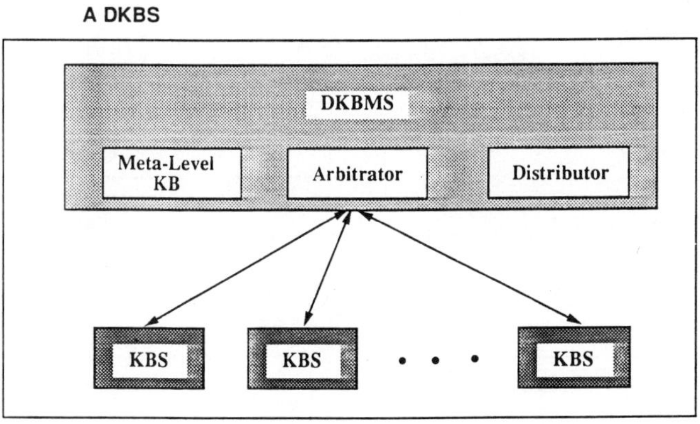

Figure 17.5 *A typical architecture of DKBS.*

messages used for communication between the external user and the system. There are two types of external messages: query and response. The query message can be sent to the DKBMS or an individual KBS. The message contains the goal. When the distributor of the DKBMS receives the query, it analyzes and decomposes the goal and then determines how many KBSs should be involved to solve the goal. If the message is received by a local KBS, the message will be forwarded to the DKBMS to invoke other KBSs if the local KBS cannot solve it alone. The response message will be received at the site where the query message originates.

Internal messages are the query messages sent by the distributor and the response messages collected by the arbitrator.

Construction of a FRASES-based DKBS FRASES is sufficient to describe metalevel knowledge for a GKB. A simple example of FRASES representation in describing a metalevel KB is shown in Figure 17.6. For instance, when the knowledge cluster (or EIF) of the distributor node is inferenced, the information about each KBS's location is needed to trigger associated rules. A message, *get-children-locations,* is sent to node KBSs from the node distributor. The format is

$$(ask\ KBSs\ 'get\text{-}children\text{-}location)$$

where, *ask* is a message-passing function. Thus, the procedure (or method) about location is activated.

Within a DKBS, multiple knowledge representation languages could be used to describe the KBSs; that is, heterogeneous knowledge bases can be adopted [29]. This would increase the complexity of a DKBMS due to migration of knowledge from one KBS to other KBSs. It it useful to consider a "common protocol" for those KBSs to communicate with the DKBMS.

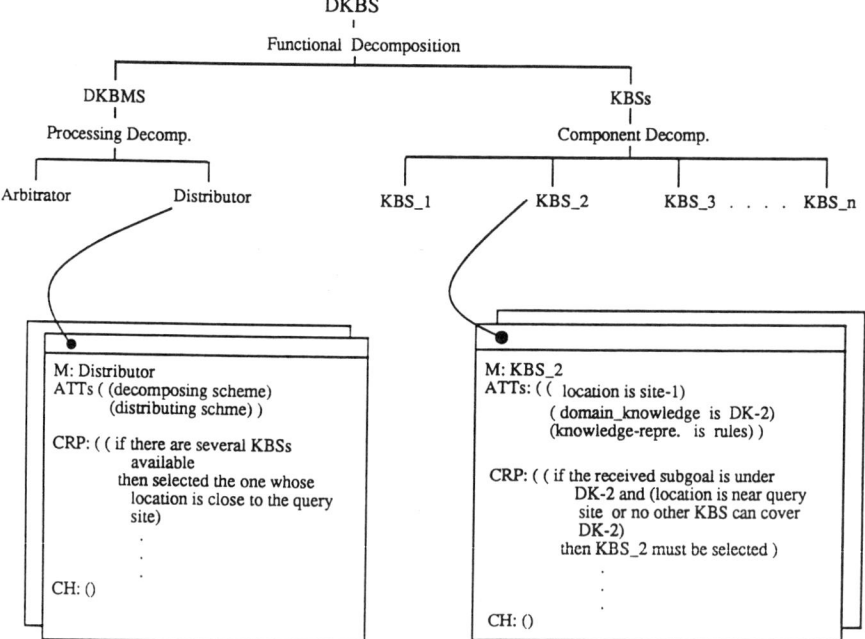

Figure 17.6 *High-level description of metaknowledge in FRASES.*

Therefore, a message unifier is built in the front of the KBS. It not only interprets the message for the KBS but "encapsules" the migrated knowledge. Since the object-oriented paradigm does not favor any knowledge representation, this is a good choice. The production rules, frames, and even some procedure information can form an individual object with a unique object-id name.

Thus, for the communication unit of a local KBS, a protocol translation from some representation to object-oriented representation can be constructed to reduce the system overhead in encoding and decoding knowledge. If every chunk of knowledge is assigned a name or object-id, then it will facilitate the update of the global metalevel KB. The FRASES representation organizes the object-ids to indicate their locations and relationships with other objects; that is, it serves as a tool in building a *dictionary* for the system. The facility of a dictionary is necessary for a distributed data base or a knowledge-base system [29,34]. The architecture for a FRASES-based DKBS is shown in Figure 17.7.

17.6 CONCLUSION

With the increasing application of knowledge-based systems, it is nontrivial to integrate a variety of systems to achieve an information-sharing network.

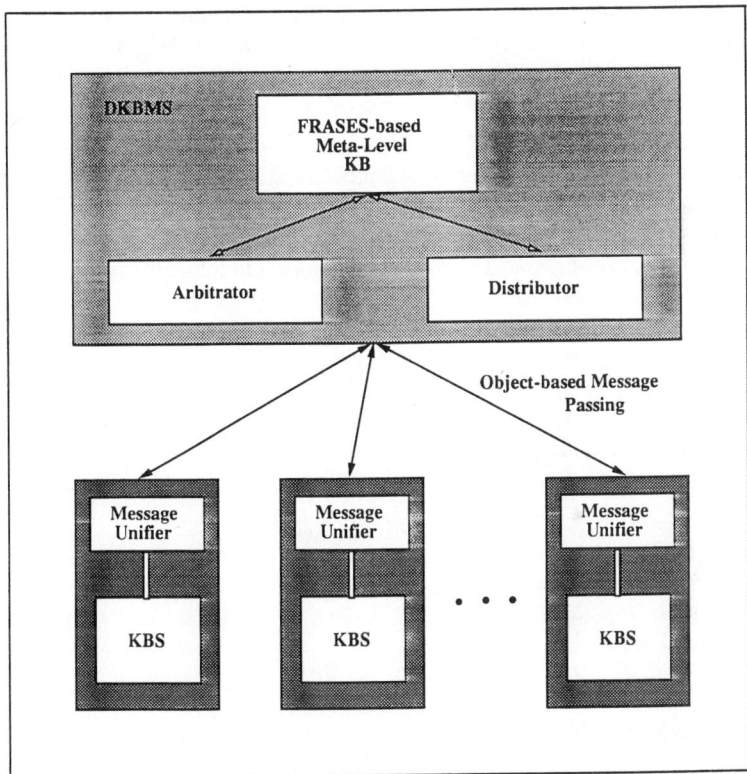

Figure 17.7 *The block diagram of FRASESbased DKBS.*

In fact, many information systems have requirements that are logically or physically distributed. In this chapter, we have investigated tightly coupled and loosely coupled architectures as a growing trend in distributed knowledge processing. The main point of our considerations is that an integrated scheme should serve as a representation for knowledge exchange. The proposed scheme, FRASES, not only facilitates representation for the varieties of problem domains but also possesses the features of the object-oriented paradigm.

Two prototype models for distributed knowledge processing using FRASES representation have been presented. Possessing the stronger semantic relationships, FRASES has facilitated partitioning a knowledge base, as well as building a metalevel knowledge base.

REFERENCES

[1] A. V. Aho, J. E. Hopcroft, and J. D. Ullman, *The Design and Analysis of Computer Algorithms,* Addison-Wesley, Reading, Mass., 1974.

[2] B. W. Wah and G. J. Li, A survey on the design of multiprocessing systems for artificial intelligence application, *IEEE Trans. on Systems, Man, and Cybernetics,* 19(4), 667–691, (1989).

[3] M. N. Huhns, Ed., *Distributed Artificial Intelligence,* Pitman, London, 1987.

[4] L. Gasser, Distributed artificial intelligence, *AI Expert,* 26–33, (July 1989).

[5] K. Oflazer, Partitioning in parallel processing of production systems, *Proc. Internat. Conference on Parallel Processing,* 92–100, (1984).

[6] V. V. Dixit and D. I. Moldovan, The allocation problem in parallel production systems, *J. Parallel and Distributed Computing,* 8, 20–29, (1990).

[7] Y. M. Huang, "Building an Expert System Shell for Design Model Synthesis in Logic Programming," Master's thesis, Dept. of ECE, University of Arizona, Tucson, 1987.

[8] J. W. Rozenblit, et al. Knowledge-based design and simulation environment (KBDSE): foundational concepts and implementation, *J. Operation Research,* 41(6), 475–489, (1990).

[9] B. W. Wah, G. J. Li, and C. F. Yu, Multiprocessing of combinatorial search problems, *IEEE Computer,* 18(6), 93–108, (1985).

[10] C. L. Forgy, Rete: a fast algorithm for many pattern/many object match problem, *Artificial Intelligence,* 19, 17–37 (1982).

[11] A. Gupta, Implementing OPS5 on DADO, *Proc. of Internat. Conference on Parallel Processing,* 83–91 (1984).

[12] D. E. Schreiner and G. Zimmermann, PESA I—A parallel architecture for production systems, *Proc. of International Conference on Parallel Processing,* (1987).

[13] A. O. Oshisanwo and P. P. Dasiewicz, A parallel model and architecture for production systems, *Proc. Internat. Conference on Parallel Processing,* IEEE Press, pp. 147–153, (1987).

[14] D. I. Moldovan, An associative array architecture intended for semantic network processing, *Proc. ACM84 Annual Conference,* 212–220 (1984).

[15] E. Durfee, V. Lesser, and D. Corkill, "Cooperation through Communication in a Distributed Problem Solving Network," in M. Huhns, Ed., *Distributed Artificial Intelligence,* Pitman, London, 1987.

[16] V. R. Lesser and D. E. Lee, Distributed interpretation: a model and experiment, *IEEE Trans. on Computers,* C-29(12), 1144–1163 (1980).

[17] S. Cammarata, D. Mcarthur, and R. Steeb, Strategies of cooperation in distributed problem solving, Vol. 2, William Kaufmann, Inc., Los Altos, Calif. *Proc. 8th IJCAI,* 767–770 (1983).

[18] M. Fehling and L. Erman, Report on the 3rd annual workshop on distributed artificial intelligence, *SIGART Newsletter,* 84, 3–12 (1983).

[19] N. J. Newell and H. A. Simon, *Human Problem Solving,* Prentice-Hall, Englewood Cliffs, N.J., 1972.

[20] M. Minsky, *Frame-system Theory in Thinking,* Laird and Wason, Eds., University Press, London, 1977.

[21] M. R. Quillian, Semantic memory, in *Semantic Information Processing,* MIT Press, Cambridge, Mass., 1968, pp. 216–270.

[22] R. P. Lippman, An introduction to computing with neural nets, *IEEE ASSP Magazine*, 4, 59–69, (1987).

[23] K. Niwa, K. Sasaki, and H. Ihara, An experimental comparison of knowledge representation schemes, *AI Magazine*, 29–36, (Summer 1984).

[24] B. P. Zeigler, *Multifaceted Modelling and Discrete Event Simulation*, Academic Press, London, 1984.

[25] K. S. Leung and M. H. Wong, An expert-system shell using structured knowledge, *IEEE Computer*, 38–47, (March 1990).

[26] C. Zaniolo, et al., Object oriented database systems and knowledge systems, in Kerschberg, Ed., *Expert Database Systems*, 1986.

[27] Y. M. Huang, "Knowledge-based Design Model Generation: Toward Object-oriented, Multiprocessing Architecture," Doctoral dissertation, Dept. of ECE, University of Arizona, (1990).

[28] A. Corradi et al., Distributed environments based on objects: upgrading smalltalk toward distribution, *Proc. IPCCC 90*, Scottsdale, Arizona, IEEE Press, 332–339 (1990).

[29] D. Carlson, and S. Ram, An object-oriented design for distributed knowledge-based system, *Proc. 22nd Hawaii Internat. Conference on System Sciences*, Honolulu, Hawaii, 55–63, (1989).

[30] B. Liskov, Overview of the argus language and systems, *Programming Methodology Group Memo 40*, MIT Lab., Cambridge, Mass., 1984.

[31] D. Tsichritzis, Active object environment, *Tech. Rep. Centre Universitare D'Informatique*, Université de Geneve, (June 1988).

[32] C. C. Hsu, S. M. Wu, and J. J. Wu, A distributed approach for inferring production systems, *Proc. of 10th Internat. Joint Conference on Artificial Intelligence*, 62–67, (1987).

[33] J. S. Conrey and D. Kibler, AND parallelism in logic programming, *Proc. of Internat. Conference on Parallel Processing*, 13–17, (1984).

[34] I. Watson and J. R. Gurd, A practical data-flow computer, *IEEE Computer*, 15(2), (1982).

[35] S. Ceri and G. Pelagatti, *Distributed Databases Principles and Systems*, McGraw-Hill, New York, 1984.

CHAPTER 18

Current Results in Japanese Fifth-Generation Computer Systems (FGCS)

TAKAO ICHIKO
TAKASHI KUROZUMI

18.1 INTRODUCTION

This chapter presents the Japanese fifth-generation computer systems (FGCS). FGCS is being developed to satisfy the demands of advanced information processing in the 1990s by eliminating various limitations of conventional computer technology. The new-generation computer will be based on novel concepts and innovative theories and technologies to perform knowledge information processing effectively. The system will have three major functions: problem solving and inference, knowledge base management, and intelligent interface. ICOT aims to develop software and hardware systems for each of these functions using logic programming and very large scale integration (VLSI) technologies. Basic software and hardware development environments have been built up so far. An experimental parallel inference machine with an effective software paradigm will be developed in the near future, and a totally integrated demonstration system will be implemented in the final stage. Thus, its performance will be upgraded to several hundred times more than the value of conventional inference mechanisms, thereby realizing a feasible environment for many pragmatic uses in knowledge information processing.

Material from technical report No. 300 published by Takao Ichiko and Takashi Kurozumi in Japanese Institute for New Generation Computer Technology was adapted for this chapter. Courtesy and copyright ICOT (Director Dr. Kazuhiro Fuchi).

Neural and Intelligent Systems Integration, By Branko Souček and the IRIS Group.
ISBN 0-471-53676-8 ©1991 John Wiley & Sons, Inc.

Ten years were allotted for effective R&D of the FGCS project. The project was then divided into three stages to make the best use of research results: a three-year initial stage for basic technology development; a four-year intermediate stage for subsystem development; and a three-year final stage for total system development (Fig. 18.1).

The intermediate-stage plan based on the results from the three-year initial stage continues to focus on the following four core research areas, with substantial results:

1. Hardware system
   a. Inference subsystem
   b. Knowledge base subsystem
2. Basic software system
3. Development support system

Hardware System Research and development of the hardware system in the intermediate stage will be based on the results of the initial-stage R&D activities on the following items: the functional mechanism for the inference subsystem, the relational data base mechanism and other mechanisms for the knowledge base machine, and the sequential inference machine. The specifications of the parallel-type kernel language version 1 (KL1) designed in the previous stage will be used as a guideline for research on the hardware system.

The intermediate-stage work on the hardware system aims to establish a machine architecture for the inference subsystem with advanced parallel processing capabilities and a parallel machine architecture for the knowledge base subsystem. Emphasis will be placed on a study of general-purpose machine architectures suitable for inference and knowledge base processing to form the core of the software system for knowledge information processing.

Inference Subsystem The inference subsystem is a machine for parallel logic programming languages and can efficiently execute KL1.

Research and development of the inference subsystem in the intermediate stage aims to establish an architecture for a parallel inference machine for large-scale parallel processing consisting of about 100 processing elements. A prototype hardware for the machine to check its operation will also be developed. The processing element will perform detailed parallel processing within the unit. Data flow, reduction, and other mechanisms studied in the initial stage will be evaluated and reorganized according to the specifications of KL1 to provide the foundations for the intermediate stage R&D.

Knowledge Base Subsystem Research and development of the knowledge base subsystem in the intermediate stage aims to define technology to

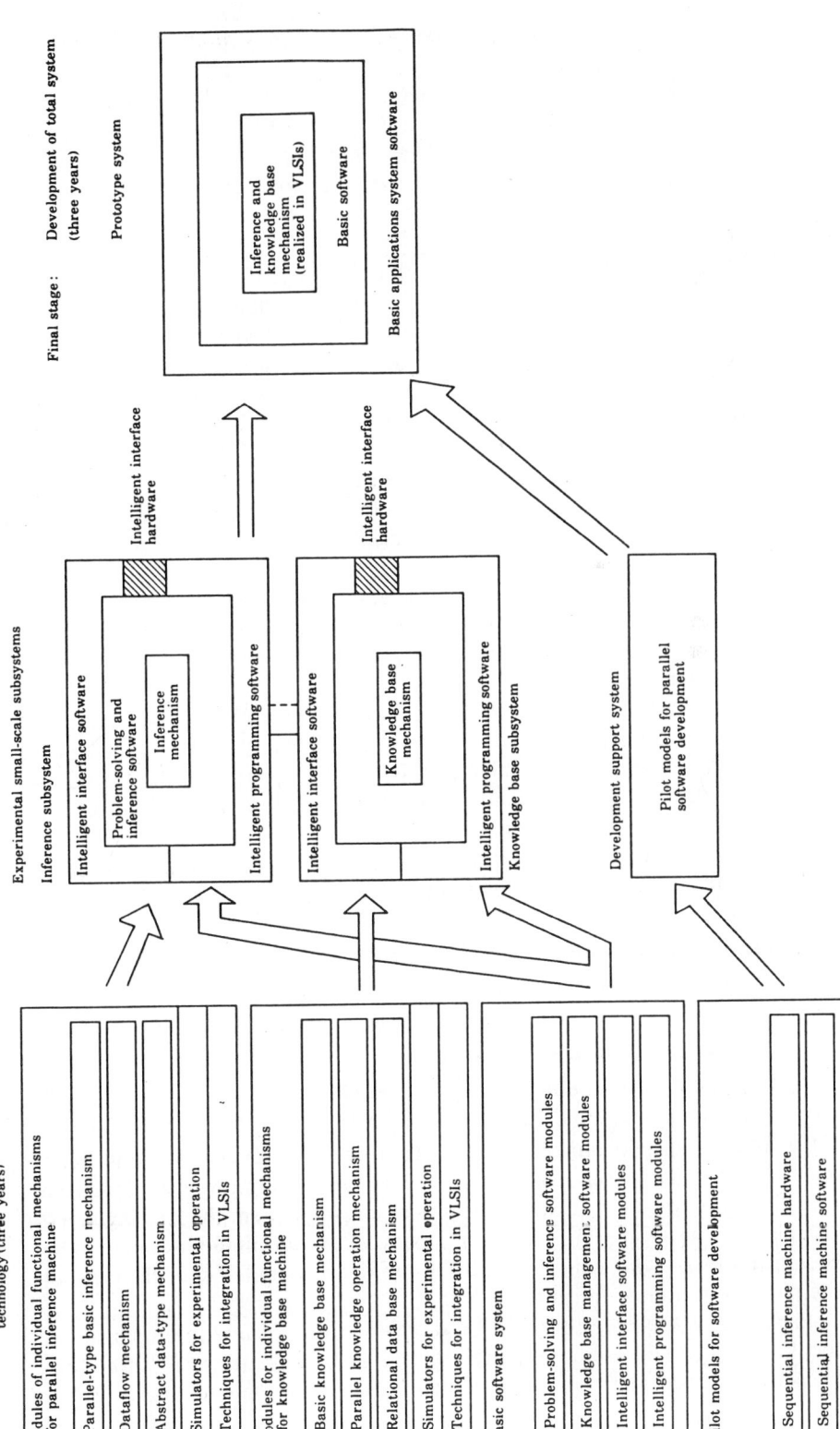

Figure 18.1 Stages of fifth-generation computer research and development.

implement a knowledge operation mechanism required by the knowledge base machine, establish the parallel architecture for the knowledge base machine, and develop prototype hardware. Mechanisms to support a distributed knowledge base will also be studied, and the technology required to implement the mechanisms will be clarified.

Basic Software System   Research and development of the basic software system will be based on the results of the basic technology developed in the initial stage. This R&D aims to provide a second version of the kernel language (KL2) by enhancing KL1 through new software technologies (e.g., parallel inference control based on parallel processing), the evaluation of the component technologies developed in the initial stage, and the development of more practical prototype systems.

Fundamental studies will be conducted on metalevel inference, distributed knowledge base management, and other functions to lay the groundwork for the cooperative problem-solving system, the final target of basic software system R&D.

A major research goal for the intermediate stage is establishing the technology to implement parallel-processing-based software modules for controlling and managing the hardware system of the inference and knowledge base subsystems.

Other points to be stressed are the clarification of the implementation technology for knowledge base management functions involving many unknown factors (e.g., a knowledge acquisition function), and the establishment of techniques to implement practical intelligent interface and programming functions in logic programming languages.

Research and development of the basic software in the intermediate stage will be conducted in the following five areas:

1. Fifth-generation (5G) kernel language
2. Problem-solving and inference software module
3. Knowledge base management software module
4. Intelligent interface software module
5. Intelligent programming software module

Fifth-Generation Kernel Language   Two kernel language versions, KL1 and KL2, will be developed. Work on KL1 aims to improve its language specifications, speed up its language processor, implement the language, and complete its programming system in a parallel execution environment. The specifications of KL1 will be established to define detailed specifications of the parallel inference machine (PIM).

Preliminary specifications will be developed for KL2 with emphasis on the knowledge representation function. This version of KL2 and other knowledge programming languages will be used throughout intermediate-stage R&D. The experience obtained using these languages will be used to remove bottlenecks in KL2 and identify functions to be supported by hardware for making KL2 faster, mainly its knowledge representation function. The final specifications of KL2 will be determined based on this research.

Problem-Solving and Inference Software Module The target of R&D in this area is to develop the cooperative problem-solving system. Work in the intermediate stage mainly aims to clarify the basic technology for developing this system (i.e., techniques to implement problem solving in a parallel environment). Different levels of parallelism, ranging from the system programming level to application systems, will be investigated.

Parallelism will be studied in a wide variety of application fields. This work will be required to develop techniques for modeling systems capable of achieving an average parallelism factor of about 100 and methods of hierarchically configuring the systems.

In addition, inductive inference, analogy, and other sophisticated inference functions necessary for implementing systems will be developed.

These basic technologies will be combined to develop prototypes of the cooperative problem-solving system.

Demonstration software will be prototyped for the problem-solving and inference software module. It will be used to apply the results in this area to real-world problems for demonstration and feed the evaluation results back to each software module.

Knowledge Base Management Software Module R&D in this area aims to establish the basis for knowledge information processing. The activities planned in the intermediate stage focus on (1) the development of a distributed knowledge base control system, an extension of the large-scale relational database management system, (2) the establishment of distributed knowledge base utilization methods to allow the distributed knowledge base control system to solve problems in parallel processing, and (3) the construction of systems to support the more sophisticated functions of knowledge acquisition. A knowledge programming language and its language processor (e.g., knowledge base editor) will provide the common base. In addition, an experimental system for distributed knowledge base utilization will be prototyped to integrate the results of these activities. This system will be able to manipulate knowledge obtained from dozens of experts.

Intelligent Interface Software Module Semantic analysis will be the primary focus of natural language processing research, while a preliminary study will be conducted on context analysis. Linguistic data will be analyzed and organized. Text-understanding systems are also planned capable of understanding, for example, junior high school science textbooks with a vocabulary of about 3000 words.

Research and development will be carried out on the construction of interactive models that permit mutual understanding between man and machine.

Intelligent Programming Software Module This theme aims to pursue various software engineering problems in logic programming language. Research and development will focus on the design of specification description languages to merge logic-based formal approaches with more natural approaches based on natural languages and graphics.

The major subject of software engineering is how to speed up, with computers, not the coding stage itself but the stages before and after coding; that is, design, maintenance, refinement, and debugging.

Rapid prototyping is considered the key factor for design support; a support system will be developed for this technique. Work will be conducted on (1) a technique to implement programs as parts; (2) a software base to achieve this technique; and (3) expert systems for design support by choosing appropriate application areas.

Development Support System Research subjects include the language processor for KL1, prototyping of software for the KL1-based parallel inference mechanism, and the parallel knowledge base management mechanism. A pilot model for a parallel software development machine will be developed to support these activities. This development machine will be a small-scale multiprocessor system tightly coupling several processors. The central processor unit (CPU) of the sequential inference machine developed in the initial stage will be used as the processor. It should achieve process-by-process parallel processing at an early stage and serve as a tool for developing parallel-processing-based prototypes software and evaluating the effectiveness of parallel processing. The machine is planned to be available as a tool from the middle of the intermediate stage. The software and hardware of both the sequential inference machine and the relational data base machine developed in the initial stage will be refined and enhanced so that these machines can be used as tools. The local area network (LAN) will also be expanded and refined to facilitate software development. In addition, the wide area network (WAN) will be arranged so that research groups can share data, software, and devices for more efficient research and development.

18.2 INFERENCE SUBSYSTEM

18.2.1 Parallel Inference Machine Architecture

An architecture for the parallel inference machine was established as follows.

Overall Structure For communications between parallel processes distributed among multiple processing elements (PEs), models appropriate for parallel software should have no difference or only a small difference in "distance" expressed by logical communication cost. However, some difference in distance will inevitably occur between the intermediate-stage PIM hardware, consisting of about 100 PEs, and the final-stage PIM, consisting of about 1000 PEs. Therefore, the entire structure of hardware for the parallel inference machine was designed hierarchically, taking into consideration the distance between PEs in hardware as shown in Figure 18.2. Parallel processing in this structure is performed at three levels: within a PE, within a cluster, and between clusters.

Processing Element   The processing element used in the intermediate-stage PIM has a KL1-oriented instruction set whose data width is around 40 bits. The PE has a parallel cache memory and a dedicated lock mechanism to permit a faster access to global memory (GM) and perform exclusive control on access. The PE must be compact, has low power consumption, and operates fast. To achieve this goal, a single PE will be installed on a PC board, and the instruction execution unit of each PE will be compact so that it can be implemented by a single large-scale integration (LSI) chip.

Parallel Cache Memory Mechanism   Data must always be consistent for all parallel cache memories. Therefore, there must be a way of informing other PEs of an address written to a cache by its PE. One key to cache memory design is to make each PE use the global bus (GB) less frequently, because the maximum number of PEs that can be connected to GM through the GB increases as each PE uses the GB less frequently. Therefore, the

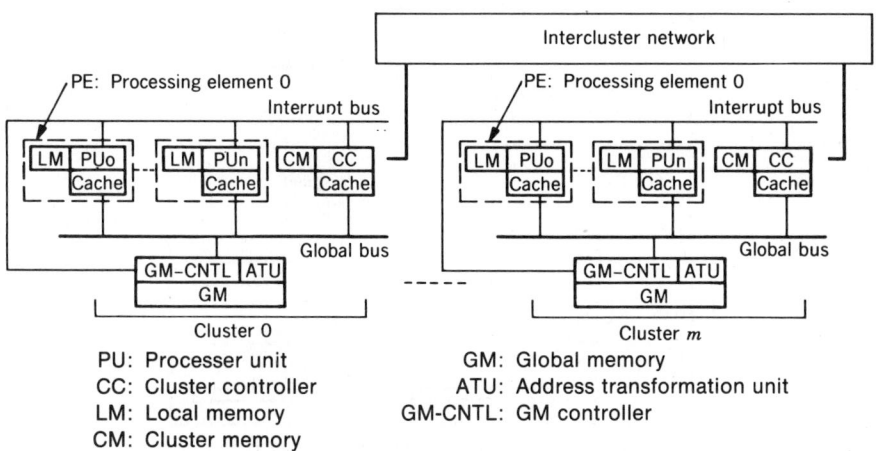

Figure 18.2 *Hardware structure of intermediate-stage PIM.*

PIM has a parallel cache mechanism in which the cache protocol is optimized by the parallel execution of KL1.

Cluster Controller and Network A cluster controller provides a single point of control over references to data coming into or out of each cluster in cluster-to-cluster message communications. Any communications request issued from a PE in a cluster to a PE in another cluster is handled exclusively by the cluster controller. The cluster controller in a cluster has a table to manage pointers between clusters; while referring to or updating the table, the controller generates a message to be sent to another cluster and sends it via the network. The controller in the destination cluster receives the message and performs the processing specified by the message or delivers the message to the processor.

18.2.2 Detailed Study of Processing Methods of KL1

KL1 is based on the parallel logic language GHC and appears to the user as a three-layered structure as shown in Figure 18.3.

KL1-u is a user language and can be regarded as a higher-level language of GHC. A concept for incorporation in KL1-u to help programmers to write parallel programs is under investigation. KL1-c provides the framework for KL1. It is a restricted version of GHC, allowing only built-in predicates in the guard part. KL1-b is a machine language for parallel inference machines. Abstract machine-language specifications (machine-independent specifications of machine languages) were investigated. KL1-p is information to be added to programs to run them efficiently on parallel machines. The specifications of KL1-b (abstract machine language), the basic machine language

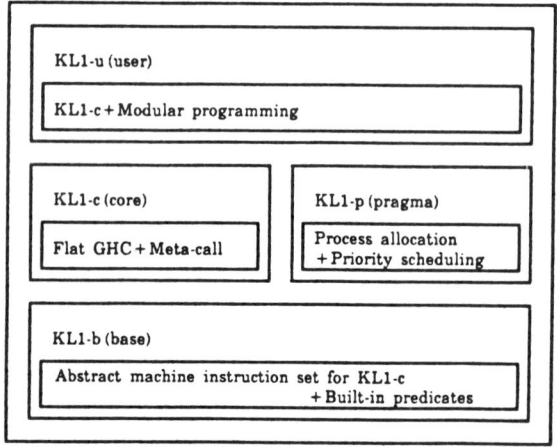

Figure 18.3 *Hierarchical structure of KL1.*

for parallel inference machines, were established by assuming the abstract machine shown here. In the assumed abstract machine, each processor will have an arithmetic unit and registers capable of handling tagged data. The memory will be accessible from all processors. Goal arguments to be reduced will be stored in the registers before unification. That is, instructions similar to those in the Warren abstract machine (WAM) will be issued for unification in the passive part.

18.2.3 Functional Specifications of Processing Methods of Parallel Inference Machines

These features correspond to the functional specifications of the abstract machine-language instructions described in the previous section.

Parallel processing in parallel inference machines is conducted in units of goals. Goals are distributed among PEs in a cluster and among clusters. Within a cluster, communication via the GM are used to perform goal tree management, scheduling, synchronization control, and load distribution. Functional specifications for these operations were studied to make use of locality in processing. For intercluster communications, functional design was produced for, for example, goal tree management based on the message communications method, and part of the design was subjected to a detailed study.

18.2.4 Functional Specifications of Hardware for Parallel Inference Machines

The functional specifications of the intermediate-stage PIM hardware were examined in detail. The PE for the intermediate-stage PIM must be designed for compactness, low power consumption, and fast operation. To achieve this goal, a single PE will be installed on a single PC board and the instruction execution unit of each PE will be made compact so that it can be implemented by a single LSI chip.

It was decided to introduce the idea of both reduced instruction set computers (RISC) and complicated instruction set computers (CISC) in the processor design to decrease the amount of hardware logic and to adopt low-power complementary metal-oxide semiconductor (CMOS) VLSI chips.

The performance of the PE must be sufficiently enhanced to increase the impact of parallel inference machines on software development. The targeted performance of a single PE is more than 250 KLIPS on average and more than 500 KLIPS for simple bench-mark programs like APPEND. The single PE can be made faster in two ways: (1) use of the pipeline scheme, which divideds an instruction into several cycles to reduce the clock cycle of the machine, and (2) addition of a hardware mechanism to support the execution of KL1 programs. This hardware mechanism uses tag architecture and KL1-specific instructions, and enables conditional interruption.

Faster access to the GM and faster inter-PE communications and synchronization are crucial for intracluster parallel processing. This led to the design of the PIM cluster with the emphasis on the parallel cache and lock mechanisms of the GM. The parallel cache was introduced to make access to the GM faster. It was optimized to fit the parallel execution characteristics of KL1. Lock operation, particularly one-word lock, frequently appears in KL1 processing. For faster execution of one-word lock, it was decided to perform distributed control by adding a hardware mechanism to each PE.

Address-by-address interleaving and the reduction of the bus cycle time were investigated for the GB to prevent the throughput of the GB from degrading the performance of the parallel inference machine. Assuming that each word of the GM consists of an 8-bit tag and a 32-bit data section, it was decided to establish an address transformation mechanism based on 32-bit word addressing and a reference recording mechanism to support virtual storage for address transformation. An interprocessor interrupt bus will be built among PEs separately from the GB to synchronize communications among PEs.

18.2.5 Control Software

The prototype control software described here was developed to investigate KL1 processing methods and architectures for the parallel inference machine.

The development of the prototype aims to

- Verify processing algorithms of KL1 processing methods for the parallel inference machine;
- Establish a basis for comparing processing methods;
- Investigate the behavior of language processors;
- Collect data to design pilot machines.

The prototype control software developed consists of a compiler, assembler, linker, and experimental simulator for parallel processing with a cluster.

The compiler reads a program written in KL1 and generates an abstract machine-language program. The abstract machine-language program is converted by the assembler into a relocatable machine-language program, which is then converted by the linker into an executable machine-language program. The compiler, assembler, and linker are written in PROLOG and C.

The experimental simulator is a pseudoparallel processing system to emulate a cluster in the intermediate-stage PIM. It was developed with the SIMASM, a multiprocess simulator support tool developed. The system simulates 4 to 16 PEs. It can operate as a parallel processing system and provide various statistical data and memory access information for hardware design. When not providing statistical and memory access information, the simulator can execute the 8-queen program in 27 s for four PEs.

The SIMASM was equipped with a utility to describe queue simulation and other tasks.

18.2.6 Parallel Inference Machine Model and Component Module

Research and development of the parallel inference machine component modules in the intermediate stage aims to obtain information on parallel execution methods and mechanisms and load-distributing mechanisms in parallel inference machines. So far, functional design was produced for the data-flow, reduction, sequential, and goal-rewriting modules. The experimental machine for the parallel inference machine model prototyped and enhanced was used in evaluation experiments to investigate the parallel inference machine component modules in detail.

18.3 KNOWLEDGE BASE SUBSYSTEM

18.3.1 Special-Purpose Unit for Knowledge Base Operation

Development of a Conceptual Model of Knowledge Base Machine Pilot System The conditions required for the knowledge base machine and methods of implementing the pilot system were determined, and the requirements for the knowledge base operation engine were defined.

1. The prototype of the relational data base machine Delta in the first stage showed the similarity between PROLOG and relational data bases. The research in the preceding years confirmed that the parallel relational data base engine (RE) implemented in LSI permits rapid execution of ordering and relational algebra operations, such as join and restriction, while data flows in stream form.
2. Investigation disclosed that the knowledge base machine must be able to
   a. Efficiently process structured variable-length data;
   b. Perform secondary-storage-to-main-storage on-the-fly processing;
   c. Efficiently process simple unification that handles a large number of facts;
   d. Perform computations for a set of knowledge.
3. Figure 18.4 shows the conceptual configuration of the knowledge base machine pilot system with the necessary capabilities described before.
   a. Stand-alone knowledge base machine.
   b. PROLOG has built-in KBM predicates for access to knowledge bases.

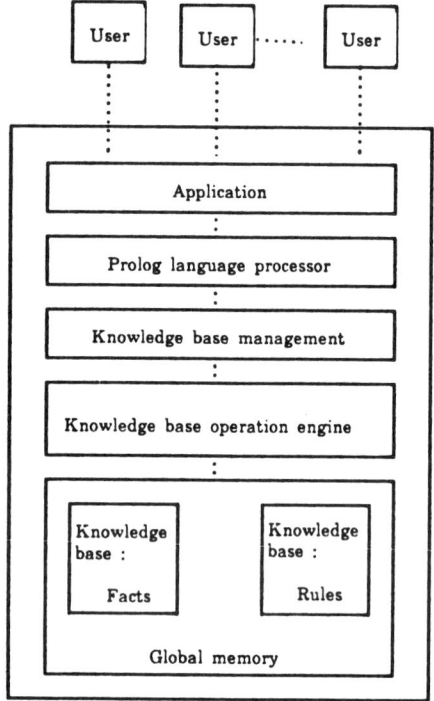

Figure 18.4 *Conceptual configuration of pilot system.*

    c. Supports the knowledge base management system (KBMS).
    d. Uses the knowledge base operation engine to perform rapid ordering and relational algebra operations on variable-length data.
    e. Adopts the on-the-fly method.
    f. Application-specific user interface and deductive processing.

Development of Prototype Knowledge Base Operation Engine Figure 18.5 is an outline of the hardware structure of the knowledge base operation engine (KE), a component of the knowledge base machine pilot system. The hardware of the pilot system is built around a general-purpose computer and has a KE. The KE receives an operation instruction from the CPU and reads the data necessary for the operation from the secondary storage linked to the KE. The operation, such as ordering or a relational algebra operation, is performed by the engine core in the KE, and the result is sent to main storage via the channel.

Research into the knowledge base machine pilot system will be carried out using a more advanced model than in the initial stage. The initial-stage model loosely coupling the inference engine and data base will be replaced

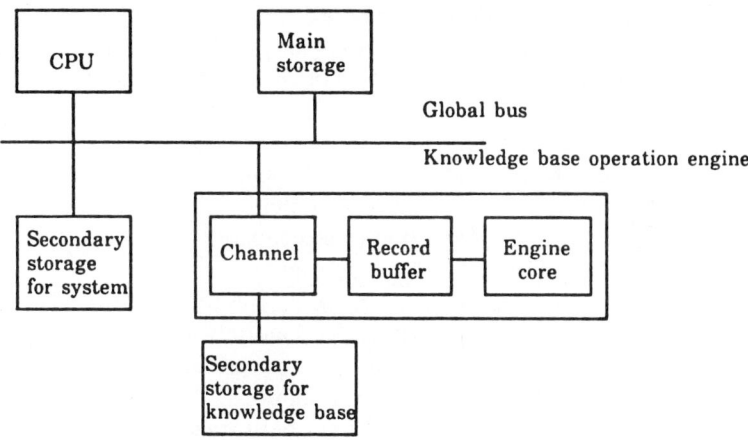

Figure 18.5 Outline of hardware structure.

with a model where the inference function and knowledge base is tightly coupled over a shorter distance. This will allow the pilot system to handle a wider variety of more complicated knowledge data in a general representation format.

Research and development activities consisted of the examination of the work; evaluation of the initial-stage engine and establishment of the functions of the knowledge base machine pilot model; and the development of the conceptual model. In addition, the knowledge base engine was implemented in hardware. The hardware engine was directly connected to the internal bus of the host computer to form a tightly coupled model. The model was then used to develop a prototype for a basic knowledge base operation engine that will provide the basis for research on smaller, faster systems built using advanced component technology in the future.

The knowledge base machine pilot system was studied as a basic experimental system for the knowledge base machine parallel control mechanism to be integrated into the hardware system of the FGCS. The results described will be incorporated into research on the knowledge base machine parallel control mechanisms.

18.3.2 Control Mechanism for Parallel Access and Processing

Research on control mechanism for parallel access and processing aims to develop a knowledge base machine that will perform a rapid search on a large amount of knowledge stored in secondary storage and be used by multiple users via host machines.

The results are outlined here.

Theoretical Study of Relational Knowledge Base Model and Retrieval-by-Unification Operation The relational knowledge base and retrieval-by-unification (RBU) operations were theoretically defined, followed by a study of the characteristics of the RBU operations.

A relational knowledge base model was formulated, as a candidate for a common knowledge base model capable of responding to various users and applications. The model is a conceptual model for the knowledge base and consists of simple structures and operations capable of dealing with various knowledge representations presented by users.

A term, a certain type of structure including variables, was used as a basic component to represent knowledge, and a relation was used as a basic container to hold terms. The RBU operation, an integration of relational algebra and unification, was formulated for term relations.

If unification-join, an RBU operation, is performed simply, the machine must perform huge amounts of processing, because it must check all the term relations to see whether the specified tuples in one attribute can be unified with any of the possible combinations of the other term relations. To make unification-join more efficient, a method to give an ordering to terms was defined, and the number of combinations reduced by ordering terms with the characteristics of the method.

Design of Unification Engine The unification engine is special-purpose hardware to perform unification-join and other operations on term relation data stored in the multiport page memory (MPPM) in the knowledge base machine.

Major R&D activities included a detailed study of the hardware structure proposed, incorporation of new processing schemes, development of detailed software to simulate the hardware, and evaluation of the engine with the hardware.

The unification engine consists of about 10 modules. A detailed study was conducted on the functions and processing schemes for each of the modules proposed. The unification engine is a dedicated unit for unification in the knowledge base machine. Software simulation was performed to evaluate the performance and features of the unification engine, by measuring two factors concerning data transfer within a sort cell and among sort cells in the engine; the amount of data transferred and transfer time. The simulation was conducted prior to implementing the unification engine in hardware. The major components of the unification engine are the sorter, pair generator, and unification unit. These are structured to permit pipeline processing. The simulation results showed that the amount of processing in the unification unit almost entirely agreed with that of the entire unification engine for uniformly distributed comprehensive data. The processing in the unification unit depends on the amount of output from the pair generator, while the output amount of the pair generator is determined by the generator's function to eliminate term pairs on which unification must fail. Because this function works less effectively for comprehensive data, the amount of processing required to order terms in the sorter and generate pairs is smaller than that of unification processing. The simulation revealed that the performance of the unification engine depends on the elimination ratio of the pair generator and the unification processing capability of the unification unit. Additional investigation of the operation of the unification unit and pair generator led to the discovery of various ways to reduce the processing time. The knowledge base machine system (Fig. 18.6) is a computer system devised to process relational knowledge. The core components of the system are the unification engine and MPPM described in the previous section. Software simulation was conducted and a detailed design for the system configuration and control method proposed was produced. These activities were helpful in understanding the behavior of the system when running specific problems.

Further investigation of the system configuration and control method revealed that the performance of the architecture depends on the type of problem. The relation between the architecture and the characteristics of specific problems was partly clarified. The knowledge base machine is a back-end knowledge base machine linked to its host, the inference machine. It provides a knowledge processing function simultaneously for multiple inference machines. It has basic functions to store relational knowledge representations and perform operations on them. The operations include unification-restriction and unification-join in addition to the conventional relational algebraic operations. When the host machine sends a request to

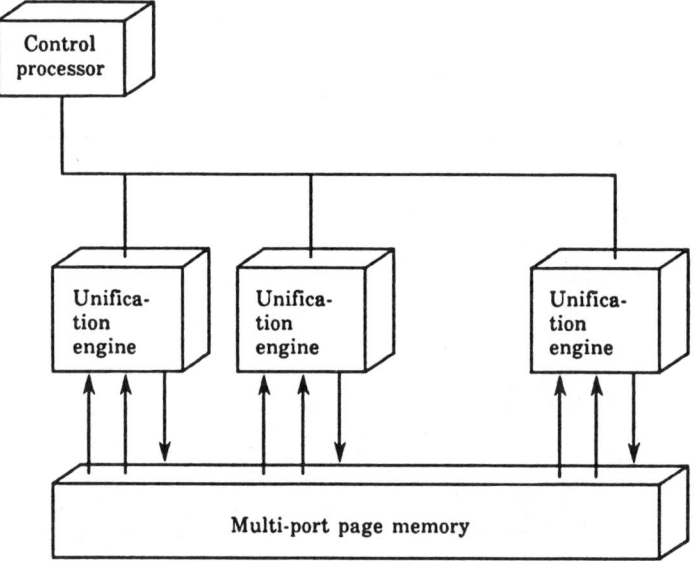

Figure 18.6 Configuration of large-scale knowledge base machine.

perform processing on relational knowledge to the knowledge base machine, the control processor analyzes the request and translates it into operations on relational knowledge. The knowledge base machine can divide an operation or group several operations page by page. The machine simultaneously processes a set of problems on multiple unification engines. To investigate control schemes and other problems of the parallel processing on the knowledge base machine, software simulation was performed by developing a model for the knowledge base machine. The simulator was planned so that the number of unification engines, page size of the MPPM, method to divide a problem, method to assign subproblems to unification engines, and other factors could be specified by parameters. The simulation showed that the page size does not much influence the processing time, that the effective utilization ratio of the page decreases as the number of unification engines increases, and that the scheme for restoring data in pages planned for other systems is not necessary, because the effective utilization ratio exceeds 50 percent for up to 1 kbyte.

18.3.3 Distributed Knowledge Base Control Mechanisms

Research into this subject aims to develop technologies to integrate multiple knowledge base machines, interconnected via high-speed buses, local area networks, and other links to form a distributed knowledge base system which will appear to the user as a logically single knowledge base machine. The development of the distributed knowledge base system, predicate logic based hierarchical knowledge management system (PHI), began as part of the intermediate-stage R&D to provide a means to develop and verify the various control and management mechanisms. Figure 18.7 shows the configuration of the PHI system. Host machines, inference machine PSIs, are linked to knowledge base machine PHIs via the ICOT-LAN. The knowledge base system in the PHI is hierarchically structured; the knowledge bases are managed by dividing them into an intensional data base (IDB) section and an extensional data base (EDB) section. This approach allows the construction

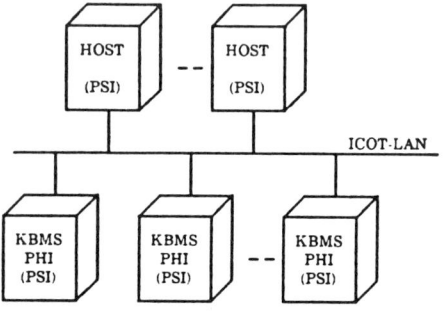

Figure 18.7 *Configuration of PHI system.*

of a system that provides the features of both the knowledge representation function of logic language processors and the management function of relational data base systems.

The relational data base management system (RDBMS) forms the nucleus of the knowledge base management unit, supporting the functions of the relational data base machine, Delta, developed in the initial stage. The knowledge management system in the knowledge base management unit stores and searches for knowledge via the RDBMS. Prototype software capable of standalone operation was developed as the first step of the RDBMS development. The RDBMS was installed on one of the PSIs interconnected via a LAN. The knowledge base engine (KBE), a hardware option for the PHI, is an operation unit that performs comparison-based processing such as search very quickly. Figure 18.8 shows the structure of the KBE. The KBE receives an operation command from a PHI, executes it, and returns the results of the operation to the PHI. The KBE repeats this sequence of processing. The KBE uses the superimposed code method, an enhanced superimposed-code-based search technique, to rapidly perform not only search but also relational operations used in knowledge processing.

18.3.4 Knowledge Representation Processing Methods for Knowledge Base Machines

A knowledge base machine must have a mechanism that allows the users and manager of a knowledge base to store, manage, and use knowledge easily. Here, the users of the knowledge base are assumed to use logic-programming-based knowledge representation suitable for various problem domains. It is also assumed that the knowledge base is shared among more than one user. If the knowledge base has information on knowledge representation to be used by users, the knowledge base manager can understand how the users obtain access to the knowledge base, and therefore can take measures

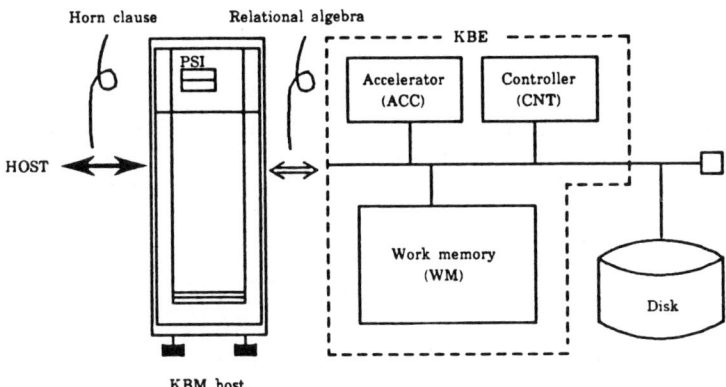

Figure 18.8 *Structure of KBE.*

against dangerous access which may spoil the uniformity and consistency of the knowledge base.

Assuming that the interface functions for individual knowledge representation have some part in common, a system was devised as a means to implement these interface functions. It will compile knowledge representations into intermediate code in some form and store the code in the knowledge base machine to efficiently process requests to the knowledge base, such as retrieve and updating, as operations on the intermediate code.

Research into knowledge representation processing methods for knowledge base machines investigated the problems involved when such interface functions of the knowledge base system were implemented as a knowledge compiler.

The effect of the implementation was also studied.

Knowledge Compiler

Knowledge Compilation Control Model   Figure 18.9 shows the execution process of knowledge compilation as the total image of the knowledge compilation control model.

The Study in Detail   Investigation was performed on program transformation, partial computation, input to partial computation, and partial computation in PROLOG. The results led future focus on the knowledge compiler and the interpreter for intermediate representation as follows:

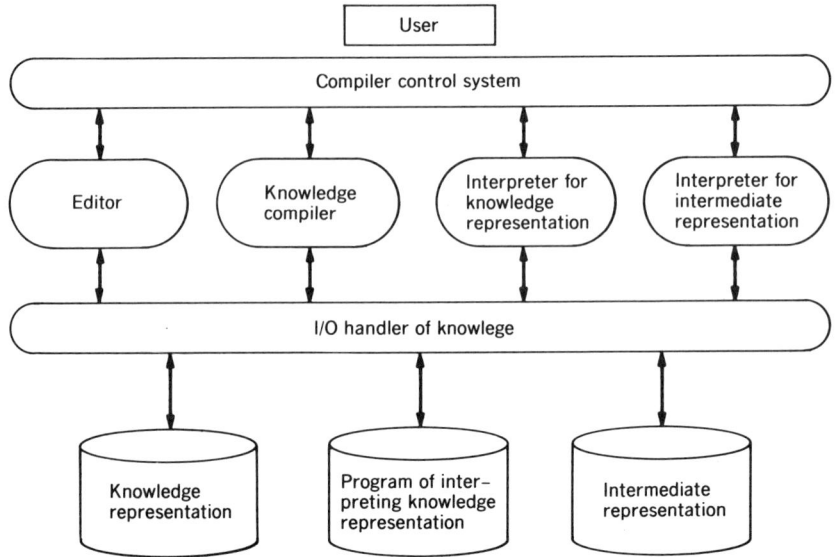

Figure 18.9 *Total image of knowledge compilation control model.*

Knowledge Compiler. A study will be conducted of methods to compile the following into PROLOG programs: (1) knowledge representations shown by object knowledge, and (2) metaknowledge and a metainterpreter to execute and interpret the knowledge representations.

Intermediate-Representation Interpreter. PROLOG will be used to describe the intermediate representation, and PROLOG-based representation methods will be studied. An investigation will also be conducted on RBU for clause search and an interface between PROLOG and RBU.

Study of Knowledge Representation. In the knowledge representation for planning, the initial state, operators to change the initial state, and the final expected condition are represented, and the represented results are used to infer a series of operators beginning with the initial state and ending with the expected condition. Typical applications of this knowledge representation include design-related tasks, such as logic design and layout, and plan-related work, such as planning of an operation sequence of robots and process control. Production systems are often used for the knowledge representation for planning. The existing production systems, including production rule representation, were surveyed and production systems suitable to the knowledge representation for planning were proposed. Future ways to research the knowledge compiler, production systems, modality-symbol-based control, and other problems were also investigated.

Study of Intermediate Representation Knowledge representation explicitly and declaratively expresses human thought. To process knowledge representation efficiently on computers, the idea of introducing intermediate representation with the emphasis on fast processing on computers and using the knowledge compiler to convert knowledge representation into intermediate representation was devised, as described earlier. The intermediate representation was examined in terms of the functions required and the architecture of computers.

Knowledge processing differs considerably from conventional information processing in that it mainly consists of pattern matching between pieces of knowledge expressed in symbols with some structure. Unification of PROLOG is a powerful pattern-matching function.

The knowledge base system is also characterized by the large knowledge bases involved. To manage large amounts of data efficiently, the system uses the schema model of the relational data base, which is based on handling several data items as a set.

The architecturally preferred approach is to store a large-scale knowledge base in a disk unit and pass only the necessary data from the disk unit to main memory. Few successful examples, however, have been reported, even for data base machines in which a filter was introduced to a disk controller or disk unit to extract necessary data. The functions of such filters

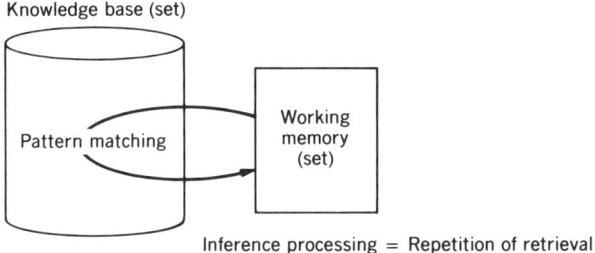

Figure 18.10 *Knowledge base processing as a set.*

have yet to be defined for knowledge bases. Unification appears to be an effective filter function, providing an approach to a new architecture.

The knowledge base processing described can be conceptually illustrated as in Figure 18.10.

Processing in the knowledge representation system appears to consist of a repetition of pattern matching between temporary states in the working memory and the knowledge base as shown in the figure. That is, a repetition of retrieval can be considered as inference processing. Some inference processing mainly changes the state. Retrieval-based processing mechanisms do not always increase efficiency. The filter mechanism described here seems effective for knowledge base processing focusing on shallow and wide inference.

18.4 BASIC SOFTWARE SYSTEM

18.4.1 Fifth-Generation Kernel Language

Detailed Study of GHC Specifications Guarded Horn clauses (GHC) is a parallel logic programming language. The grammar rules of the language were refined and enhanced to provide the foundation for KL1 in the basic software system. The semantics of GHC were defined from the following two aspects: (1) parallel input resolution neglecting the difference between guards and bodies, and (2) restriction on the parallel input resolution to execution in a single environment.

Development and Evaluation of a Prototype KL1 Language System

Implementation on a General-Purpose Machine A basic study was conducted on a GHC-based parallel programming environment to give it wider acceptance*. The results of the study were used to develop a GHC system

---

* GHC is also regarded as the effective way to merge message passing—a standard parallel—programming construct with the declarative style, which makes more portable and easier to debug, and deductive capabilities of logic programming.

on a general-purpose machine that runs programs efficiently. The system is based on FGHC (flat GHC: a subset of GHC that balances the existing compiling technology with the descriptive power of the language). It executes a sequence of guard goals sequentially and a sequence of body goals on a pseudoparallel basis.

Sequential Implementation KL1 consists of four components is hierarchically structured as shown in Figure 18.3: KL1-c, KL1-p, KL1-u, and KL1-b. Two software modules had already been developed for the sequential processing system: a compiler that converts KL1-u and KL1-c into KL1-b, and an abstract machine emulator that directly executes KL1-b. The emulator is implemented in ESP on the PSI. The system was further enhanced by the addition of the following debugging support tools: a style checker that uses the grammar and characteristics of the language to perform static analysis as debugging prior to program execution, a tracer with which to debug a program while running it, and an algorithmic debugger with which to debug a program based on its specifications after execution. These additional functions proved effective in parallel programs where debugging is far more difficult than in conventional sequential programs.

Pragma Evaluation System KL1-c forms the core of KL1. KL1-p is the pragma language that indicates goal distribution. These two components work together to extract the distribution processing capability inherent in a parallel machine. A pragma evaluation system was developed as an experimental environment (support tool) to advance the tentative specifications of the pragma to practical specifications applicable to real-world problems. The system is a support tool to examine the pragma specifications, and directly compiles a KL1-c program with the pragma into PROLOG, a sequential logic language.

Distributed Implementation A distributed implementation of KL1 was developed on the multi-PSI according to the basic design of the components of KL1 (tentative specifications were used for KL1-p). The specifications of KL1-u in the distributed implementation are based on the KL1-u specifications on the PSI established, and are added with a pragma that allows programs to specify load distribution among processing elements. For KL1-b, the abstract machine language of the system, the syntax of body goals was expanded by introducing the pragma, to permit, for example, display of the PEs that should execute goals and the priority of goal execution.

Execution Environment of Programs and Enhancement of Program Transformation Technique A GHC execution environment program has a hierarchical structure in which an abstract machine layer is sandwiched between the hardware layer and the user layer. The abstract machine layer contains differently shaped multiple abstract machines. The execution control program for user processes is installed on an abstract machine. The user selects the machine whose shape is most suitable for solving a goal and executes the goal on that machine.

The GHC execution environment program consists of an abstract machine unit, an execution control program unit, and a supervisor unit to coordinate the two units.

The supervisor/abstract machine layer has a supervisor and differently shaped multiple abstract machines. The execution control program layer is an execution control program to execute user processes in the user layer on abstract machines. It consists of a user interface, a shell, a server for user processes, and a device server. If the GHC execution environment program, the PSI was used as the hardware, and the pseudomulti-PSI, a simulator for the multi-PSI, was used as the distributed processing system.

KL2 Research into this subject was continued to investigate kernel language version 2 (KL2) basically and accumulate experience at an application level. The results are given here.

KL2 is an extension of KL1 and will also have a hierarchical structure ranging from a user language to a core language including a pragma, or to a base language. KL2 is being studied in the following areas to improve programmability and implement functions not offered by KL1.

1. Knowledge representation,
2. Parallel problem solving,
3. Support for increased software productivity,
4. Description of real-time systems,
5. Integration with knowledge bases.

These areas can be positioned in the structure shown in Figure 18.11.

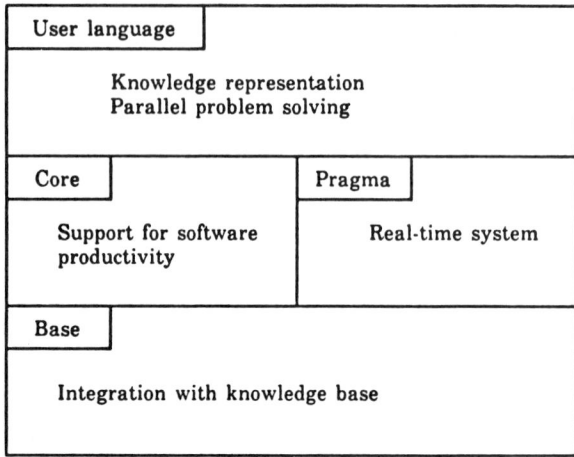

Figure 18.11 *Various aspects of KL2.*

The core language should be essentially an extension of GHC. In GHC, however, all programs have a guard, making program transformation (unfolding) more difficult. Improvement on GHC and design of a new language are under investigation for easier program transformation and partial evaluation.

The pragma language does not describe logical execution of programs. It mainly describes utilization and control of program execution. Three areas must be investigated for the pragma language: implementability of high-degree parallelism, control on search, and unified handling of meta.

18.4.2 Problem-Solving and Inference Software Module

Major research subjects in this area are parallel inference software, basic metainference software, basic coordinative problem-solving software, and demonstration software.

Parallel Inference Software An experiment on methods to describe various parallel applications and parallel algorithms was conducted for evaluation. It had been planned that the experiment would feed the results back to give the execution environment and specifications of KL1. A detailed design was developed for PIMOS, an operating system for parallel inference machines. PIMOS is designed as parallel inference control software to provide users with an environment of parallel languages. Application-based evaluation of the parallel inference software was conducted in two areas: (1) languages to describe problems concerning a parallel operation system and methods to compile the languages into KL1 (FGHC), and (2) applications of KL1 such as LSI layout design and syntactic analysis for natural language processing.

Parallel Inference Control Software As with conventional sequential computers, parallel inference machines cannot control program execution smoothly without an operating system, the basic software to manage program execution.

PIMOS is designed as the operating system for the multi-PSI version 2, intermediate-stage PIM consisting of 100 PIMs, and final-stage PIM, or fifth-generation machine, consisting of 1000 PIMs. A revised version of the original PIMOS will be installed in the final-stage PIM.

Application-Based Evaluation of Parallel Inference It is almost confirmed that the committed choice type parallel logic language, the base for KL1, has sufficient descriptive power for clear algorithms. PROLOG is easy to use for general problem solving. It is not clear that the committed choice type parallel logic language can be characterized likewise. The application capability of the language was evaluated, aiming at solving problems in a parallel operation system where multiple components run in parallel. Problems on problem solving for a parallel operation system were defined, the committed choice type parallel logic language was shown to be too primitive as a language to represent problems, and an approach based on a problem-oriented language was proposed. The approach consists of a language simply representing problems and an inference mechanism including search and simulation based on the representation. The language contains both AND and OR parallelism. The parallel operation system description language, ANDOR-II, is a parallel logic language with an all-solution search function. The function is implemented by adding OR parallelism to FGHC, which has AND parallelism. A given problem can be described declaratively and easily using AND parallelism to represent parallel events and OR parallelism to represent objects with multiple possibilities (possible world).

Programming of Search Problems An effective search technique must be established for problems in which the search space is very large and the object to be searched for has a complex structure.

Large-scale integration layout is one such problem. The search space is very large because a vast number of components must be laid out, and problem solving is complicated because the layout problem is on the border between logic design and device design and is therefore restricted by both areas. This makes it necessary to search for solutions quickly and intelligently. LSI layout differs considerably between design methods (e.g., custom LSI, silicon compiler, standard cell, gate array). This research focused on the standard cell LSI, because its design is automated to some degree by existing computer-aided design (CAD) systems and can therefore be evaluated relatively easily, and because fairly large scale LSI can be designed.

The results of the investigation described suggest that interactive search in which a designer and a system cooperate to solve problems is possible for the building block layout, whereas a parallel search for increased speed is effective in cell layout. The CIL (complex indeterminate as language element), which can describe constraints positively, and IQ (incremental query system), which provides an interactive execution environment, have a function to execute a sequence of PROLOG goals step by step and partially change goal sequences. CIL and IQ were used to develop a prototype building block layout system, which was then evaluated. A prototype cell layout system for evaluation using GHC and capable of describing parallelism positively was also built.

Parallel Syntactic Analysis A parallel syntactic analysis technique suitable for implementation in parallel logic languages was proposed. The technique exploits the features of committed-choice language, a parallel logic language that describes given natural-language grammar categories (nonterminal symbols) as processes able to run in parallel. The technique can handle in a unified manner all of the logic-language-based grammar description formats proposed so far, such as definite clause grammars (DCG), extraposition grammars (XG), and gapping grammars (GG). It also works as an extremely efficient parsing algorithm for these logic grammars even in sequential execution environments, because it does not repeat the same processing during analysis and a program is completely compiled into a logic language.

The parallel syntactic analysis method for logic-language-based grammar description languages is an extension of the parallel parsing technique for DCG, which was designed to basically handle the structure of context-free grammar (CFG). The structure of the grammar rule to be processed was extended so that it could also handle context-sensitive grammar (CSG) and O-type grammar with a higher-level descriptive power and XG and GG, which can describe a relation between nonadjacent elements. A method to convert a grammar written in these grammar description formats into committed choice parallel languages such as GHC, PROLOG, and Concurrent PROLOG was given. In the converted program, all grammar components, such as words and grammar categories, are expressed as predicates of a logic language, each moving as a process able to run in parallel. Each process corresponds to a partial tree of a parsing tree, making it unnecessary to store an interim result during analysis as a side effect. These features permit parallel syntactic analysis to be used as an extremely efficient parsing system even in sequential execution in PROLOG. The technique enhanced seems to work as an effective parsing system for GG and other grammar description formats for which efficient algorithms have not been found. The algorithm of the technique is described in the PROLOG-based metaprogram; however, it can also be easily implemented in parallel logic languages such as GHC. Parallel parsing seems effective for various grammar description formats, particularly GG, because even no efficient sequential algorithms have been found for the syntactic analysis of GG. It also provides a practical parsing algorithm for CSG, because it ensures termination unless cyclic rules are included.

Basic Metainference Software Research in this area seeks to implement the sophisticated inference ability possessed by human beings in software and to develop an environment and tools to support the implementation.

Automated Partial Evaluation of PROLOG Programs The partial evaluation of PROLOG programs serves as an effective optimization technique in metaprogramming but it suffers from poor execution efficiency. PEVAL, a goal-directed partial evaluation program, was implemented to increase the efficiency of the partial evaluation. The goal for the present was to implement all layers of partial evaluation programs (Peval, Peval', Peval'') with the same basic material. As the first step toward this goal, the aim was to develop an automated powerful partial evaluation program in PROLOG. The most crucial problem in this research is to what extent recursively defined predicates must be expanded. An automated program was successfully developed on a multi-phase-structure basis. This was accomplished by statically analyzing the partial evaluation program to introduce conditions that prevent recursively defined predicates from being expanded infinitely in the program.

Proof Support System The first step to construct a logic model is to establish a logic system suitable for the characteristics of the domain involved. Then a model for the domain is developed by establishing an axiom system in the logic system. Finally, the characteristics of the defined model are verified through the configuration of proof within the model. The investigation was conducted extensively from various aspects, such as interface and description language, by introducing actual and virtual keyboards and a logic expression input editor, a structure editor corresponding to the tree structure of a logic expression.

Computer Algebra System Research into this subject was conducted using the algorithm approach. The survey and investigation were carried out on algorithms for computation of polynomial greatest common divisors (GCDs) and factorization, canonical simplification with respect to algebraic relations, and integration in finite terms, and prototypes were developed for arbitrary precision arithmetic to be used as the basic parts of the computer algebra system. A further survey was conducted on factorization and integration. The polynomial factorization and computation of GCDs were studied. A new concept called p-factor and p-decomposition was introduced for the factorization of univariate polynomials over finite fields. The concept of separability was also introduced with respect to the roots of a Zassenhaus' minimal polynomial to distinguish the irreducibility of a factor obtained from the roots. A theoretical study was also made of the lattice algorithm, a factoring algorithm with polynomial time complexity, and it was demonstrated that lattice algorithms can generally be applied to the computation of GCDs and factors of elements in a Euclidean ring that satisfy given conditions. (In lattice algorithms, computation can be performed in polynomial time with respect to the degree of the polynomial to be factored.)

Particularly, for univariate polynomials over integers, experimental factoring algorithms are programmed according to the foregoing results. In R&D on the computer algebra system, survey and research were conducted to achieve a system capable of advance integration of elementary functions in finite terms with system implementation. Programs required as basic parts are being prototyped one by one from the bottom.

Basic Coordinative Problem-Solving Software Basic coordinative problem-solving software is an important application of parallel problem solving. A language useful for describing coordinative problems was designed, and R&D was continued on programming methods for coordinative problem solving and applications using parallel kernel language KL1. Mandala, an object-oriented knowledge representation language, had been developed as a descriptive language for coordinative problem solving, but the object-oriented part of Mandala had not been completely implemented in KL1.

Afterwards, an object-oriented language implemented in KL1 alone was proposed and a language was designed that can execute procedures in an object in parallel, with a few restrictions as possible imposed on its parallelism.

A software reutilization system for parallel coordinative problem solving has been under study. Program reutilization is a program development method that saves existing software as a part so that it can be reused in developing similar software. Mandala has been selected as the language to develop a prototype system.

Research into a KL1-based application system in the logic device CAD field aimed to define a procedure for implementing the application system in a parallel programming environment through program development. Investigation was conducted on specification representations suitable for the style of KL1-based programming, and the results were used to develop a prototype system.

Demonstration Software Research in demonstration software aims to apply the results of R&D on problem-solving inference to real-world applications for general evaluation and feed the evaluation results back to each research subject.

An intelligent electronic device computer-aided engineering (CAE) system and a computer Go and Shogi system (CGS) were selected as actual applications for research.

The intelligent electronic device CAE system was chosen to study the requirements of a true CAE system for electronic devices. R&D on this subject is being conducted to clarify methods for problem-solving inference, knowledge base development, and knowledge base utilization applicable to troubleshooting for common carrier systems and layout design of analog LSI as sample problems.

A simplified CGS prototype system was developed for determining moves. Various functions and tools used in the system were also developed. The results almost achieved one of the goals of this theme: clarifying and implementing the knowledge required to determine moves in Go and a technique called *sakiyomi* (guessing moves ahead). The system is satisfactory as a total system, because it can determine moves from the start to the end.

18.4.3 Knowledge Base Management Software Module

Knowledge Representation Utilization System Research in the utilization system aims to investigate knowledge representation and inference engines for representing a dynamic system and conducting qualitative reasoning on the system. The investigation will be necessary for evaluating the knowledge representation capability of Mandala and the knowledge base management system. The investigations are outlined here. Functional design of the knowledge base management system in Mandala is discussed from the standpoint of functional design in the section on the cooperative problem-solving system and other related sections.

The first step of the research was investigating a reasoning system to make a computer understand the world of physics. With knowledge similar to laws in physics, the system will infer temporal transition of the state of a physical system from the given initial information. It was assumed that the knowledge is based on physical laws in textbooks of physics. The qualitative physical reasoning system (Qupras) uses physical laws to infer

1. Relation between the objects forming a physical system;
2. Temporal transition of the state of the system—that is, what will happen to the entire system at the next moment.

Qupras infers the relation between objects and transition of a state while performing qualitative reasoning using knowledge of physical laws and objects.

Basic Knowledge Acquisitions System The basic knowledge acquisition system has been studied as a basic software technology for knowledge base management. The results of the activities in this area are given next.

Hypothesis Generation and Verification System Conventional deduction-based reasoning systems start with existing general knowledge and proceed to specific knowledge. Therefore, these systems cannot draw an effective inference for facts not included in the general knowledge. This problem can be solved with a system capable of performing nondeductive inference, which requires a sort of logical leap such as induction, analogy, or common reasoning. An inference result is essentially a hypothesis in such a system; the system can generate hypotheses. Investigation on such inference has just been started. In the investigations, hypotheses were given as a set of possible hypotheses in a system; the system did not actually reach a solution for generating hypotheses.

A theoretical study was conducted to gain perspective on the implementation of systems with such inference capability. The results of the study were used to propose a new theory called *Ascription,* a general basic theory on inference involving logical leaps.

Analogy System Implemented in Metaprogram The general meaning of analogy is to find a similarity between two or more events and draw some inference based on the similarity. This discussion assumes that analogy is an inference category for obtaining unknown facts which cannot be obtained by conventional deduction.

Inference rules were implemented in metaprograms using logic programs and incomplete logic data bases as examples. An investigation was conducted on ways to increase the efficiency of similarity-based prediction of facts by performing partial evaluation on a given similarity. Future research subjects include the establishment of techniques to discover a similarity between attributes and the interrelationship between similarities.

Increased Efficiency of Logic Program by Program Transformation
Programs declaratively written are easy to understand. In general, however, they run inefficiently under the standard depth-first search strategy for PROLOG in which goals are executed from left to right. Research into this subject aims to improve the execution efficiency of declaratively written easy-to-understand programs by transforming and synthesizing them into equivalent but more efficient programs.

A program transformation technique for typical programs in the generate-and-test type was studied. This study has some features not found in research based on conventional program transformation.

Knowledge Base Management by Parallel Logic Language To endow computers with problem-solving capabilities possessed by professionals, knowledge used in solving problems must be stored and managed in data bases. A knowledge base management system provides functions to represent experts' knowledge in a format that can be understood by computers, and acquire knowledge and save it in data bases systematically. A knowledge data base was devised to accumulate and manage knowledge expressed in a set of Horn clauses and ways were studied to implement the knowledge data base in GHC. The possibility of constructing a knowledge base management system that will process rules and facts was checked, and it was confirmed that data bases consisting of rules and facts, system control programs, and knowledge base management programs can be handled with a single framework of logic programming. Three methods for parallel control over GHC-based relational operation processes were also investigated. The investigation showed that parallelism differs between the types of commands when the processing granularity is made smaller by dividing data.

The problems had logically good characteristics. Actual knowledge base management systems involve a number of more complicated problems.

Learning Function in Algebraic Manipulation Mathematicians solve problems using expert knowledge in mathematics. They choose an appropriate method that seems to help solve problems and sometimes try to reach a solution on a trial-and-error basis. The expert knowledge they use includes object knowledge, which directly transforms a problem, and metaknowledge, which represents know-how to select object knowledge. Research into this subject aims to study the learning process of mathematicians by focusing on learning from examples.

Learning can be roughly divided into two types: one obtains knowledge on a domain (in this example, the domain of mathematics), the other obtains knowledge on how to learn. The following three levels of learning in the domain of algebraic manipulation were investigated and part of the detailed design developed.

- *Rule Learning:* This type of learning in the algebraic manipulation draws general terms from a series (mathematical progression). It corresponds to a problem in conventional inductive inference.
- *Concept Learning:* This type classifies rules in the algebraic manipulation.
- *Strategy Learning:* This type learns heuristics to apply rules in the algebraic manipulation.

Basic Distributed Knowledge Base Management Software Research into this theme aims to provide a dictionary data base for natural language processing application systems and a proof data base function for theorem-proving systems. Two subsystems are currently under study and development: the knowledge base management subsystem and the knowledge base manipulation subsystem.

Distributed Knowledge Base Management Subsystem (Kappa) The distributed knowledge base management subsystem Kappa uses the non-first normal form (NF^2) model instead of the relational model as the model of the lowest layer to efficiently handle knowledge with a complicated structure. The NF^2 model can be considered an extension of the relational model. Models for knowledge representation capable of handling terms and taxonomy are positioned. A prototype system was partly developed to confirm whether Kappa can be implemented.

Basic Distributed Knowledge Base Manipulation Software (SIGMA) SIGMA is a knowledge base management and manipulation system built on Kappa. It is designed as a support tool for expert system development. SIGMA will use KSI, a knowledge representation language based on the concept of sets, to represent knowledge and to manage knowledge bases, perform inference, explain an inference process, and keep knowledge bases consistent. A prototype of SIGMA was partly developed to confirm the feasibility of the software. Investigation was also conducted on the interconnection between SIGMA and Kappa which will be implemented.

Demonstration Software Work on this subject aims to check the usefulness of expert systems using knowledge bases and knowledge base management technology by applying them to real-world applications. An expert system for the logic design of computer hardware was chosen as an application. A detailed design was produced for the expert system, and a prototype system was partly developed.

18.4.4 Intelligent Interface Software Module

Pilot Model for Discourse Understanding System A basic study of the pilot model was made and its prototype partly developed. The results of these activities were evaluated, and detailed theoretical and technological studies were conducted on anaphora processing, a crucial technology for context processing.

The discourse understanding system DUALS (discourse understanding aimed at logic-based systems) is an experimental system to build a logic-programming-based computational model for discourse understanding. To construct such a model, it is necessary to review problems, such as identification of ellipsis and anaphora, speech actions, and planning, from a computational standpoint, and to formulate inference. DUALS version 1 is the first attempt to implement the situation semantics theory as a foundation for discourse understanding models. It has been developed in PROLOG. PROLOG's term description alone, however, is not always sufficient to express complicated structures such as semantic structures of sentences, making it difficult to gain perspective on the entire system. This problem is also a stumbling block in problem domains, such as context understanding, where a number of unknown factors are involved.

DUALS version 2 was developed based on this experience. Design of the second version aimed to further clarify the structure of the system and provide a basis for future experiments on various discourse understanding. CIL was used to provide a unified description of the entire system. CIL's functions to represent and unify partial terms were used to clarify the system structure effectively. The latest work in situation semantics was also introduced to represent semantic structures of sentences and expand the framework of situation semantics.

DUALS version 2 consists of the six modules shown in Figure 18.12; text analysis, object identification, discourse analysis, problem solving, text generation, and I/O routine.

The text analysis unit performs morphological, syntactic, and semantic analyses on a sentence made up of kanji (Chinese characters) and kana (Japanese syllabary), and generates an intermediate representation (event type) of the sentence. The unit uses a variant of Watanabe grammar and implements rapid parsing using the SAX analysis algorithm. The object identification unit matches objects (generally noun phrases) which appear in the intermediate representation of a sentence to objects in the real world. It also identifies anaphora and ellipsis. The discourse analysis unit interprets speech actions and stress discourse structures. The current unit is very primitive; it only determines whether the input sentence is a query, and stores the description situation of each sentence in the list structure. The

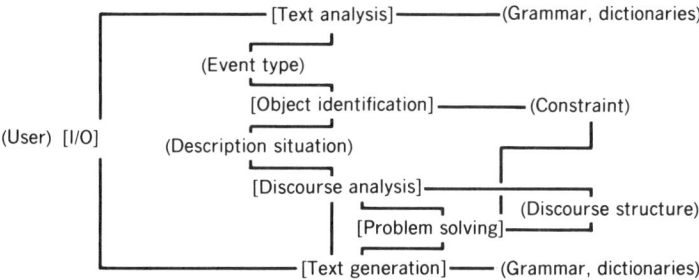

Figure 18.12 *Structure of DUALS version 2.*

problem-solving unit obtains a solution to the query by performing a search or inference. The text generation unit generates a Japanese sentence from the intermediate representation (descriptive situation) of the input sentence. When the sentence to be generated is a solution to a question issued by the user, the sentence generation unit first produces the intermediate representation of the output sentence from the intermediate representation of the question and the solution from the problem-solving unit, then generates a surface sentence. The I/O routine provides a user interface. It uses the Pmacs editor on SIMPOS to permit user-friendly data entry. DUALS version 2 is written in CIL and ESP, and installed on the PSI. As in the first version, sample sentences were extracted from tests designed to measure the reading skill of Japanese students in elementary schools.

The results of the design and installation of DUALS version 2 showed that the semantic representation based on situation semantics and the processing modules will provide a basic system in discourse-understanding experiments in the future.

The model proposed by Barwise was introduced as the framework for identification on anaphora relations in the anaphora treatment. In this model, indexing, which represents anaphora relations, is performed on a subset of English, and a model-theory-like interpretation is given by considering it a logical expression. Unlike interpretations in ordinary predicate logic, this interpretation is regarded as procedural, and assignment of a value to the logical expression is considered a partial function. Therefore, this framework seems suitable for implementation in computer programs; especially, it appears to provide excellent compatibility with CIL, which can represent a partial value assignment as an associative list. It was shown that Barwise's model based on situation semantics can be easily installed in CIL. An example of detailed design of the model was demonstrated.

The model, of course, does not provide a framework capable of treating all anaphora phenomena. Particularly, the interpretation rule shown is directly established based on the syntactic structure of English (L(SS)). This seems quite different from the framework of Kamp, which is very closely related to the model. That is, intermediate structures such as Kamp's discourse representation structure (DRS) are removed from Barwise's model. Neither framework, however, can be used directly as a model for Japanese, because its syntactic structure differs from that of English. Still, the idea of partial assignment and dynamic interpretation shown in Barwise's model provides a basic approach to anaphora treatment in Japanese.

Basic Sentence Analysis and Synthesis Software Research in this area seeks to develop Japanese phrase structure grammar (JPSG), a grammar system to describe basic Japanese structure and its language processor.

A basic framework to describe grammar was designed and used to describe one of the most basic structures. The framework was enhanced by expanding the area of linguistic phenomena involved; new features and principles were introduced according to the expansion. It was demonstrated that the framework can also describe and formulate Japanese morphology. The sentence-level concept was introduced, and the priority of the elements forming a sentence, such as tense, aspect, and modality, was established. A study was performed on whether the structure of a sentence consisting of these elements could be governed by any rule. The same investigation was performed on complicated sentences involving these elements and conjunctive postpositional particles. The resulting framework was then used to experimentally describe sentences extracted from Japanese readers for sixth-grade students in elementary schools.

A prototype parser based on the idea of JPSG has been under development to demonstrate the feasibility of applying JPSG to language understanding. An attempt was made to install some of the functions of JPSG for evaluation using two schemes: conditional unification and CIL-based GALOP. A grammar description language was also designed to provide an interface between linguists and the parser, and experimentally used. The results of the evaluation are summarized here.

The conditional unification technique allows most of the grammar descriptions of JPSG to be converted easily into programs. This facilitates the development of an analysis system to verify JPSG, a goal set up for this research subject. The technique does not depend on an analysis algorithm.

The processing efficiency of the parser was not very high, because the conditional unification was implemented as PROLOG programs. The parser implemented in C-PROLOG on the VAX 11/8600 took about 20 s to analyze the sentence, "Ken was made to read a book." The processing speed is relatively fast, considering that the conditional unification simultaneously searches all possible solutions.

18.4.5 Intelligent Programming Software Modules

Under the theme of intelligent programming software modules, research and development was conducted on the experimental program transformation, proof, and synthesis system, software knowledge management system, and experimental program transformation, analysis, and verification system. Demonstration software was developed for their systems.

Experimental Program Transformation, Proof, and Synthesis System

Theorem-Proving System The theorem-proving system must have functions to process a huge amount of mathematical knowledge. The system must also have utility functions for proving theorems such as proof editing, two-dimensional I/O, and formula manipulation functions.

This system is used primarily as a notebook by mathematicians. It may also be used as a computer-aided instruction (CAI) system for mathematics education, an expert system, or a problem-solving and inference system. It will be particularly useful for programming support; thus, this application of the system will be an important theme in the future.

When development of the theorem-proving system started, the following areas were chosen to be covered by the system:

- Linear algebra (LA),
- QJ,
- Synthetic differential geometry (SDG).

A prototype of the system, named the CAP-LA system, was developed using the DEC10-PROLOG machine, and its first version was developed. Based on these experiences, the second version of the CAP-LA system was completed.

The CAP-LA system deals with linear algebraic theories appearing in textbooks for first-year university students. Tsuchiya's textbook was used to design and evaluate the proof description language (PDL) and system. As a checker of proofs written in PDL, the CAP-LA system supports evolution of linear algebraic theories. Development of the second version of the CAP-LA system emphasized improvement of the following features:

- Proof speed,
- Interline interpolation function,
- Verification process display functions for system developers and experts,
- Structure editing function and syntax guidance function,
- Representation of proofs in a natural language.

The objectives were achieved satisfactorily.

Term Rewriting System The heaviest mathematical work load on researchers is formula manipulation. Generally, the work requires complicated and profound inference concerning equal signs.

Therefore, the term rewriting system is an important subsystem for the theorem proving system. Having both proving and computing functions, this system can be used alone or can support many other systems.

As a theorem-proving subsystem, the term rewriting system is typically applied to equational logic. It is also useful for proving theorems based on predicate logic. The latter is a typical application of a SDG theorem-proving system. Other applications of the system may be program synthesis and verification.

When its computing functions are brought to the fore, the term rewriting system is a programming language and its executor. Although term rewriting is regarded as a functional programming technique, the system can also handle logic. Thus, it is a candidate logical language making up for its weak points in representing mathematical functions. The kernel technique for realizing a logical as well as a functional programming language may be term rewriting.

An experimental system called Metis was developed. It has the following functions based on the term rewriting technique:

- Determining directions of equations and converting equations to rewriting rules;
- Removing ambiguity from rewriting;
- Executing rewriting.

Moreover, Metis was improved as described here. The biggest problem for Metis is processing equations such as the commutative law because, once this law is converted to a rewriting rule, the possibility of terminating the rewriting cannot be guaranteed. To solve this problem, an AC-unification algorithm and an AC-completion procedure into which associative and commutative laws had been incorporated were first implemented. An unfailing completion procedure that does not necessarily convert equations to rewriting rules was then implemented.

Another problem is how to treat not-equal signs (i.e., unequal relationships). However, this was completely solved by implementing Jieh Hsiang's S-strategy.

Program Construction System (PCS) There is a lot of similarity between proving mathematical theorems and computer programming. This fact is recognized from mathematical and other perspectives. The realizability interpretation given by Kleene in his theory of constructive mathematics is a principle of extracting programs from proofs. The concept of programming constructive proving was researched. Since a number of systems based on this concept had already been reported, their features were investigated.

Sample programs were then studied to seek techniques for developing such a system. For this work, Satoh's QJ was used as the logic system and

Quty as the programming language. Partly modified PDL was used as a specification and proof description language in CAP-LA, and its affinity with the theorem-proving system was checked.

Figure 18.13 shows the image of PCS. The proof editor and system manager, proof parser, and proof verifier have the same functions as the CAP system. The module knowledge base corresponds to the proof data base in the CAP system and is used to store and retrieve created modules. The proof compiler is the module extracting programs by analyzing verified proofs.

Program Design, Development, and Maintenance System The R&D results of software design support system using a knowledge base are outlined here. Fundamental measures covering the whole software life cycle are now needed to improve software productivity and maintainability. Reusing software resources and accumulating and transferring software development techniques are becoming particularly important to cope with the shortage of software engineers. The focus is on developing a system with a data base, supporting software transfer techniques and analysis/verification of designs. The techniques to be transferred are closely associated with knowledge information processing techniques such as knowledge representation and application techniques.

Models of design activities were investigated to clarify the role of a knowledge base in designing software. The concept of a three-source software design model was then implemented. In this concept, a software design model is a set of interactions between three systems: the processing system (designers); the resource system (know-how, design information, and program parts); and the system being processed (specifications and programs to be designed). According to the concept, a software design activity is defined as "an activity of the processing system to obtain necessary information from the resource system and use it to convert the processed system from specifications to refined specifications to a program."

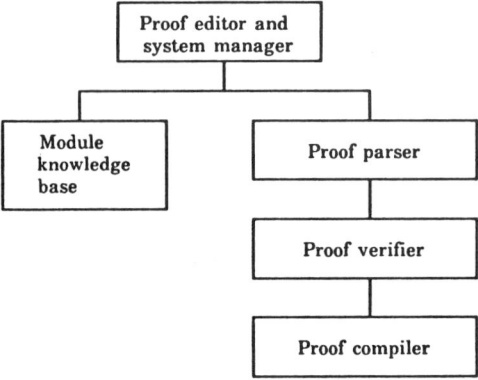

Figure 18.13 *Outline of PCS.*

Software Knowledge Management System The R&D goal of SIMPOS is to provide a better environment for logic programming. The development placed importance on the following items:

- Unifying interfaces between subsystems,
- Management of multiple PSI machines connected by a network,
- Flexible handling of abnormal conditions (errors).

The internal algorithms and execution efficiency of modules were also improved. The logic language, ESP, having object-oriented functions, was used to describe the system. As a result, the following objectives were achieved:

- Compatibility between the system and application programs through the logic language;
- Accessibility to the system through an object-oriented paradigm.

Demonstration Software

Specification Design Expert System This system solves basic problems in automatic software design. As an example of telephone exchange software with parallel job processing, high reliability, and real-time processing, an expert system for specifications representing user requirements in the initial stage of software development is to be developed.

Japanese-Language Specification Description System This system supports the description of software and software documents. The system processes software specifications to help system programmers maintain and manage software easily. It also facilitates the creation of manuals from basic specifications.

Recently, enlargement of software, including an operating system, has given rise to maintenance and management problems. To solve these problems, a system was designed, consisting of a software describer for software and its documents and an editor managing the descriptions.

Knuth's WEB is a well-known system describing both programs and their specifications. The system creates easy-to-read specification sheets from descriptions of specifications and programs, edits descriptions of programs into programs, and creates an input file for the compiler.

In the ICOT system, a more advanced method than that in WEB is used to refine specifications step by step and systematically. A concept of restricted Japanese language has also been introduced into the system to remove ambiguity from descriptions in Japanese. The system describes software and software documents in almost the same format as TELL. Created programs are expressed in the object-oriented logic programming language ESP.

Experimental Image-Based Software Production System In pursuit of an intelligent software specification creation system, on-line banking software was studied. A specification description support system with functions to understand application-dependent nomenclature and programming knowledge, and a specification design support system featuring functions to use knowledge about the application were designed and partly developed.

18.5 DEVELOPMENT SUPPORT SYSTEM

18.5.1 Parallel Software Development Machine

Software The software system consists of the connection control software, which runs on the parallel software development machine, multi-PSI, and part of the SIMPOS programming and operating system installed on the PSI, the processing element of the multi-PSI.

The connection control software links and coordinates multiple PSIs. It provides a function to control a dedicated network and a user-interface function to enable a host PSI to control execution smoothly on the networked PSIs.

SIMPOS running on the PSI is vital for smooth operation of the multi-PSI, system. The PSI, serving as the PE of the multi-PSI, is also designed to work as a standalone sequential inference machine. Therefore, the PSI offers a user-friendly interface for the multi-PSI or true parallel inference machine systems in the future, because it can be used as a front-end processor or a workbench to provide a cross compiler for program development. Basic development and improvement were completed for SIMPOS. The subsystems of SIMPOS were improved and enhanced so that the PSI could be used more easily as the processing element or front-end processor of the multi-PSI. The PSIs have become widely used in many kinds of research areas with the improvement of available software tools.

Hardware The multi-PSI system is being developed in two stages; the first version and the second version (Figs. 18.14 and 18.15). The first version has a structure that can be completed in a short time to enable research on parallel software to be conducted as soon as possible. It can act as an experimental system for the design of a larger, more advanced version of multi-PSI, multi-PSI version 2. The first version is a small-scale multiprocessor system made up of six to eight PSIs linked via the connection network hardware developed.

In the development of the multi-PSI version 2, the PSI itself is being improved and reduced in size, because the second version should have more CPUs and higher performance. The connection network is also being refined to link up to 64 PSIs. The entire part of the KL1 processor will be implemented in firmware to achieve high performance. An improved, smaller

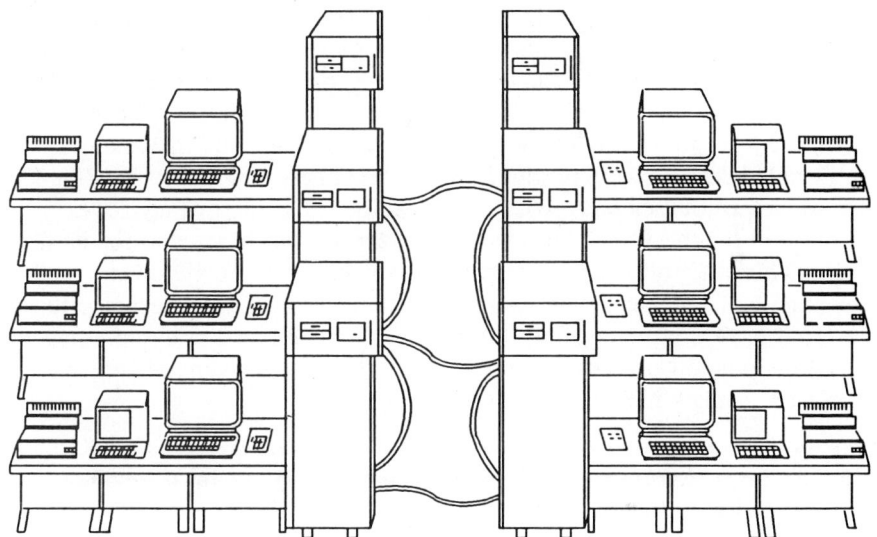

Figure 18.14 *Configuration of multi-PSI version 1 (six PEs).*

processing element is also being configured as a separate standalone system to be used as the front-end processor for the multi-PSI version 2 or as a simulator. This system, a smaller version of the PSI, is called the PSI-II. A prototype of the multi-PSI version 1 was developed for evaluation. The connection network hardware and processing element hardware of the multi-PSI version 2 were designed in detail, and a prototype PE was developed. The PE hardware of the multi-PSI version 2 was used to conduct the partial prototyping of the PSI-II.

The R&D results of the hardware system for the parallel software development machine are outlined here separately for the connection network hardware and PE. Various items disclosed in the evaluation and the results of the detailed investigation will be used to develop prototypes of the multi-

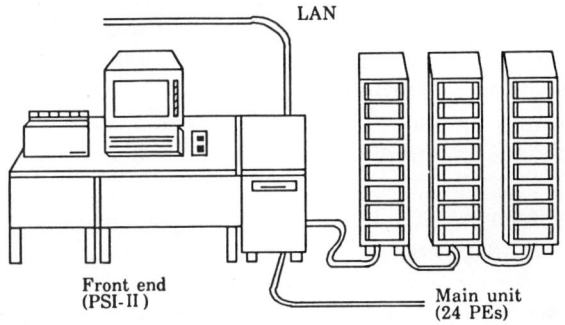

Figure 18.15 *Configuration of multi-PSI version 2 (24 PEs).*

PSI version 2 in the future. In this development, prototypes will be characterized more as small-scale parallel inference machines than as parallel software development machines. This shift in approach is expected to facilitate R&D of parallel software considerably, including parallel inference control software PIMOS, and R&D of the intermediate-stage and final-stage PIMs.

Detailed designs were produced for the hardware and firmware of the PE, and their prototypes were partially developed. Detailed designs were also developed for the PSI-II hardware using the PE hardware and for its firmware. This was followed by partial prototyping. The PE hardware used as the engine of the parallel software development machine is designed to run fast and have a large-capacity memory. The design also stressed compactness and low power consumption so that a system consisting of 16 to 64 PEs could be configured relatively easily.

18.5.2 Improvement and Enhancement of Sequential Inference Machine (High-Speed Processor Module)

The high-speed processor module is an extension of the sequential inference machine. It is a dedicated processor capable of rapidly processing programs in a PROLOG-based inference language and has a large-capacity memory. Research and development on the module consisted of designing improved and enhanced software and producing functional designs of smaller hardware. Three types of software were developed to make full use of the power of the smaller version of the module, called the smaller CHI (cooperative high performance sequential inference machine): programming system software, basic software for the CHI, and an ESP translator to provide functional compatibility with the basic part of the SIM. A detailed design of the smaller hardware and its prototype were developed. The firmware of the high-speed processor module was also designed in detail and coded to complete the R&D of the module.

The development results are outlined next for the software and hardware systems.

Software System The R&D activities consisted of (1) improvement and enhancement of the programming system software, (2) improvement and enhancement as well as design and prototyping of the basic software for the CHI, (3) basic design and prototyping of the ESP translator, and (4) development of a firmware simulator for the CHI, a development tool for the smaller CHI firmware. These activities provided the expected results.

Hardware System A detailed design and circuit design were produced for the smaller CHI. A prototype was then developed and checked for performance. The firmware of the smaller CHI was also designed in detail and coded. The development activities for the smaller CHI consisted of the following: (1) research, development, and manufacture of two types of

CMOS gate arrays, and (2) research and development, circuit design, and manufacture of processor modules. These were required to replace CML devices used in the CHI developed with CMOS and transistor-transistor logic (TTL) devices, thereby making the hardware more compact. (3) Circuit design and manufacture of the main memory module and memory interface module; these were conducted to introduce higher-density memory devices (1 Mbit) for smaller hardware. (4) Circuit design and manufacture of the host interface module; these were conducted because the I/O control module was changed to make the system smaller.

18.5.3 Improvement and Enhancement of Network System

Development of the network system in the FGCS project is being conducted in two areas: a SIM network system and an international network system.

SIM Network System   The SIM network system is designed to facilitate R&D of the FGCS project. It connects the sequential inference machine PSIs installed at ICOT and its related research institutions via Ethernet-type LAN and DDX. The network functions designed and partially prototyped were improved and enhanced to build a network system. The system is currently in operation, providing communications services, such as electronic mail between 18 sites including ICOT.

The structure and functions of the system will now be outlined.

Network System Structure   Figure 18.16 shows the entire configuration of the SIM network.

LAN Interface Adaptor (LIA). This device performs interface conversion and protocol conversion to connect a host to a LAN.

LAN Interface Board (LIB). This board provides on IEEE488 interface between a host and a LIA. The board is installed in the host.

NCLIA + Network Console. This pair provides the network manager with functions such as management on diagnostic information and network structure information and collection of log information. Network structure and other information is sent to hosts by communications between the LIA and NCLIA.

Bridge (BRG). Unit LANs are connected to each other by the BRG, which allows inter-LIA communication across the different unit LANs.

Gateway (GW). The GW is a public network connection unit that uses the DDX-packet network and allows LIAs linked to a composite network to communicate with LIAs linked to another composite network. A composite

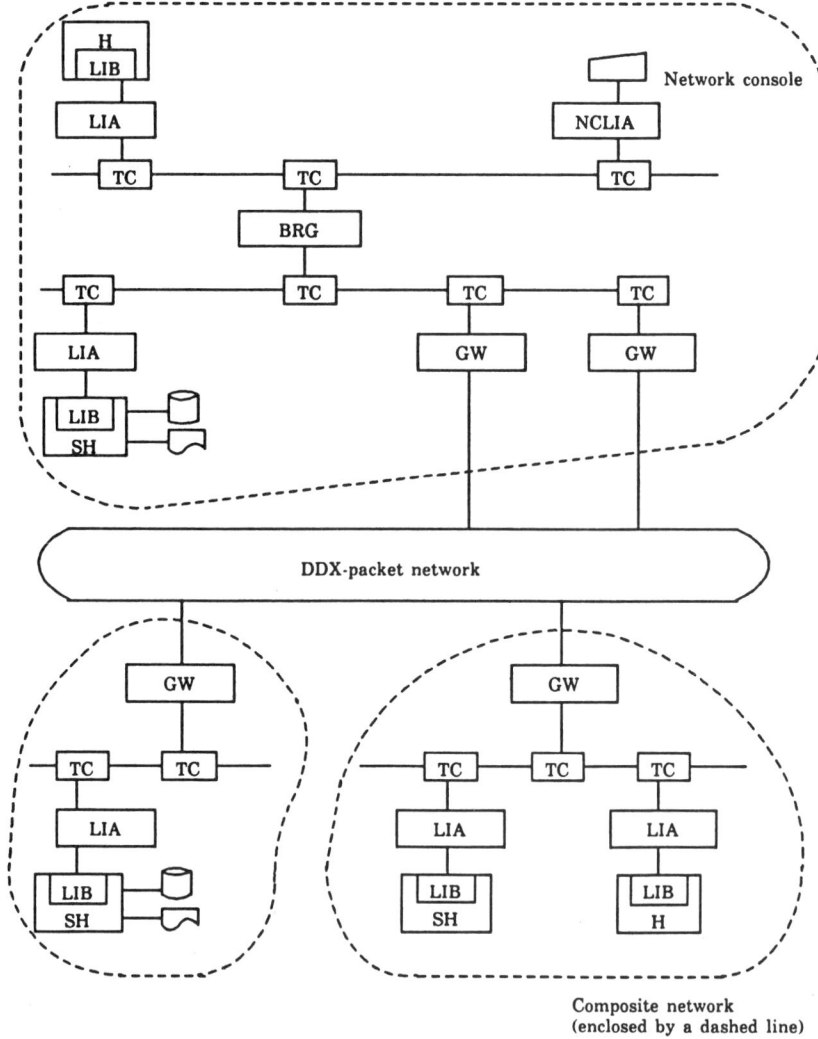

Figure 18.16 *Entire structure of SIM network system.*

network consists of one or more unit LANs interlinked with a BRG. Generally, one site consists of one composite network.

Transceiver (TC). This unit is connected to a coaxial cable, the transmission line of the LAN. It sends and receives signals via the cable, and provides electrical insulation.

H/SH. H represents general host computers, such as the PSI and general-purpose computers, while SH means server hosts. In principle, a server host is used in each composite network and works as a mail server, print server, and file server for general hosts.

Network Functions The network functions of the SIM network system are divided into basic and application functions. Application functions use the basic functions to provide users with various useful services.

Basic Functions. The basic functions provide the network-based interprocess communications function, offered by the LIA, to programs running on the PSI. The functions are supported by the network control program (NCP).

A new NCP was developed based on the logic design produced previously. The new NCP has the following features:

1. Provides a $1:n$ communications function;
2. Provides commitment mode;
3. Provides datagram mode;
4. Provides value-added mode enhanced in function and reliability in $1:1$ communication;
5. Allows a heap and string to be used as a data transfer buffer;
6. Enhances simplified communications interface (n line);
7. Coexistence with user-specific NCP.

Application Functions. The following application functions of the SIM network provide users with a variety of useful services:

1. *Remote object:* This function provides a basic mechanism which allows the user of a PSI to perform object generation or deletion, method calling, and other operations for another PSI. The operations, however, are restricted. It was decided to develop various systems forming the application functions based on remote objects.

2. *File server system:* This system provides a remote file and directory access function and a manipulator function. The remote file and directory access function permits the user program to access files and directories of all PSIs and general-purpose machines connected to the SIM network. The manipulator function uses the remote file and directory access function and the functions of other server systems. It allows the user of a host to copy, print, and perform other operations on files and directories of another host. This system was developed based on remote objects. It is provided to users as part of the file system function.

3. *Mail server system:* This system provides a mail function and an electronic bulletin board function. The mail function permits users to exchange electronic messages. It has various optional functions such as forwarding

mail and producing printout. The bulletin board is something like a blackboard that can be shared by several users. Users can exchange their ideas via the bulletin board. Mail and bulletin boards are stored in the mailbox of the server PSI. An auxiliary server function was added to the system. This function stores mail and other information when the server PSI is disconnected from the network. It allows users to send mail when the server PSI malfunctions or is disconnected from the network.

4. *Print server system:* This system provides a remote print function and a manipulator function. The remote print function allows the user program to use the printers linked to all PSIs on the SIM Network. The manipulator function enables the user at a terminal on the network to cancel remote print, display output conditions, and perform other operations. This system was developed based on remote object. It is offered to the user as part of the print system function.

5. *Talk system:* This system provides a full-duplex, real-time interuser communications function via a bitmap display (BMD). The system has an additional function. When a user wants to communicate with another user on the network, this function lists the names of users currently using the PSIs on the network and allows the user to select the appropriate name. This function has substantially increased the ease of use of the network.

International Network System The computer network plan of the FGCS project has been made around the PSI. At present, communications between PSIs and general-purpose computers are limited; the only service supported is file transfer with VAX/VMS via Ethernet. Both overseas and domestic research institutions have long been requesting electronic-mail-based communications with ICOT over a computer network. The FGCS project therefore earmarked the problem of networking general-purpose machines as an urgent task. The SUN workstation was introduced to link computers at ICOT with other institutions. It has been used as a gateway to connect with computers at universities and other research groups. Connection to overseas sites was considerably enhanced by, for example, becoming a regular member of CSnet. At present, ICOT is officially linked to ARPA Internet via CSnet. A stable connection has also been established via the UUCP network with MIT and the following gateways: UKC in Britain, INRIA in France, and ENEA in Sweden. Connection to other countries will be implemented as the need arises in the future. ICOT is currently connected to domestic research institutions through JUNET and DECnet. Both networks support an increasing number of research groups. JUNET also supports electronic-mail-based communications with countries which have no direct link with ICOT. Development and enhancement of these networks have given ICOT sophisticated electronic mail link comparable to that of leading research institutions anywhere in the world.

Networking still involves a number of technical problems. Services available on the international network are currently restricted to electronic mail.

It is imperative to continue to enhance the network to provide the same level of reliability and service matching as the state-of-the-art ARPA Internet.

18.6 CONCLUDING REMARKS

Current R&D activities at ICOT are also being used to pursue more sophisticated possibilities and to realize them so that the project can move on the final stage smoothly. Some advanced subjects previously untouched will be handled. It was decided to start with these subjects as important issues to be studied.

Under these circumstances, the guidelines for the R&D activities were set as follows:

1. R&D will be conducted to develop basic research from the beginning and to achieve effective integration of individual research areas.
2. The R&D system established on the basis of the component and overall system technologies will be more solidly organized in gradual readiness for research in the final stage.

Since ICOT is involved in the development of state-of-the-art technology, it is considered imperative to exchange information and research results with scientists working in related fields both at home and abroad. A research infrastructure in cooperation with universities and industry, which gives ICOT useful advice and supplements its activities, has been built up for basic research advancement.

Some researchers in advanced research areas have the cross-sectional view that the generation analogy will not hold very well in advanced information processing systems of the 1990s and that, because of the diversified research involved, the research base is seen more as *branching rather than as a movement from one generation to the next*. ICOT always listens to the pros and cons of the matter. From the beginning, ICOT has tried to verify the hypothesis that parallel architecture based on logic programming will be a next generation computer oriented toward advanced information processing in the 1990s. The main focus is to research computer science based on logic programming and to develop symbol crunching super parallel computer and knowledge information processing technology within this framework.

Another national project in Japan, less widely publicized, is reported to be challenging the U.S. lead in numeric supercomputers. As a result, it can also be said that both those qualities offer an excellent foundation on which to raise AI applications. Finally, an observer says, "While initiatives worldwide all have problems, they have also led to progress." He is quite right. We are looking forward to a good solution in the FGCS.

CHAPTER 19

An Advanced Software Paradigm for Intelligent Systems Integration

TAKAO ICHIKO

19.1 AN APPROACH FOR HIGH-QUALITY SOFTWARE

In this chapter, a method for modeling human concept in computer software design is discussed from the viewpoints of highly parallel processing and effective artificial intelligence technologies oriented toward the computer applications designer in hardware and software.

This section is concerned with new software design approach for achieving high-quality technology. It describes a more integrated software design method enabling a designer to more flexibly design and more easily apply application design concepts with high concurrency in components and parallelism in components networks. This compares favorably with design based on conventional mainframe computer under machine constraints. Besides reporting experiments, future design issues and possibilities are also discussed. This design method is investigated by the introduction of a VLSI component (32-bit microprocessor) on the basis of a fundamental paradigm composed of a high-performance module component and high-quality software tool engineering bridging between the two.

This section was based on a new design paradigm for high software quality, differing from the conventional design paradigm. In this section, the

---

Part of this chapter is © 1989 IEEE. Reprinted, with permission, from *Proceedings of* IEEE International Workshop on Tools for Artificial Intelligence (Architectures, Languages, and Algorithms), Hyatt–Dulles Hotel, Fairfax, Virginia, USA, October 23–25, 1989, pp. 709–716: T. Ichiko, K. Takeuchi, N. Nango, "An Entity Model for Conceptual Design"

Neural and Intelligent Systems Integration, By Branko Souček and the IRIS Group.
ISBN 0-471-53676-8   ©1991 John Wiley & Sons, Inc.

differences between the new paradigm and the conventional one are discussed in detail, mainly from the viewpoint of the interface between hardware and software in overall system realization, and some concrete application examples are referred to, demonstrating the enhancement of high-quality software. In particular, this enables an application-specific expert, who is not an expert computer designer, to design a pertinent application computing mechanism easily, according to a specified design objective. Full consideration is taken of the fact that the designer is not required to gain further specialized knowledge of computer hardware and electronics. At the same time, parallel application is possible by high-quality software implementation on functional module components network by using a high-level language (HLL). This means that by using an open-ended computing network, the designer can be released from conventional machine constraints.

19.1.1 The Conventional Software Design Process

Conventional software design, which has now spread over the world on the general-purpose mainframe computing system with the operating system (OS), is briefly depicted in Figure 19.1.

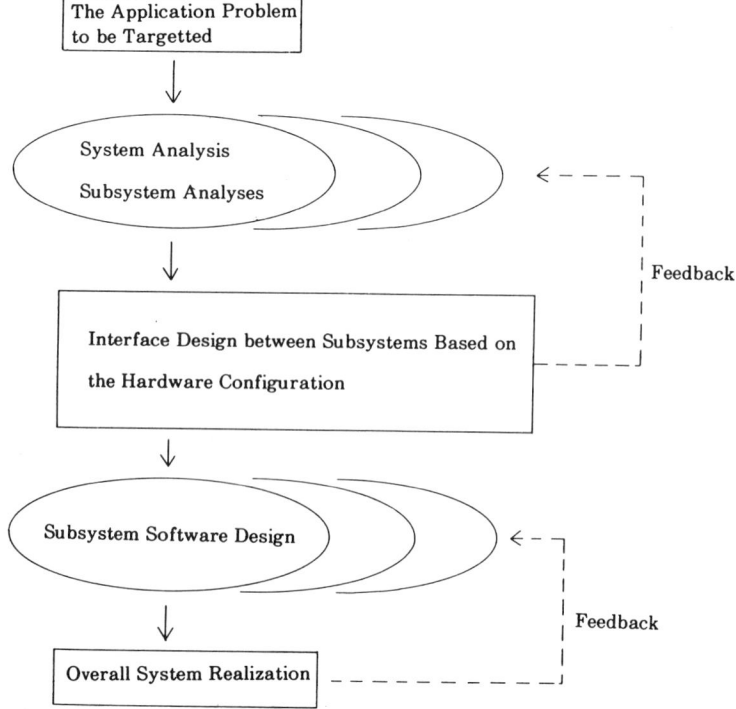

Figure 19.1 *The process of conventional software design.*

System Analysis/Subsystem Analyses First of all, the application problem to be software-systemized is broken down into a number of subsystems, according to the system analysis and subsystem analyses. In the design process, the description of the subsystems should be clarified, and their functionality and descriptive range should be specified in detail from the logical point of view.

Interface Design between Subsystems Based on the Hardware Configuration Following the analysis of the system and subsystems, the computing hardware configuration should be determined for the construction of the subsystems, and the interface between them should be designed in detail. The performance limits of these subsystems depend on the constraints of the hardware configuration. The relationship between the subsystems, the independence of them, the load balance/partiality, and the data flow quantity/velocity are to be fixed according to the interface specification. In the design process, an overall image of the software system is uniformly specified, and the result is verified and fed back to the subsystem design, according to the design objective. This fed-back information can specify the nature of the subsystems. Here, the analytical results essentially include the previous hardware constraints.

Subsystem Software Design After the hardware configuration and the interface have been determined in the subsystem, only the specific contents of the software implementation are assumed to be variables. This means subsystem functional design is realized assuming that the hardware is given. Such software design is performed within the closed subsystems.

Operational Test Next, individual operational tests of the subsystems, of the interface between them, and of overall system functionality are performed according to the design objective. At this design stage, it is difficult not only to return to the design process in system analysis/subsystem analyses, but also to alter the hardware configuration.

From the point of view of software design, it is more desirable to easily feedback the operational test results into the previous system analysis and subsystems design according to the design objective, especially in the case of a complex computing system. Examples of drawbacks in such design feedback are as follows: (1) difficulties in the hardware design change; (2) difficulties in the interface design change; (3) repetitive design process is infeasible. The third point means that it is difficult to both construct the whole from partial units without inconsistency, and also to optimally design a software system specific to an application under very severe constraints on the hardware configuration. In a sense, these difficulties are derived from the design origin itself. This is oriented toward overcoming these difficulties by the software enhancement of the effective functional module component, based on an advanced very large scale integration (VLSI) concept [1].

19.1.2 Functional Module Component

The functional module component is conceptually composed of a processor and a memory. The interconnection of the components is easy to implement from the viewpoints of both hardware and software by the usage of links. In the field of design simulation, the component concept is extended from its original image and a more optimal computing network can be constructed [1] by the introduction of the component-based software design method, enabling a software designer to overcome the previously mentioned difficulties. Some examples of conceptual extensions are as follows: (1) repeatability of software functional design units; (2) expandability of the computing network according to the design objective; (3) load balance evolution among components based on a more effective interconnection of components; (4) communication between two components based on a simple mechanism. Here, a sender waits till the receiver can receive the information, and also the receiver is waiting till the information arrives, on the basis of a message passing scheme. For an orientation toward overcoming the problems of conventional hardware design such as the positioning of the central processing unit or peripheral units, the components are assumed to be embedded in the individual hardware units, and the connection between two hardware units is done on the components' basis. Compared with the conventional method, the design method enables a designer to obtain more easily computing mechanisms specific to the pertinent application, realize a processing speed proportional to data flow volume, evolve a higher speed response time, and enhance the ease and flexibility of the design, even in the case of complex computing mechanisms (e.g., numerical, business, real-time, or process control applications). It is important that this is oriented toward the realization of a more optimal computing mechanism specific to an application without the need for any specialized knowledge of hardware and electronics. This means that with the higher-quality software implementation and advanced VLSI hardware technology, the easy design method can even be applied by a nonexpert computer designer. The designer has only to be conscious of data flow from the point of view of functionality. Here, on the background of the extended software conceptual design for software reutilization with high quality developed so far, the present study accompanied by the introduction of advanced artificial intelligence (AI) technology is expected to contribute to effective parallel process of computing mechanisms. See Section 19.2.

19.1.3 The Component-Based Software Design Process

The introduction of the component concept means the 'softening' of hardware. In other words, system realization is aiming at software not on conventional hardware with constraints but on component-based hardware with on open-endedness, as pointed out in Section 19.1.2. Therefore, the conventional approach of computing mechanism design centered on hardware, can be changed to a software-based one centered on software. Figure 19.2 outlines an example of the new design approach.

System Analysis/Functional Design   An application problem to be targetted is analysed from the viewpoint of functionality. The results are specified by the subsystems in terms of functionality. In the conventional design functional ("logical") correction should have been done by the fed-back contents at the time of hardware configuration, from the hardware ("physical") point of view. In contrast, such a design correction is not necessary for the new approach; in this case the logical design is more important. For example, hierarchical subsystems can be realized.

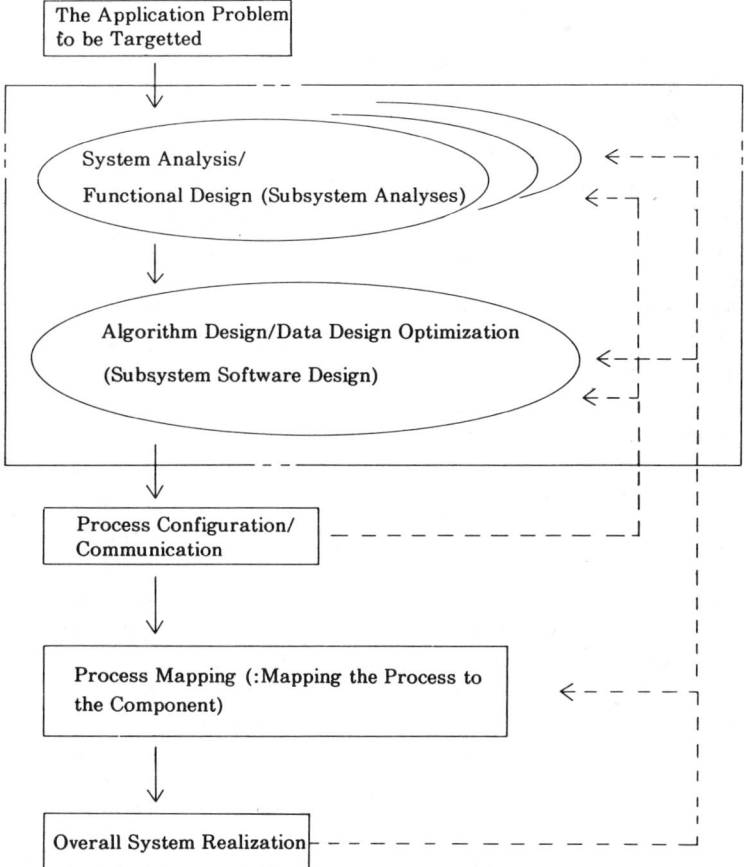

Figure 19.2 *The process of component-based software design.*

Algorithm Design/Data Design Optimization This is a refinement process of the functional design results. Compared with the conventional design process, this algorithm design is almost equal to the interface design and subsystem software design. Thus, software design is preceded in the new design process.

Process Configuration/Communication Functions are regarded as processes, and the data interface between the functions means process communication. In the new design process, attention is paid to the data flow between processes from the following points of view:

Concatenation: direct data input/output may or not be found.
Independence: some processes may or not be parallelized for parallel paths.
Dependency: some processes may or not be sequentialized for serial paths.
Load balance on link: communication volume/communication time.
Load balance on process: computation volume/computation time.

These results are fedback to the algorithm design with regard to process/data design, functional design, process division, and so forth. Here, feedback in design tasks is not automated, but performed manually from the viewpoint of the algorithm design.

Process Mapping Following the previous process configuration/communication, processes should be physically mapped onto the components. This method differs according to the hardware constraints on component number, cost, and other factors. Examples of the mapping factors include the number of components, system throughput (dependent on parallelism, load distribution, communication load), ease of design, better understanding, flexibility and so on. From the viewpoint of VLSI technology maturity, the component count factor may be secondary. Whether the throughput is primary or secondary depends on the design objective. In this design process, ease of design and a better understanding of process mapping are primary factors. This is also effectual for software documentation with high quality. Flexibility may be secondary if the processes are easily modified for design changes. In the case where automated process mapping can be done, process mapping becomes easier even if the mapped results are difficult to understand.

Operational Test and Throughput Evaluation The overall software system results obtained should be evaluated according to the design objective. Functional performance and throughput are mainly restricted by hardware in the conventional design, but at the same time, as expected, are based on the plan at the design stage in a completed system. However, in the new design approach, they cannot be guaranteed during the design process, and therefore logical refinement becomes important in the design. As a result, flexibility for changing the design, throughput evaluation, and feedback to the preceding design stage for this high-quality software design process must be stressed. In the design process, it is regarded as very important to be able to easily feedback to any stage of the design, depending on the previous points stressed.

It is possible to more optimally design by the repetition of the following steps: (1) functional division · merge → process/data (:link interconnection) reconfiguration; (2) process mapping → overall system realization results evaluation for higher quality. The problem-solving procedure is shown in Figure 19.3 as an example.

A component is a VLSI-oriented single-chip computing mechanism with high repeatability, like the transputer [1,2], although different in terms of hardware modeling. It can complete processing by sending/receiving information to/from the neighboring component. This concept can be extended to a general-purpose off-the-shelf component. (Component-link interconnection means a diagram or physical interconnection of the previous transmission components through their links, as shown in Figures 19.5 and 19.6.)

19.1.4 Automation of Process Mapping

In this component-based software design process, design factors such as subsystems, functions, data, and so on, are mutually crisscrossing in com-

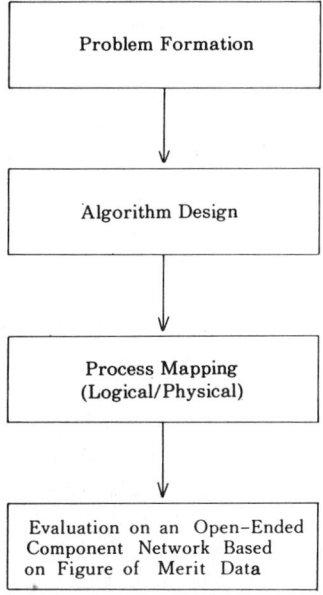

Figure 19.3 *Macroscopic problem-solving procedure.*

plex forms that must be logically matched. Various kinds of careless mistakes and inconsistent errors are apt to intrude in design tasks, and therefore it is desirable to automate part of the design process, such as process mapping. An example of this is a design simulation method. A process mapping can be divided into two categories:

1. Process mapping based on functional division and/or merge (logical process mapping);
2. Process mapping on component-link interconnection (physical process mapping).

(1) Analyzed functions can have a hierarchical structure. Some processes are operational in parallel with others. They can be executed on a component. They are related to each other only through communication. Functionality on a certain level can be defined as a process. One of the most simple method for logical process mapping is to develop the functionality into the lowest-level functions and to express it in terms of functions without any hierarchy as processes.

(2) One of the simplest methods for physical process mapping is to map one process to one component with the introduction of arbitrary components count according to the data input/output (I/O) channels. Actually, physical process mapping method with restricted components count is taken into consideration in 19.1.4.

19.1.5 On the Mapped Results Evaluation

As described previously, more optimal software design process is available for higher quality from the initial problem formation, algorithm design up to process mapping. One of the primary goals is to feed back the mapped results into the preceding design stage. From the viewpoint of flexibility at the preceding design stage, the feedback from the mapped results is relevant for higher software quality. For example, load concentration in a certain process or in a certain communication channel depends on the preceding functional division. Of course, such a fact cannot often be fully foreseen at the design stage, so therefore some design changes are needed. Examples of feedback of typical mapped results are the introduction of pipeline or parallel operations into the overloaded processes by more divisions of the process so as to distribute the load, or the reduction of the communication overload in the necked channel by its division. This intends to realize the design feedback for the gradual enhancement of higher-quality software, from the point of view of higher performance and improved functionality. In particular, compared to the conventional approach, the cheaper computation cost/more expensive communication cost [3] should be taken into consideration in a component-based design approach. In the design simulation of a component-based design process for more optimal design feedback, it is intended to introduce figure-of-merit data, enabling a designer to realize gradual higher-quality software enhancement, according to the design objective.

19.1.6 Application Example

Figure 19.4 shows an example of the transmission delay time relationship on components-link in a simple computing mechanism for component-based design [4].

The horizontal line shows a delay in the parallel processing between transmission components (TC) in terms of the time dimension, and the vertical line shows the delay propagation status of the components (TC_1–TC_4)-link (L_1–L_4). In the simulation, the maximum value of the previous transmission delay time can be calculated for critical performance in parallel processing. This maximum value is some input link arrow is shown at the component reached first by the link. The value is directly related to the initiation of operation in the nearest component as indicated in Figure 19.4, and it is necessary to delay the initiation by the value [3].

Thus, as an approach for a component-based design process, such figure-of-merit data-based evolution (e.g., minimizing the maximum values of the individual components for higher quality from the point of view of performance and functionality) is effective for gradually optimizing a design process. It should be noted that this approach enables a designer to obtain optimization with ease according to the design objective, with the introduction of preprocessing syntax specification enhancement for the simulation.

According to the context of this section mentioned so far, some concrete application examples such as sorting, fast Fourier transform (FFT), and the complex stock control problem of the Information Processing Society of Japan (IPSJ) [5], have been studied over a long period. Due to the regularity

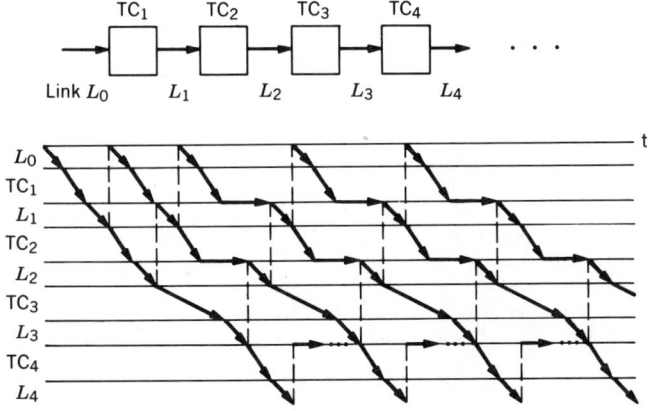

Figure 19.4 An example of parallelism in components-link interconnection.

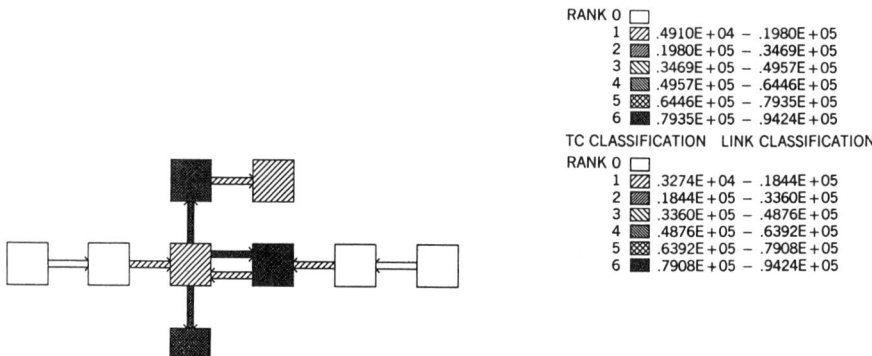

Figure 19.5 A fundamental parallel processing component network for stock control.

of processing, the first two examples are not so difficult. Therefore, the third example is more suitable from the viewpoint of high-quality software as an application using a component-based design process. This design problem is typically expressed in BNF [5]. Figures 19.5 and 19.6 show the concrete application results according to the descriptions of the previous 19.1.3–19.1.5. (The logical process mapping involving intelligent design aid technology from the viewpoint of software conceptual design can be referred to in detail in Section 19.2.)

As shown in the remarks, the two figures of merit data on the components (TC) and links are depicted graphically, on an open-ended component network, in various shades, using an on-line CRT or plotter (via a mouse input). Figure 19.6 is the result of a gradual evolution of Figure 19.5 for higher-quality software in the normal operation of stock control with high concur-

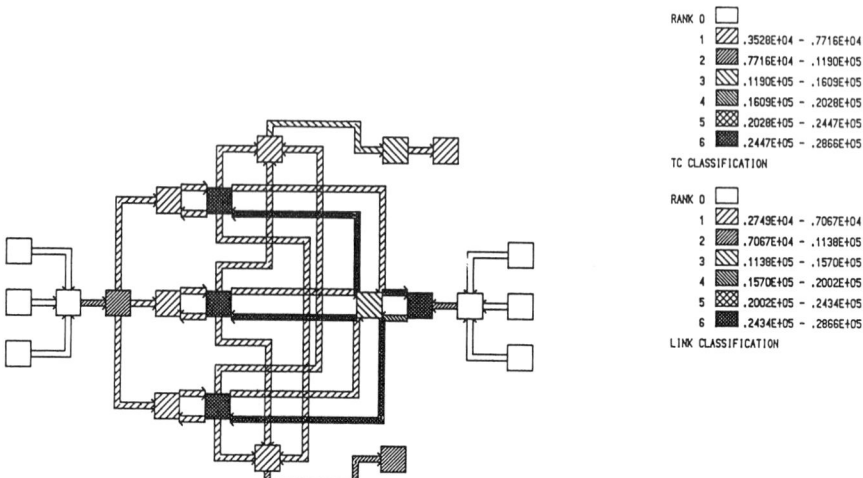

Figure 19.6 A highly parallel processing component network for stock control.

rency and parallelism, from the point of view of performance and functionality. It can be seen that the maximum value in Figure 19.6 is reduced from that in Figure 19.5, by the effective introduction of components. In Figure 19.6, the synchronization of data and procedures should be taken more fully into consideration than it is in Figure 19.5 from the high-level design point of view, which can be absorbed in the logical mapping, at present. In more detailed terms, gradual parallelizations of both the processing in the horizontal line and the vertical line in Figure 19.6, which refer to order, container acceptance, and shipping management (top line) and order answering/out of stock management (bottom line), were performed. Except for the alteration in the number of I/O processes, there is little difference between the logical processes (including their number) in Figure 19.5 and those in Figure 19.6 from a functional point of view. At this stage, it is possible to reduce the components (cost) according to the designer's objective. In particular, it is also useful to realize a distributed control of stock data base. Moreover, some kinds of design evolution can be expected (e.g., parallelization of fundamental hierarchy structures).

19.2 AN ENTITY MODEL FOR CONCEPTUAL DESIGN

Semantic description of the initial stage in computer software design is conceptual and highly abstract. The design method is too heuristic from the original nature [6]. The aim is to construct an interactive design aid environment with highly semantic description facility and a logically effective verification facility for increasing software productivity oriented toward reutilization. This needs a mental process in the design [7,8] using expert shell technology. The method for design simulation presented here is a semantic technique based on an effective human interface [9] in the entity model constructive process. The system helps humans understand and verify during the design process. This can be achieved by semantically representing an entity model and performing predicate calculus on its entity characteristics, using rule-based inference from the data base for software conceptual description.

19.2.1 The Entity Model

For the purpose of this section, an entity model is defined as follows: An entity model is a model that discriminates different functions and specifies them in terms of nodes with links that enable them to be referred to in objective space.

The entity to be represented depends on the objective space itself. The semantics in the space has to be extracted at the interaction between the real human world model and the entity model. From the point of view of practical research on design process by human beings, it can be assumed that human beings dynamically create entities for discrimination and logically verify on the basis of predicate calculus generated by entity-based recognition, according to the interactive consideration status. With regard to the conventional human interface based on predicate calculus which results from inference, such deep mental process cannot be fully realized just by a better understanding of the semantics of the conceptual description. In this study, generation, understanding and verification of the descriptions are based on mental processes determined by the entity model for conceptual design (refer to Figure 19.7). The verification in Figure 19.7 can be divided into two categories:

1. *Validity Verification:* This means the verification based on the interaction between that which human beings want to express and that which the predicate calculus performed in an entity model.
2. *Semantic Verification:* This means the verification based on a comparison between the predicate calculus to be performed in an entity model and the predicate calculus in the data base.

From the following points, it can be seen that the conceptual design is highly abstract.

1. No description of the data instance is needed.
2. Data are conceptually specified, and their meaning is interpreted from the viewpoint of the categorized and structured relations. No description of the data characteristics is needed.

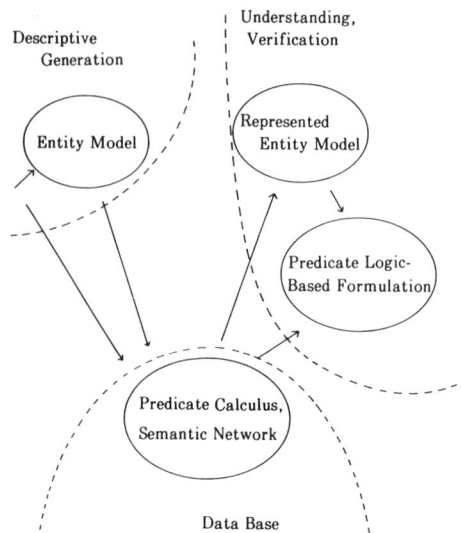

Figure 19.7 *Macroscopic basis for this research.*

3. Processes are specified from the viewpoint of the temporally causal relations between data. No feedback operation to instantiation is needed.

19.2.2 Method for Description

An entity represented in a semantic description for conceptual design falls into one of the following categories:

1. Data concept
2. Process concept
3. Data status
4. Event
5. Control status

Here, 1 and 2 are used in static situations, and the objective space is a definite space where its characteristics are to be discriminated for attributive classification. On the other hand, 3, 4, and 5 are used for expressing dynamic behavior [10], and the objective space allows entity creation, deletion and also the dynamics of reciprocal action. This space corresponds to time-space in the real world. When the symbols that represent data concepts and process concepts are to be used in objective space, they are transformed into the data status and event respectively as entities with a time interval, appearing under a certain dynamic condition. The reason why control status is also introduced as an entity that appears under a certain dynamic condition, is that it plays a role in the creation of and relation between other entities, and that the structural description and temporally causal relation description between one entity and another result in a fully semantic expression in a conceptual sense (refer to item 2 under "Process Organization Refinement."

Hierarchical Data Concepts Here, data concepts can be specified hierarchically from the point of view of similarity or differences identified among the data concepts. This is done in the following way.

1. *Hierarchy by Classification:* This is the case when a more constrained concept is connected to a higher concept. A set of the former instances is a partial set of the latter. This hierarchy is apt to be made independently of process design, and such introduced data concepts compose an objective space specific to a particular area of application.
2. *Hierarchy by Status Introduction:* This is done when the intermediate symbolized entity to be introduced for data status identification on a specified process is connected to a higher concept.

Structured Data Concepts   It is possible to express data concepts by their components and method of organization. This method involves the following three categories.

1. *Composition:* This is the Cartesian product of the components.
2. *Selection:* This is the discriminated union of the components.
3. *Repetition:* This is a set of the same concept components.

Hierarchical Process Concepts   Here, process concepts can be specified hierarchically from the point of view of the similarity or difference among the process concepts. This is done in the following ways.

1. *Specialization of Process Functions:* This is the case when a more specialized process is connected to a generalized function. The former input and output data concepts are more constrained than the latter input and output data concepts, respectively.
2. *Specialization by Assigning a Specified Status to Input and Output Data or by Setting Motive Condition/Status Output:* This is done when the process whose input and output data are the intermediate entities to be introduced for data status identification on a specified process, or when the process that has a motive condition controlled by a trigger signal at a specified time, are connected to the same functional process.

Process Organization Refinement   The process organization is refined so as to be semantically synthesized from the lower processes. The synthesis results specify temporally causal relations between the data and time-space positional relations, at the time and space where the lower processes are to be initiated.

The temporal location is assigned at the time when an entity is existing in objective space. Next, a description of the process entity model is given.

1. Under the identification condition for process internal status assignment, data are identified according to conceptual difference, identification difference and status transitive difference at each identification.
2. Control status is introduced as the intermediate symbolized entity for an identification node for controlling the initiation of lower processes and its entity structure is described in the entity model. The control function is classified on the basis of its structure as shown in Figure 19.8.
3. Based on causal relations, the data to be identified in item 1 are mutually linked and expressed in terms of input and output of the lower processes. Under this condition, processes take on event characteris-

AN ENTITY MODEL FOR CONCEPTUAL DESIGN

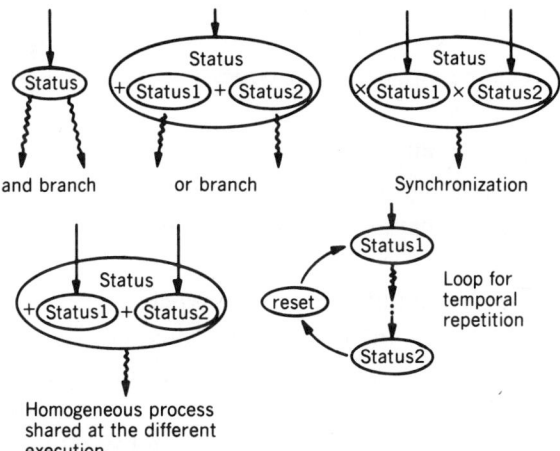

Figure 19.8 *The structures of control status and their function: (→) output; (⤳) trigger signal; (+) selection; (×) composition.*

tics, and therefore their functions can be specialized, based on specialized input and output data.

4. If a constraint is found in the ordering process initiation, the constraint is reduced to the existence of an entity and described as a trigger signal.

19.2.3 Constructing the Entity Model

Methods for constructing the entity model and performing predicate calculus on the entity characteristics obtained from the description of the entity structure and temporally causal relations are formulated as follows; the results of an application are then shown in a simple example.

Constructing the Entity Model from Data

1. The entity model can be constructed from the point of view of identifying the similarities or differences in the data entities assigned as input and output. Basically, entity identification can be performed according to the following rules (where i, i', and k are integers).

1) A particular entity is assigned in terms of both (data 1, data 2, . . . , data i) and (data 1, data 2, . . . , data i, data $i + 1$, . . . , data j), where data k is the lower structure of data $k + 1$.

2) When data i has a more constrained concept than data i', the same entity is assigned by both (data 1, data 2, . . . , data i) and (data 1, data 2, . . . , data i', data $i + 1$, . . . , data j). Here, their structural relations can be expressed in tree form.

2. The entity obtained in 1 can be classified in the following way.

1) Data assigned as input to the process
   → external input
   Data assigned as output from the process
   → external output
   Data assigned as input in the lower process
   → internal input
   Data assigned as output in the lower process

2) These features are propagated to the lower data (downward propagation).

3) When the lower data concept is a selection from, or repetition of, the upper data concept, the preceding features are propagated to the upper data. When the lower data concept is a composition of the upper data concept, and all of the lower data have the same features as just described, the features are propagated to the upper data (upward propagation).

4) When an external input (or output) and an internal input (or output) are competitive, an external input (or output) can be assumed.

5) When an input (or output) is internal, but the internal output (or input) found (i.e., is not referred), then an inconsistency is assumed.

Figure 19.9 shows a visual description of a simple application example, and Figure 19.10 shows the construction of the data-based entity model as just described.

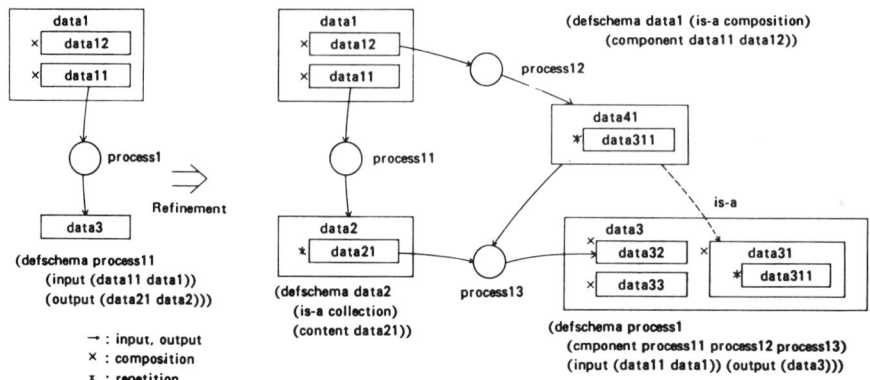

Figure 19.9 *A visual description of a simple example.*

AN ENTITY MODEL FOR CONCEPTUAL DESIGN   519

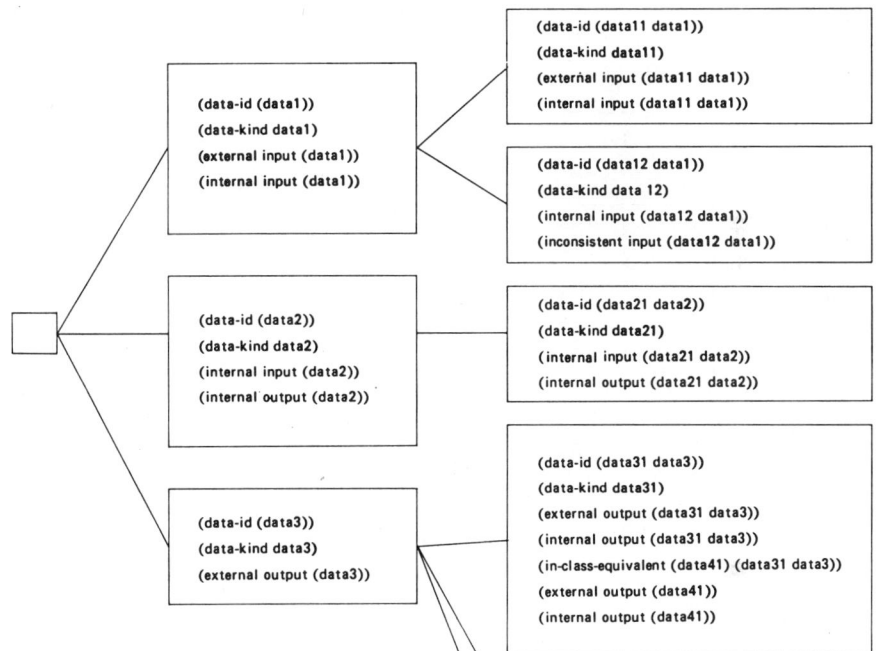

Figure 19.10 *The data structure of the example in Figure 19.9.*

Constructing the Entity Model Based on Events Process execution can be simulated and as a result an entity model based on events can be constructed. From this, temporally causal relations between entities can be seen by tracing the directional graph, which is shown in the data entity model with input, output data and trigger signal in the lower process (refer to the preceding subsection, "Constructing the Entity Model from Data"). Each node on the graph represents the context, that is data status, process on, or trigger signal. Causally related nodes can be connected by an arc. Basically, this is done according to the following rules.

1. Initially, nodes representing the data status are generated, assuming that the external input data are set to the status of temporal parameter value 1.
2. The data status can be instantaneously passed to the data to be structured (i.e., the same temporal parameter value).
3. New data status can be specified in the following way.
   - Process output on the basis of process on (the temporal parameter value is the sum of the temporal parameter value at the time of process on and 1).

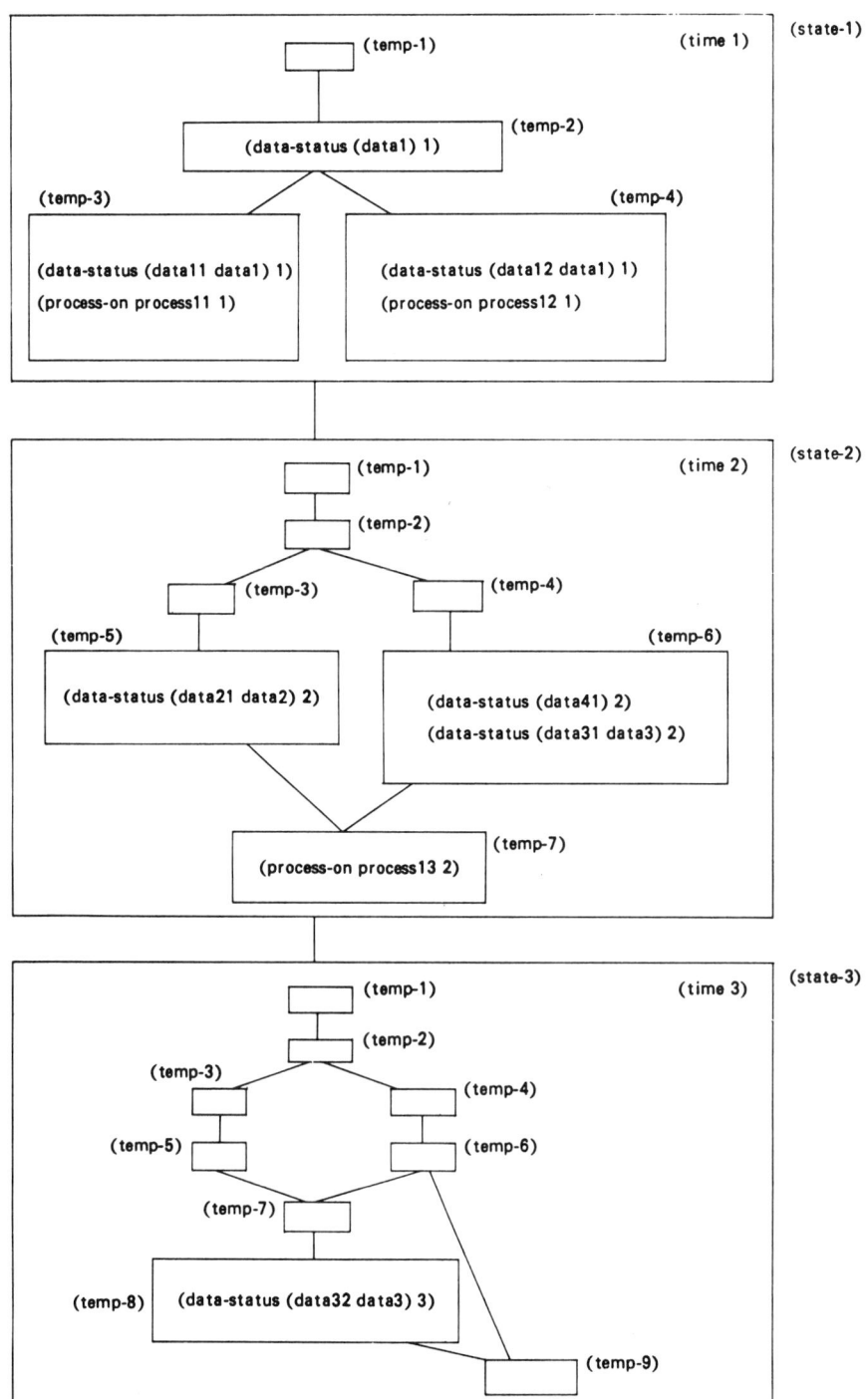

Figure 19.11 *The results of constructing an event-based entity model.*

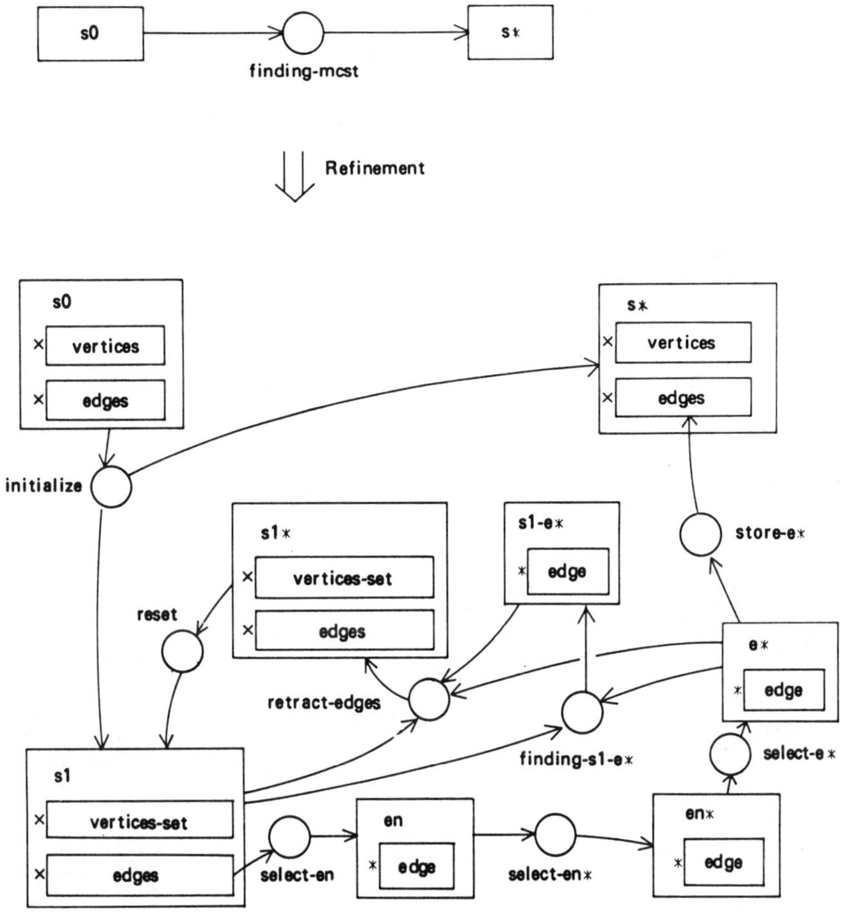

AN ENTITY MODEL FOR CONCEPTUAL DESIGN   521

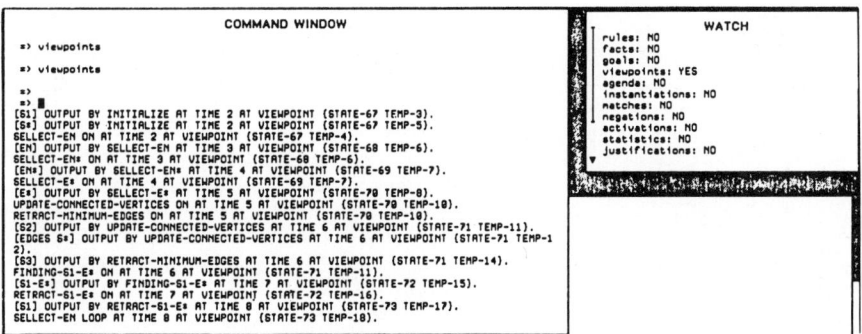

Figure 19.12 *A visual description of a minimum-cost spanning tree (mcst) and its data structure: s0 = initial-cost network; s1 = intermediate-cost network; s* = minimum-cost tree; en = sorted edge; en* = minimum-cost edge on each vertex; e* = true minimum cost edge.*

Figure 19.13 *The simulation results of the event-based entity model of mcst.*

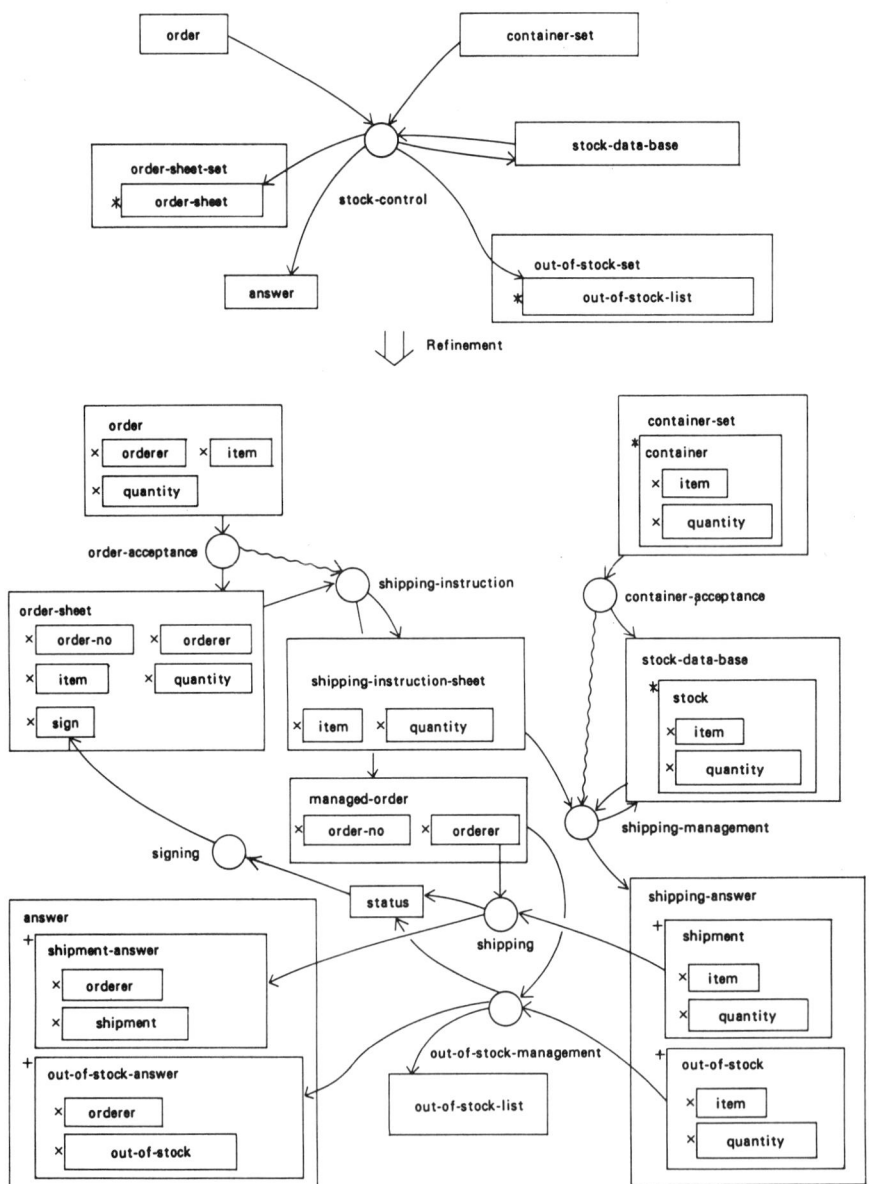

Figure 19.14 A visual description of stock control.

```
                         COMMAND WINDOW
[OUT-OF-STOCK ANSWER-OUT-OF-STOCK ANSWER] IS EXTERNAL OUTPUT DATA. VIEWPOINT (STATE-57)
[ITEM OUT-OF-STOCK ANSWER-OUT-OF-STOCK ANSWER] IS EXTERNAL OUTPUT DATA. VIEWPOINT (STATE-58)
[QUANTITY OUT-OF-STOCK ANSWER-OUT-OF-STOCK ANSWER] IS EXTERNAL OUTPUT DATA. VIEWPOINT (STATE-5
9)
[OUT-OF-STOCK ANSWER-OUT-OF-STOCK] IS EXTERNAL OUTPUT DATA. VIEWPOINT (STATE-57)
[ITEM OUT-OF-STOCK ANSWER-OUT-OF-STOCK] IS EXTERNAL OUTPUT DATA. VIEWPOINT (STATE-58)
[QUANTITY OUT-OF-STOCK ANSWER-OUT-OF-STOCK] IS EXTERNAL OUTPUT DATA. VIEWPOINT (STATE-59)
[STOCK-DATA-BASE] IS EXTERNAL OUTPUT DATA. VIEWPOINT (STATE-60)
[STOCK STOCK-DATA-BASE] IS EXTERNAL OUTPUT DATA. VIEWPOINT (STATE-61)
[ITEM STOCK STOCK-DATA-BASE] IS EXTERNAL OUTPUT DATA. VIEWPOINT (STATE-62)
[QUANTITY STOCK STOCK-DATA-BASE] IS EXTERNAL OUTPUT DATA. VIEWPOINT (STATE-63)
[STOCK-DATA-BASE] IS INTERNAL INPUT DATA. VIEWPOINT (STATE-60)
[STOCK STOCK-DATA-BASE] IS INTERNAL INPUT DATA. VIEWPOINT (STATE-61)
[ITEM STOCK STOCK-DATA-BASE] IS INTERNAL INPUT DATA. VIEWPOINT (STATE-62)
[QUANTITY STOCK STOCK-DATA-BASE] IS INTERNAL INPUT DATA. VIEWPOINT (STATE-63)
[ORDER-SHEET] IS EXTERNAL OUTPUT DATA. VIEWPOINT (STATE-65)
[ORDER-SHEET ORDER-SHEET-SET] IS EXTERNAL OUTPUT DATA. VIEWPOINT (STATE-65)
[ORDER-SHEET-SET] IS EXTERNAL OUTPUT DATA. VIEWPOINT (STATE-64)
[ORDER-SHEET] IS INTERNAL INPUT DATA. VIEWPOINT (STATE-65)
[ORDER-SHEET ORDER-SHEET-SET] IS INTERNAL INPUT DATA. VIEWPOINT (STATE-65)
[ORDER-SHEET-SET] IS INTERNAL INPUT DATA. VIEWPOINT (STATE-64)
[ORDERER ORDER-SHEET ORDER-SHEET-SET] IS EXTERNAL OUTPUT DATA. VIEWPOINT (STATE-66)
[ORDERER ORDER-SHEET ORDER-SHEET-SET] IS INTERNAL INPUT DATA. VIEWPOINT (STATE-66)
[ORDERER ORDER-SHEET] IS EXTERNAL OUTPUT DATA. VIEWPOINT (STATE-66)
[ORDERER ORDER-SHEET] IS INTERNAL INPUT DATA. VIEWPOINT (STATE-66)
[ITEM ORDER-SHEET ORDER-SHEET-SET] IS EXTERNAL OUTPUT DATA. VIEWPOINT (STATE-67)
[ITEM ORDER-SHEET ORDER-SHEET-SET] IS INTERNAL INPUT DATA. VIEWPOINT (STATE-67)
[ITEM ORDER-SHEET] IS EXTERNAL OUTPUT DATA. VIEWPOINT (STATE-67)
[ITEM ORDER-SHEET] IS INTERNAL INPUT DATA. VIEWPOINT (STATE-67)
[QUANTITY ORDER-SHEET ORDER-SHEET-SET] IS EXTERNAL OUTPUT DATA. VIEWPOINT (STATE-68)
[QUANTITY ORDER-SHEET ORDER-SHEET-SET] IS INTERNAL INPUT DATA. VIEWPOINT (STATE-68)
[QUANTITY ORDER-SHEET] IS EXTERNAL OUTPUT DATA. VIEWPOINT (STATE-68)
[QUANTITY ORDER-SHEET] IS INTERNAL INPUT DATA. VIEWPOINT (STATE-68)
[ORDER-NO ORDER-SHEET ORDER-SHEET-SET] IS EXTERNAL OUTPUT DATA. VIEWPOINT (STATE-69)
[ORDER-NO ORDER-SHEET ORDER-SHEET-SET] IS INTERNAL INPUT DATA. VIEWPOINT (STATE-69)
[ORDER-NO ORDER-SHEET] IS EXTERNAL OUTPUT DATA. VIEWPOINT (STATE-69)
[ORDER-NO ORDER-SHEET] IS INTERNAL INPUT DATA. VIEWPOINT (STATE-69)
[SIGN ORDER-SHEET ORDER-SHEET-SET] IS EXTERNAL OUTPUT DATA. VIEWPOINT (STATE-70)
[SIGN ORDER-SHEET ORDER-SHEET-SET] IS INTERNAL INPUT DATA. VIEWPOINT (STATE-70)
[SIGN ORDER-SHEET] IS EXTERNAL OUTPUT DATA. VIEWPOINT (STATE-70)
[SIGN ORDER-SHEET] IS INTERNAL INPUT DATA. VIEWPOINT (STATE-70)
No applicable rules.

=> ▮
[ANSWER-OUT-OF-STOCK ANSWER] IS EXTERNAL OUTPUT DATA. VIEWPOINT (STATE-55)
[ORDERER ANSWER-OUT-OF-STOCK ANSWER] IS EXTERNAL OUTPUT DATA. VIEWPOINT (STATE-56)
[ORDERER ANSWER-OUT-OF-STOCK] IS EXTERNAL OUTPUT DATA. VIEWPOINT (STATE-56)
```

Figure 19.15 *The simulation results of the data-based entity model of stock control.*

- Propagation from the upper data.
- Propagation from the lower data.

If the generated data status already exists and propagation from the lower data is found, nodes are connected to generate a new node, creating a new data status. Data that have not yet been generated are generated at the time of alignment of all the statuses propagated from the lower data. The lower data status nodes are connected thus generating a new node, and then this becomes the generated data status. Here, the nodes that represent the data status generate only the necessary node (process input or trigger signal).

4. A trigger signal can be specified as process output at the time of process on. (The temporal parameter value is the sum of the temporal parameter value at the time of process on and 1.)
5. The process can be switched on at the generation time of a pair of all

COMMAND WINDOW

SHIPPING-MANAGEMENT ON AT TIME 3 AT VIEWPOINT (STATE-67 TEMP-13).
[STOCK-DATA-BASE] OUTPUT BY SHIPPING-MANAGEMENT AT TIME 4 AT VIEWPOINT (STATE-68 TEMP-14).
[SHIPPING-ANSWER] OUTPUT BY SHIPPING-MANAGEMENT AT TIME 4 AT VIEWPOINT (STATE-68 TEMP-15).
OUT-OF-STOCK-MANAGEMENT ON AT TIME 4 AT VIEWPOINT (STATE-68 TEMP-18).
SHIPPING ON AT TIME 4 AT VIEWPOINT (STATE-68 TEMP-19).
[ANSWER-OUT-OF-STOCK] OUTPUT BY OUT-OF-STOCK-MANAGEMENT AT TIME 5 AT VIEWPOINT (STATE-69 TEMP-20).
[STATUS] OUTPUT BY OUT-OF-STOCK-MANAGEMENT AT TIME 5 AT VIEWPOINT (STATE-69 TEMP-21).
[ANSWER-SHIPMENT] OUTPUT BY SHIPPING AT TIME 5 AT VIEWPOINT (STATE-69 TEMP-22).
[STATUS] OUTPUT BY SHIPPING AT TIME 5 AT VIEWPOINT (STATE-69 TEMP-23).
SIGNING ON AT TIME 5 AT VIEWPOINT (STATE-69 TEMP-23).
SIGNING ON AT TIME 5 AT VIEWPOINT (STATE-69 TEMP-21).
[SIGN] OUTPUT BY SIGNING AT TIME 6 AT VIEWPOINT (STATE-70 TEMP-25).
[SIGN] OUTPUT BY SIGNING AT TIME 6 AT VIEWPOINT (STATE-70 TEMP-26).
No applicable rules.

(a)

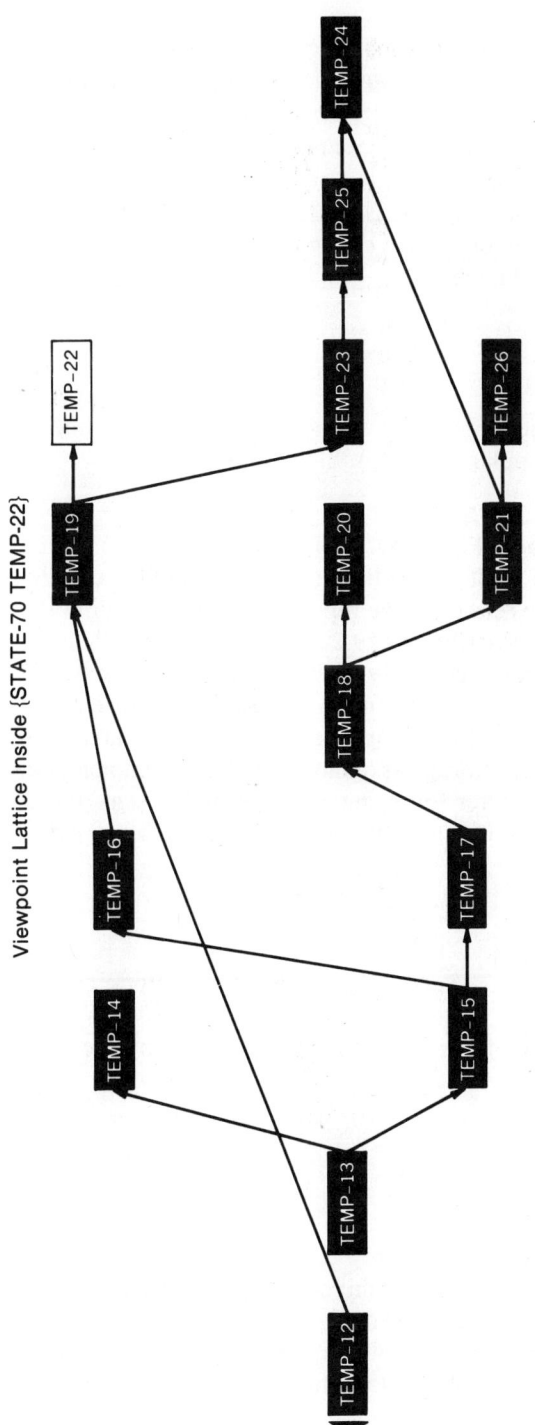

Figure 19.16 *The simulation results of the event-based entity model of stock control;* (a) *nodal description;* (b) *graphic representation.*

the input data status and trigger signal (if necessary). Here, the temporal parameter value is the maximum temporal parameter value of the related data status and trigger signal.

6. In the case where a causal relation occurs between a certain output and the process, a loop in the process is assumed to have been generated and an output node is not generated. Next, Figure 19.11 shows the result of constructing the event-based entity model as described, with regard to the example in Figure 19.9.

19.2.4 Examples of Application

Two typical kinds of real applications are now shown. One is a well-known problem, and the other is a popular design problem in IPSJ [5].

Minimum-Cost Spanning Tree Figure 19.12 is a visual description of the algorithmic process of a solution. It shows the result generated by a similar inquiry on entity and the causal relation to be represented during the interactive consideration process, as described in 19.2.1 to 19.2.3. Here, the symbols representing the data do not necessarily correspond to the data storage assignment on a computer. In such a conceptual design, the representation of an entity needs to be based on their properties because data have them according to their roles. The data storage assignment is another implementation-specific problem on a computer following a conceptual design. Figure 19.13 shows the result of representing this application example by an event-based entity model.

Stock Control Design Problem This is another example of a practical application. Figure 19.14 shows a visual description of this stock control design problem.

Figures 19.15 and 19.16 show the results of representing it by the data-based entity model and the event-based entity model, respectively.

ACKNOWLEDGMENTS

I should like to thank the following company for giving me permission to adapt material from their publication: T. Ichiko, "An Approach for High Quality Software," Proceedings of Int'l Symposium on Information Technology Standardization, Elsevier (North-Holland), pp. 177–188, 1990. Courtesy and copyright Elsevier Science Publishers B. V. North-Holland Mathematics and Computer Science.

REFERENCES

[1] I. Barron et al., Transputer does 10 or more MIPS even when not used in parallel, *Electronics* Nov. 17, 109–115 (1983).

[2] C. A. R. Hoare, Communicating sequential processes, *Comm. of ACM,* **21**–8, 666–677 (1978).

[3] T. Ichiko, Integrated design simulation on component network oriented toward non expert designer, *Proc. of ICCI '89,* 2, 122–126 (1989).

[4] T. Ichiko, VLSI oriented structural analysis and synthesis on an integrated system device, *Proc. of MIMI '84 Bari,* 35–38 (1984).

[5] Y. Morisawa, BNF expression on stock control design problem, *J. Information Processing Soc. Japan (IPSJ)* **25**–11, 1255–1260 (1984).

[6] S. Tamura, et al., On a system planning and design method, *Proc. of IEEE Internat. Conf. on Systems, Man and Cybernetics,* 1202–1206 (1983).

[7] B. Chandrasekaran, Deep models and these relations to diagnoses, *Lecture Notes* (1986).

[8] D. Brown et al., Knowledge and control for a mechanical design expert system, *J. IEEE Expert,* 92–100 (1986).

[9] J. Baalen et al., Overview of an approach to representation design, *Proc. of AAAI '88,* 392–397 (1988).

[10] D. Harel et al., STATEMATE: A working environment for the development of complex reactive systems, *Proc. of IEEE Internat. Conf. on Software Engineering,* 396–406 (1988).

CHAPTER 20

Integrated Complex Automation and Control: Parallel Inference Machine and Paracell Language

PETER FANDEL

This chapter discusses how complex automation and control problems are solved today with distributed control systems, and how they are handled using the parallel inference machine (PIM) and intelligent Paracell language. It will be shown that today's solutions are inadequate for very large and complex problems, and that the use of massively parallel controllers programmed in a parallel language can adequately solve these problems, as well as problems currently considered unsolvable. Integration tiles and cells are described in detail.

20.1 THE INADEQUACIES OF LARGE-SCALE DCS SOLUTIONS

It must be emphasized that, for applications in process control and factory automation applications, traditional solutions using programmable controllers and small scale distributed computer systems (DCS) are quite adequate. Although the technology introduced in this chapter is fully capable of solving these small- to medium-sized problems, there is no great need for new solutions. It is only when, the size of DCS solutions grow to include more than 10 or so nodes, that the inherent communications and memory requirements of these systems render them impractical. See Figure 20.1.

Paracell, SoftScope are Trademarks of Flavors Technology, Inc.
Many thanks to all Flavors asked to proof this document.

Neural and Intelligent Systems Integration, By Branko Souček and the IRIS Group.
ISBN 0-471-53676-8   ©1991 John Wiley & Sons, Inc.

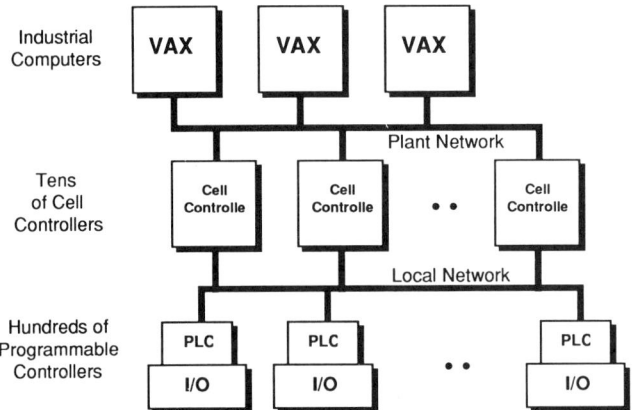

Figure 20.1 Conventional solutions to massive automation problems consist of a mixture of many different programming environments.

20.1.1 DCS System Architecture: Theory versus Practice

In a distributed computer system, each computer is dedicated to solving a portion of the problem directly. All of the computers, or *peers*, in the system are networked together so that each may have access to the state of the entire system.

Theory In order to keep system optimization, in theory, each peer in a distributed computer network must contain all the information that all the other peer elements in the distributed computer system have available to them. If they do not replicate the state of the entire system at full sampling rate, then each segment of this system is optimized presumably at the degradation of system optimization. A single element system can optimize its own performance and system performance because they are one and the same. In order for a multiple element system to optimize performance, each element must maintain, at full sampling rate, a copy of all the other elements' states in addition to its own.

Thus each peer in a DCS must have n times the memory required to perform its local task, where n is the number of peer elements in the DCS (see Figure 20.2). Further, the memory requirements of the entire system are equal to n^2 times local memory requirements. Both this theoretical model and an actual implementation (CM* at Carnegie Mellon) indicate that it is folly to think about more than 10 peer elements and maintain system optimization for a tightly coupled DCS system.

Practice In actual implementation, of course, system optimization is only a goal and not a realization, and although the peer-to-peer elements in a DCS

20.1 THE INADEQUACIES OF LARGE-SCALE DCS SOLUTIONS

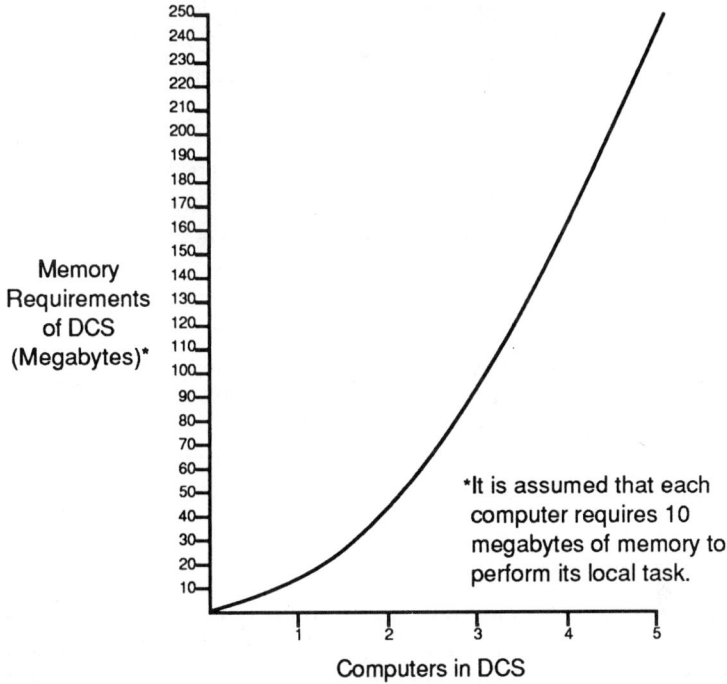

Figure 20.2 DCS memory requirements. (It is assumed that each computer requires 10 Mbytes of memory to perform its local task.)

must contain more memory and high bandwidth, they need not contain the memory discussed in the theoretical model discussed above. Systems with up to 30 to 40 peer-to-peer relationships (9 peers fully connected have 36 relationships) have been effective when designers have paid attention to bandwidth limitations and memory selection. Any more than 10 peers brings into question the design and the system optimization related to a DCS system.

The important thing to realize is that a single node on a DCS network has little information about the system, and can only optimize its own performance, not system performance. Theoretically at least, to optimize entire system performance, each node must know "everything" about the entire system. Engineering practicality, of course, leads to an interim analysis and consideration. For example, hierarchical DCS that adds one or more layers of computers in order to consolidate information. Under any circumstances, of course, DCS should be avoided for large-scale systems.

20.1.2 DCS Programming

As the scope of DCS solutions increase, so do the difficulties in programming these systems. Major difficulties include software overhead, and applications specific programming. Due to these difficulties, 15 to 25 percent of the cost of building a typical new factory is for software. These costs are prohibiting the process control and factory automation industries from implementing desired control strategies.

Software Overhead In a typical DCS solution, 50 percent of the software effort is for systems software (i.e., peer to peer communications, data base software, system status software—that which is not part of the control problem) (see Figure 20.3). This varies, of course, with the size of the system much in the same way as memory requirements vary.

Applications Specific Programming Distributed computing would be more tenable were it generically programmed. However this is rarely the case. In DCS every programmable logic controller (PLC) and every cell controller is programmed separately with application specific programs. Maintenance of these systems is a major problem for the control engineers.

20.1.3 DCS Reliability

The name *distributed computer system* implies a certain degree of fault tolerance. "If one computer fails, my production won't stop because I have a hundred computers." This, however, is far from the reality encountered in the industry. Based on Detroit sources, Flavors estimates that fully one third of the computers in factories are single points of failure, which means the failure of any one of those computers results in the stoppage of the entire factory.

Taking a small factory of 100 PLCs and cell controllers, of which the failure of any 30 will halt production, it becomes clear that DCS is not inherently fault tolerant, but is in fact fault-sensitive.

20.2 APPLYING PIM/PARACELL

The PIM/Paracell solution to control and automation problems is a revolutionary rather than an evolutionary advancement in control technology. Not

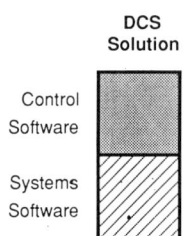

Figure 20.3 *Fifty percent of DCS software is systems software.*

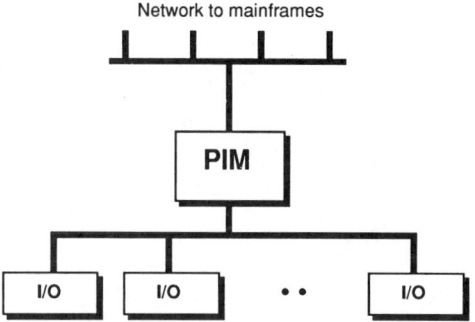

Figure 20.4 *The PIM/Paracell solution to massive control and automation problems.*

only does the PIM/Paracell system overcome the inadequacies of DCS, but goes well beyond the intended capabilities of DCS. (See Figure 20.4.)

This section of applying the PIM and Paracell glosses over technical details presented in the latter part of this chapter. In short, the PIM is a parallel processing industrial computer capable of implementing the control code of hundreds of programmable controllers and their cell controllers, and providing this code a single view of the system via a massive global memory. A typical PIM installation will replace a layer of cell controllers, most of the PLCs, and several industrial computers. Any remaining PLCs take on the role of concentrators. Concentrators are PLCs that have been generically programmed to pass data back and forth between the PIM and the factory floor. The pyramid concept still exists within the PIM In software. However, the pyramid resides in one box.

20.2.1 Incorporating the PIM

The PIM Is a plantwide controller, that interfaces directly to industry standard I/O on one end, and mainframes on the other. All that should be done in real-time is done within the PIM. The management information system (MIS) and other data-base-intensive operations are done by area mainframes. Communication to the factory floor is via industry standard input/output (I/O) networks such as those developed by Allen-Bradley and Modicon. Communication to mainframes and other administrative computers is via shared-memory links (via bus to bus coupling). At the time of publishing, 10 different I/O protocols and bus architectures are supported by the PIM. (See Figure 20.5.)

Development of control code for the PIM is done on Apple Macintosh development workstations. Flavors has fully capitalized on the easy-to-learn and fun-to-use human-machine interface of the Macintosh.

20.2.2 Outcome

The outcome of incorporating PIM and Paracell solution into a factory are mostly derived from the simplification of the control solution. The following

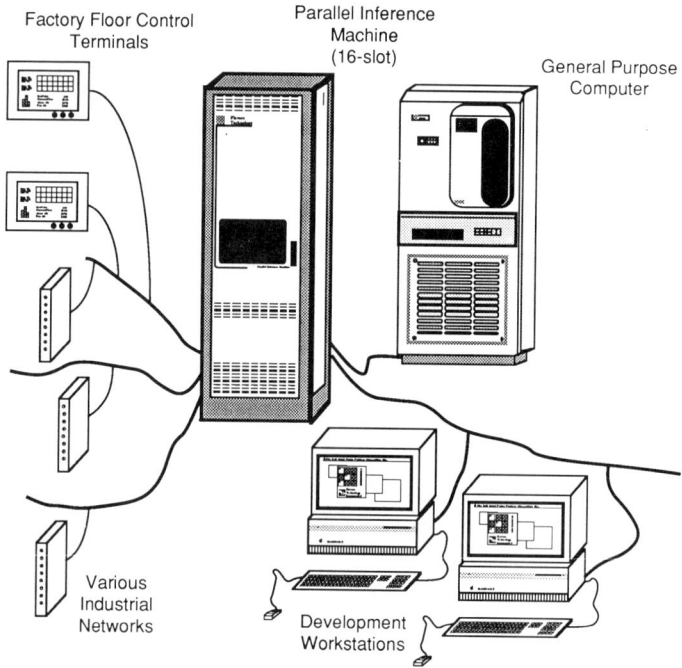

Figure 20.5 *Industry standard I/O links and high-speed shared memory links to mainframes and the development workstations tie the PIM to the factory or plant.*

are important aspects of PIM/Paracell installations from the point of view of the user.

Productivity The Paracell language and Navigator environment form a highly productive package. Using the PIM and Paracell will reduce software effort by at least an order of magnitude, and resulting control code can be read and understood by nontechnical personnel. This is due to one box replacing many, and thus removing integration overhead. This is due to the high-level nature of Paracell and the capacity of the PIM. And, this is due to the Navigator project management tools delivered with Paracell. (See Figure 20.6.)

Project Management and the Mythical Man Month The more programmers involved in a project, the less effective each programmer will be due to the coordination difficulties (for reasons similar to those that make DCS inadequate for large problems). Because of the 10-fold decrease in software effort afforded by PIM/Paracell solutions, less control engineers are required to code the system, and their individual productivity increases. The coordination of control engineers is directly enhanced through the Navigator project management tool.

20.2 APPLYING PIM/PARACELL   535

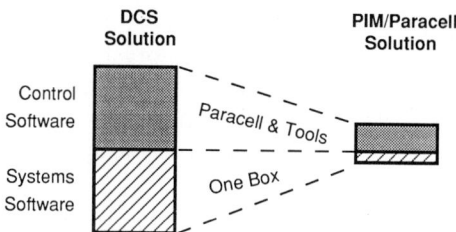

Figure 20.6 *Systems software is virtually eliminated, and the development of control software requires much less effort.*

Running on a development workstation, the Paracell Navigator provides a three-dimensional view of a control system, organizes the application in a natural manner, helps the system architect divide the program among multiple developers, helps individual programmers organize their work, and automatically documents authors, revision dates and code/data relationships for each segment of code and each variable created.

Real-time Figure 20.7 most clearly illustrates how the PIM's architecture results in a performance behavior very unlike most other computers. Every aspect of the PIM design was designed for the execution of Paracell programs in real time. Our criteria for real time in order of importance are

1. Predictable response time
2. Reliable response time
3. Fast response time

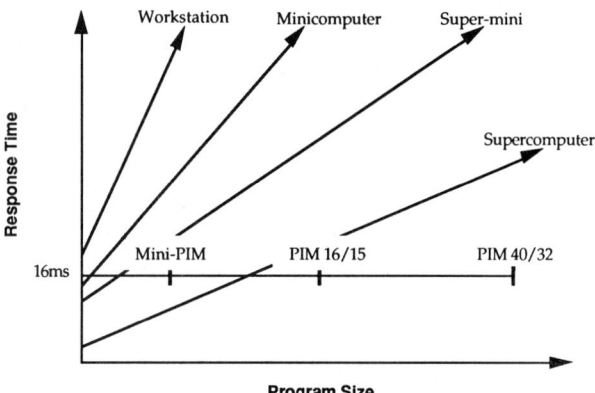

Figure 20.7 *For most computers, as the amount of software to execute increases, response time decreases. Traditionally, this is addressed (if at all) by getting a faster computer or trying to tie together multiple computers. With the PIM, the size of the program does not affect response time, but affects the number of processors needed to perform the task. Adding processors to the PIM has no effect on the performance of existing code. The 40-slot PIM may contain up to 32 processor boards, which support 16,384 virtual processors.*

Cell controllers, programmable controllers, and industrial computers, the programmable devices that form DCS, do not have a constant response time. As code is added, response time (scan time) degrades. When the scan time of a PLC reaches a certain point, the PLC shuts down, and the response time of a real-time central processing unit (CPU) seriously degrades as I/O is added.

Post Tuning Post tuning is the ability to modify and improve the control of a process or factory after production has been begun. Individual pieces of control code may be modified and loaded into the PIM from the development workstation while the PIM is controlling a process, without affecting the rest of system and without interrupting even the control where the change takes place.

Dynamic Technology Factories are built to last many decades, whereas computer technology advances continually. The incorporation of new software technologies such as fuzzy sets and neural networks was foreseen by the PIM/Paracell design team. An additional processor board may be installed, and a fuzzy-logic-based package may be programmed that makes use of the global view of the factory available to all processors.

The way in which the PIM was designed enables Flavors Technology to incorporate the latest processors at minimum effort. Motorola 68040 processor boards are expected within a couple of years that will significantly boost performance. Specialized processor boards are also feasible that are optimized for neural network performance for example. Implementation of these new technologies would be nonintrusive.

Expandability The PIM is a modular computer, and processors and/or memory may be added in chunks as needed. All computer systems, of course, can be expanded. Unlike most other systems, however, adding processors and/or memory to the PIM up to the maximum configuration does not affect the performance of existing Paracell code running in existing processors.

20.3 PARACELL

When Flavors undertook to design and develop the PIM/Paracell control system, they elected to begin with the language (Paracell). This removed any hardware platform biases that might have affected the design of the language. The PIM was then developed to optimally execute Paracell and only Paracell. The Paracell language was conceived in 1981 by Richard Morley and Doug Currie as an extensible, truly parallel language with an emphasis on programmer productivity and real-time performance. Detailed development work began in 1987, and a beta version was released in August 1990.

20.3.1 Attributes

Introducing a new language is not a task to be taken lightly. However, we feel that the promise of an order of magnitude increase in performance justified the decision. Other parallel processing computers have elected to coerce naturally serial languages to execute in Parallel. We designed Paracell from the beginning to be a parallel language. Some attributes of Paracell, and a description of the performance gains we expect will precede a description of some Paracell details.

An Intelligent Language The productivity of Paracell is due to the high IQ of each statement, and the fact that it is written in English. It is widely accepted that the average code produced per programmer per year is a constant around 2000 lines. Because 2000 lines of FORTRAN code do ten times the work of a similar number of lines of assembler code, higher-level languages clearly result in higher net programmer productivity. Paracell is a very high level language, and the net productivity of Paracell statements is at least an order of magnitude higher than traditional languages such as FORTRAN, and C. With this in mind, projects previously thought to be untenable may be reconsidered. Much of this increase is due to the organization of Paracell programs, in that once a statement is written and loaded into the PIM, it is automatically executed every 16 ms by default. There is no need for extra lines of code to decide when or when not to execute the statement (e.g., subroutine calls, branches, etc.). (See Figures 20.8 and 20.9.)

Some examples of Paracell code that illustrate the net productivity per line of code:

```
Every five seconds add 1 to X.

If pressure is greater than 5000 then
    set release valve to open.
```

	Development Cost	Lines of Code	Cost/Line of Code	Man Years	Lines/Man Year	Cost/Man Year
IBM Checkout Scanner	$3M	90K	$33	58	1551	$51.7K
Lotus 1-2-3 Version 3	$22M	400K	$55	263	1520	$83.6K
Space Shuttle	$1,200M	25.6M	$47	22,000	1163	$54.5K
Lincoln '89	$1.8M	83.5K	$21	35	2327	$51.4K
Citibank Automatic Teller Machine	$13.2M	780K	$17	150	5200	$88.0K

Source: Fortune - Sept. 25 '89

Figure 20.8 *Programmer productivity, in lines of code, is relatively constant regardless of the size of the project. (From* Fortune, *September 25, 1989.)*

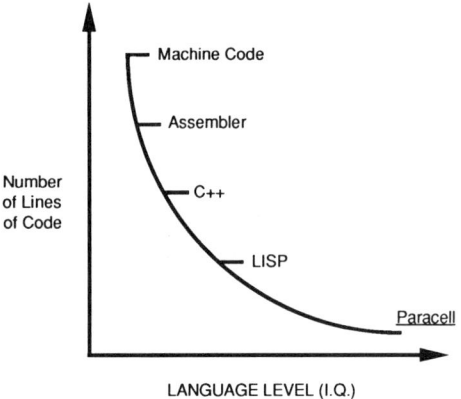

Figure 20.9 *The key to increased productivity is the power per line of code.*

The two statements automatically execute 60 times a second. The former continually checks to see if 5 s have passed since it was last reset, and automatically resets and adds one to the value X every 5 s. The latter monitors pressure continually, and sets release valve accordingly. The user is not required to write any more code to ensure their execution.

A Rich Language Part and parcel of a high-level language is its richness. In order to save the user the trouble of fussing with pointers in implementing structures such as stacks, queues, and other "collections," Paracell has included all of these. Paracell also supports automatic timers, and has full trigonometric, logarithmic, exponential, and other advanced mathematical functions built in.

A Real-time Language Paracell is a real-time language chiefly because of the manner in which programs are organized into "tiles," and the predictable hardware resources available to those tiles. Every frame (16 ms) every piece of Paracell code has a guaranteed and predictable amount of time to execute, and guaranteed access to any global data, even if all other tiles are accessing the same location in the same frame.

Another aspect of Paracell that aids real-time performance is the numerical type called Paracell numbers. Single-precision Paracell numbers are rational numbers with four digits after the decimal point. Unlike floating-point numbers (which are also supported), the minimum distance between representable numbers does not vary with the magnitude of the numbers. Further, Paracell numbers offer better arithmetic performance than floating-point numbers.

20.3.2 Hardware Resources

In order to write Paracell programs, it is helpful for the programmer to have a basic understanding of the hardware resources made available by the PIM.

20.3 PARACELL

The programmer does not need to know how these resources are made available, just that they are there.

Cells The basic resources available to Paracell programs are *cells*. Cells are virtual processors with fixed and guaranteed resources. As hardware is added to a PIM, the number of cells increases, but the resources of a cell never change or differ from one another. The software view is of 16,384 cells sharing a large common (global) memory. In each PIM frame (16 ms) all 16,384 cells simultaneously import data from global memory, then execute their Paracell code in parallel, and finally simultaneously export data to global memory. Every frame, every cell can import up to sixteen 32-bit words from global memory, and export up to eight 32-bit words to global memory. (See Figure 20.10.)

Global Memory All cells have equal and coherent access to a common data memory. Any or all cells may *import* any variable from this global memory any or every frame. Likewise, any or all cells may *export* any variable into this global memory any or every frame. The user understands that, if more than one cell exports to the same global variable in the same frame, the programmer may specify which cell has priority.

Intelligent memory operations may be performed by the global memory with each import or export. This is so that multiple cells may have cumulative affects on individual global variables within a single frame. This will be described in more detail later.

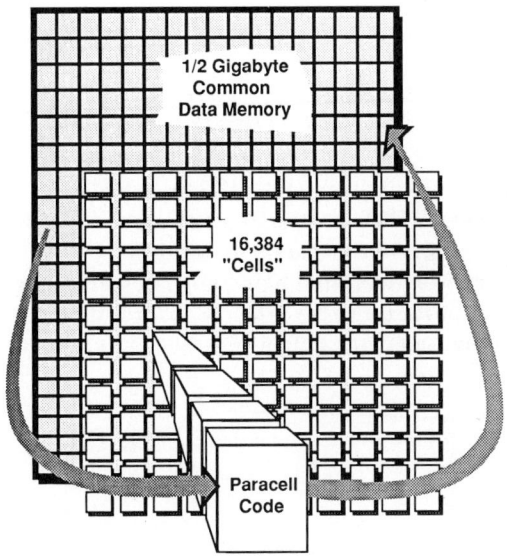

Figure 20.10 *The software view: thousands of independent cells simultaneously repeating the read-execute-write cycle.*

Local Memory All cells have their own local memory wherein the tile code is stored, as well as local variables.

20.3.3 Paracell Program Organization

Paracell programs are written in statements, statements are grouped together into tiles, and tiles are assigned to cells for execution. An average tile will contain two to four statements, and, by default, a cell will have one tile assigned to it for execution.

Tiles Paracell programs are organized into tiles. A tile will contain a memory to memory transform that may take the form of a function, rule(s), or short program. When loaded in the PIM, each "tile object" holds the code for the tile, and information about how the tile is scheduled. By default a tile has a one to one relationship with a cell such that all of the cell's resources are devoted to that one tile all of the time. (See Figure 20.11.)

If a tile contains to much code for one cell to execute in one frame, the state of the tile is saved at the end of the frame, all exports are held, and execution will continue during the next frame with the original import values. The user will be informed that one of the tiles is too big, and the user may take one of four actions:

1. Break the code in to multiple independent tiles.
2. Break the code in to multiple tiles in a tile sequence.
3. Create a cell group (not to be confused with a tile group).
4. The user may choose to leave it alone if it does not need to execute every frame.

Groups and Sequences If all tiles were allocated to cells one to one, 16,384 tiles could be loaded into the PIM. However, many tiles will not need to execute at the frame rate (60 times per second). Thus provisions have been made for allocating multiple tiles to single cells. When a tile does not need to execute every frame, it may be placed in a *tile group*. Tiles in a tile group take turns using the cell's resources in a round robin fashion. For example, if two tiles were assigned to one cell, one of the tiles would execute on odd frames, the other on every frames, each tile executing 30 times per second. By grouping together control tiles that do not need 16-ms response

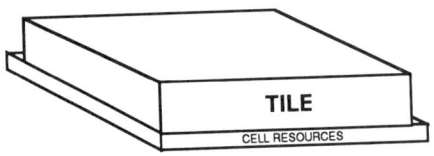

Figure 20.11 *One tile, one cell.*

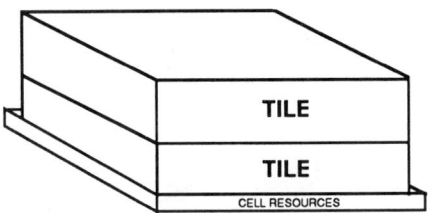

Figure 20.12 *Many tiles, one cell.*

(such as temperature control), and allocating them to one or a few cells, many more tiles may be loaded into a PIM than there are cells. (See Figure 20.12.)

Tile sequences are a flexible way of implementing larger pieces of code as a single unit. A tile sequence consists of two or more tiles linked via *piped variables*. A piped variable is a local variable that is shared by two or more tiles in a tile sequence. Piped variables are initialized by the first tile in the tile sequence, and successive tiles will view the value of piped variables as set by previous tiles in the tile sequence. (See Figure 20.13.)

Some problems are not decomposable enough to match the resources of cells, and although tile sequences may resolve many of these problems, large tiles will occasionally be needed. In order to accommodate large tiles, provision have been made to join cells together to form *cell groups*. (See Figure 20.14.)

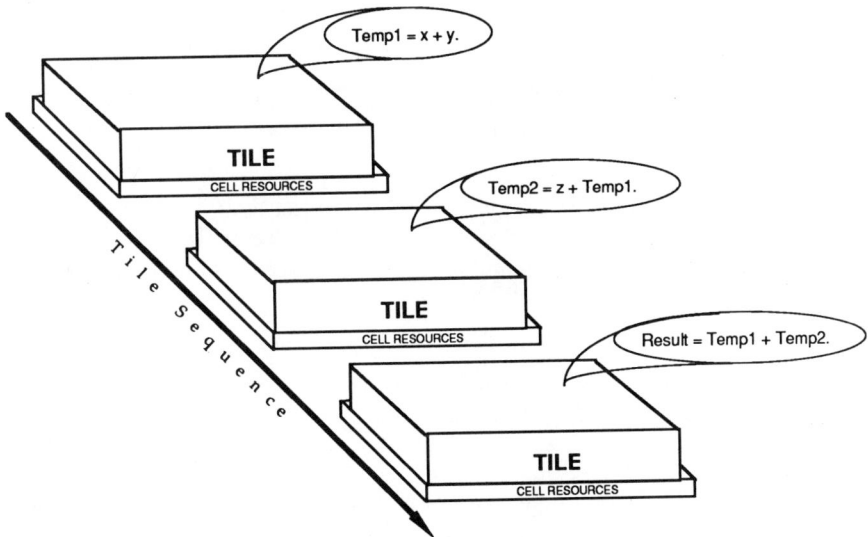

Figure 20.13 *Tile sequence example ($Temp_1$ and $Temp_2$ are piped variables).*

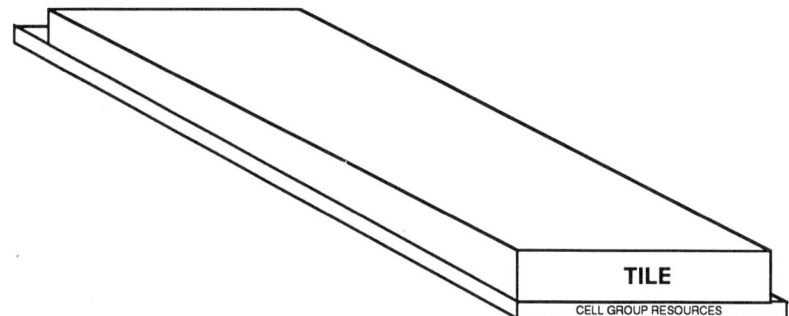

Figure 20.14 *A large tile in a cell group.*

20.3.4 Paracell Syntax

Paracell syntax is based on the English language (it can be easily modified to German, French, Japanese Kanji, or other languages), and uses English punctuation. Just as there are many ways in English to state something, there are many ways in Paracell. Among the concepts Paracell relies upon to ensure the readability of control code are alternate syntax, noise words, aliasing, automated variable declaration, multiple word variables, automatic type conversion and context-dependent parsing.

In the following equivalent examples, which illustrate the flexibility of Paracell, the first is verbose, the second terse.

1.
```
Big circle is a boolean.
Pi is the constant 3.1415927.
Circumference and radius are paracell numbers.

Set circumference to 2 times pi times radius.
If circumference is greater than 10, then big circle is
    true, else,
  Big circle is false.
```

2.
```
Circumference = 2 * pi * radius.
If circumference > 10, then big circle = 1, else
big circle = 0.
```

In the latter example, the compiler will search the data base for the variable names used, assign definitions when found, and offer default definitions based on context when not found.

Alternate Syntax Alternate syntax simply means that the programmer often has one or even two alternate but equivalent means of constructing a statement. For example, the following two statements are equivalent, but adhere to different syntaxes:

```
Set inventory to inventory plus surplus.
Add surplus to inventory.
```

Noise Words Noise words include parts of speech such as "the," and "a," and their omission have no effects on the meaning of Paracell statements. Specific words such as "and," are only sometimes considered noise-dependent on context. For example, the following two pairs of statements are equivalent, but the noise words have been removed from the latter pair:

```
Maximum temperature is the constant 3000.
Add A, B, and C to D.

Maximum temperature is constant 3000.
Add A, B, C to D.
```

Aliasing The programmer may define aliases for any variable or keyword so as to enrich the vocabulary of Paracell. Once the alias is defined, it may be used throughout the system. For example, a programmer may prefer the word "otherwise" to the word "else." Aliasing may also be used to simplify the naming of variables. For example, a tile could create the alias "tank" for the variable "Ammonia tank 37," in order to simplify the code, or to use generic tank control code.

It is further possible to have the word "tank," for example, to be an alias for two different variables in two different tiles. If "tank" were used subsequently in a new tile, the programmer would be asked to define it according to one of the two existing aliases, or to create a new variable and alias.

At any time, the user may view any control code with all aliases removed, and replaced with the root variable names, or keywords.

Automated Variable Declaration Variables need not be declared in Paracell statements. When the Paracell compiler comes across a variable name in a Paracell statement for which it cannot find a definition, it offers the user a default definition based on the context in which the variable name was used. The user may accept the definition or override the compiler. When the compiler finds multiple definitions for the same name due to aliasing, the compiled offers the list of possible definitions for the user to choose from, and again the user may override and create a new definition.

Multiple Word Variables Multiple word variables are supported. For example:

```
Maximum temperature is the constant 3000.
If temperature exceeds maximum temperature then
    set red alert to on.
```

In a multiple word variable is first used embedded in a Paracell statement, it must have underscores between the words in order for the compiler to successfully understand the word. These underscores are removed automatically once the compiler has entered the word into the variable dictionary.

Automatic Type Conversion Most languages require type conversion statements to be performed prior to mathematical operations involving mixed types such as integers and floating-point number. In Paracell, it is not necessary to coerce variables for mixed type operations. At compile time all mixed-type operands are automatically coerced, and operations are checked for potential precision loss. Warnings are given to the programmer accordingly.

Context-Dependent Parsing Certain keywords can have different meanings depending on how they are used. For example, the keyword "and" can be both a conjunction and a logical operator. In the following examples, the former "and" is a logical operation to be performed with the Boolean variables *B* and *Z*. The latter "and" indicates that two assign operations are to be performed, setting both *A* and *B* to *Z*.

```
Set A to B and Z.
Set A and B to Z.
```

20.3.5 Paracell Semantics

Semantics refers to the way in which Paracell code is implemented—what the computer does when you tell it to do something. Only those aspects of Paracell semantics that are nonintuitive are described here. These differences relate to the relationship of Paracell statements within a tile when global variables are used, and when local variables are used.

Global Variable Semantics Statements within a tile execute in parallel with respect to global variables. This means that a tile containing the pair of similar statements:

```
Add 1 to inventory.
Add 2 to inventory.
```

whereby inventory is a global variable, would *not* add 3 to inventory. Instead, the last statement that modifies the value of inventory "wins," so that in this case 2 is added to inventory.

To ensure the parallel effect of statements in a tile, even when there are more statements that can be executed in one frame, all results (exports) are held until all statements have had a chance to execute. Exceptions occur when sequences are specified and when intelligent memory operations are used.

Statements in multiple tiles (except statements with intelligent memory operations—see later) also execute in parallel with respect to global variables. This means that two tiles containing the similar statements:

```
Tile A-    Add 1 to inventory.
Tile B-    Add 2 to inventory.
```

whereby inventory is a global variable, would *not* add 3 to inventory. Instead, the programmer is asked to specify which tile will "win," such that 1 or 2 is added to inventory, according to the programmer's priorities.

There are cases where global variables need to be modified every time a tile so specifies, regardless of the number of such operations specified in parallel—a job counter, for example. Those cases are satisfied through the use of *intelligent memory*. Intelligent memory permits tiles to perform operations on global variables as the variables are imported and/or exported. For example, 10 tiles containing the intelligent memory statement

```
Read and increment job counter.
```

whereby *job counter* is a global variable, would result in 10 being added to *job counter*. These increment operations are carried out within global memory.

Local Variable Semantics The statements in a tile execute serially with respect to local variables. This means that a tile containing the statements

```
Add 1 to inventory.
Add 2 to inventory.
```

whereby inventory is a local variable, would add 3 to inventory.

Piped Variable Semantics The statements in a tile, and in multiple tiles, execute serially with respect to piped variables. In the example

```
Tile A-    Set piston count to L4 count times 4.
Tile B-    Add V6 count times 6 to piston count.
Tile C-    Add V8 count times 8 to piston count.
```

in which *piston count* is declared as a piped variable by all three tiles, the value of *piston count* at the end of tile A's execution is "piped" in as the initial value of *piston count* for tile B's execution, and so on for tile C. Thus, the final value of the piped variable, *piston count,* is affected by all tiles that declared it as a piped variable. Note, that tiles containing piped variables must have an execution order explicitly specified by the programmer.

Mixed Type Operations When operations involve sources and targets of mixed type, the sources must first be converted to the type of the target in order for the operation to be performed. When an obvious conversion is not evident, such as adding a floating-point number to a character, Paracell will still try to accommodate the user.

20.4 THE PIM

The PIM was developed by Flavors Technology Inc., in Amherst, New Hampshire, from 1987 to 1989. The goal of the development team, headed by Richard Morley, father of the programmable controller, was the creation of a new generation of industrial computer capable of addressing the needs of today's large scale automation and control problems. The architecture of the PIM was designed to execute Paracell programs in real time.

The basic architecture of the PIM is that of a systolic parallel processing machine. The PIM has two basic building blocks, processors (equipped with local memory), and global memory. The PIM may be configured to meet a wide range of needs by matching the right amount of processors and global memory to fit an application. For future requirements, processors and/or global memory may be added to meet expanding needs. (See Figure 20.15.)

The up-to-16,384 virtual processors in the PIM import data, execute tiles, and export data in regular cycles called frames. The frame time of the PIM is typically set at about 16 ms.

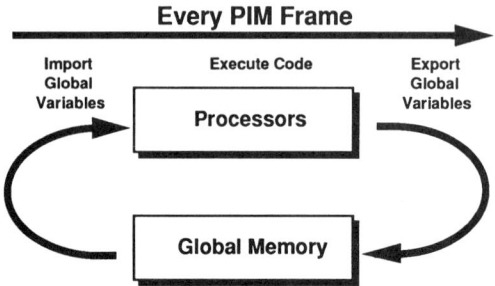

Figure 20.15 *The basic architecture of the PIM.*

20.4.1 Basics

A PIM consists of a mixture of four kinds of boards plugged into a high-speed bus, and various peripherals and network interfaces. The global memory is distributed over three kinds of boards, two of which are dual-ported to allow communications via I/O networks, or bus coupling. The fourth kind of board contains all the user processors.

The boss board (one per PIM) controls the PIMbus and provides the interface to the development workstations. The boss also contains 4 Mbytes of dual-ported global memory accessible by the multiprocessor (MP) boards and the development workstations. Memory boards contain from 32 to 128 Mbytes each of global memory. Input/output boards contain from 4 to 16 Mbytes each of dual-ported global memory accessible by the MP boards and by other computers or peripheral devices. (See Figure 20.16.)

The processors are distributed four per MP board, and access the global memory on the other boards via the PIMbus. Each processor supports 128

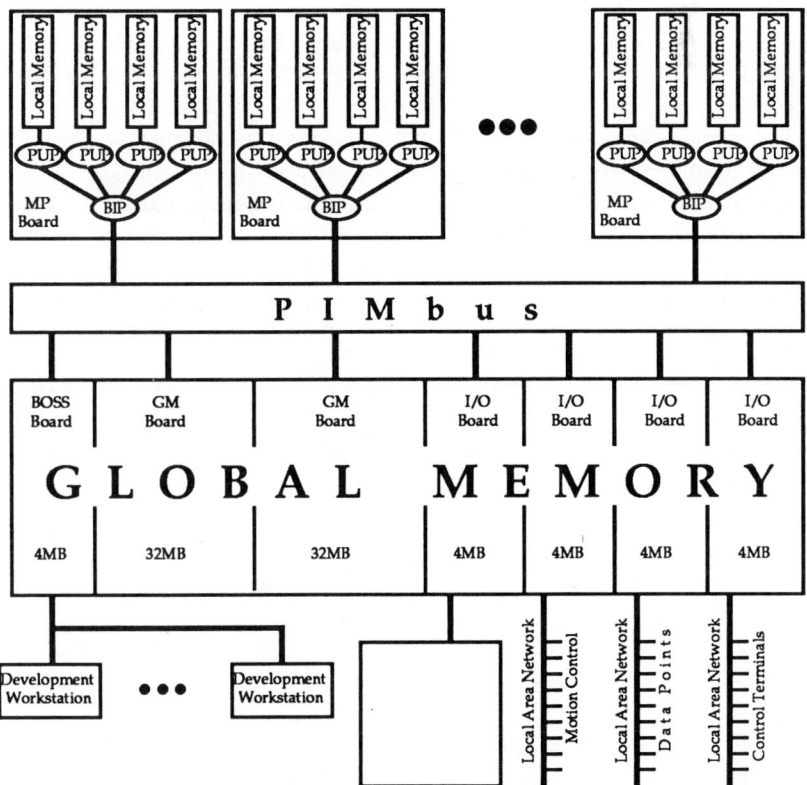

Figure 20.16 *The PUPs and the BIPs make up the brains of the PIM. Processors are also located in the boss board and I/O boards, and arithmetic logic units are located in all global memory.*

548 INTEGRATED COMPLEX AUTOMATION AND CONTROL

virtual processors or cells (512 cells per MP board). The minimum number of processors is 4 (one MP board). The maximum number of processors is 128 (32 MP boards).

A minimum-configuration PIM comes equipped with one boss board (containing 4 Mbytes of global memory), and MP board (supporting 512 cells), and a development workstation. A PIM may be configured with up to 40 boards, one of which must be a boss board. The remaining 39 boards must include at least one MP board, and may include up to 32 MP boards. These limitations aside, any combination of memory, I/O, and MP boards are acceptable. A typical maximum configuration includes one boss board, 32 MP boards, two memory boards, five I/O boards, and four development workstations (the development workstations are not required for the operation of the plant). This configuration contains a total of 16,384 cells and 340 Mbytes of global memory.

20.4.2 The PIMbus and the Boss Board

The PIM uses an advanced, proprietary, bus architecture to realize its 100-Mbyte/s performance. The electrical interface used by the PIMbus conforms to the IEEE 896 Futurebus+ standard. The PIMbus is driven by the boss board.

The user bandwidth of most asynchronous bus designs is significantly smaller than the gross bandwidth due to arbitration overhead. The PIMbus is a synchronous 25-MHz bus. There are no bus requests and there is no bus arbitration. Rather, the boss board grants 40-ns bus cycles to the MP boards according to a precise rotating schedule. The number of grants per PIM frame that an MP board receives is independent of the number of MP boards, thereby guaranteeing predictable performance (the user bandwidth of the PIMbus is 94.37 Mbytes/s with 32 MP boards).

To take full advantage of the synchronous design of the PIMbus, all reads (272,480 PIMbus cycles) are done during the first portion of a PIM frame and all writes (136,240 PIMbus cycles) during the second portion, requiring only two mode switches (8 PIMbus cycles each) per PIM frame. PIMbus reads and writes are never interspersed within a frame. (See Figure 20.17.)

$$272{,}480 + 136{,}240 + 16 = 408{,}736 \text{ PIMbus cycles per PIM frame}$$

$$408{,}736 \times 40 \text{ ns per PIMbus cycle} = 16{,}349{,}440 \text{ ns per PIM frame}$$

$$16{,}349{,}440 \text{ ns} \approx 16 \text{ ms} = 1 \text{ PIM frame}$$

Each processor has a cell 0 and a cell 1 and so on. The frame begins by satisfying all the imports for all cell 0's. Then all the imports for all the cell 1's, then all the cell 2's, and so forth. This is done because the import/export process is going on at the same time all the cells are executing. So it is necessary that by the time the cell 1's are executing, all of the imports for

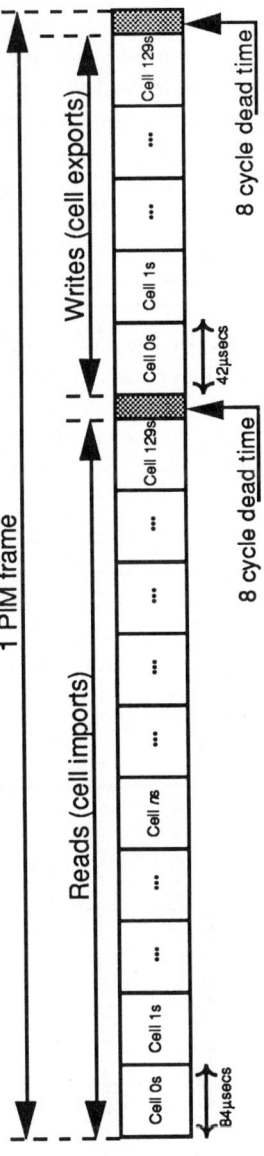

Figure 20.17 *Dissection of a PIM frame.*

those cells have already been transferred to local memory, and by the time the cell 2's are executing all the imports for those cells are in. The exports are then satisfied in the same way (except that there are half as many) and the whole cycle is repeated. Of the 130 cells, the first and last are reserved and are not available to the Paracell programmer.

The *dead time* is required for the bus to switch read/write mode. Eight PIMbus cycle are required for each mode switch in order to empty the bus of requests due to the eight-way interleaved memory (see global memory).

The PIM is not dependent on any particular type of chips. The processors we require (cells) are "virtualized" from whatever processor chips currently being used.

Reads Because PIMbus cycles (40 ns) are much faster than dynamic RAM (about 200 ns), the global memory boards are eight-way interleaved and cycle at 320 ns.

The boss board grants cycles for each of 32 MP boards in order, regardless of whether or not they are all present. During the read phase of a frame, any data traveling on the data lines is a response to a request made eight cycles previously and has nothing to do with the signals on the address lines at that moment. (See Figure 20.18.)

Writes The boss board gives a 40-ns grant to an MP board. The MP board puts data and address on the PIMbus and the memory board receives the request and begins cycling. The boss board gives a grant to the next MP board in line which puts another datum and address out on the bus directed to a different memory bank on the memory board. The boss board grants cycles for each of 32 MP boards in order, regardless of whether they are all present.

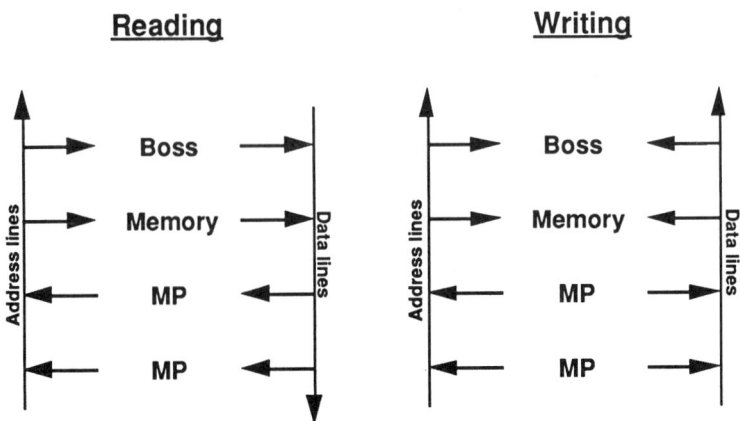

Figure 20.18 *Traffic during reads and writes.*

The Boss Board The boss board is the central bus controller of the PIM. Each PIM must have one of these boards plugged into the highest numbered slot of the PIMbus back plane. The boss takes care of general system utility functions and acts as the workstation interface. The boss board also contains 4 Mbytes of global memory (shared by the workstations). A minimum PIM configuration is one boss board, and one MP board. (See Figure 20.19.)

20.4.3 The Multiprocessor Boards

The MP boards are the main processing elements of the PIM. A PIM comes equipped with a minimum of one and a maximum of 32 MP boards. An MP board contains four identical PIM user processors (PUPs), each with its own floating-point unit, local memory and shared memory. A fifth microprocessors, called the bus interface processor (BIP), controls access to the PIMbus for global imports and exports. The BIP has its own local memory and I/O, plus dedicated hardware for controlling the bus accesses (each MP board receives a grant every 32 PIMbus cycles). The BIP does not directly access global memory. Instead, the BIP sets up transfers (imports and exports) between the PUPs and the PIMbus. These transfers are set up in the BIP's export/import memory, which is continually being loaded with export and import requests from the four PUPs. (See Figure 20.20.)

The PUP Each PUP is a compact microprocessor system, optimized for rapid real-time processing. The heart is a Motorola 68030 running at 20.0 MHz, coupled with a 68881 floating-point coprocessor also running at 20 MHz. Each PUP is time-sliced into 130 virtual processors. A PUP has 256KB of fast static random-access memory (SRAM) and 4 Mbytes of dy-

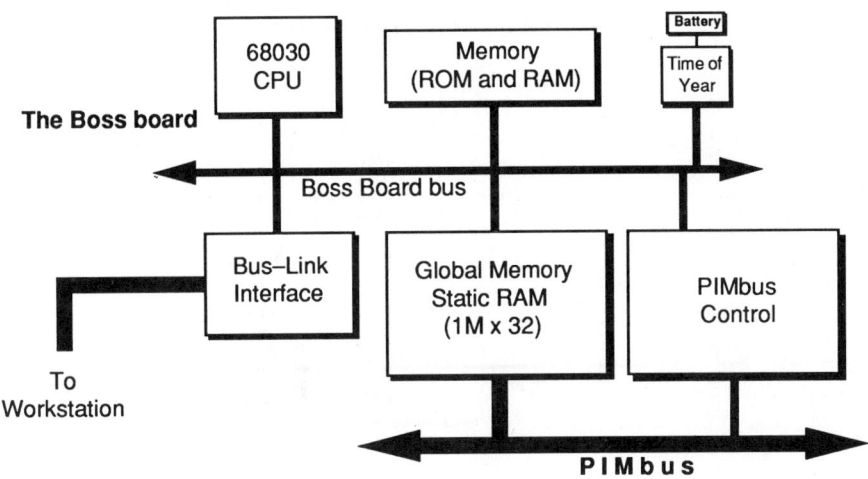

Figure 20.19 *Boss board detail.*

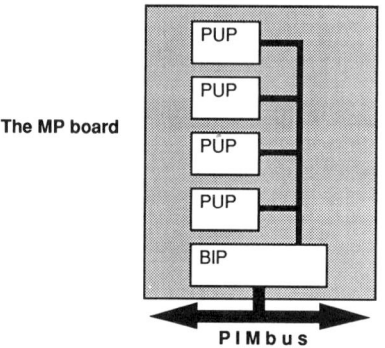

Figure 20.20 *Each MP board has five Motorola 68030 processors.*

namic RAM (DRAM) with ECC (Error Correction Code) which is shared with the BIP. Error correction allows any one memory chip to be removed from any or all memory banks without any loss of data. The user is informed of any memory chips that have failed. Tile code is stored in the local memory. (See Figure 20.21.)

20.4.4 Global Memory

All global memory, whether it resides on memory boards, I/O boards, or the boss board, is divided into eight banks accessible in overlapping cycles by the PIMbus. Error detection and correction is applied to all accesses. Global memory is smart memory. An on-board arithmetic/logic unit (ALU) per bank allows simple arithmetic and logic functions to be performed in a single memory cycle. (See Figure 20.22.)

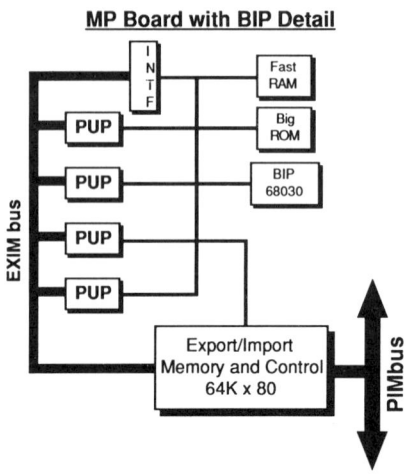

Figure 20.21 *The BIP processor off-loads housekeeping and I/O overhead from the PUPs.*

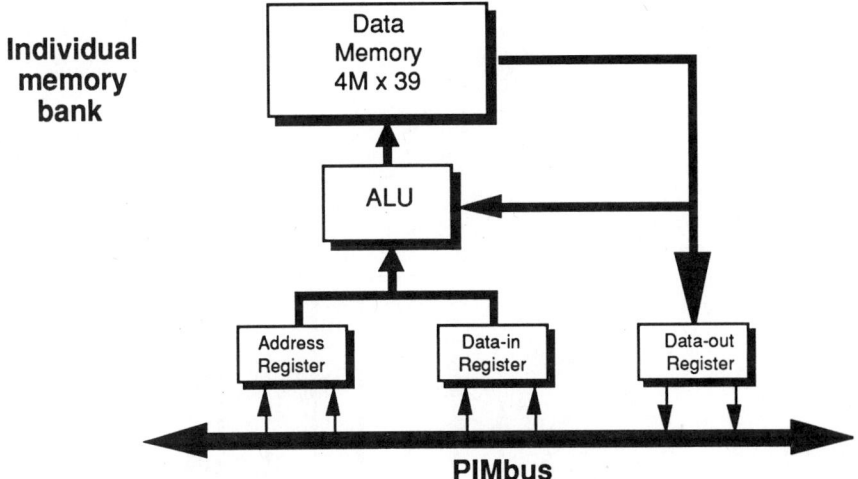

Figure 20.22 *All global memory is capable of performing arithmetic as instructed by the cells via a 3-bit operation code in the address. The memory is 39 bits wide to allow for the 7-bit error correction code.*

Interleaving Each bank of memory can cycle every 320 ns (eight PIMbus cycles). The eight memory banks are interleaved at 40-ns offsets. The entire memory board cycles every 320 ns; however, each memory board cycle overlaps the next. This results in a dead time of eight PIMbus cycles at the end of each read and write phase in order to empty the pipeline, during which no MP board places requests on the bus. (See Figure 20.23.)

Memory ALUs Each memory bank has its own ALU that performs operations on data in accordance with three dedicated signals on the PIMbus. This capability alleviates PIMbus traffic and processing cell load by allowing simple operations to be performed locally on the memory board. For example, in order for a cell to add a number to a variable in global memory, it need only export the address of the variable with the number to be added and set the dedicated ALU signals on the PIMbus accordingly. The variable does not have to be imported by the cell, and the add operation is performed by the ALU on the memory board. Without memory ALU's, the operation would require two PIMbus cycles instead of one, and more instructions would be required of the cell.

One of four operations may be performed with each global memory read, and one of four operations may be performed with each global memory write.

When a cell requests to import a variable, the cell may specify an operation to be performed after the read takes place, such that the value received by the cell reflects the previous value of the variable.

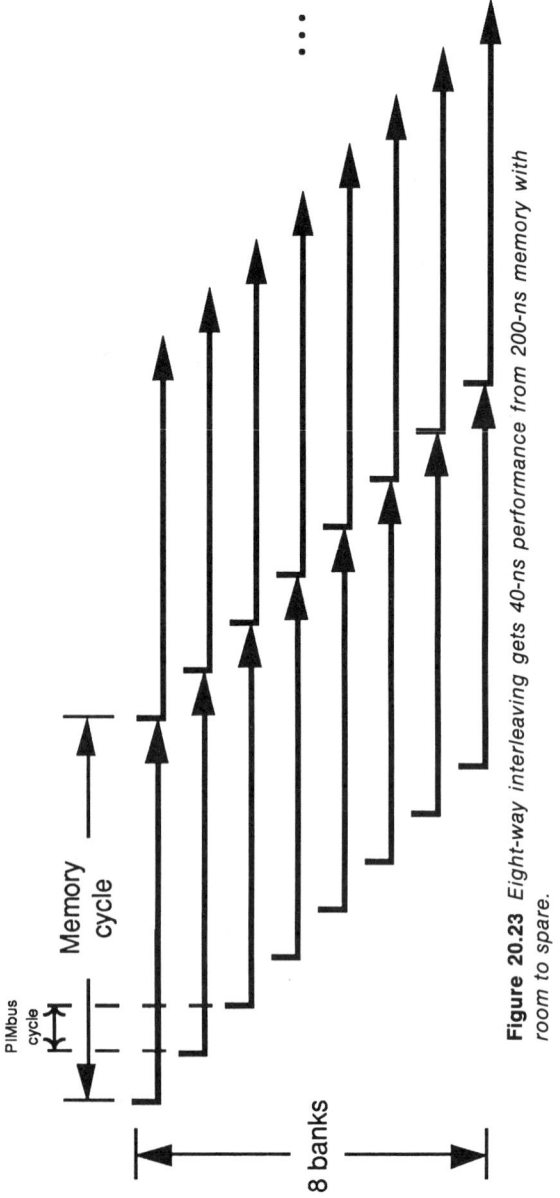

Figure 20.23 Eight-way interleaving gets 40-ns performance from 200-ns memory with room to spare.

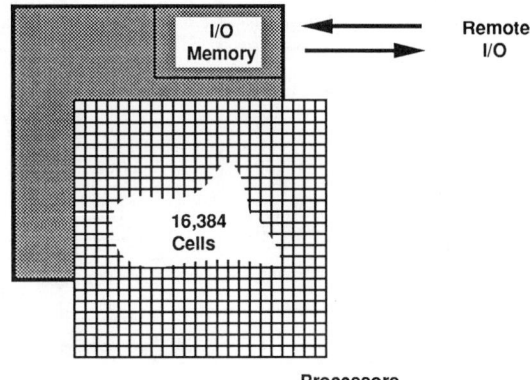

Figure 20.24 From the cells point of view, I/O memory is no different from the rest of global memory.

Decrement	Subtracts one from the variable
Increment	Adds one to the variable
Set	Sets all bits of the variable to 1's
Clear	Sets all bits of the variable to 0's

A cell may request to export an operation to a variable. In this case, the data sent on the data lines of the bus does not represent a new value to be placed at the specified address, but represents a modifier to the existing value.

Add	Adds a positive or negative number to the existing value
Exclusive OR	Performs an XOR logical operation on the existing value
OR	Performs an OR logical operation on the existing value
AND	Performs an AND logical operation on the existing value

20.4.5 I/O

In a typical automation/control application, one PIM can replace up to 25 minicomputers used as cell controllers, and as many as 100 PLCs providing a tightly coupled operation. The PIM can interface to many tens of thousands of I/O points through industry standard local area networks. The PIM must also be able to interface to the general-purpose computers that oversee the non-time–critical aspects of a factory. To that end, the PIM comes equipped with powerful and flexible distributed I/O subsystems.

All PIM I/O (with the exception of I/O to the development workstations) is performed through I/O memory, a subset of global memory. (See Figure 20.25.)

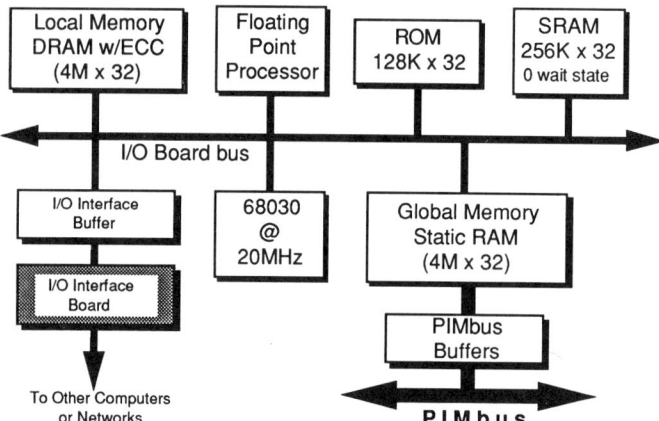

Figure 20.25 *Each I/O board comes equipped with a dedicated Motorola 68030 processor whose job it is to make I/O transparent to the cells.*

As explained in earlier chapters, the raw control resources available to Paracell programs consist of up to 16,384 cells communicating to each other through global memory. That global memory resides in memory boards, I/O boards, and the boss board. The global memory residing in the I/O boards (referred to as I/O memory) contains I/O variables associated with external data points. These associations are done in software when the system is configured.

The I/O Board The I/O board contains a microprocessor with a floating-point unit, various types of memory, and a buffer for an I/O interface board. These all connect to the 32-bit I/O board bus. The I/O processor is a Motorola 68030 running at 20 MHz. The I/O processor may access all of the memory on the I/O board bus, and communicates with the I/O interface board through the I/O interface buffer. The main function of the I/O processor is the transfer of data between global memory, local memory, and external devices through the I/O interface board. The bandwidth of every I/O board is approximately 21 Mbyte/s, and a PIM configured with 16 I/O boards is capable of 336 Mbyte/s of throughput. (See Figure 20.26.)

The secondary function of the I/O processor and the sole function of the floating-point coprocessor is the processing of data in transit. For example, some external devices may not format data according to the IEEE standards adhered to by the PIM. This could necessitate compute-intensive floating-point conversions.

Input/output board global memory contains 4 or 16 Mbytes of RAM, depending on the chips used. The global memory is dual-ported to the PIMbus, which always has priority over the I/O board bus. Unlike memory elsewhere on the PIM, global memory on the I/O board uses SRAM, which is four times as fast as the DRAM. Static RAM is used on the I/O board

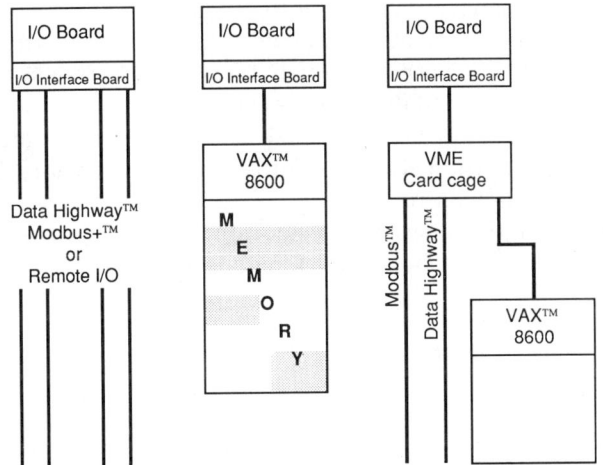

Figure 20.26 *Three identical I/O boards each connected to a different I/O interface board.*

global memory because it is dual-ported and allows the I/O processor access in between PIMbus accesses.

The I/O Interface Board With every I/O board, and I/O interface board is plugged in the opposite side of the backplane. The I/O interface board varies in design according to the device(s) to which it is connected. I/O Interface boards connect to networks such as Modicon's Modbus+, Allen-Bradley's Data Highway, and remote I/O networks. Other I/O interface boards act as bus couplers, connecting the I/O board bus to standard bus architectures such as VMEbus, Qbus, Multibus, and others. (See Figure 20.27.)

20.4.6 The Development Workstation

Paracell programs are developed on the development workstation, currently an Apple Macintosh IIfx. All of the tools used to develop Paracell programs

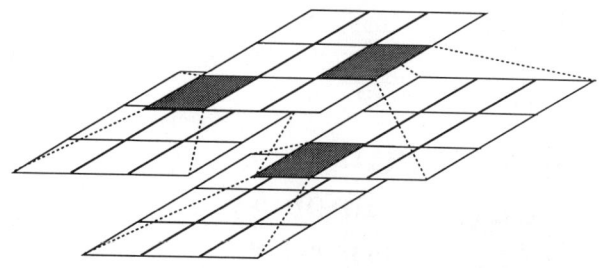

Figure 20.27 *A three-deep example of the Navigator view of a Paracell program. This view, taken 10 levels deep, can organize massive programming projects.*

have been written in accordance with the Macintosh human-machine interface standards. Those familiar with Macintosh will be instantly productive, and others will quickly learn as with all Macintosh applications.

The development workstation is connected to the PIM via a shared memory bus coupler, enabling the Macintosh to view the global memory in the boss board as an extension of the Macintosh's internal memory. The result is a net 1-Mbyte/s bandwidth.

Multiple workstations may be connected to one PIM enabling a team of programmers to simultaneously develop code for the PIM.

20.5 THE PARACELL ENVIRONMENT

In addition to the Paracell language, tools are provided on the macintosh workstation for creating, modifying, monitoring, and debugging Paracell

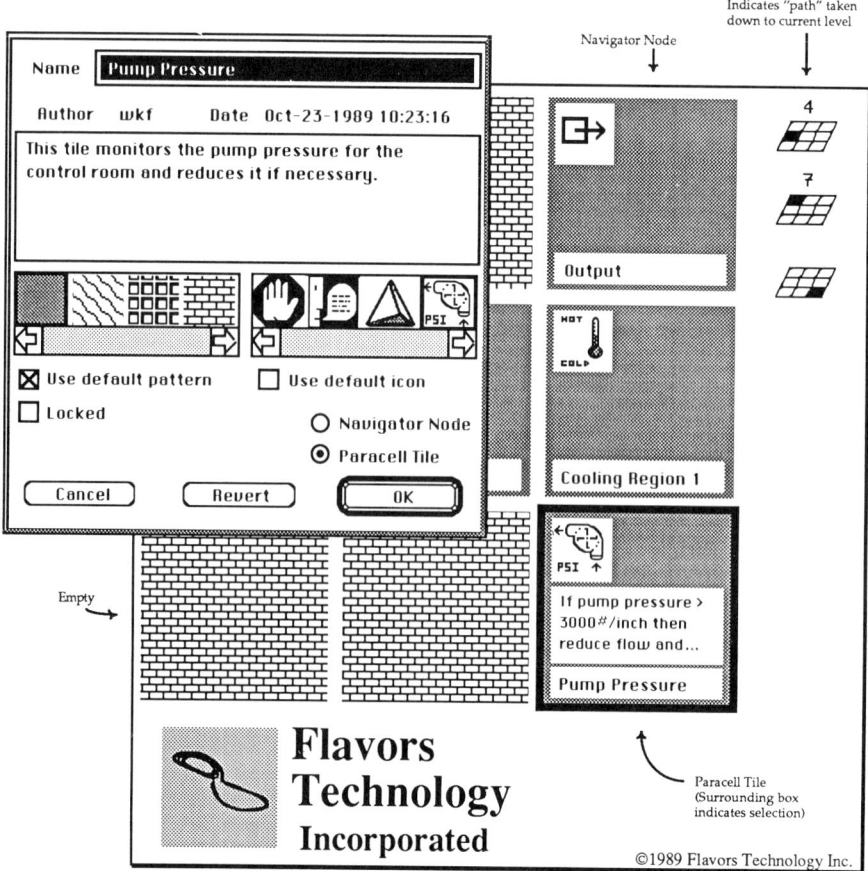

Figure 20.28 The Navigator window and a dialogue box as seen by the user.

20.5 THE PARACELL ENVIRONMENT

programs. Central to these tools is the Paracell Navigator, which is used to organize and browse the tiles. The other window-oriented tools which may be accessed from the Navigator include the Paracell Editor, Paracell Soft-Scope, and Paracell Debugger.

20.5.1 The Paracell Navigator

This tool is first and foremost a management tool. The Navigator allows the project manager to conceptualize, organize and view many tens of thousands of tiles three-dimensionally, and provides a natural means of dividing a problem into manageable pieces. (See Figures 20.28 and 20.29.)

The Navigator displays a 3 × 3 matrix of squares. Each square is either a Paracell tile, a Navigator node (a link to another 3 × 3 matrix), or empty. Squares are selected either by keypad, or by moving a mouse-controlled pointer to the square and "clicking" the mouse button. Whether a tile or a node, clicking the icon or name portions of the square displays a dialogue box (above left). Selecting other areas in the square displays the Paracell code for that tile or "enters" another 3 × 3 matrix one level deeper. On the right side of the window is a "map" showing the present location in the hierarchy. Clicking in the map will select a new location higher in the hierarchy. (See Figure 20.30.)

Two small round buttons in the lower right corner of the dialogue box enable the user to specify whether a tile or node is being created. If the user creates a Paracell tile, the Paracell editor is called, and the users may enter the Paracell control statements for the tile. (See Figure 20.31.)

The dialogue box is used to define and classify nodes and tiles. Users may choose from libraries of patterns and icons to better organize projects. For example, each major subtask could be assigned a different pattern and each programmer within the team could have a personal icon, or as in the example, an icon may be used that represents the machine being controlled.

Browsing Accompanying the Navigator are facilities which allow the user to browse through tiles and variables. Tiles may be viewed by author, creation date, imports, exports, and tile groupings. Variables and aliases may be

Figure 20.29 *Empty square, Navigator node, and Paracell tile.*

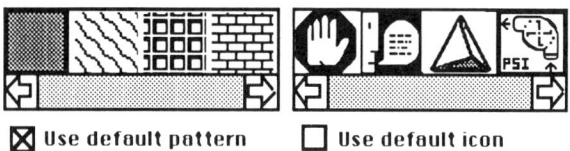

Figure 20.30 *Pattern and icon selection mechanisms.*

viewed by name, and when requested, a detailed description of a variable may be viewed.

The detailed variable description includes all the aliases for the given variable (or if it is an alias, the variable to which it is mapped), all tiles that import the variable, all tiles that export the variable, the type of the variable, whether it is an I/O variable (see Chapter ?), the units of the variable (e.g., kelvins), the scope of the variable, if it is an I/O variable the location of its associated data point, when the variable was created and by whom, and the documentation notes for the variable.

Documentation In addition to documentation that may be imbedded within Paracell code, which is available in most computer languages, the Navigator automatically documents the creation and modification of all tiles and variables with author, and time stamp. Revision histories for tiles and variables are updated, and are always available through the Navigator.

20.5.2 The Paracell Editor

The Paracell editor is an environment for the creation and modification of tiles. Included in the environment are various tools to assist in the organization and viewing of tiles. The chief benefit that the Paracell editor provides the user is *power typing*. While the user enters code, the editor is working behind the scenes, checking spelling, searching the dictionary for alias matches, and performing small-scale analysis or parsing like plurals and possessives. When the editor cannot understand something the user has written, the user is not interrupted. Rather, when the user finishes a session, the editor will highlight ambiguous words or questionable operations.

Although aliases provide a powerful flexibility to individual programmers, the system manager or an outside viewer may wish to view Paracell code without aliases. To this end, the editor has the facility for replacing aliases with the standard system-assigned variable names at the option of the user. This does not modify the source code, but is merely a display option.

The Translator Once Paracell code has been created with the editor, the next step is to translate it to Flavors implementation language (FIL). The translator completes the parsing begun by the Editor and replaces all aliases with their root definitions. The translator is fully incremental at the tile level.

This means that a single tile, or several tiles, may be created or modified without compiling and other tiles or inhibiting any running tiles.

The Loader The Paracell loader takes the FIL code, binds tiles to cells in the PIM, binds variables to global memory addresses, loads the code into the local memory of the cells, and initiates the final conversion from FIL to machine code. The conversion to machine code is performed locally by the processors in the PIM so that more powerful processors may be easily integrated into the PIM when they become available.

When a user specifies a tile or group of tiles to be executed on the PIM, it is automatically loaded. Running tiles in cells not being loaded are not affected in any way by the loading of tiles into other cells so as to minimize downtime for software changes. When a tile is loaded over another tile to replace it, or when a tile is loaded to be added to a tile group, the replacement or addition happens synchronously such that the performance of other tiles is not affected.

Normally, the loaded is fully incremental at the tile level. A single tile or several tiles may be loaded without reloading or otherwise inhibiting any other tiles. An option is provided to reload a number of tiles, up to the full number of tiles in the Navigator pyramid. This option would be used to optimize variables' physical memory allocation.

The loader maintains maps of cell and memory utilization. These maps are in symbolic form on the workstation for access by the debugger, and in binary form on the PIM for debugging.

20.5.3 The Paracell Debugger

The Paracell debugger provides mechanisms for debugging and controlling the execution of Paracell programs. Various views and controls are provided at the variable, import, export, tile, frame, and hardware levels. Built-in debugger tiles handle user interaction so that the debugger has no visible impact on the performance of running tiles. The debugger tiles also perform run time hardware diagnostics. These tiles use import and export slots which are needed to synchronize the PIM frame with memory banks and do not impose a performance penalty on user programs.

Variable Monitoring The main debugging facility is the real-time display and control of global variables. Imports may be set, and exports may be monitored for any tile. SoftScope (see next section) provides a user-friendly interface for variable monitoring.

Performance Diagnostics The debugger also can monitor the execution of tiles and provide performance data for any specified tiles. The available performance data for a given tile includes the number of times the tile has completed, and the minimum and maximum cell time left over each time the tile completes.

20.5.4 SoftScope

SoftScope provides a user friendly interface monitoring Paracell program execution. Global PIM data from the debugger are displayed in real time and formatted so that users do not have to develop data display programs for debugging new Paracell applications. SoftScope does not control the execution of PIM; it reports what the PIM is doing.

SoftScope Displays SoftScope allows users to construct arbitrary display windows freely and easily. There are various types of display elements from which to choose:

Digital display	Looks much like the display of a multimeter
Binary display	An on/off display (e.g., PIM cabinet door open/closed)
Bar chart display	An analog version of the digital display
Strip chart display	An analog display that keeps a recorded history of values

20.6 SUMMARY

The continued advancement of the process control and factory automation industries is not being hindered by any lack of progress in software technology, nor in microprocessor or memory technologies. The performance of processors and memories is steadily improving as always, and the software industry is providing more and more innovative and easy to use packages that place to power of computers in to the hands of the people with the problems. However, these improvements are constrained by the control model into which they must fit. A model which has been shown inadequate for the large and complex control problems being put forth by heavy industry around the world.

The decision makers in automation and control need to step back and take a look at the jungle of networked programmable devices controlling their operations. This jungle is a larger version of the one made up of relays in the late 1960s. Just as the PLC and ladder logic put control under the umbrella of one machine and one language in 1969, there is a need again, for one computer and one language capable of encompassing the much more complex control systems of today.

CHAPTER 21
Robotics and Flexible Automation Simulator/Integrator

MIOMIR VUKOBRATOVIĆ

21.1 INTRODUCTION

Adequate study of robotic mechanisms starts with the development of computer-oriented methods for forming the mathematical models of kinematics and dynamics of spatial active mechanisms. The rising rate of development of robotic mechanism mathematical models went the way from the numerical-iterative computer methods, over the numeric-symbolic ones, and, finally, to the forming of mathematical models in symbolic form. While the first version, even the second, is predominantly a research version of the computer procedures in forming mathematical models of robotic mechanisms, the third one, symbolic model form, is a typical implementation version. Mathematical models of robot kinematics and dynamics are not the goal, but tools for the synthesis of dynamic control of these mechanisms. Thus, the customized software for the synthesis of dynamic control laws of various complexity is a significant step toward the realization of control system, the performances of which satisfy the evermore complex tasks of industrial manipulation, as well as the manipulation being performed in some unconventional working environments. The further and evermore complex role of robots should be sought in the scope of flexible production cells, and broader, of intelligent technological systems. Hence the customized software should be expanded toward the simulation and control of flexible models, by means of which a specific, new, and more extensive promotion of

---

Neural and Intelligent Systems Integration, By Branko Souček and the IRIS Group.
ISBN 0-471-53676-8 ©1991 John Wiley & Sons, Inc.

robotics technology is carried out. The number of robots applied within various flexible automation cells (FMC) is increasing. One of the major problems arising in these applications is the fast and efficient selection of adequate robots for each FMC. On the other hand, it is often necessary to redesign the robot controller in order to cope with the specific requirements in the FMC. To solve these problems, various software systems for simulation of robotic FMC have been developed. Most of these software systems simulate all subsystems within the FMC as finite automata, neglecting their kinematic and dynamic characteristics. There are only a few software packages that enable simulation of the robot kinematics. Such an approach is usually quite acceptable if NC machines, conveyers, and so on, are in question: These subsystems usually behave uniformly in all working regimes, and therefore their performance may be represented by very rough models, such as finite automata models.

21.2 CUSTOMIZED SOFTWARE FOR MANIPULATION ROBOT DYNAMICS

21.2.1 Symbolic Models of Robot Dynamics

In the last several years, many papers were published, rendering an important contribution to the development of computer methods for the forming of mathematical models of robotic mechanisms, as well as the computer synthesis of nonadaptive and adaptive dynamic control of robotic systems of various complexity and dedication. The modeling methods may be classified with respect to the laws of mechanics on the basis of which motion equations are formed. One may distinguish methods based on Lagrange, Newton-Euler, Appel, and other formalisms for dynamic modeling of interconnected multibody systems. The other criterion is whether the method yields the solution to both the direct and the inverse problems of dynamics or only to the inverse problem (computation of driving forces for a given desired motion of manipulator). The number of multiplications/additions is yet another criterion of comparing the methods. This criterion is the most important one from the point of view of their real-time applicability.

The common property of the numerical-iterative methods is that they do not depend on specific manipulator design. They employ general laws of kinematics and dynamics of joint-interconnected rigid bodies. For real-time applications it is suitable to customize the algorithm for specific manipulator design.

Although the manipulator modeling methods based on Lagrangian and Newton-Euler equations have been developed so rapidly, it was shown that, no matter from which equations the model originates, it may be simplified further by taking into account the specific manipulator configuration. Because of that, the necessity for the model optimization algorithm arose.

During last several years a number of papers dealing with such a problem appeared, as well as program packages for automatic model generation: SYM [1], SYMB [2,3], EMDEG [4], SYMORO [5], ARM [6,7]. They all have more or less the same structure; The optimized model is generated in two steps, where the output of the first one, the suboptimal model, is the input for the second.

Khalil and Kleinfinger [5] utilize the Newton-Euler model both for serial-link and tree-structured robots. Using the fact that driving torques depend linearly on joint masses and tensor of inertia, they regroup system parameters with an intention to minimize the overall number of system parameters. Identification of the parameters that should be grouped is done by the Lagrangian equations, which clearly express linear dependency in closed form. When all parameters are assigned to some group, all group members are set equal to zero, except one, which carries the contribution of all group members. That parameter is introduced in Newton-Euler equations, obtaining the minimal number of operations. This algorithm is not iterative, which guarantees relatively short model generation time, as well as the possibility of applying the algorithm for redundant manipulators with more than 6 degrees of freedom (DOF). Also, it is applicable only for inverse dynamics model generation.

Neuman and Murray [7] gave a different approach to the multistep manipulator model generation. Starting from equilibrium equations that originate from an arbitrary method and written in the form of trigonometrical polynomials, they reduce them due to algebraic/trigonometrical rules. In [7] the 6-DOF PUMA manipulator, driving torques expressions are given. In the form of polynomials, the expressions contain as many as 12,000 multiplications! After reduction, the number of operations is almost 100 time lower. The reduction algorithm is based on a searching technique, which means that the model generation time grows very fast with increasing number of degrees of freedom. On the other hand, in opposite to [7], the general algebraic laws are applied to any set of trigonometrical polynomials describing not only the robot dynamics but also the dynamics of other mechanical systems. Many other modeling techniques have been developed [8–11], being directed at reduction in computational requirements. Now we are able to summarize the characteristics that any multistage model generation algorithm should have

1. Ability to generate wide class of models, which means general optimization algorithm for a given model type;
2. Ability to generate redundant manipulator models—avoiding time explosion in relatively complex systems.

21.2.3 Example

A simple example is given here. Figure 21.1 shows the input symbolic model for the computation of driving torques of a planar manipulator with two

```
*ROBOT_DU : OFF-LINE program          continued
SUBROUTINE ROBOT_DU$OFF               *Max frequency
INCLUDE 'M$SYM:ROBOT_DU.CMM'          OMD_2_1_3= QDD_1+QDD_2
RRR_1_3= RR_1_1*RR_1_1                UR_1_1=-WW_1_3_3*A_1
BJZZ_1= AJZZ_1+AM_1*RRR_1_3           UR_1_2= QDD_1*A_1
RRR_2_3= RR_2_1*RR_2_1                UR_2_1=-WW_2_3_3*A_2
BJZZ_2= AJZZ_2+AM_2*RRR_2_3           UR_2_2= OMD_2_1_3*A_2
AMS_1_1= AM_1*RR_1_1              VD_2_2_1= UR_2_1+C_2*UR_1_1+S_2*UR_1_2
AMK_1= AMS_1_1/A_1                VD_2_2_2= UR_2_2-S_2*UR_1_1+C_2*UR_1_2
AMS_2_1= AM_2*RR_2_1                  UMS_1_2= AMK_1*UR_1_2
AMK_2= AMS_2_1/A_2                    AFA_1_2= AM_1*UR_1_2
END                                   AFC_1_2= AFA_1_2+UMS_1_2
                                      UMS_2_1= AMK_2*UR_2_1
                                      UMS_2_2= AMK_2*UR_2_2
                                      AFA_2_1= AM_2*VD_2_2_1
                                      AFC_2_1= AFA_2_1+UMS_2_1
*ROBOT_DU : ON-LINE program           AFA_2_2= AM_2*VD_2_2_2
SUBROUTINE ROBOT_DU$ON                AFC_2_2=AFA_2_2+UMS_2_2
INCLUDE 'M$SYM:ROBOT_DU.CMM'          ANC_1_3= QDD_1*BJZZ_1
IF (F$_1) THEN                        ANC_2_3= OMD_2_1_3*BJZZ_2
                                      AFF_2_2_1=-SPF_1+AFC_2_1
*   Min frequency                     AFF_2_2_2=-SPF_2+AFC_2_2
    C_2=COS(Q_2)                  AFF_2_1_2=
    S_2=SIN(Q_2)                  S_2*AFF_2_2_1+C_2*AFF_2_2_2
END IF                                AFF_1_1_2= AFF_2_1_2+AFC_1_2
IF (F$_2) THEN                        AMKV_1_2= AFF_1_1_2+AMK_1*UR_1_2
                                      RF_1_3= A_1*AMKV_1_2
*   Medium frequency                  AMKV_2_2= AFF_2_2_2+AMK_2*VD_2_2_2
    OM_2_1_3= QD_1+QD_2                RF_2_3= A_2*AMKV_2_2
    WW_1_3_3= QD_1*QD_1               ANN_2_2_3=-SPN_3+ANC_2_3+RF_2_3
    WW_2_3_3= OM_2_1_3*OM_2_1_3       ANN_1_1_3= ANN_2_2_3+ANC_1_3+RF_1_3
END IF                                P_1= ANN_1_1_3
                                      P_2= ANN_2_2_3
                                      END
```

Figure 21.1 *Customized symbolic model.*

revolute joints moving in a horizontal plane (input for the optimization algorithm). Figure 21.2 shows the output of the optimization algorithm—the generated optimized model.

Table 12.1 shows the number of multiplications (*) and the number of additions (+) for several six joint manipulators. The UMS-2 robot has TTR-RRR structure, while the other three robots have RRR-RRR structure. The column SYM shows the number of multiplications and additions in the models obtained after customization of the Newton-Euler equations [1]. These models are used as input models for the optimization algorithm presented in this chapter. The column SYM-OPT shows the numerical complexity after the optimization. Thus comparing the efficiency of the input and output model, we see that it is improved from 25 to 75 percent.

One floating-point (FP) operation takes about 20 μs on a low-cost micro-

```
*ROBOT_DU_OPT : OFF-LINE program     *ROBOT_DU_OPT : ON-LINE program
SUBROUTINE ROBOT_DU_OPT$OFF          SUBROUTINE ROBOT_DU_OPT$ON
INCLUDE 'M$SYM:ROBOT_DU_OPT.CMM'     INCLUDE 'M$SYM:ROBOT_DU_OPT.CMM'
RRR_1_3= RR_1_1*RR_1_1               C_2=COS(Q_2)
BJZZ_1= AJZZ_1+AM_1*RRR_1_3          S_2=SIN(Q_2)
RRR_2_3= RR_2_1*RR_2_1               QD_1_2   = QD_2+QD_1
BJZZ_2= AJZZ_2+AM_2*RRR_2_3          QDD_1_2  = QDD_2+QDD_1
AMS_1_1= AM_1*RR_1_1                 QDD_1_1_2 = QDD_1+QDD_1_2
AMK_1= AMS_1_1/A_1                   VAR_1  = -QD_1_2*QD_1_2
AMS_2_1= AM_2*RR_2_1                 VAR_2  = QDD_1_2*CONS_2
AMK_2= AMS_2_1/A_2                   VAR_3  = QD_1*QD_1
CONS_1= A_2*(AM_2*A_1+AMK_2*A_1)     VAR_4  = QDD_1_1_2*C_2
CONS_2= BJZZ_2+A_2*(AM_2*A_2+        VAR_5  = CONS_3*QDD_1
         AMK_2*A_2+AMK_2*A_2)        VAR_6  = S_2*VAR_3
CONS_3= BJZZ_1+A_1*(AM_2*A_1+        VAR_7  = QDD_1*C_2
    AM_1*A_1+AMK_1*A_1+AMK_1*A_1)    VAR_8  = VAR_3+VAR_1
END                                  VAR_9  = VAR_2+VAR_5
                                     VAR_10 = VAR_7+VAR_6
                                     VAR_11 = VAR_10*CONS_1
                                     VAR_12 = VAR_8*S_2
                                     VAR_13 = VAR_12+VAR_4
                                     VAR_14 = VAR_11+VAR_2
                                     VAR_15 = VAR_13*CONS_1
                                     VAR_16 = VAR_15+VAR_9
                                     END
```

Figure 21.2 *Optimized model.*

computer. About 100 FP operations are distributed on one processor board. Thus, the dynamic model computations take about 2 ms on a six-processor computer. Much lower computing times are achievable by the use of array-processor-oriented controllers. Preliminary results [12] of implementation on FRT-300 array processors for PUMA-560 robot show that computing time is reduced to 0.1 ms.

The main problem with control of robotic systems is: To what extent it is necessary to include dynamic terms in the control law?

21.2.4 Nonadaptive Dynamic Control of Robots

Requirements for dynamic control are most accurate tracking of "fast" trajectories, simplest control structure, and control robustness to robot parameters variations. Dynamics approach can be realized via optimal control synthesis, which is realized via minimization of some criterion (minimization of movement execution time) [13]; control by linear optimal controller (synthesis based on linearized model) [14,15]; control via "inverse method" demanding "on-line" calculation of the complete robot dynamics model [16]; control via force feedback, using the fact that by means of force measurement on the robot direct information about the robot dynamics is obtained [17-19]; decoupled control, via "on-line" calculation of robot dynamics compensates the same and decouples it into subsystems [20]; decentralized control stabilizing each joint (subsystem) of robot independently from the rest of system; the stability of the whole system is investigated and, if necessary, global (cross-coupling) feedback loops are introduced [21,22].

21.2.5 Computer-Aided Control Synthesis [23]

Modules of general software are nominal trajectory-forming block, kinematic constraint block, mechanism dynamic parameter setting block, actuator parameters setting block, synthesis of nominal dynamics, control quality setting, local control synthesis, nonlinear model analysis, global control synthesis, and synthesis of control in discrete form simulation block, and control laws choice of microcomputer realization. A global flowchart of the package for control synthesis is presented in Figure 21.3. The algorithm enables the control synthesis for an arbitrary manipulator configuration and arbitrary manipulator type. Thus, the user can synthesize control either for (1) an already built manipulator (evaluation of manipulator) or (2) for a manipulator that is under design. In case (1), the algorithm can help to improve control for the existing manipulator and determine the equipment necessary for this improvement. In case (2) the user can investigate the influence of the choice of various manipulator parameters upon the control synthesis and the performance of the manipulator.

21.2.6 Hierarchical Control Structure

Usual approach in control of manipulation robots included in flexible manufacturing cells is by hierarchical decomposition of control task. This means that the robot control system is decomposed in hierarchical levels. This control structure comprises three control levels [24]:

Strategical control level concerns the problem of trajectory planning in the given complex environment. Using the information from various sensors (visual, tactile, etc.), the strategical control has to plan trajectory of the robot hand in order to accomplish the robotics task given by higher control level or from operator. Output of this control level is a specified path in external (hand) coordinates. *Tactical* control level has to perform mapping from external (hand) coordinates into internal (joint) coordinates of the robot (i.e., to compute joint trajectories of the robot). In doing this the tactical control level has to take into account the state of the sensors. *Executive* control level has to ensure tracking of the joint trajectories specified by tactical control level. This level might include various control algorithms, which take into account various dynamic effects of the robotics system in order to compensate for the dynamics of the robot.

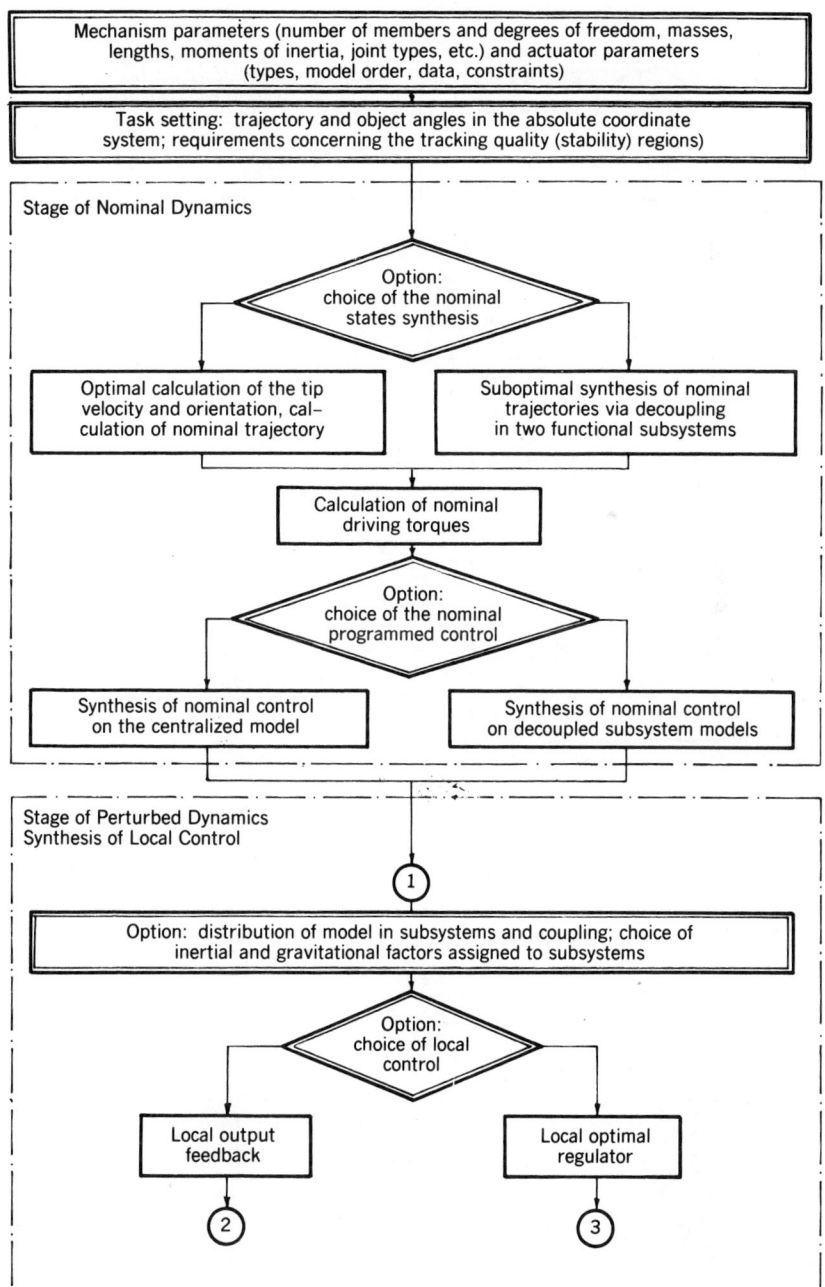

Figure 21.3 *Software for computeraided control synthesis: (━━) algorithm; (══) user's option.*

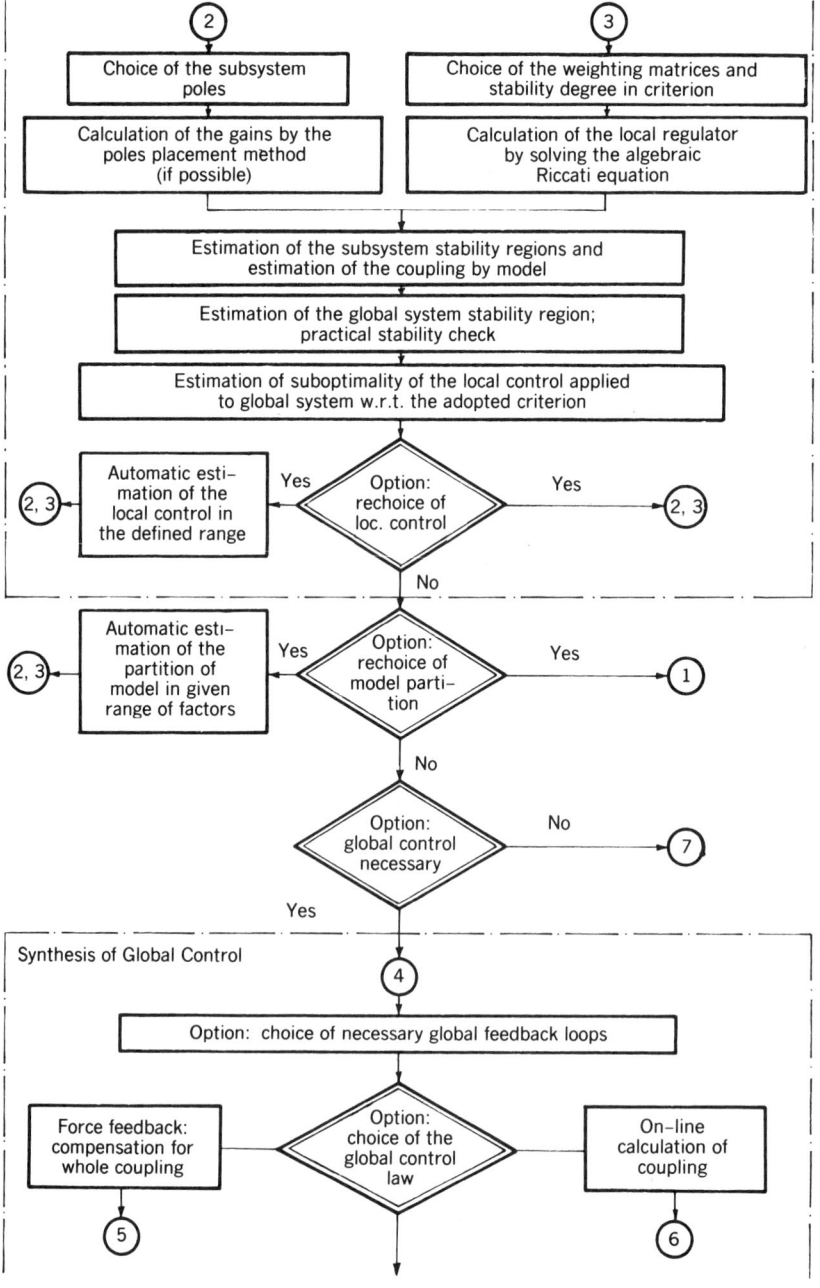

Figure 21.3 (Continued)

21.2 CUSTOMIZED SOFTWARE FOR MANIPULATION ROBOT DYNAMICS

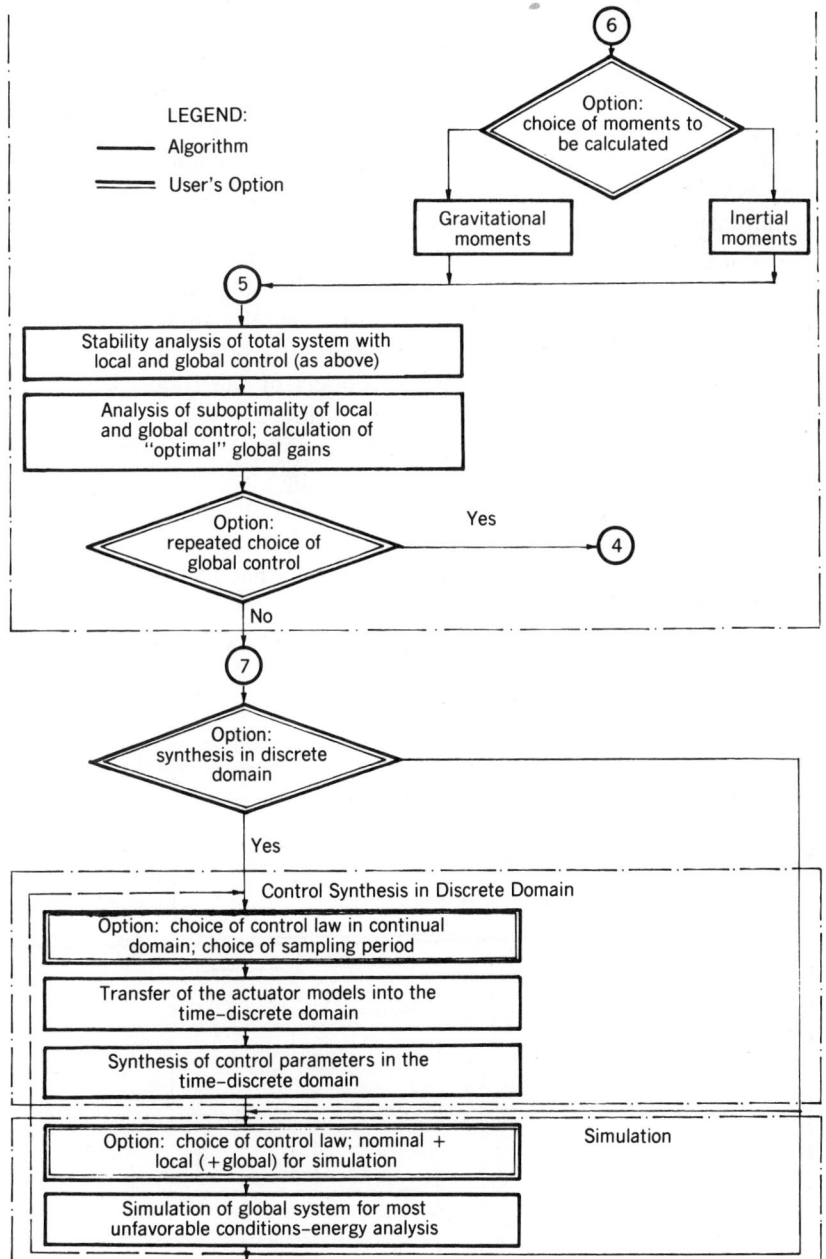

Figure 21.3 (*Continued*)

21.2.7 The Software Development System (SDS) for Robotic Controllers [25]

The software development system for robotic controllers (RC) is primarily aimed at generating robot models appropriate for implementation on microprocessors. The SDS is usually located on the host computer, although it might also be put directly on the controller itself. In the last case, upon imposing robot parameters, the robot programs would become "self-generated," implying that such a robot controller may be called *universal*. The global SDS organization is shown in Figure 21.4. Upon imposing initial requirements, such as kinematic configuration and dynamic parameters, the model "design" begins. It involves the generation of symbolic expressions describing the kinematic and dynamic models of the robot. Besides, the model "design" includes the optimization of the trigonometric expressions in order to evaluate them with the minimal number of numeric operations. The last part of the SDS is the model "coding" module. It produces an output source program appropriate for compilation, linking, and location into the robot controller. The SDS provides for the generation of the linearalized dynamic model. Linearalized models are important in the application of highly developed linear control theory, while the models of sensitivity with respect to the variation of robot parameters are directly applicable in

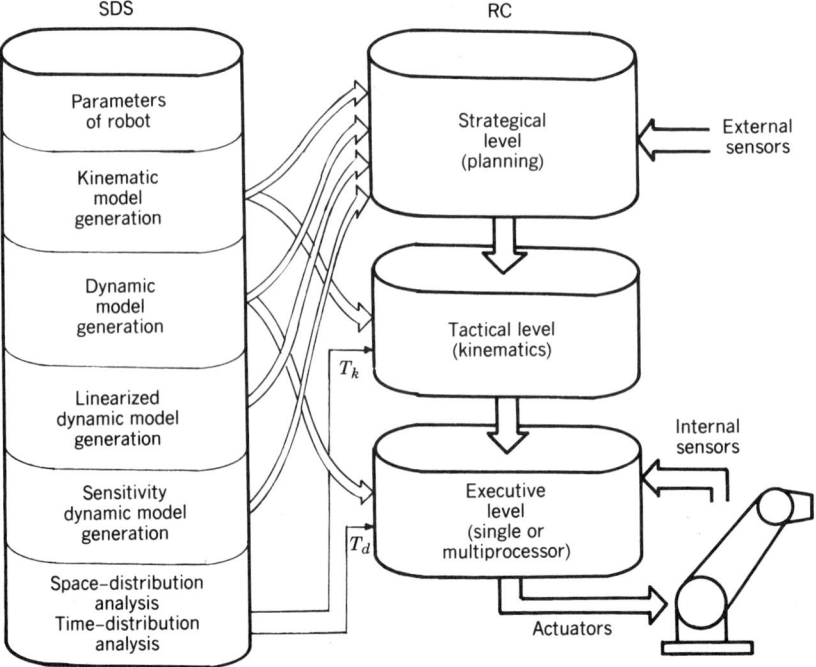

Figure 21.4 *Software development system.*

adaptive control algorithms and examination of the control robustness. The tactical control level involves the transformation of external (world) coordinates into the joint coordinates using the kinematic model. The execution level usually performs the function of digital servo systems enabling tracking of the desired trajectories. This level, however, usually includes the compensation of the robot model nonlinearities, and therefore requires the approximate or even exact dynamic model. The time distribution analysis (TDA) module determines the kinematic models, taking into account parameters of the mechanism and the actuators, as well as representative manipulation tasks (with given payload mass, velocities, and accelerations). The space distribution analysis (SDA) module determines how to distribute the model components on several parallel processor units. The input of this module is the elements of dynamic model matrices generated in symbolic form as subroutines, which are called *subtasks*. Since the subtasks are independent of each other, the search problem (which subtask to adjoin to which processor unit) is not very complicated. Given the number of processors n_p and the execution time of each subtask τ_l ($\tau = 1, \ldots, L$, with L being the number of relevant model element), the SDS determines the optimal distribution of subtasks among processors so that their execution times become as equal as possible.

21.2.8 Adaptive Control Algorithms [26–29]

A number of papers have been published dealing with the problem of dynamic control of manipulation robots with variable parameters. Two general approaches are well known: robust nonadaptive control and adaptive control. Especially, in last few years, many papers have been published on adaptive control for robotic systems. These adaptive controls have to solve the problem of parameter variation (i.e., to ensure satisfactory performance of the robotic system if its parameters change).

Analysis of various forms of adaptive indirect algorithms from the standpoint of numerical complexity and quality of trajectories tracking: local adaptive decentralized feedback with various forms of adaptive (decentralized or centralized) feedforward control, local nonadaptive decentralized feedback control (fixed feedback gains) with adaptive feedforward control, development of implicit self-tuning control strategy as one possibility for simplified and efficient procedure for adaptive control synthesis.

Indirect decentralized adaptive control includes [30]

1. Estimation of unknown dynamic parameters of robot payload using recursive least square method:
    a. Complete estimation model using sensitivity robot model with fast symbolic computations,
    b. Complete estimation model using dynamic model of robot terminal link with fast symbolic computations;

2. Synthesis of decentralized self-tuning control:
   a. Main control structure: dynamic compensator ("feedforward" control) + local feedback control; Decentralized adaptive control with adaptation in feedback loop; Decentralized adaptive control with adaptation in feedforward loop; Decentralized adaptive control with simultaneous adaptation in feedback and feedforward loop;
   b. Synthesis of feedforward and local feedback control is accomplished with: Calculation of regressive matrix in estimation model depending on real-time sensor data; Calculation of regressive matrix ϕ in estimation model depending on nominal data.

Implementation of decentralized adaptive control:

- Possibility of using two different methods for compensation of driving torques: driving torque measurement in terminal link by torque sensor and calculation of approximative control variation;
- Different possibility of calculation of internal robot acceleration: approximation by nominal acceleration, accelerometer, velocity signal differentiation with filtration, filtration of estimation model by stable first-order filter;
- Algorithms of decentralized adaptive indirect control (specially with driving torque measurement and off-line calculation of regression matrix) absolutely satisfy real-time condition for implementation of modern microprocessor control systems.

Except in very special tasks, we can conclude that recent industrial applications do not require adaptive controllers of industrial robots.

21.2.9 Control of Manipulation Robots in Operational Task Space (Cartesian Space)

Cartesian space control of manipulation robots in tasks such as arc welding and contour following requires high precision in position of manipulator's gripper.

Dynamic model of robotic manipulators in Cartesian space is based on dynamical model in joint space. By this, problem is mapping from one to the other space. Dynamic equations of robot's gripper movement in Cartesian space correspond to the dynamical movement equations in joint space. This relation is essentially nonlinear, and it is realized by means of inverse Jacobian matrix [31].

$$P = H(q)\ddot{q} + h_c(q,\dot{q}) + h_g(g) \qquad (21.1)$$

$$F = \overline{H}(x)\ddot{x} + p_c(x,\dot{x}) + p_g(x) \qquad (21.2)$$

$$\overline{H}(x) = J^{-T}(q)H(q)J^{-1}(q) \qquad (21.3)$$

$$P_c(x,\dot{x}) = J^{-T}(q)h_c(q,\dot{q})$$

$$p_g(x) = J^{-T}(q)h_g(q)$$

where
- n = number of degrees of freedom
- $q = [q^1 \ q^2 \ q^3 \ \cdots \ q^n]^T$ = state vector in joint space
- $x = [x^1 \ x^2 \ x^3 \ \cdots \ x^n]^T$ = state vector in Cartesian space
- $H(q), \overline{H}(x)$ = corresponding inertial matrix in joint and Cartesian space
- $h_c(q,\dot{q}), p_c(x,\dot{x})$ = centrifugal and Coriolis force vector in corresponding spaces
- $h_g(q), p_g(x)$ = gravitational force vector in corresponding spaces
- P = active torque vector in all manipulator's joints
- F = generalized force vector acting on a manipulator gripper

Cartesian space control of manipulation robots relies on the dynamic model (21.2), which represents balance of forces on a robot's gripper. Relation between active torque vector in all joints of manipulator and generalized force vector on the robot's gripper is

$$P = J^T(q)F \qquad (21.4)$$

Last few years, the control law proposed in [31] is cited in literature very often. It fully compensates influence of centrifugal, Coriolos, and gravitational force effects:

$$F = \overline{H}(x)[-K_p \,\Delta x - K_v \,\Delta\dot{x} + \ddot{x}_0] + p_c(x,\dot{x}) + p_g(x) \qquad (21.5)$$

By appropriate choice of position and velocity gain factors of local feedback loops (K_p and K_v, respectively), the qualitative gripper's trajectory tracking is attained. Drawbacks of this algorithm lies in sensitivity on dynamic parameter's variation of last mechanism's link and in the necessity of on-line inverse Jacobian calculation.

In order to improve the proposed algorithm in [31] the nominal decentralized control law is introduced:

$$F = F_0 - K_p \,\Delta x - K_v \,\Delta\dot{x} \qquad (21.6)$$

$$F_0 = H(x_0)\ddot{x}_0 + p_c(x_0,\dot{x}_0) + p_g(x_0) \qquad (21.7)$$

By this control law we attain very qualitative trajectory tracking. Sensitivity on dynamic parameter's variation of last segment is decreased, and on-line inverse Jacobian calculation is avoided. That is the reason why this control algorithm is more suitable for the microcomputer implementation.

21.3 SOME SPECIFIC TASKS IN APPLIED ROBOTICS

21.3.1 Mathematical Model of Manipulator with Constraints of Gripper Motion

The manipulator position can be defined by the generalized position vector X_g, which consists of n independent parameters. For a manipulator with 6 DOF we adopt the position vector [32,33].

$$X_g = [x_A \quad y_A \quad z_A \quad \theta \quad \varphi \quad \psi]^T \quad (21.8)$$

where x_A, y_A, z_A are Cartesian coordinates of the gripper point that determine its position, and θ, φ, ψ are Euler angles that determine its orientation. Imposed gripper motion reduces the number of degrees of freedom. Let n_r be this reduced number of DOF. It holds that $n_r < n$ (equality holds when there is no constraint). Introduce n_r free and independent parameters u_1, ..., u_{n_r} that define the constrained position of the gripper. The reduced position vector X_r is introduced:

$$X_r = [u_1 \cdots u_{n_r}]^T \quad (21.9)$$

We express the constrained motion by the second-order Jacobian form connecting the position vector (21.8) and the reduced position vector (21.9):

$$\ddot{X}_g = J_r \ddot{X}_r + A_r \quad (21.10)$$

where J_r is the reduced Jacobian and A_r is the associated reduced vector of dimensions $J_r(n \times n_r)$ and $A_r(n \times 1)$, respectively. For motion without constraints:

$$\ddot{X}_g = J\ddot{q} + A \quad (21.11)$$

Combining (21.10) and (21.10), we obtain

$$\ddot{q} = J^{-1}J_r\ddot{X}_r + J^{-1}(A_r - A) \quad (21.12)$$

The constraints produce reactions, forces, and moments that are introduced into the dynamic model developed by general theorems, which can be presented in the following form:

$$H\ddot{q} + h = P + D_1 F_A + D_2 M_A \quad (21.13)$$

or

$$H\ddot{q} + h = P + DR_A \qquad (21.14)$$

where the matrix D and reaction vector R_A are

$$D_{(n\times 6)} = [D_{1\,(n\times 3)} \mid D_{2\,(n\times 3)}], \qquad R_A = \begin{bmatrix} F_{A\,(3\times 1)} \\ M_{A\,(3\times 1)} \end{bmatrix}$$

where $H = H(q) = n \times n$ inertial matrix
$h = h(q,\dot{q}) = n \times 1$ vector consisting of gravity, centrifugal, and Coriolis effects
$P = P(t) = n \times 1$ vector of driving torques (forces) in joints
$\ddot{q} = n \times 1$ acceleration vector of internal coordinates
$D = n \times 6$ matrix associated with the reaction vector
$R_A = n \times 1$ reaction vector

Depending on the constraint imposed and the manipulator configuration, there are some conditions that should be satisfied by the six-component reaction R_A. Namely, there are $6 - (n - n_r)$ scalar conditions that can be expressed in matrix form:

$$ER_A = 0 \qquad (21.15)$$

where E is a matrix of dimension $(6 - n + n_r) \times 6$.

Now, Eqs. (21.12), (21.14), and (21.15) define the complete mathematical model of closed-chain configuration.

We will consider only the problem of calculating the nominal dynamics. The nominal dynamics (prescribed manipulation task) assumes that the forces, which we want to realize during motion, are given. Thus F_A, M_A, and, accordingly, R_A are known. Figure 21.5 presents a manipulator with additional force and moment acting on the gripper. These values must be prescribed so that they satisfy (21.15). Now, the necessary driving torques (forces) can be solved from (21.14). However, if we want to calculate the unknown motion and reactions, then substituting (21.12) into (21.14), we obtain

$$HJ^{-1}J_r\ddot{X}_r + h - DR_A = P - HJ^{-1}(A_r - A) \qquad (21.16)$$

Combining (21.16) and (21.15), we obtain matrix equation

$$\begin{bmatrix} HJ^{-1}J_r & -D \\ \hline 0 & E \end{bmatrix} \begin{bmatrix} \ddot{X}_r \\ \hline R_A \end{bmatrix} + \begin{bmatrix} h \\ \hline 0 \end{bmatrix} = \begin{bmatrix} P \\ \hline 0 \end{bmatrix} + \begin{bmatrix} -HJ^{-1}(A_r - A) \\ \hline 0 \end{bmatrix}$$

$$(21.17)$$

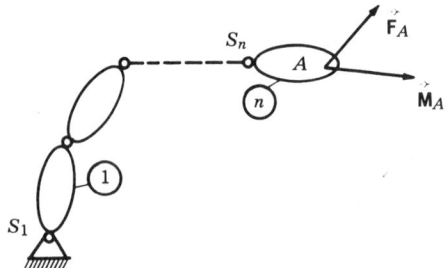

Figure 21.5 Manipulator with additional force F_A and moment M_a acting on gripper.

Equation (21.17) represents a system of $n_r + 6$ equations that can be solved for $n_r + 6$ unknowns $\ddot{X}_r$ and R_A. Now, by using (21.12) we obtain $\ddot{q}$; that is, we may calculate the manipulator motion.

21.3.2 Adaptive Hybrid Position/Force Control of Manipulation Robots

Model of constrained motion of robot:

$$P = H(q)\ddot{q} + h(q,\dot{q}) + J^T F \qquad (21.18)$$

where $P = n \times 1$ vector of driving torques
$H = n \times n$ inertia matrix
$h = n \times 1$ vector of gravity, centrifugal, and Coriolis forces
$J = n \times n$ Jacobian matrix with respect to the point of contact of the end-effector and the surface
$F = n \times 1$ vector of the Cartesian forces and moments by which the end-effector is acting upon the surface, and vice versa

Cartesian force F by which the end-effector is acting upon the contact surface in the first approximation may be modeled as the stiffness force

$$F = K_E(p - p_E) \qquad (21.19)$$

where $K_E = n \times n$ matrix of the surface stiffness (including stiffness of the end-effector and force sensor)
$P = n \times 1$ vector of the coordinates of the contact point between the end-effector and the surface
$p_E = n \times n$ vector of the coordinates of the point of impact between the end-effector and the surface

The idea of hybrid control [19,34] is simple: By introduction of the $n \times n$ selectivity matrix S, the directions in Cartesian space in which only force control is applied are distinguished from those directions in which only position control is applied. The selectivity matrix S is the diagonal $n \times n$ matrix whose ith element at the diagonal is 1 if the ith DOF of the end-effector with respect to the Cartesian frame is force controlled, and is 0 if it is position controlled. Hybrid control schemes usually include explicit force feedback loops. The forces at the end-effector are measured by force transducers, and the error between the desired force and actual force is used to compute control signals that are directly applied to actuators inputs. Here, $(I - S)p^0(t)$ are desired position trajectories of the end-effector, $SF^0(t)$ are desired "trajectories" of the Cartesian forces.

The basic problems are control of contact force between a robot and a constraint surface, and interaction between position and force control. Based on considerations of these problems it can be concluded that

> Fixed force feedback gains cannot ensure stable contact force with various surfaces; adaptive control is necessary.
>
> Since the robot in essence is a "position" device, implicit force control (in which the equivalent position of the end-effector corresponding to the desired Cartesian force is determined and realized) is more stable and robust and therefore more applicable.
>
> As the position control of robot behaves as low-pass filter, implicit force control cannot respond to sharp variations of forces; therefore, explicit force feedback has to be also applied.

Proposed control scheme [35,36]:

Combination of implicit and explicit force feedback,
Adaptive force feedback gains.

If we assume that the force F is directly measured, the stiffness of the environment may be identified by

$$K_E^{EST} = F/(p - p_E), \qquad p \neq p_E \tag{21.20}$$

Based on identified stiffness we may determine the "equivalent position" p that corresponds to the desired force $F^0(t)$ as

$$p_E^0(t) = (S_E^{EST})^{-1}F^0(t) + p_E \tag{21.21}$$

The actuator control signals are computed according to the following control law:

$$\begin{aligned}
u(t) = &-K_p J^{-1}(q)[(I_n - S)(p(t) - p^0(t)) + S(p(t) - p_E^0(t))] \\
&- K_v J^{-1}(q)(I_n - S)(\dot{p}(t) - \dot{p}^0(t)) + J^T(q)SK_F(F^0(t) - F(t)) \\
&+ K_{FI} \int (F^0(t) - F(t))\, dt - K_D \dot{p}(t)\overline{K}_{PR}
\end{aligned} \tag{21.22}$$

where $u = n \times 1$ vector of actuator input signals
$p^0(t) = n \times 1$ vector of the desired nominal trajectory of the end-effector
$K_p = \mathrm{diag}(k_p^i) = n \times n$ matrix of local servo position feedback gains
$K_v = \mathrm{diag}(k_v^i) = n \times n$ matrix of local servo velocity feedback gains
$K_F = \mathrm{diag}(k_F^i) = n \times n$ matrix of force feedback gains
$K_{FI} = \mathrm{diag}(k_{FI}^i) = n \times n$ matrix of gains in feedback loops with respect to the integral of force errors
$\overline{K}_{PR} = \mathrm{diag}(\overline{k}_{PR}^i) = n \times n$ matrix of the coefficients of proportionality between the actuator torque and input signal
$K_D = \mathrm{diag}(k_D^i) = n \times n$ matrix of damping feedback gains in directions in which forces are controlled

The fixed feedback gains cannot ensure satisfactory maintenance of the desired forces in the contact with various environments. Therefore, the feedback gains have to be updated according to the identified stiffness of the environment. The force feedback gains and damping gains are selected to ensure certain bandwidth and overcritical damping of the system.

21.3.3 Cooperative Manipulation Models

In some industrial application of robotics systems it is sometimes more convenient or even necessary to use more than one robot to accomplish a given task. An example of this situation is when a single manipulator cannot handle the object either because it is beyond manipulator's load capacity, or when the geometrical properties of the object make it difficult to manipulate. In the coordination of several robots we can recognize four cases:

  a. Robots coordination but without common workspace (no contact)—robot synchronization in time;
  b. Robots share common workspace (trajectory planning for collision avoidance);
  c. Robots must come into mutual contact (problem of robot synchronization in time and space and robustness problem);
  d. Robots coordination in time and space performed on higher control level.

Two Cooperating Robot Arms [37,38] Two cooperating arms with 6 DOF each are considered in cooperative work. They handle a rigid object permitting no relative motion between grippers and manipulated object. Using the invariance of the relative position and orientation between the grippers, the kinematic constraints are determined. In order to obtain the most convenient form of kinematic relations among internal coordinates of manipulators the relations in terms of internal velocities are considered. Internal coordinates of one manipulator are taken as independent variables for the description of the dynamic model of cooperative work. Internal coordinates of the other manipulator are determined by the constraint relations. When the motion of one manipulator is given, the motion of the tip of other manipulator is defined and inverse kinematics uniquely determines the motion of other manipulator in its internal coordinates. The set of six independent variables has to be chosen, which will serve as generalized coordinates, to describe the dynamic behavior of the system. We adopt that set to be the set of internal coordinates of M_1. In other words, when internal coordinates (or velocities and accelerations) of M_1 are known, internal coordinates of M_2 can be determined via constraint relations. The dynamic model of cooperative work is obtained using the individual dynamic models of the manipulators and the model of interaction. The model of interaction is represented by the balance of forces and torques (Fig. 21.6).

Object Motion The procedure of dynamic model forming can be summarized in the following steps [37]:

Step 1: Determine the constraints posed on the internal coordinates of manipulators.

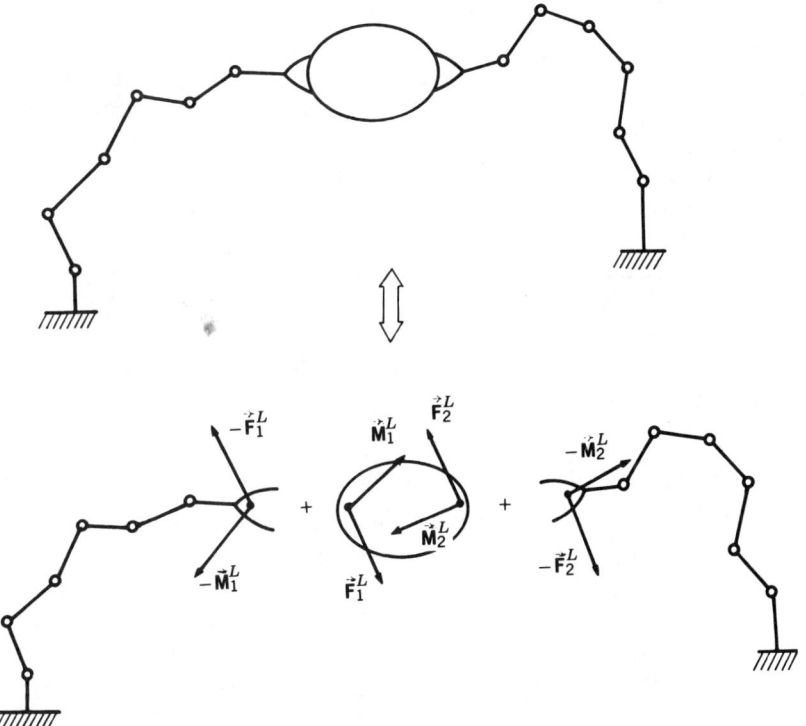

Figure 21.6 *Dynamics decomposition (n = 2).*

Step 2: Choose the set of generalized coordinates as independent variables.

Step 3: Break down the problem of the system motion in the problem of subsystems motion (manipulators and the object) under the influence of the rest of the system, represented by corresponding forces and torques of interaction.

Step 4: Determine the dynamic model of the object motion under the influence of manipulators. The dynamic models of manipulators under the influence of external forces and torques are known.

Step 5: Map all equations in the space of generalized coordinate and eliminate introduced forces and torques of interaction.

Dynamic Model of the Cooperative Manipulation

$$X(q_1,q_2)\ddot{q}_1 + \chi(q_1,q_2,\dot{q}_1,\dot{q}) = P_1 + \phi(q_1,q_2)P_2$$

where q_1, q_2 are the vectors of internal coordinates of manipulators M_1 and M_2, respectively. Strong similarity between this model and the model of a single manipulator can be noticed. The term $X(q_1,q_2)\ddot{q}_1$ corresponds to the inertial forces, the term $\chi(q_1,q_2,\dot{q}_1,\dot{q}_2)$ corresponds to the sum of Coriolis, centrifugal, and gravitational forces, P_1 is the vector of driving torques produced by actuators of M_1; and the term $\phi(q_1,q_2)P_2$ is the vector of driving torques produced by M_2 and mapped to the joint space of M_1. The matrix $X(q_1,q_2)$ acts like an operator that maps forces from the joint space of M_2 to the forces in the joint space of M_1. The matrix $X(q_1,q_2)$ is an equivalent inertial matrix and is obtained as a sum of corresponding inertial matrices of M_1, M_2 and the object, multiplied by necessary mapping operators. When we compare the system of one manipulator and the system of two cooperating manipulators, we note the common property considering number of degrees of freedom, but the difference arises when number of inputs is considered. The direct dynamic problem, it means determination of the system motion when input torques are known can be resolved for both manipulators, while inverse dynamic problem cannot be resolved directly due to the redundancy of driving system (more actuators than the number of degrees of freedom).

21.3.4 Dynamics of Flexible Manipulators [39,40]

During deformation all points of the body move relative to their underformed position, changing by that the relative positions (i.e., mutual distances and angles). Motion of elastic bodies consists of these components: translation, rotation and pure deformation. The first two correspond to ideally rigid-body motion while the third one represents the relative displacements (extension and shear) in the vicinity of the point considered. The deviation vector is now depending not only on time, but also on the position of the considered point on the body $\mathbf{u}_i = \mathbf{u}_i(x,y,z,t)$:

$$\tilde{\mathbf{u}}_i(\tilde{x}_i, \tilde{y}_i, \tilde{z}_i, t) = \tilde{\mathbf{u}}_i(\tilde{x}_i, \tilde{y}_i, \tilde{z}_i, u_i^1, \ldots, u_i^{N_i}) \tag{21.23}$$

According to the supposition of small deformations, (21.23) can be expanded in a power series with respect to small parameters u_i^α ($\alpha = 1, \ldots, N_i$). Taking the first-order terms only gives

$$\tilde{\mathbf{u}}_i(\tilde{x}_i, \tilde{y}_i, \tilde{z}_i, t) = \sum_{\alpha=1}^{N_i} \tilde{\mathbf{f}}_i^\alpha(\tilde{x}_i, \tilde{y}_i, \tilde{z}_i) u_i^\alpha(t) \tag{21.24}$$

where $\tilde{\mathbf{f}}_i^\alpha$ = interpolation functions
  u_i^α = generalized elasticity coordinates
  $\tilde{x}_i, \tilde{y}_i, \tilde{z}_i$ = coordinates of the position vector of the considered point before deformation with respect to the local coordinate system

In the lumped-mass method the system is substituted by a finite number of concentrated masses in chosen points, interconnected by springs. In the method of finite elements the interpolation functions are defined within a subdomain of finite-dimension bodies (finite elements), while u_i^α represent displacements of the points of their connections (nodes).

Dynamics of Robots with Complaint Joints First suppose that the mechanism links are rigid; that is, the mechanical structure compliance is due only to the elasticity of the driving transmission system. The actuators, reducers and transmissions are represented by rigid bodies interconnected by torsional or linear springs (Fig. 21.7). The variable q^i is the small angular (or linear) deflections of the equivalent spring

$$q^i = q^{io} + \Delta q^i, \qquad P_i^M = P_i^{Mo} = K_{i-1,i}\, \Delta q^i \tag{21.25}$$

where q^i = internal coordinate of complaint joint,
  q^{io} = internal coordinate of "rigid" joint,
  P_i^M = load on output shaft of complaint joint,
  P_i^{Mo} = load of "rigid" joint,
  $k_{i-1,i}$ = spring stiffness.

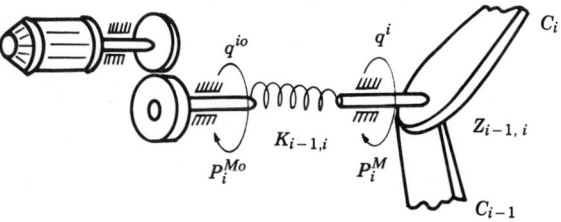

Figure 21.7 Model of joint compliance.

Dynamic Model of a Manipulation Robot with Complaint Joints

$$P = H(q,\theta)\ddot{q} + h(q,\dot{q},\theta), \qquad P = K_z \Delta q \qquad (21.26)$$

where $\Delta q = [\Delta q^1, \ldots, \Delta q^n]^T$ = vector of internal coordinates elastic deviations, and $K_z = \text{diag}[K_{i-1,i}] = n \times n$ diagonal stiffness matrix.

Dynamic Model of Robot with Elastic Links The vector of elastic angular displacements $\tilde{\varphi}_i(\tilde{x}_i, \tilde{y}_i, \tilde{z}_i, t) = \tilde{\varphi}_i(\tilde{\mathbf{r}}_i, t)$ defines the relative rotations with respect to the local trihedron of axes, connected to the rigid skeleton. The vector $\tilde{\mathbf{r}}_i$ is the position of the arbitrary point before deformation. To define total displacements and orientation with respect to reference coordinate system, it is necessary to form the linear and angular displacements of the ith link tip of joint Z_1 (Fig. 21.8):

$$\tilde{\mathbf{u}}_i = \tilde{\mathbf{u}}_i(-\tilde{\mathbf{r}}_{i+1,i}, t) \quad \text{and} \quad \tilde{\varphi}_i = \varphi_i(-\tilde{\mathbf{r}}_{i+1,i}, t) \qquad (21.27)$$

causing the linear and angular displacements of the rest of the mechanism chain. Define the elastic rotations matrix of the ith link tip, as a transformation matrix from the trihedron $\overline{Z}_i \overline{\xi}_i \overline{\eta}_i \overline{\gamma}_i$ into the corresponding trihedron $Z_i \xi_i \eta_i \gamma_i$, connected to the rigid skeleton, with its axes parallel to the axes of the internal coordinate system at O_i. Dynamic equations of motion as function of generalized coordinates, and their derivatives are derived by means of the virtual-work method. The complete dynamic model of the manipula-

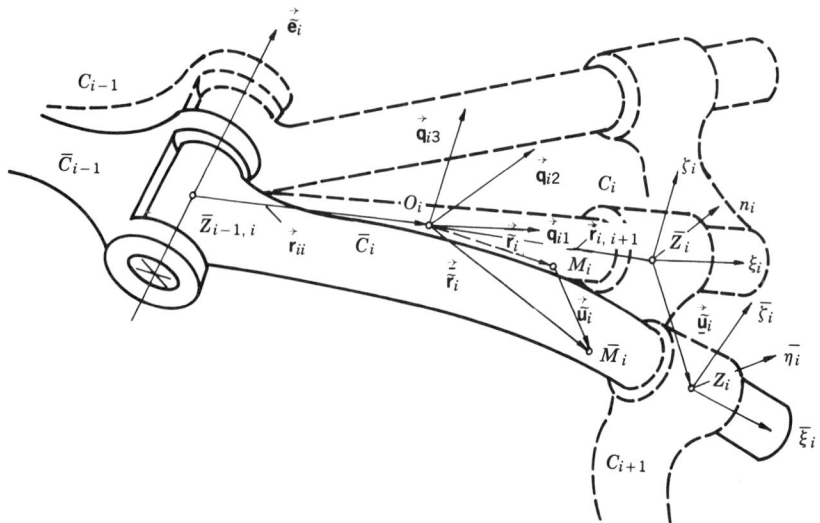

Figure 21.8 Local elastic displacements.

tion robot with elastic links can be presented in the form

$$H^{qq}(q,\theta)\ddot{q} + h^q(q,\dot{q},\theta) + H^{qu}(q,\theta)\ddot{u} + C^{qu}(q,\dot{q},\theta)\dot{u}$$
$$+ K^{qu}(q,\dot{q},\ddot{q},\theta)u = p^q \qquad (21.28)$$

$$H^{uu}(q,\theta)\ddot{u} + (C^{uu}(q,\dot{q},\theta) + D^{uu}(t))\dot{u} + (K^{uu}(q,\dot{q},\ddot{q},\theta) + K_s^{uu}(\theta))u$$
$$+ H^{uq}(q,\theta)\ddot{q} + h^u(q,\dot{q},\theta) = p^u \qquad (21.29)$$

where θ is the vector of parameters.

The first equation is describing the motion of the manipulation system, and the second one its structural dynamics. In deriving the model, all values are expressed in the form of a sum of nominal and additional components. Matrix H^{qq} and vector h^q have the same structure like the corresponding matrices of the equivalent rigid system. The remaining matrices of the system express the dynamic effects of elasticity. Their elements depend on system parameters, internal coordinates and their derivatives as well as on so-called generalized inertial and stiffness parameters. These parameters qualitatively represent the effects of deformation motion upon the system dynamics. The vector $u = [u^1, \ldots, u^n]$ is the $N \times 1$ vector of the mechanism elastic coordinates ($N = \Sigma_{i=1}^n N_i$), sets of $n \times N$ matrices $\{H^{qu}, C^{qu}, K^{qu}\}$ and $N \times N$ matrices $\{H^{uu}, C^{uu}, K^{uu}\}$ express the inertial generalized gyroscopic effects and the effects of geometric rigidity, D^{uu} is a $N \times N$ matrix of structural damping and K_s^{uu} = block diag K_i is a block matrix of structural stiffness. The $N \times n$ matrix H^{uq} and the $N \times 1$ vector h^u express the influence of inertial, gyroscopic, gravitational, and external forces onto the mechanism oscillatory motion. The vectors p^q and p^u of dimensions $n \times 1$ and $n \times 1$, respectively, represent the vector of generalized active forces in the mechanism joints and the microelastic control forces, acting onto the mechanism links (forces of active damping/active stiffness forces, etc).

21.3.5 Complex Mobile Robot [41]

Equivalent Manipulator Concept [42] A mobile platform consists of a drive unit with several actuators, driving wheels and a damping system preventing transmission of jerks from the wheels to the upper level of the platform. For the sake of simplicity we shall consider two-dimensional (2D) case when the platform moves in xy plain of a fixed $Oxyz$ frame. The torques that are to be produced about wheel shafts can be reduced to F_x, F_y forces along x and y axes and the torque M_z about the z' axis of the local coordinate frame of the platform (Fig. 21.9b). Thus, the rotation and the translation of the platform can be modeled by a 3-DOF manipulator with two sliding and a revolute joint (first three joints in Fig. 21.9b). The damping system is described by three additional joints to describe oscillations along z axis (sliding joint) and about x' (roll) and y' (pitch) axes (Fig. 21.9). The maximal number

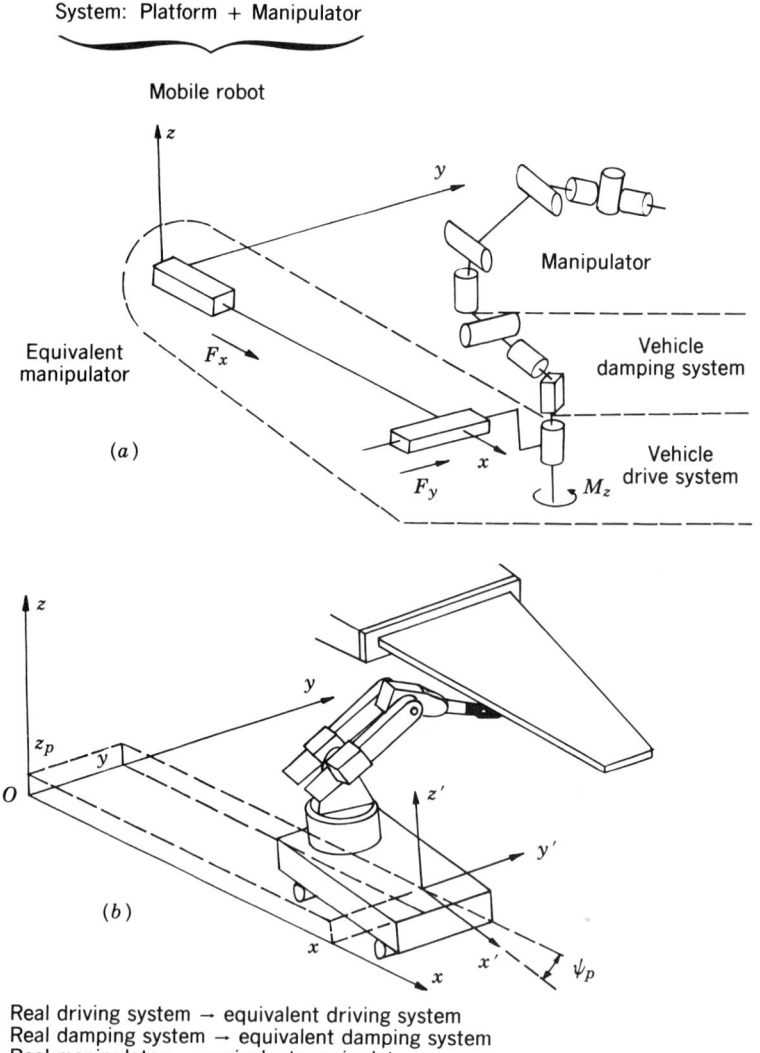

Figure 21.9 *Mechanical scheme of complex mobile robot.*

of equivalent joints describing drive unit is 6 for the platform moving along 3D surfaces. Since the damping system can generally be extended to 6 equivalent joints, we see that up to 12 joints of equivalent manipulator are necessary for describing kinematic and dynamic behaviour of an arbitrary platform. Notice that the first three joints in Fig. 21.9 are powered by an equivalent drive mechanism, while the next three joints are passive, noncontrollable joints consisting of equivalent spring-damper mechanisms. One of the motivations for developing the equivalent manipulator representation was extensibility of model complexity. Now, it is quite simple to add several joints and links of a desired robot mounted on the platform to the platform links (the last six joints in Fig. 21.9).

Dynamic Model, Simulation, and Control The equivalent manipulator representation is very useful to avoid the derivation of complicated dynamic equations for each particular mobile robot. However, we have to find the transformations between real drive system of the platform and the equivalent one. After that it is easy to form the model because we can apply any modeling method developed for open-chain linkages. The dynamic model of the equivalent manipulator is of the form

$$P = H\ddot{q} + h(q,\dot{q}) \tag{21.30}$$

where $q = [q^1 \cdots q_r]^T$ is the vector of joint coordinates,
$H = n \times n$ positive-definite inertia matrix,
$h(q,\dot{q}) = n$-vector due to Coriolis, centrifugal, and gravitational forces,
$P = n$-vector of driving torques/forces in joints.

The equivalent joints corresponding to the actuators driving the platform as well as robot joints are powered. For the sake of simplicity we shall accept that the actuators are direct current (dc) motors modeled as

$$u_i = a_i P_i + b_i \dot{q}_i \tag{21.31}$$

where u_i = input voltage
P_i = produced torque in joint i
q_i = joint coordinate
a_i, b_i = parameters

The inductive effects are neglected. The equivalent joints representing damping effects are modeled by

$$c_i(q_{mi} - q_i) - v_i \dot{q}_i = P_i \tag{21.32}$$

where q_{mi} = constant equilibrium position of the spring in joint i
c_i = the spring constant
v_i = damping friction coefficient

From (21.31) and (21.32) it follows that the vector $P = [P_1 \cdots P_r]^T$ can be expressed as

$$P = Au - B\dot{q} + Cq_m + Cq \tag{21.33}$$

where A, B, and C are diagonal constant matrices. Here $u = [u_1 \cdots u_{n_p}$

$0 \cdots 0 \ u_{n_p+n_d+1} \cdots u_n]^T$ is the input vector. The number of powered equivalent joints of the platform is denoted by n_p, and the number of equivalent joints corresponding to damping effects by n_d. The q_m vector has the form $q_m = [0 \cdots 0 \ q_{m_I} \cdots q_{m_J} \cdots 0 \cdots 0]^T$, where $I = n_p + 1$, $J = n_p + n_d$.

Combining (21.30) and (21.33), we obtain the model of the entire system

$$H\ddot{q} + h(q,\dot{q}) + B\dot{q} + Cq = Au + Cq_m \qquad (21.34)$$

Since H is a positive-definite $n \times n$ matrix and H^{-1} always exists, we can express $\ddot{q}$ as function of q and $\dot{q}$. Thus, both direct and inverse dynamics are provided. The system (21.34) can be unstable even when $u = 0$ if the elastic modes are not damped. In reality, viscous frictions are always combined with elastic effects, and the oscillations are damped. The control system for both the platform and the manipulator consists of feedforward term u_{FF} and feedback term u_{FB}:

$$u = u_{FF} + u_{FB} \qquad (21.35)$$

Various types of control laws are applicable. We shall adopt the following algorithm:

$$u_{FFi} = k_{vi}\dot{q}_i, \qquad u_{FBi} = k_{Pi}(q_i - q_i^0) + k_{vi}(\dot{q}_i - \dot{q}_i^0) + k_{Fi}P_i \qquad (21.36)$$

for controllable joints $i = 1, \cdots, n_P + n_d + 1, \cdots, n$. Thus we introduce position and velocity feedback as well as force feedback. This control law would make the overall system decoupled and linear if no uncontrollable joints exist. The mobile robot in that case still exhibits nonlinear oscillatory effects due to the elastic effects in the damping system of the platform. Further improvements of the controller (21.35), (21.36) can be achieved by introducing new sensors measuring the platform oscillations. The proposed control scheme is shown in Figure 21.10.

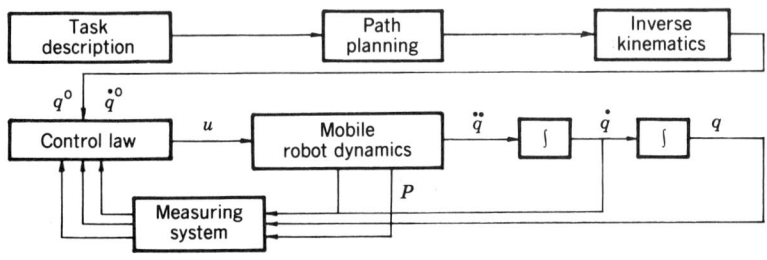

Figure 21.10 *Control system for a mobile robot.*

21.3.6 Control of Manipulation Robots by Using Neural Network Model

In robot control, neural network approach represents upgrade of classic nonadaptive and adaptive control algorithms with possibility to design a system with acceptable performance characteristics over a very wide range of uncertainty. Since the exact dynamics of the controlled object (robot manipulator) are generally unknown, we use a neural network model and appropriate learning rules, by which an internal model of inverse robot dynamics is acquired during execution of movement. Also, neural networks have great potential for application in intelligent systems, where robot would be able to learn from its experience, recognize changes in environmental conditions, reach to them, or to make decisions based on changing manufacturing events. In summary, a neural network controller should exhibit several important features as adaptation based on learning and generalization mechanism, utilization of large amounts of informations from internal and external robot sensors, and fast collective processing capability.

Using decentralized control and computed torque control algorithm [23], models of neural networks as feedforward robot controllers are developed. Based on repetitive trails (i.e., repetitive execution of the working task and position and velocity errors during execution), training of neural network controller can be realized. In this way, control algorithm is redistributed between feedforward and feedback loop, and as a result we have faster response and improved performance of system.

Neural Network Approach Manipulation robot models (21.1) are highly nonlinear systems with strong coupling between subsystems. For this type of plant in practice conventional and common way of local PID regulators for each subsystem (degree of freedom of robot mechanism) is applied. However, this type of control is not adequate for advanced industrial robots with requirements for high precision and speed in complex working environments. In that case the coupling between subsystems is strong, and we have to apply "dynamic" control—that is, control based on dynamic model of robot mechanism in the form of feedforward control [23]:

$$u = u^0 = KP\varepsilon - KD\dot{\varepsilon} \qquad (21.37)$$

where u^0 = nominal centralized control, which is synthesized using dynamic robot model and model of actuators but for nominal values of internal coordinates
KP = position feedback gain
KD = velocity feedback gain
$\varepsilon = q - q^0$ is feedback error
q, q^0 = actual and nominal internal coordinates
n = number of degree of freedom

Also, as another solution we can apply computed torque method for calculation of driving torque [23]:

$$P = H(q,\theta)[\ddot{q}^0 + KP(q - q^0) + KD(\dot{q} - \dot{q}^0)] + h(q,\dot{q},\theta) \quad (21.38)$$

where P = vector of driving torques
$H(q,\theta)$ = inertial matrix of system
$h(q,\dot{q},\theta)$ = vector depends gravitational, Coriolis, and centrifugal effects
θ = vector of parameters

However, in the procedure of controller design we must cope with structured uncertainties (inaccuracies of model parameters and additive disturbances), unstructured uncertainties (unmodeled high-frequency dynamics as structural resonant modes, actuator dynamics, sampling effects) and measurement noise. In this case, classic nonadaptive algorithms are not robust enough, because these algorithms compensate only small part of mentioned uncertainties. Also as a solution, adaptive control techniques can tolerate wider range of uncertainties, but in the presence of sensor data overload, heuristic informations, limits of real-time applicability and very wide interval of system uncertainties, application of adaptive control is not enough for satisfactory performance.

Hence, as a solution of these problems, neural networks as compensation elements in intelligent control systems represent serious possibilities:

1. *Learning:* neural network model with time-varying weighting factors enables us to learn internal model of robot dynamics and in this way we can compensate all robot uncertainties,
2. *Generalization:* if neural network once learned some movement, it could control quite different and faster movements,
3. *Fast Computational Capability:* collective and parallel way of information processing enables real-time applicability.

Dynamic Feedforward Controller based on Network Models Here, neural network model is applied as a form of feedforward robot control using decentralized control algorithm and computed torque method.

The proposed neural network model can be regarded as example of autonomous driving torque generator. Let us explain the problem of robot control in a computational framework. There are causal relations between robot driving torque and the resulting robot movement coordinates. Let $P(t)$ denote the time history of driving torque and $q(t)$ denote the time history of the robot internal coordinates. Also we can denote causal relation between P and q using functional F; that is, $F(P(\cdot)) = q(\cdot)$. If we want the robot to track desired trajectory q_d, the problem to generate a desired driving torque P_d, which realizes q_d is equivalent to finding an inverse of the functional F.

Hence, our primary interest is learning of inverse dynamic model of robot mechanism, because exact dynamics are generally unknown.

A neural network (NN) can be considered as continuous nonlinear dynamic system with two different forms of network topology:

1. NN model as fixed nonrecurrent single-layer network;
2. NN model as fixed nonrecurrent multilayer network with intermediate hidden units.

In the case when we consider single-layer network and decentralized algorithm, output of artificial neurons generates driving torques in robot joints based on time-varying weighting factors and nominal desired coordinates, velocities, and accelerations in the form

$$P_i^0 = \sum_{j=1}^{n} w_{ij} H_{ij}(q^0,\theta)\ddot{q}_j^0 + w_{i,n+1} h_i(q_i^0,\dot{q}_i^0,\theta), \quad i+1,\cdots,n \quad (21.39)$$

where P_i^0 = joint driving torque generated by neural network, and w_{ii} = adaptive weighting factors neural network $\{W = W(w_{ii})\}$.

Using model (21.38) and model of robot actuators, we can easily calculate nominal modified feedforward control in the form

$$u_i^0 = f(q_i^0,\dot{q}_i^0,\ddot{q}_i^0, P_i^0(q^0,\dot{q}^0,\ddot{q}^0,w_{ij})), \quad i = 1,\cdots,n, j = 1,\cdots,n+1 \quad (21.40)$$

so that complete control law has the form according to Eq. (21.1).

In the case when we consider single-layer and "inverse" method controlled computed torque is calculated according to

$$P_i^0 = \Sigma w_{ij} H_{ij}(q^0,\theta)[\ddot{q}_j^0 + KP\varepsilon_j + KD\dot{\varepsilon}_j] + w_{i,n+1} h_i(q,\dot{q},\theta) \quad (21.41)$$

Using Eq. (21.40) and model of robot actuators we can easily synthesize complete control law.

The topology of neural network in this case is specified as deterministic model, because we divide internal dynamic model of robot mechanism in $n^2 + n$ separate units with $n^2 + n$ tuning weighting factors. But this choice is not limited, and we can divide dynamic model in greater number of separate units. As was mentioned, this is a deterministic approach, because we use any a priori knowledge about the dynamic structure of the controlled object and in this way we have better scale-up properties. The limitation of this approach is in the case when we do not know the exact form of the dynamic equation to be known a priori.

The output of artificial neuron is bounded with hard-limiter signum function (activation function of neuron) due to the real physical constraints.

Based on the multi-layer (three-level) network with 18-36-6 formation (18 units = input level, 36 units = hidden level, 6 units = output level), special topology of neural network can be considered. The number of neural network units in the input and the output layer are determined according to the number of robot internal coordinates, velocities, accelerations, and robot driving torques [44].

This type of feedforward neural network with sufficient number of units in hidden layer can approximate any continuous mapping. Thus, neural networks can be used as very general computation models. Furthermore, using distributed representation scheme in neural networks can give a high level of fault tolerance. Also, in this multilayer neural network, we use a black-box approach, because we did not use any a priori experience and knowledge about inverse dynamic robot model. All we need to do is to feed in the neural network necessary informations (desired trajectory, desired driving torque, feedback signal) and let it learn by test trajectory.

During the training phase we adjust 864 weighting factors in proposed neural network. As activation function for neuron in multilayer neural network special squashing function is adopted, so we calculate output of neural network (driving torque) according to the formula

$$P_i^0 = P_i^{\max} \frac{1 - \exp\left[-0.5 \sum_m w_{HO}^{mi} \left(1 - \exp\left\{-\sum_t w_{IH}^{tm} I_t\right\}\right) \big/ \left(1 + \exp\left\{-\sum_t w_{IH}^{tm} I_t\right\}\right)\right]}{i + \exp\left[-0.5 \sum_m w_{HO}^{mi} \left(1 - \exp\left\{-\sum_t w_{IH}^{tm} I_t\right\}\right) \big/ \left(1 + \exp\left\{-\sum_t w_{IH}^{tm} I_t\right\}\right)\right]}$$

$i = 1, \ldots, 6; m = 1, \ldots, 36; t = 1, \ldots, 18$

(21.42)

where $P_i^{\max}$ = maximal permitted value for driving torque in ith joint
w_{HO}^{mi} = weighting factor between ith neuron in output layer and mth neuron in hidden layer
w_{IH}^{tm} = weighting factor between mth neuron in hidden layer and tth neuron in input layer
I_t = network input (robot internal coordinates, velocities, and accelerations)

Application of these networks in robot control are divided in the *learning phase* and *generalization (pattern associator) phase*. In the learning phase we propose some type of *unsupervised learning,* when by using of trial- and-error method and external inputs from trajectory generator, neural network acquires knowledge about real internal dynamic robot model as a result of interaction with real robot and real environment.

Training and learning of one-layer network is accomplished using error-correcting method based on well-known Delta rule (Widrow-Hoff least mean square algorithm):

$$\tau \frac{dw_{ij}(t)}{dt} = D_{ij}(t)E_i(t) = D_{ij}(t)[P_i - P_i^0]$$

$$= D_{ij}(t)\left[P_i - \sum_{j=1}^{n+1} w_{ij}(t)D_{ij}(t)\right],$$

$$i = 1, \cdots, n, j = 1, \cdots, n+1 \qquad (3.36)$$

where τ = the learning time constant
E = the complete feedback error, and
D = the selected part of dynamic robot model.

The basic principle of the proposed single-layer approach (modified Kawato approach [43]) is similar to the theory of least mean square adaptive filter and theory of identification; that is, this neural network represents neural identifier of nonlinear inverse dynamic robot model.

Complete scheme of control laws with neural networks are shown in Figures 21.11 and 21.12.

Advance features and potential for future exploitation of neural network modeling in robot control includes

a. Internal pattern associator substitutes external feedback loop as dominant part of control and in this way system is faster;
b. Control and learning are done simultaneously, which enables possibility of adaptation to sudden change of robot dynamics;
c. As form of adaptive control neural network does not require exact parameter model nor parameter estimation;
d. This type of neural network models can be easily implemented in a parallel distributed processing machine with short time of processing.

21.3.7 Industrial Robots Controllers [45,46]

During the last years development of controllers was dictated by automating more and more complex tasks from industrial practice and conducted to the following requirements: needs for achieving higher and higher work speeds and compensation of dynamics, necessity for synchronization of manipula-

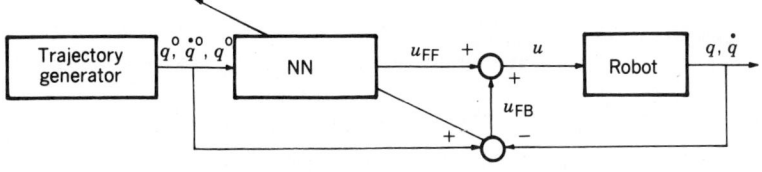

Figure 21.11 Neural network as element of decentralized control.

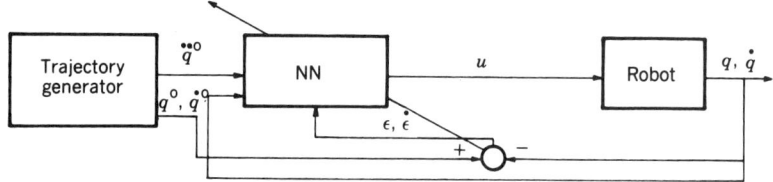

Figure 21.12 *Neural network as element of computed torque method.*

tor with the technological environment, needs for higher accuracy and robot self-sufficiency in locating the workpieces (external sensors: tactile, proximity, vision), needs for studying complex tasks (i.e., tasks in which it is unavoidable to use programmed variables, using commands for assigning values to the programmed variables, calculation of arithmetic expressions, etc.

Basic Program Support The contemporary controller (Fig. 21.13) most often consists of robot program interpreter, trajectories generator calculating the desired trajectories of manipulator tip, trajectories transformer calculating the necessary displacements in the manipulator joints, control synthesis module calculating on basis of current manipulator state, and the wanted joint displacements the actuator command signals in the individual joints.

Contemporary controllers are realized using microcomputer technology, wereby usually one microprocessor is used for program interpretation and calculation of the necessary joints displacements, while one or more microprocessors is used for calculating the actuator control signals.

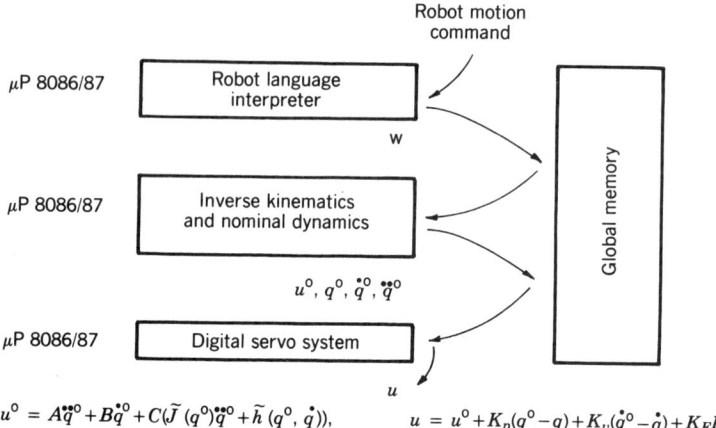

Figure 21.13 *One robot control system implementation.*

21.4 SIMULATION AND DYNAMIC CONTROL OF THE ROBOTIZED MANUFACTURING CELLS

21.4.1 Introduction

Many software packages for simulation of flexible manufacturing cells (FMC) have been developed in the last several years. The benefits of using such software packages are numerous. For example, the simulation of FMC may be used to speed up design of FMC, selecting NC machines and robots; the simulation is also used to test control of FMC at the higher control level (i.e., to test synchronization of all subsystems in the FMC, and so on. In the majority of the existing software packages for simulation of FMC, robots and NC machines are usually modeled in the form of finite automata [47,48]. Such models of subsystems are useful for fast testing of the highest control level of FMC; that is, by such models we may check relatively simply synchronization of all machines, conveyers, buffers, and robots. This approach is usually satisfactory for NC machines, conveyors, and soon, since their dynamics do not have significant influence upon the complete system performances. However, this does not hold for robotic systems, due to their highly nonlinear dynamic behavior. Actually, robot behavior highly depends on the dynamic regime, and it varies with various positions, speeds, and accelerations of the robot joints [49]. On the other hand, it is often required that the software package for simulation of FMC can answer the following questions: could each robot in the particular FMC reach all desired positions in the working space (i.e., are all subsystems properly located in the FMC), are structure of the robots properly selected, could each robot find its paths, and so on. This means that the software packages for simulation of FMC have to include kinematic models of the robots and geometric models of all subsystems. Many packages enable testing of FMC's "geometry," but there are quite a few that include kinematic models of robots [47,50].

21.4.2 Hierarchical Control of Flexible Manufacturing Cell

The control of FMC is always hierarchical. The basic hierarchical control levels are as follows [24]:

  a. The highest control level has to distribute the tasks (operations) to subsystems in FMC: to robots, NC machines, conveyors, and so on. The task of the highest control level is to synchronize all machines and robots in time and in space. One of the tasks of this control level might be to coordinate two or more robots in the space if they share common work space, but often this task is associated to lower control levels. The output of the highest control level are the tasks that have to be implemented by robots and NC machines in FMC.

b. Strategical control level of robot has to plan the robot paths in such a way to realize the task imposed by the highest control level. The strategical level has to ensure robot motion between the obstacles in its work space. Practically the strategical control level has to plan the paths of the robot gripper. One of the tasks of the strategical control level is to minimize the time required for the travel of the robot hand between two imposed positions in the work space or along the desired path. This level also plans the motion of the robot in some complex tasks (e.g., assembling). The strategic control level also plans the forces which the robot has to realize upon the objects in the workspace, and so on.

c. Tactical control level of robots has to realize mapping of the desired trajectory of the robot given in external coordinates into so-called internal coordinates of the robot (the internal coordinates represent coordinates of the robot joints). Namely, the desired motion of the robot is realized by movements of the robot joints driven by corresponding actuators. Therefore, it is necessary to compute movements of the robot joints that correspond to the imposed motion of the gripper (which has been defined at the strategical control level).

d. Executive control level has a task to realize the trajectories imposed by the tactical control level. At this control level, based on the imposed trajectories of the joints coordinates and based on information on actual state of the robot (obtained via corresponding sensors at the robot), the control signals for actuators are generated.

This hierarchical control structure is simulated in the software package as a control of any FMC imposed by the user.

21.4.3 Description of Software Package

The software package for simulation of FMS enables simulation of various FMS consisting of up to two robots, several NC machines, conveyors, buffers and various sensors. In the so-called initialization phase the user has to prepare input data on all subsystems which he or she wants to include in particular FMC. The software package for simulation of FMS consists of several modules. Here we shall briefly describe these modules.

a. The program module for simulation of the highest control level simulates the control of complete FMC (i.e., synchronization of all subsystems in FMC). In the first phase of simulation package development, we have adopted concept of so-called nondynamic control of FMC, which is realized by Petri nets. Using Petri nets, it is possible to achieve synchronization of all subsystems within FMC in a relatively simple way. However, in future work we intend to include so-called dynamic control of FMC at the highest control level; that is, we shall include control of FMC, which takes into account

21.4 CONTROL OF THE ROBOTIZED MANUFACTURING CELLS

dynamic characteristics of subsystems (especially robots). However, other modules are independent of the module for simulation of the highest control level.

b. The program module for simulation of NC machines, conveyors, and buffers simulates these subsystems in simple manner by their models in the form of finite automata. Actually, NC machines are simulated by the time periods that they require for execution of various programs and operations within each program. The package enables imposing arbitrary number of various NC machines and conveyers with arbitrary location of these subsystems in the working space. During simulation the user may impose various failures of each machine.

c. The program module for simulation of sensors enables arbitrary location of the sensors at the working scene. In the first phase of simulation package development we have included three types of external sensors (sensors that are not mounted on robots and NC machines): tactile sensors, proximity sensors, and cameras. We have also included various types of internal sensors (e.g., potentiometers or shaft encoders in robot joints, tachometers, accelerometers, tactile sensors, on/off sensors, etc.). The user also may impose failures of each sensor in the system to test behavior of the control system in such situations.

d. The package includes program module for graphic presentation of FMC. By this model user gets visual presentation of the scene at the 3D graphics. Each subsystem (e.g., robot, NC machine, conveyer, etc.) may be represented by a st of regular geometric bodies. The user commands drawing of a scene at arbitrary moment during the simulation, or he or she may require graphical representation of the scene in discrete moments (e.g., the scene may be represented each 0.5 s during simulation).

e. The main program module in the package is the module for simulation of robot. The package enables simulation of up to two robots of arbitrary type and structure. The module for simulation of robots consists of several submodules. Actually, the package enables simulation of the complete control unit for each robot. Each submodule simulates one level in hierarchical control [51] of manipulation robot:

e.1. The package includes various algorithms for path planning of the robot hand at the strategical control level [52]. By these algorithms the robot controller (i.e., its strategic control level) plans the paths of the robot hand in the presence of obstacles in the working space. The output of this control level is the path of the hand that has to be realized by the lower control levels.

e.2. The next submodule simulates the so-called tactical control level of a robot controller. At this control level, hand trajectory is mapped into joint trajectories. This control level has to solve so-called inverse kinematic model of a particular robot. For the selected structure of the robot the package generates kinematic model in symbolic form and automatically solves the inverse kinematic problem [53]. The output of the tactical control level are desired trajectories of robot joints.

e.3. The submodule for simulation of executive control level enables simulation of various nondynamic and dynamic control laws for manipulation robots. The task of the executive control level is to realize desired trajectories of the robot joints that are imposed by the tactical control level. Using the separate program package for synthesis of control for manipulation robots [54], the user may synthesize various control laws for a particular robot and then simulate robot performance with the selected control law and control parameters, within the particular FMC (in which the robot should be implemented). The package includes synthesis and simulation of nondynamic control laws consisting of local servos around the robot joints [51]. However, various dynamic control laws (i.e., control laws that take into account robot dynamics) may be synthesized and simulated. For example, the synthesis and simulation of global control in the form of force feedback or on-line computation of robot dynamics is included [49,55], as well as the so-called controller based on computed torque method [56], robust control [57], and others. The package simulates microprocessor implementation of control; that is, the sampling period of the controller is simulated. The output of the executive control level are control signals for actuators driving the robot joints.

e.4. The submodule for simulation of the robot performance includes various models of the robot. These models are discussed in the next section.

In Figure 21.14. a global flowchart of the described software package is given. In the text to follow we shall concentrate on the models for simulation of robot performance, since this module plays a central role in this package.

21.4.4 Models of Manipulation Robots in Simulation of FMC

In the software package for simulation of FMC, three general types of robot models are included. Here we shall present these types of robot simulation models.

Finite-Automata Model of Robot The simplest model of a robot may be taken in the form of finite automata. This means that the model of robot accepts commands from the highest control level (e.g., commands like "move", "grasp", "stop," etc.) and it responds, after a certain period of time, that the command has been executed. The model also may respond that the command has not been executed since some failure in the robot system has occurred. The user of the package may impose failure of the robot to execute certain command. However, the problem is how to determine the time that the robot requires to execute a certain command. For example, the duration of execution of the command "move from one posi-

21.4 CONTROL OF THE ROBOTIZED MANUFACTURING CELLS

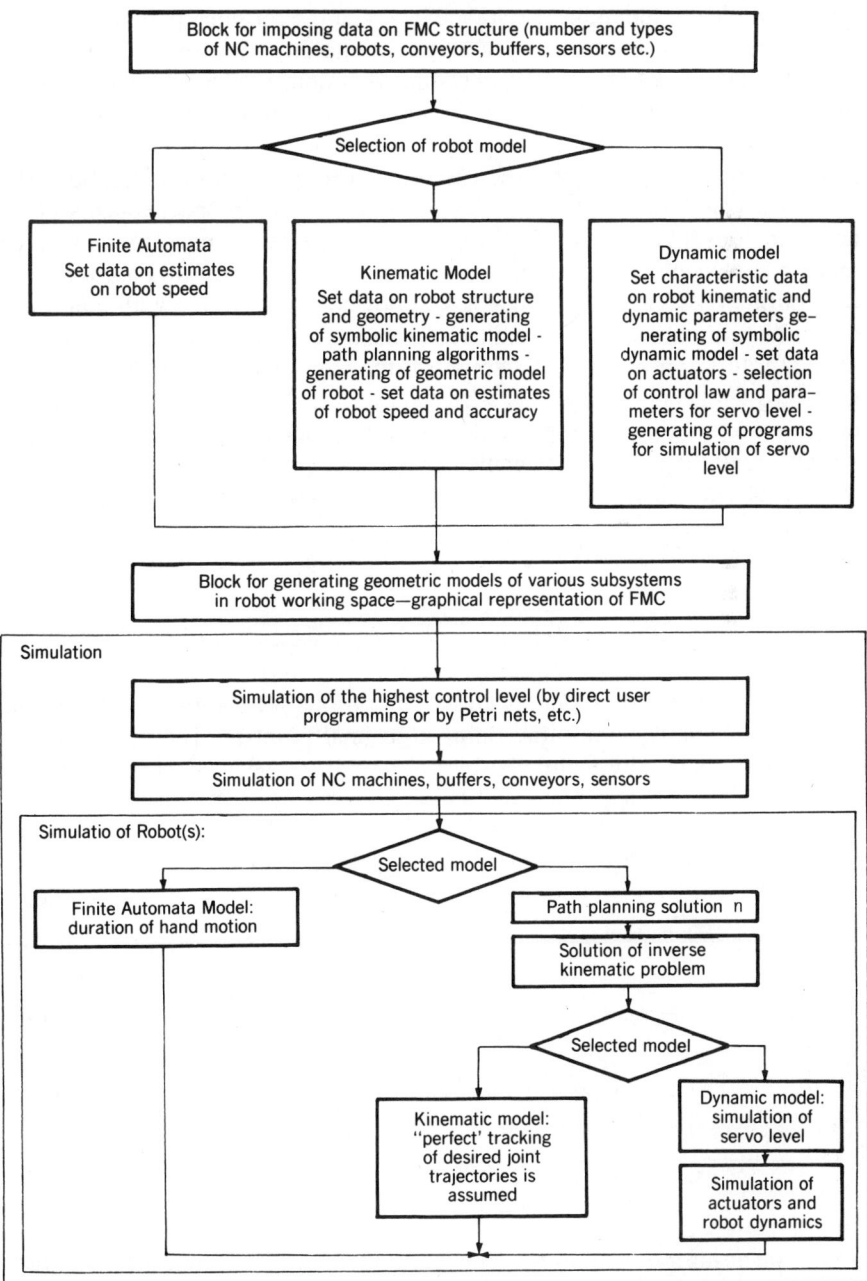

Figure 21.14 *Global flowchart of simulation or robotic FMC.*

tion to another'' depends on distance between the terminal points and on the robot capabilities. In order to get rough estimation of time period required to execute certain command, we introduce a simple model that describes the movement of the robot hand. The motion of the robot hand is described by velocity and acceleration of the hand center of mass (or some other selected point on the hand) and by three rotational velocities and accelerations (for example, velocities of Euler angles of the hand with respect to axes of the absolute coordinate system). Actually, we describe the hand motion by velocities and accelerations of the so-called external (Cartesian) coordinates of the robot [58]. We suppose that the robot hand can move with some prescribed maximum velocities and accelerations, and using these velocities we estimate duration of each movement of the robot. If robot external coordinates are $s = (x_c, y_c, z_c, \theta, \phi, \psi)^T$, where x_c, y_c, z_c are the Cartesian coordinates of the selected point on the robot hand with respect to absolute coordinate system, and θ, ϕ, ψ are Euler angles of the robot hand, then we assume that the maximum velocities and accelerations of external coordinates are

$$(\dot{x}_c^2 + \dot{y}_c^2 + \dot{z}_c^2)^{1/2} < v_m, \qquad \dot{\theta} < \dot{\theta}_m, \quad \dot{\phi} < \dot{\phi}_m, \quad \dot{\psi} < \dot{\psi}_m$$

$$(\ddot{x}_c^2 + \ddot{y}_c^2 + \ddot{z}_c^2)^{1/2} < a_m, \qquad \ddot{\theta} < \ddot{\theta}_m, \quad \ddot{\phi} < \ddot{\phi}_m, \quad \ddot{\psi} < \ddot{\psi}_m$$

(21.43)

Using these maximal velocities and accelerations we estimate duration t_d of certain movement from initial point s_0 to final point s_f. The motion of the robot hand is actually simulated by

$$x(t) = x_0 + \frac{a_m(t - t_0)^2}{2}, \qquad t < t_1$$

$$x(t) = x_0 + \frac{a_m(t_1 - t_0)^2}{2} + v_m(t - t_1), \qquad t_1 < t < t_2$$

$$x(t) = x_0 + \frac{a_m(t_1 - t_0)^2}{2} + v_m(t_2 - t_1) \qquad (21.44)$$

$$+ v_m(t - t_2) - \frac{a_m(t - t_2)^2}{2}, \qquad t_2 < t < t_3$$

$$t_1 = \frac{v_m}{a_m} + a, \quad t_2 = \frac{x_f - x_o}{v_m}, \quad t_3 = t_2 + \frac{v_m}{a_m}$$

under the assumptions that $t_2 > t_1$ (i.e., $v_m^2/a_m < x_f - x_0$), that the movement starts at the moment t_0, and where $s_0 = (x_0, y_0, z_0, \theta_0, \phi_0, \psi_0)^T$ are the coordinates of initial point, and $s_f = (x_f, y_f, z_f, \theta_f, \phi_f, \psi_f)^2$ are the coordinates of the final (desired) point of the robot hand. Equations (21.44) hold for the first coordinate x_c in the vector of external (hand) coordinates s. The similar equations hold for the rest five external coordinates. Using Eq. (21.44) we get time required t_d for execution of the desired movement from point s_0 to point s_f. Thus, in our simulation package, when the highest control level sends the command that the robot has to move from its present (initial) point s_0 to some desired point in external coordinate space s_f the package computes time duration according to (21.44), and after this time period has passed it sends information to higher level that the desired movement has been executed. (Obviously, this holds if the user has selected this simple model to represent robot in the simulation). In the model internal (joint) coordinates of the robot are not computed at all. The complete robot is replaced by a single body (hand), the movement of which is described just by maximal linear and rotational velocities [e.g., by Eqs. (21.43) and (21.44)].

The main advantage of this model is its extreme simplicity. The time required for simulation of such model is very short, since its computation at every integration interval requires only a few multiplications and additions. This model is useful if the user wants to test the highest control level of the FMC—in other words, if he or she wants to check synchronization of all subsystems in FMC. By this model of robot the user may quickly test behavior of the complete FMC if some failures in subsystems appear, he or she can check information flow between the highest control level and other subsystems, and so on. However, the main drawback of this robot model is that it gives extremely rough estimation of the robot performance. Time required for certain motion estimated by (21.44) may be very rough so that synchronization with other subsystems may be very suboptimal (from the standpoint of time required for execution of complete task). On the other hand, this model can be used just to test the highest control level of FMC. We cannot test controller of the robot nor can we get any information of the robot capabilities. It should be noted that this model may be used to check roughly path planning algorithms but in the case if these algorithms treat robot as one point (point at the robot hand) which should move between the obstacles. This means that all NC machines, conveyors, and other obstacles in the working space are modelled by the regular geometric bodies and the path planning algorithms (at the strategical level of the robot controller) plan the movement of the robot tip in its working space. However, such algorithms are extremely suboptimal since they neglect the robot geometrical structures [52]. It also should be noted that the model (21.44) may be improved if instead fixed values for the maximum velocities of the robot hand $v_m, \dot{\theta}_m, \dot{\phi}_m, \dot{\psi}_m$ we take variable values that vary depending on the robot position s. Also it is possible to take into account acceleration and deceleration phase of each movement (i.e., to include maximum accelerations of the robot hand).

Kinematic Model of the Robot In order to enable testing of strategic and tactical levels in the robot controllers and to enable obtaining information on robot kinematic capabilities, we introduced kinematic model of the robot. The kinematic model of the robot describes relation between the angles (displacements) of the robot joints and the robot hand (external) coordinates. If by q we denote the $n \times 1$ vector of the robot internal (joint) coordinates (n is the number of the robot simple joints), we can write the relation between the vector s of external coordinates and q as

$$s = f(q) \qquad (21.45)$$

where f is a function that maps the $n \times 1$ vector q to the vector s (the number of external coordinates could be less than, or equal to 6). Equation (21.45) represents the direct kinematic model of the robot. Since the highest control level usually imposes desired movement of the hand (i.e., it sets the trajectories of the external coordinates), the tactical control level has to compute the joints coordinates q by solving the inverse kinematic problem:

$$q = f^{-1}(s) \qquad (21.46)$$

Once the trajectories of the joints coordinates are computed they have to be realized by executive control level. In our package, if the user selects to simulate the robot behavior just by kinematic model, it is assumed that the desired trajectories of the robot joints (which are computed at each sampling interval by the tactical level of the robot controller) are perfectly realized. The tactical control level is completely simulated (as well as strategic control level), and it is assumed that the realized joints trajectories $q(t)$ are equal to desired (nominal) trajectories. The package just checks whether the desired accelerations $\ddot{q}^0(t)$, velocities $\dot{q}^0(t)$, and angles (displacements) $q^0(t)$ could be realized by the actuators in the robot joints. In other words, the simulator checks if

$$|\dot{q}_i^0(t)| < \dot{q}_{iM}, \quad |\ddot{q}_i^0(t)| < \ddot{q}_{iM}, \quad q_{im} < q_i^0(t) < q_{iM}, \quad i = 1, 2, \cdots, n$$

$$(21.47)$$

where $\ddot{q}_{iM}$ = maximum acceleration of the ith joint
$\dot{q}_{im}$ = maximum velocity of the ith joint
q_{iM} and q_{im} = minimum and maximum angle (displacement) of the ith joint

If the conditions (21.47) are met, it is assumed that the desired trajectories of the robot joints are perfectly realized. Otherwise, the desired movement of the robot must be slow down (or it must be changed if required angles are out of the allowable range). Using computed trajectories of the robot joints coordinates $q^0(t)$ and using limits (21.47) upon the joints accelerations, velocities, and angles the simulation model gives answers whether the motion set by the higher control level can be met or not, and estimates time duration of the movement. Therefore, using this model the user can test whether the robot hand can reach all required positions in the cell and whether all required movements of the robot are feasible. In other words, the user may test whether the robot can find collision-free paths between the desired positions. By this one can select the "optimal" location of the robot and other subsystems in the cell. One can check the capabilities of the selected robot (from the "geometric point of view") to perform certain tasks within specific FMC. Therefore, this model may be used to select appropriate robot structure for specific FMC (i.e., to select which of the robots available at the market is the most appropriate for the specific FMC, or to design a new robot, if necessary). Also the kinematic model may be used to redefine the subtasks dedicated to a robot. The problem of a robot redundancy may be solved by this model, as well as the problem of how to avoid singular points of a robot kinematics (i.e., how to find singular-free paths of the robot hand).

It is obvious that the kinematic model gives better estimates of the robot performance and capabilities than the finite-automata model. However, the estimates of the robot speed to accomplish certain movements may be very rough, since it is based on inequalities (21.47) in which rough estimates of the allowable joints velocities and accelerations are used. Namely, since the constraints upon the joints velocities $\dot{q}_{iM}$ and accelerations $\ddot{q}_{iM}$ must cover all possible configurations of the robot, they may be too conservative. These constraints do not account for actual capabilities of the robot actuators and actual dynamic load upon them, but only take some rough estimates of the allowable joints velocities and accelerations. Therefore, the estimates of the time required by the robot to execute certain commands imposed by the higher control level may be rough, and thus, the coordination between the subsystem within the FMC may be very suboptimal. The kinematic model of the robot can be used to test in detail both strategic and tactical levels of the robot controller. However, the servo level of the robot controller cannot be tested. It should be noted that the above model may be improved; instead fixed values for joints maximal velocities and accelerations we introduce variable values which vary depending on robot configuration q. This model may be also used for test interpolation of the robot trajectories (i.e., the primitive level of the robot controller [59]).

Dynamic Model of the Robot In order to test complete controller of robots and FMC, we have introduced simulation using dynamic model of a robot [60]. The dynamic model of the robot consists of the model of mechanical part of the robot (mechanism) and the model of actuators that are driving the robot joints. The model of the mechanical part of the robot is usually assumed in the form

$$P = H(q)\ddot{q} + h(q,\dot{q}) \qquad (21.48)$$

where $P = n \times 1$ vector of driving torques in the joints
$H = n \times n$ inertia matrix of the mechanism
$h = n \times 1$ vector of centrifugal, Coriolis, and gravity moments (forces) around the axes of the joints

Various types of actuators are applied to drive robots: dc motors, ac motors, hydraulic actuators, pneumatic actuators, and so on. The models of actuators are in general nonlinear, but for the dc motors (which are still most often applied for industrial robots) linear state model may be adopted:

$$\dot{x}^i = A^i x^i + b^i u^i + f^i P_i, \quad i = 1, 2, \cdots, n \quad (21.49)$$

where $x^i = (q_i, \dot{q}_i, i_{Ri})^T = 3 \times 1$ state vector of ith actuator model
i_{Ri} = rotor current of ith dc motor
u_i = scalar input to ith actuator
P_i = driving torque (load) in ith joint
A_i = 3 × 3 matrix
b^i, f^i = 3 × 1 vectors.

The connections between the models (21.48) and (21.49) (through the state coordinates $q_i, \dot{q}_i$ and driving torques P_i) are evident. Certain constraints upon the actuators input u amplitude as well as on the allowable driving torques should be also added to these models. The robot controllers generate the input signals for the actuators according to the selected control law at the executive (servo) level:

$$u_i(t) = g_i(x(t), x^0(t)) \quad (21.50)$$

where x is the state vector of the overall system $x = (x^1, x^2, \cdots, x^n)^T$, and $x^0(t)$ is the nominal state vector that corresponds to desired joint trajectories $q^0(t)$ that are imposed by the tactical control level (according to (21.46) and robot hand trajectory $s^0(t)$ imposed by the strategic level). In (21.50) g_i denotes selected control law for the ith actuator. For example, if simple local servo is applied to control the ith joint then $g_i(x(t), x^0(t)) = -k_i^T(x^i(t) - x^{0i}(t))$, where k_i is 3 × 1 vector of local servo feedback gains. To this local servo control law we may add feedforward term (nominal control [54]), which is computed either using the entire dynamic model of the robot (21.48), (21.49) or using only the model of actuators (21.49) in which the dynamic coupling $f^i P_i$ between the joints is ignored. This control law is the most frequently implemented in practice, and it is satisfactory for robot positioning and tracking of relatively slow trajectories. However, to improve accuracy of the tracking of the desired nominal trajectories of the joints $q_0(t)$, we may implement more complex dynamic control laws. For example, we may apply global control, which is added to local servo signals, so that the control law is in the form [55,23]

$$u_i(t) = -k_i^T(x^i(t) - x^{0i}(t)) + u_i^0(t) + k_i^G P_i \quad (21.51)$$

where u_i^0 = (centralized or local) feedforward term
k_i^G = global feedback gain for ith joint
P_i = value representing information on actual coupling between joints $P_i(t)$

21.4 CONTROL OF THE ROBOTIZED MANUFACTURING CELLS

This information may be obtained in two ways: by force feedback, or by on-line computation of the robot dynamic model. Namely, by measuring torques (by force sensors) we obtain direct information on coupling between the joints. The second solution requires on-line computation of the dynamic model of the robot (21.48) or some approximate models of the robot may be adopted [23]. As mentioned above other types of dynamic control laws may be simulated by our software package. The separate software package may be used to synthesize control parameters for the selected control law [54], and then, by simulating the complete FMC we can verify whether the synthesized control law meets all specific requirements for the particular FMC. The effects of these dynamic control laws are to improve accuracy of tracking of trajectories and to increase speed of the robot. Namely the accuracy of tracking of desired (nominal) trajectories is usually constrained by the required robot speed: If the trajectories are relatively slow, even simple servo may ensure sufficiently accurate tracking. However, if the speed of the desired movement is increased, the accuracy is decreased. The dynamic control law improves accuracy of trajectory tracking even for higher speeds. On the other hand, settling time which is required by the robot to settle at required terminal point, depends on the applied control law: dynamic control laws may considerably reduce settling time. However, the relation between robot speed and accuracy highly depends on the specific movements and robot task. For each particular movement this relation between the accuracy and the speed may vary, although the control system tends to achieve uniform performance for the robot for all regimes. On the other hand, requirements regarding the accuracy and the speed of the robot vary depending on the specific task (for example, the accuracy is much lower if the robot has only to place a workpiece at the conveyor than if it has to accurately place a workpiece at certain location). It is quite obvious that the simulation of the robots within FMC does not include these dynamic effects; one cannot get the complete insight in the actual robot capabilities. This may cause at least two negative consequences. The designer of the system may not utilize all advantages of the specific robot. On the other hand, time scheduling of subtasks dedicated to robots and other subsystems at the FMC control level may be very suboptimal. Our software package supplies the user with accurate estimates of the robot speed/accuracy ratio for each particular movement of the robot desired in the specific task of the FMC. If the dynamic control of the robot is applied, we can estimate the increase of this ratio. Therefore the user can achieve optimal scheduling of the subtasks at the FMC's control level during off-line programming of the cell. On the other hand, for each particular movement (action) of the robot we obtain what is desired speed (if no maximal) and what accuracy is acceptable and therefore we may select the most appropriate controller, or we may optimize the robot performance from the standpoint of energy consumption. In other words, this package enables to precisely determine the relations between the higher and the lower control levels. This package enables to simulate dynamic model of various robots of arbitrary type and structure. The user has just to set basic data on robot structure, geometry and dynamic parameters and the package automatically generates the dynamic model (21.48) in symbolic form [3]. Simulation of the complete dynamic model of the robots and the entire controller may be time consuming, in spite of the fact that fast symbolic models of robots dynamics are used. Often it is not necessary to use complete dynamic model, but various approximate models may be used, in which certain dynamic effects may be ignored (for example, Coriolis or centrifugal forces). By this computer time required for simulation may be significantly reduced without affecting the accuracy of simulation results. It should be kept in mind that even complete dynamic models (21.48), (21.49) are some approximation of actual system behavior, since many effects (for example, elastic modes of the robot links, etc.) are not taken into account. However, the model (21.48), (21.49) is much more accurate representation of the robot performance than finite-automata or kinematic models.

21.4.5 An Example

The developed software system has been tested on various types of robots and various FMC. Here we shall briefly present an example of simulation testing of particular FMC consisting of two robots, three NC machines, two buffers, two conveyers, and several different sensors. The simulated FMC is presented in Figure 21.15. The Petri net for control of this particular FMC [61,62] is applied. First, the performances of the FMC using the rough models of robots (finite-automata models) are simulated. Then, the kinematic models of both robots to select the appropriate locations of the robots and ensure that all desired positions can be reached by the robots end-effectors are applied. The trajectories of the hand of the robot at the left side of FMC are presented in Figure 21.16. Maximum speeds and the accuracy of the robots servos is assumed and we have got an estimate of duration of one cycle of task execution (time required between the entrance of a work object at the input conveyor and its exit at the output conveyor) of about 30 s. Then the complete dynamic models of the robots when local servos including feedforward terms [23] are applied are simulated. It has been found that the robots can move faster in some movements and therefore we have been able to improve synchronization of the entire cell and to achieve the cycle time of about 24 s. This means that about 20 percent improvement has been achieved by simulation of the complete robot dynamics. In Figure 21.17 are presented the errors in tracking of the joint trajectories developed if the first robot is forced to move faster than with its nominal speed. During transfer between the desired end-positions of the robot hand the tracking errors are high, but they do not reflect execution of the entire tasks. Namely, when the robots approach their goal positions their end-effectors achieve desired posi-

21.4 CONTROL OF THE ROBOTIZED MANUFACTURING CELLS

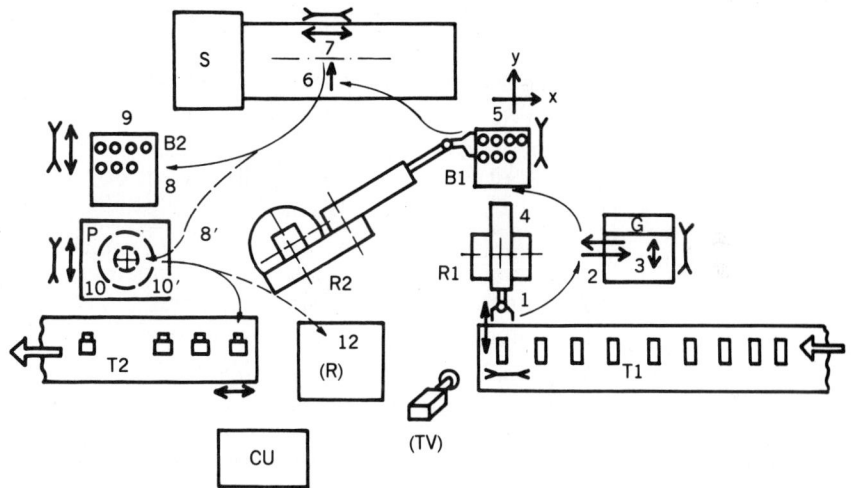

Figure 21.15 Example of FMC.

T1: INPUT TRANSPORTER
R1: MANIPULATOR
G: MILLING MACHINE
B1: BUFFER 1
S: LATHE
R2: ROBOT
B2: BUFFER 2
P: CALIBRATING PRESS
T2: OUTPUT TRANSPORTER
R: REJECT
TV: VISUEL (TV) SYSTEM
CU: CONTROLLER
GRIPS (CLOSES)
RELEVES (OPENS)

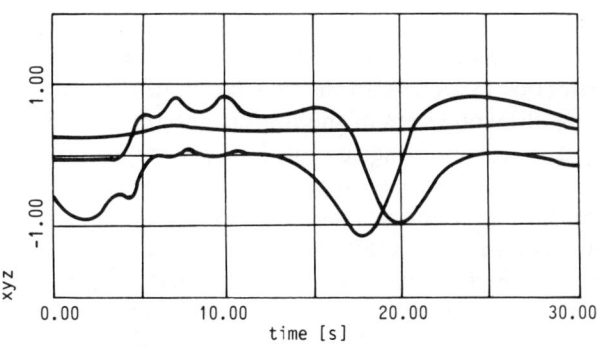

Figure 21.16 Hand trajectories of robot.

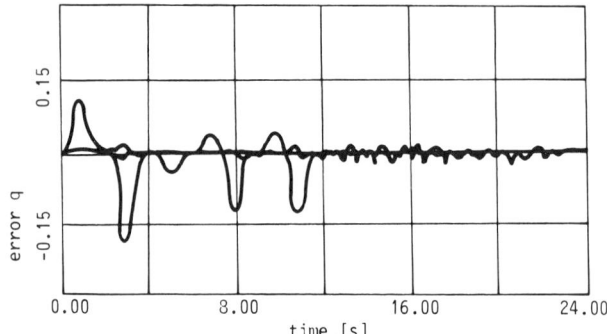

Figure 21.17 Errors in tracking of joint trajectories.

tions with sufficient accuracy in spite of the high error developing during transfers. This means that these faster movements may be accepted. Further improvement may be achieved if we synthesize dynamic control for robots using software package for robot control synthesis [54,23]. By application of dynamic control law, which includes compensation for coupling between the joints, we may achieve more accurate tracking of faster trajectories and by this to reduce the duration of one cycle of the entire cell. The effects of control which include dynamic compensations upon the robot speed and accuracy has been shown in [49].

21.4.6 Conclusion

The developed software system offers some obvious advantages over existing software tools for simulation and design of FMC [63,64]. As mentioned, this package may used either for design of various FMCs (selection of robots and other subsystems, allocation of robots in working space, etc.), or for synthesis of appropriate control at both top control level (programming of FMC) and at lower levels of robots controllers.

It is obvious that simulation of robots using their dynamic models requires much longer processor time than simulation by the finite automata model and by kinematics model. Also the simulation of dynamic models requires knowledge on all parameters of the robotic systems, which may complicate the user's job. The more precise model is applied the more specific data on robotics systems have to be supplied, but the benefits that should be expected from the simulation are also increased.

However, users may use these model options step by step. First, they may test only the highest control level using the finite automata model of the robot. This simulation is relatively fast, and users can test whether they can roughly coordinate all subsystems within their FMC. Actually, the user can program the sequence of subtasks (operations, movements of the robot hands, etc.), and then the parallel actions are recognized and better synchronization and coordination of subsystems may be achieved, but taking into account only rough estimates of the robot capabilities.

Next, the user has to specify all relevant data for modeling the FMC geometry: geometric models of all subsystems, their locations of the scene, data on robot kinematics, and so on. The execution of the FMC's tasks is simulated using kinematic robot model. By this the path planning algorithms can be tested as well as trajectory generation level in the robots controllers and the coordination between the robots and other subsystems may be improved.

At last the user has to impose data on robot dynamic parameters and actuators, as well as on the executive (servo) control level. Then the user can simulate the performance of the FMC using the dynamic model of the robots. By this a complete insight in the robot capabilities may be obtained: how fast and how accurate can the robot perform each particular movement. In this way one can test whether the selected robot and its controller are capable of satisfying all requirements in specific FMC. The control at both path planning level and at the servo level may be improved in order to meet these requirements. On the other hand, synchronization and coordination of the robot(s) with the other subsystems can significantly improved. Actually such simulation enables programming and control at the highest control level of FMC in such a way that the actual dynamic capabilities of the robots are taken into account.

REFERENCES

[1] M. Kirćanski, M. Vukobratović, N. Kirćanski, and A. Timčenko, "A New Program Package for the Generation of Efficient Manipulator Kinematic and Dynamic Equations in Symbolic Form," Third ICAR, October 1987.

[2] M. Vukobratović and N. Kirćanski, Numerical generation of closed form mathematical models of robotic mechanisms, *Acta Applicandea Mathematical,* 2(2) (1984).

[3] M. Vukobratović and N. Kirćanski, *Real-Time Dynamics of Manipulation Robots,* Springer-Verlag, Berlin, 1985.

[4] W. J. Burdick, An algorithm for generation of efficient manipulator dynamic equations, *IEEE Int. Conf. on Robotics and Automation, San Francisco* (1986).

[5] F. J. Kleinfinger and W. Khalil, Dynamic modeling of closed-loop robots, 16, *ISIR,* 1986.

[6] K. P. Khosla and P. C. Neuman, Computational requirements of customized Newton-Euler algorithms, *J. Robotic Systems,* 3 (2) 309–327, (1985).

[7] P. C. Neuman and J. J. The complete dynamic model and customized algorithms of the PUMA robot, *IEEE Trans. on Systems, Man and Cybernetics,* 4, (1987).

[8] A. Izaguirre and R. Paul, Automatic generation of the dynamic equations of the robot manipulators using a LISP program, *Proc. IEEE Internat Conf. on Robotics and Automation, San Francisco,* 220–227, (1986).

[9] S. Yin and J. Yuh, An efficient Algorithm for automatic generation of manipulator dynamic equations, *Proc. IEEE Internat Conf. on Robotics and Automation, Scottsdale, Arizona,* 1812–1817, (1989).

[10] W. R. Toogood, Efficient robot inverse and direct dynamics algorithms using micro-computer based symbolic generation, *Proc. IEEE Internat Conf. on Robotics and Automation, Scottsdale, Arizona,* 1812–1817, (1989).

[11] O. W. Schiehlen, "Computer Generation of Equations of Motion," in *Computer Aided Analysis and Optimization of Mechanical System Dynamics,* NATO ASI Series, Vol. F9, Springer-Verlag, Berlin, pp. 183–215, 1985.

[12] N. Kirćanski, A. Timćenko, Y. Jovanović, M. Kirćanski, M. Vukobratović, and R. Milunov, Computation of customized symbolic robot models on peripheral array processors, *IEEE J. of Robotics and Automation,* (1990).

[13] M. E. Kahn and B. Roth, The near minimum time control of open loop articulated kinematic chains, *Trans. ASME J. Dynamic Systems, Measurement and Control,* 164–172, (Sept. 1971).

[14] M. Vukobratović and D. Stokić, *Control of Manipulation Robots: Theory and Application,* Scientific Fundamentals of Robotics Series, Vol. 2, Springer-Verlag, Berlin, 1982.

[15] M. Vukobratović and D. Stokić, Contribution to the decoupled control of large-scale mechanical systems, *Automatica,* 1, 1980.

[16] R. C. Paul Modelling, trajectory calculation and servoing of a computer controlled arm, A.I. Memo 177, Stanford Artificial Intelligence Laboratory, Sanford University, Sept. 1972; also in Russian, Nauka, Moscow, 1976.

[17] M. Vukobratović and D. Stokić, Dynamic control of manipulators via load-Feedback, *J. Mechanism and Machine Theory,* 17(2) 107–118, (1982).

[18] Y. S. J. Luh, D. W. Fisher, and I. P. R. Paul, Joint torque control by a direct feedback for industrial robots, *IEEE Trans. on Automatic Control,* AC-28(2), (1983).

[19] H. M. Raibert and J. J. Craig, Hybrid position/force control of manipulators, *J. Dynamic Systems, Measurement and Control, Trans. of the ASME,* 103(2) 126–133, (1981).

[20] J. Roessler "A Decentralized Hierarchical Control Concept for Large-Scale Systems, Proc. of the Second IFAC Symp. on Large-Scale Systems, pp. 171–179, Toulouse, Pergamon Press, 1980.

[21] F. Miyazaki, S. Arimoto, M. Takegaki, and Y. Maeda, "Sensory Feedback Based on the Artificial Potential for Robot Manipulators," Preprint of the 9th IFAC World Congress, Vol. 6, pp. 27–32, Budapest, Pergamon Press, July 1984.

[22] D. Stokić and M. Vukobratović, Robustness of decentralized robot control of payload variation, *J. Robotic Systems,* 7(2)(1988).

[23] M. Vukobratović, D. Stokić, and N. Kirćanski, *Non-Adaptive and Adaptive Control of Manipulation Robots,* Scientific Fundamentals of Robots, Series Vol. 5, Springer-Verlag, Berlin, 1985.

[24] V. S. Medvedov, A. G. Leskov, and A. S. Yuschenko, *Control Systems of Manipulation Robots* (in Russian), Nauka, Moscow, 1978.
[25] M. Vukobratović, General structure of software for modelling and control of manipulation robots (in Russian), *Technical Cybernetics*, AN SSSR, No. 3, Moscow, 1987.
[26] M. Vukobratović, D. Stokić, and N. Kirćanski, Towards non-adaptive and adaptive control of manipulation robots, *IEEE Trans. on Automatic Control*, AC-29(9) 841–844 (1984).
[27] M. Vukobratović and N. Kirćanski, An approach to adaptive control of robotic manipulators, *Automatica*, 21(6) (1985).
[28] A. V. Timofeev and Yu. V. Ekalo, Stability and stabilization of programmed motions of robots, (in Russian), *Automatica and Remote Control*, 20, 1976.
[29] A. J. Kiovo and T.H. Guo, Adaptive linear controller for robotic manipulators," *IEEE Trans. on Automatic Control*, 28(20), (1983).
[30] B. J. Oh, M. Jamshidi, and H. Seraji, Decentralized adaptive control, *Proc. IEEE Internat. Conf. on Robotics and Automation*, 1016–1021 (April 1988).
[31] O. Khatib, A unified approach for motion and force control of robot manipulators: the operational space formulation, *IEEE J. Robotics and Automation*, RA-3(1) (1987).
[32] M. Vukobratović and V. Potkonjak, *Applied Dynamics and CAD of Manipulation Robots*, Springer-Verlag, 1985.
[33] M. Vukobratović and D. Vujić, Contribution to solving dynamic robot control in a machining process, *Mechanism and Machine Theory*, 22(5)(1987).
[34] M. T. Mason, Compliance and force control for computer controlled manipulators, *IEEE Trans. on Systems, Man and Cybernetics*, SMC-11(6) (1981).
[35] D. Stokić, Contribution to constrained motion control of manipulation robots, *Robotica*, 1990 in press.
[36] D. Stokić, M. Vukobratović, D. Vujić, and A. Rodić, "Dynamic Models in Simulation and Synthesis of Position/Force Control of Robots," Eighth CISM-IFToMM Symp. on Theory and Practice of Robots and Manipulators (Ro.man.sy '90), Cracow, Poland, 1990.
[37] M. Vukobratović, *Applied Dynamics of Manipulation Robots: Modelling, Analysis and Examples*, Springer-Verlag, Berlin, 1989.
[38] M. Djurović and M. Vukobratović, Contribution to dynamic modelling of cooperative manipulation, *Mechanism and Machine Theory*, 1990 in press.
[39] H. W. Sunada, "Dynamic Analysis of Flexible Spatial Mechanism and Robotic Manipulators," Doctoral dissertation, University of California, Los Angeles, 1981.
[40] D. Šurdilović and M. Vukobratović, Contribution to dynamics of flexible manipulation robots, *Mechanism and Machine Theory*, 1990 in press.
[41] Y. Ichikawa, and N. Ozaki, Autonomous mobile robot, *J. Robotic Systems*, 2(2) (1985).
[42] N. Kirćanski, M. Vukobratović, M. Kirćanski, and A. Timčenko, "Modelling, Simulation and Control of Complex Robotic Systems," Sixth CISM-IFToMM Symposium on Theory and Practice of Robots and Manipulators, Udine, Italy, 1988.

[43] H. Miyamoto, M. Kawato, T. Setoyama, and R. Suzuki, Feedback—error learning *neural network* for trajectory control of a robotic manipulator, *Neural Networks,* 1(1) 251–265 (1988).

[44] D. Katić, Neural network model for learning control of manipulation robots, *Proc. of the Internat. Conf. on Intelligent Autonomous Systems* 2, Amsterdam 1989.

[45] B. Karan, M. Timotijević, M. Djurović, V. Devedžić, and N. Josifović, "A System for Programming the UMS-7 Industrial Robot," 3rd Yugoslav-Soviet Symp. on Applied Robotics and Flexible Manufacturing Systems, Moscow, 1986.

[46] N. Kirćanski, Dj. Leković, M. Borić, M. Vukobratović, M. Djurović, N. Djurović, T. Petrović, B. Karan, and D. Urošević, An Advanced Distributed Control System for Education in Robotics", Eighth CISM-IFToMM Symp. on Theory and Practice of Robots and Manipulators, Cracow, Poland, 1990.

[47] M. Vojnović et al. "An ESPRIT Project in Advanced Robotics," *3rd International Conference on Advanced Robotics '87 ICAR,* Versailles, IEEE, Washington, D.C., October 1987.

[48] L. S. Zenkevich and A. A. Dimitriev, "Mathematical and Program Support for Adaptive Robots for Assembling", The Second Yugoslav-Soviet Symposium on Applies Robotics and Flexible Automation, Arandjelovac, ETAN, Belgrad, 1984.

[49] M. Vukobratović and D. Stokić, Is dynamic control needed in robotic systems and if so, to what extent?, *Internat. J. Robotic Research,* 2 (1983).

[50] G. Spur et al., "Planning and Programming of Robot Integrated Production Cells," ESPRIT Technical Conference, Brussels, 1987.

[51] M. Vukobratović and D. Stokić, *Applied Control of Manipulation Robots: Analysis, Synthesis and Exercises,* Springer-Verlag, Berlin, 1989.

[52] D. Stokić, M. Vukobratović, and V. Devedžić, "Expert System for Synthesis of Dynamic Control of Manipulation Robots," CISM-IFToMM Conference on Robots and Manipulators. "Ro.man.sy 88," Udine, Italy, 1988.

[53] V. M. Kirćanski, K. M. Vukobratović, M. N. Kirćanski, and M. A. Timčenko, A new program package for generation of efficient manipulator kinematic and dynamic evaluation in symbolic form, *Robotica* (July 1988).

[54] K. M. Vukobratović and M. D. Stokić, A procedure for interactive dynamic control synthesis of manipulators, *Trans. on Systems, Man, and Cybernetics,* (Sept./Oct. 1982).

[55] K. M. Vukobratović and M. D. Stokić, *Control of Manipulation Robots: Theory and Application,* Scientific Fundamentals of Robotics Series, No. 2, Springer-Verlag, Berlin, 1982.

[56] R. C. Paul, Modelling, trajectory calculation and servoing of a computer controller arm, A.I. Memo 177, Stanford Artificial Intelligence Laboratory Stanford University, September, 1972.

[57] H. Asada and E. J. J. Slotine, *Robot Analysis and Control,* Wiley, New York, 1986.

[58] J. J. Craig, *Introduction to Robotics: Mechanics and Control,* Addison-Wesley, Reading, Mass., 1986.

[59] J. S. Albus, H. G. McCain, and R. Lumia, NASA/NBS Standard Reference Model for Telerobot Control System Architecture (NASREM), NBS Technical Note 1235, 1987.

[60] K. M. Vukobratović, *Applied Dynamics of Manipulation Robots: Modelling, Analysis and Examples,* Springer-Verlag, Berlin, 1989.

[61] B. M. Jocković, An Application of Petri-nets in a Control System of Flexible Manufacturing Cells, Fourteenth Symp. on Microprocessing and Microprogramming, Zurich, North-Holland, 1988.

[62] A. A. Leskin, *Algebraic Models of Flexible Manufacturing Systems* (in Russian), Publ. comp. NAUKA, Leningrad, 1986.

[63] K. M. Vukobratović and M. D. Stokić, Software Package for Simulation of Flexible Manufacturing Systems, (in Russian), *Technical Cybernetics,* No. 3, Moscow, 1989.

[64] M. D. Stokić, K. M. Vukobratović, and Dj. Leković, "Modelling of Robots and their Environment in Simulation of Flexible Manufacturing Cells", IFAC Symp. on Information Control Problems in Manufacturing Technology, Madrid, Pergamon, London, 1990.

CHAPTER 22

Intelligent Data Base and Automatic Discovery

KAMRAN PARSAYE
MARK CHIGNELL
SETRAG KHOSHAFIAN
HARRY WONG

22.1 INTRODUCTION

Intelligent data bases deal with *information*, not just data. By using them, we can perform tasks— involving large amounts of information—that otherwise could not possibly be performed. Their use makes possible what we had not even considered before. They provide new viewpoints for business, science, and education. For instance, throughout history, we have conducted science by gathering data with observations, analyzing observations in our minds, and making new "discoveries." By ushering in *automatic discovery,* in contrast, intelligent data bases help us make discoveries automatically.

Intelligent data bases represent the evolution and the merger of several technologies, including

- Automatic discovery
- Hypermedia
- Object orientation
- Expert systems
- Traditional data bases

These technologies are uniformly integrated within the framework of intelligent data bases to provide the intelligent technologies for the 1990s (see Fig. 22.1). It is, however, important to note that our discussion here is not purely

---

Neural and Intelligent Systems Integration, By Branko Souček and the IRIS Group.
ISBN 0-471-53676-8   ©1991 John Wiley & Sons, Inc.

hypothetical or theoretical. The technologies discussed are available in commercial form.

In this chapter we focus on what intelligent data bases are, the technologies they use, how they work, and how they impact business, science, and knowledge.

22.2 ARCHITECTURE OF AN INTELLIGENT DATA BASE

The top-level architecture of the intelligent data base consists of three levels:

- High-level tools
- High-level user interface
- Intelligent data base engine

As Figure 22.2 shows, this is a staircase-layered architecture—that is, users and developers may independently access different layers at different times. The first level is *high-level tools*. These tools provide the user with a number of facilities such as automated discovery, data quality and integrity control, and intelligent search. The second level is the *high-level user interface*. This level creates the model of the task and data base environment that users directly interact with. The base level of the system is the *intelligent data base engine*. The intelligent data base engine incorporates a model that allows for a deductive object-oriented representation of multimedia information that can be expressed and operated on in a variety of ways.

To grasp the importance of intelligent data bases from an intuitive point of view, let us first compare traditional data bases with human memory. Then we provide a more precise definition in terms of the three-layered architec-

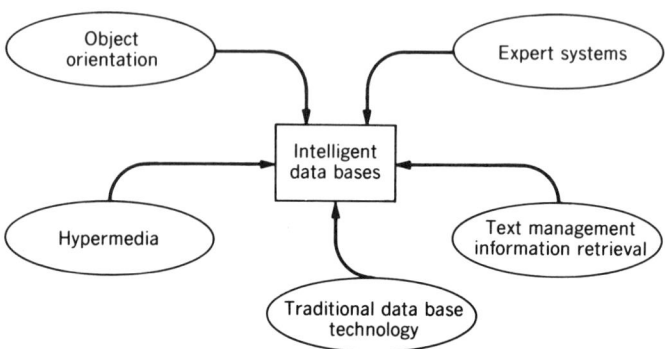

Figure 22.1 *Intelligent information technologies for the 1990s.*

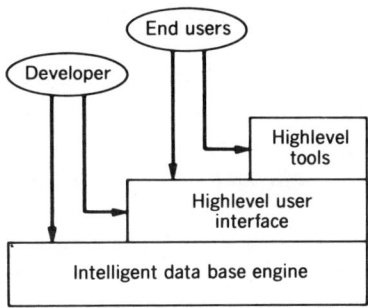

Figure 22.2 *Intelligent data base architecture.*

ture of intelligent data bases. Data base systems of the 1970s and 1980s could store many more facts (in text form) than human memory, but could not access information in a flexible manner. In contrast, our minds deal with a very small amount of information in a much more powerful way:

- We *notice* both usual and unusual patterns. We do not need to be given a specific pattern to search for, but automatically establish criteria for judging norms and anomalies.
- We can deal with partial and *inexact* information, and we can make educated guesses based on what we have seen before. We can recall someone called Johnson or Jackson who is about 30 years old and has a yellow or orange-colored car.
- We can deal with images and sounds as well as purely textual data. Moreover, we can combine information in image, text, or sound forms.
- We can not only access data, but reason with it based on its structure. We have categories of inheritance among objects such as vehicles, cars, and Fords.
- We can associate information not only in linear form, but based on associative links. For example, we can associate the term IBM with mainframes, then with corporate structure, then with a direct sales force, and so on.

It is obvious from all this that our memory is far more flexible and powerful than a traditional data base, but it can deal with far less data.

Suppose your mind could only deal with exact information: How useful would it be? You would hardly be able to function. By making data bases deal with information in as flexible and powerful a way as human minds do, we being a new era.

We now discuss each of these layers in turn, and show how they form the basis for intelligence in a data base system.

22.3 HIGH-LEVEL INTELLIGENT TOOLS

The growing size of data bases means that, while our dependence on data bases increases, it becomes harder to find information and see patterns, the chance of errors increases, and it becomes more difficult to know what is in the data base. Intelligent data base tools are extremely helpful because they

- Automatically discover patterns, rules, and unexpected knowledge in the data base;
- Automatically identify suspicious and anomalous data items, and guard against errors;
- Allow for flexible access to the data by providing inexact queries, partial matches, and best guesses.

Thus, intelligent data base tools impose order on, and extract knowledge from, the sea of data that surrounds us all. We discuss three categories of intelligent data base tools:

- Automatic discovery
- Automatic error detection
- Flexible query processing

These intelligent data base tools are integrated within a uniform environment consisting of three components (shown in Fig. 22.3):

Data Dictionary: holds information about data (i.e., metadata);

Concept Dictionary: is used to define virtual fields;

Data Analysis and Discovery: find patterns, detect errors, and process queries.

These components may be used together: for example, one may discover some unusual patterns, then ask some inexact queries to pursue the matter further.

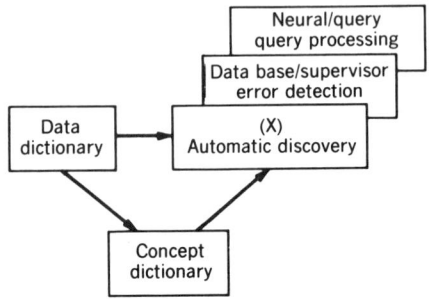

Figure 22.3 *Intelligent data base tools.*

The *data dictionary* defines the structure of the data base—it manages *metadata* (i.e., data about data). As one might expect from a data dictionary, it allows one to define data base schemas by defining fields, their lengths and types. However, an intelligent data dictionary does much more: it provides summary information about the data base so that one can quickly get a feeling for the structure of the data.

The *concept dictionary* allows one to define virtual fields on the data base. For instance, one can

- Define data types, say size = {large, medium, small};
- Define concepts based upon existing information, say define *young* to be less than 30 years old.

Defining fields and types which correspond to concepts, allows the discovery of rules in a more natural form and facilitates the expression of constraints in more intuitive terms.

22.4 AUTOMATIC DISCOVERY

Most data bases contain implicitly defined relationships, patterns, and correlations. There will never be enough analysts and time to find this knowledge. Rather than a data base crystal ball, we need tools which *generate* or *discover* this information automatically. In other words, a query language can deal with queries, but it cannot automatically form hypotheses since it does not know what to ask for. To discover knowledge, we need to

a. Form hypothesis.
b. Make some queries.
c. View the results, and perhaps modify the hypotheses.
d. Continue this cycle *until a pattern emerges*.

This series of tasks is automated with a machine learning program, which repeatedly queries the data base until knowledge is discovered. Thus a discovery program is not a query language. Instead, it *discovers* what queries should be asked, and builds the mechanisms to do so. This, in effect, provides an intelligent layer on top of a data base and its query language.

The technology for extracting knowledge from large data bases may be viewed as a merger between statistics and artificial intelligence, and from the viewpoint of *machine learning*.

Although discovery programs heavily utilize statistics, their output is not in purely statistical form. This is advantageous because most end users do not know statistics well enough to interpret the results—and there are not enough statisticians to go around. The knowledge produced by discovery programs is thus in the form of easy to read rules.

As an example of how a discovery program works and how it differs from statistics, suppose you are a disk drive manufacturer, and you have sporadic defect problems that you cannot figure out. (This, by the way, is a true story.) A sample from a list of your data (3354 total records) looks like this:

Serial No.	Step	Product Code	Lot No.	Oper. No	Code (error)
NI81400800	FSI	XMB19	24	176613	419
NI81130796	MA1	XMI14	24	32787	932
NI81809383	FSI	XMI14	24	213272	931
....	...	.....	..	......	...
....	...	.....	..	......	...
....	...	.....	..	......	...
NI81000875	CB2	X418172	37	213272	932
NI81201054	CAM	X418172	44	32787	931
NI81000878	FSI	A418172	37	51512	932
NI81201022	FSI	X418172	43	213272	932
NI81400817	MA1	XMI14	24	38978	931

Using all of this data, you can try forming hypotheses for a long time, or you can let a discovery program find out that one particular operator is not properly trained and is causing the problem (which is what happened in this case). The rule that the discovery program produced in less than 5 min to identify the source of the problem looked like this:

```
% Rule 10
CF = 72
  "ERROR CODE" = "932"
IF
  "OPER" = "213272"
AND
  "PROD" = "X418172"
;
% Margin of Error: 4.7%
% Applicable percentage of sample: 11.3%
% Applicable number of records: 379
```

It would have taken a lot longer to find the source of the problem in another way. What is more, this form of rule is better understood by most users than a statistical correlation matrix. Thus we can conclude that without a discovery program, we are only partially utilizing our data bases. Leaving this untapped knowledge in the world's data bases is, to say the least, irresponsible.

22.5 AUTOMATIC ERROR DETECTION

Data bases are growing in size and number every day. As the size of a data base grows, it becomes harder to identify errors. Undetected errors leave a trail of disaster behind them. Thus we need to detect errors and ensure data quality automatically.

Traditionally, data base systems have enforced simple integrity constrains such as range checking. An intelligent data base allows comprehensive integrity constraints to be expressed much more easily in terms of pattern-matching rules. This allows the user to change the integrity constraints without changing the application program. The system automatically checks for anomalies and deviations in values, based on a set of criteria provided by the user.

In most applications integrity is encoded in terms of procedural programs when the application is developed. Often, integrity constraints are "spread over" a program, with each constraint being coded as a specific set of instructions. As the constraints (inevitably) change, the program needs to be "reopened" for further surgery.

In an intelligent data base, constraints are expressed in terms of a set of rules that are "separate" from the program. Then as the constraints change, the rules are changed without "reopening" the program. This has two advantages:

1. The chance that the constraints are accurate increases.
2. The ability to change the constraints improves.

The system also provides a set of statistical features for measuring data quality. For instance, deflection after the standard deviation and from typical correlations are detected. These features should be particularly useful for users and developers with large data bases.

A data quality control system thus has two components:

1. It finds anomalous data items and unusual patterns by itself.
2. It enforces integrity constraints that are maintained *separately from* data bases and application programs by using easy-to-read rules.

There are many other concrete examples of a way one can apply data quality control in business. For example, suppose items are missing from your inventory, or unauthorized actions are taking place against your inventory: An automatic system can be used to call your attention to such problems. As another example, suppose you have an employee (whose signature threshold is low) issuing an unusual number of small purchase orders. He is doing this in order to buy, part-by-part, an unauthorized large-ticket item. A data quality system can detect such an anomalous use of purchase orders, and signal it to the user.

By using object orientation, we can specify an object model in terms of which constraints are enforced. Thus inheritance, attached predicates, and the full power of an object-oriented system are used for error detection.

Errors in a data base *will* cost us—it is just a question of how much they are going to cost. By using a data quality control system, we can avoid errors, and thereby save both time and money. In most cases, an automatic error-detection system will quickly pay for itself many times over. In fact, we cannot afford to be without such tools.

22.6 FLEXIBLE QUERY PROCESSING

When searching for information, a partial match is often better than no response. In traditional query systems, a data base record either matches a query, or it does not. For example, a person is either *young*, or not. A flexible query system, on the other hand, manages information the way our mind does—that is, it deals with inexact queries, and provides best guesses. It provides a degree of confidence for how well a record matches a query.

As an example of flexibility in information processing, suppose you are a detective and you have a description for a suspect: He is tall, in his thirties, and drives a yellow car. As it may turn out, the car is listed as orange, and the suspect is 37 and 6'4". Your human mind can match these descriptions (e.g., you do not have to know the suspect's exact height), but a data base query cannot.

A flexible query system can solve queries involving inexact numeric values: for example, someone is almost 18 years old; inexact words: for example, Janssen, Jensen, and other types of inexact data.

You can specify conceptual terms upon which the system will base its queries. For example, *adult* might mean those people who are older than 21, *young* might mean those people who are less than 30, and *bright color* might mean the colors red, yellow, green, blue, orange, and white. This helps you to gather potentially useful, solid information, in spite of the inexact queries you started out with.

Although a neural component may be used in such a system, the user need not have any previous experience with neural networks in order to use the system. In a traditional neural network programming system, you have to define a network structure, then program and train it before you can use it. A flexible query system is programmed to automatically install itself on the data base. Therefore, it is not even necessary to know what a neural network is.

Traditionally, the set of conditions have been two-valued: that is, given a record R and a set of conditions C, either R satisfies C or not. In a flexible query system, a degree of confidence is used to reflect how well R satisfies C. How well R matches C depends on three factors:

1. How well each field of R matches the conditions;
2. How well the fields of R match the interfield criteria;
3. How important each field is to the query.

To use SQL as a notation, an inexact query has the format:

```
SELECT Name, Age, Telephone
FROM PERSONNEL
WHERE
    Name = Dovid Smth and
    Age IS-CLOSE-TO 18;
```

In this case, the name *Dovid Smth* is misspelled, but the system matches it to *David Smith*. The condition Age *IS-CLOSE-TO* 18 is similar to "age is around 20," or "age is about 20." (As noted earlier, inexact ranges and concepts for fields are defined in the *concept dictionary*.)

The system can also provide a guess (i.e., a hypothetical answer that is not a record in the data base). For instance, if you know someone's zip code, it can try to guess the make of that person's car.

Moreover, a query may have a degree of importance (or weight) associated with each condition. The weight is a number between 1 and 100 that shows the importance of the condition; for example,

```
SELECT Name, Age, Telephone
FROM PERSONNEL
WHERE
    Name = Dovid Smth, weight = 70 and
    Age IS-CLOSE-TO 18, weight = 50;
```

Here the Name condition is more important than the Age condition. A query also has a *query output threshold* to filter all the records in the relation that have a *combined query confidence factor less* than the threshold.

There is a system-provided inexact comparator called *Near*. To use this comparator *Near*, a user must first specify a *closeness* value C, which is indeed a percentage on the field. Field values will be compared with the *Value* in the subcondition according to C. For example, a user may specify *closeness* on field *Age* of relation *Person* to be 10 percent. In this case, a subcondition "Age *Near* 30" will have the following values:

Age Value	Confidence Factors (%)
33	100
36	90
39	81
42	73
...	...

To apply *Near* on an nonscalar field (that is, with data type characters or string), a nonnumeric algorithm is used.

A similarity matrix can be used to define a user-defined comparator with a great deal of ease. A set of values can be defined in a similarity matrix by specifying the level of similarity between them. For example, a user-defined operation *similar-color* can be defined on a set of values {yellow, orange, red, green} by the following similarity matrix:

	Yellow	Red	Orange	Green
Yellow	—	30	70	10
Red	30	—	60	0
Orange	70	60	—	5
Green	10	0	5	—

This matrix suggests that yellow is more similar to red than to green. The matrix need not be symmetrical.

To sum it all up, if we do not use inexact queries, we are not using the full information available in our data bases.

22.7 INTELLIGENT USER INTERFACES

It is a commonly accepted truism that the user interface is critical to any application or environment. Early computer interfaces seemed to confuse *intelligent* with *esoteric*. One not only had to be smart to use early interfaces, one also had to learn a complex and somewhat arbitrary syntax.

The solution was, and still is, to make interfaces less esoteric and more forgiving. If a typical user can sit down at the computer to perform a task, and at each stage of the program it is obvious what to do next (within the context of the task), then the interface is satisfactory.

The human mind seems to be inherently associative in nature. There are many sources of evidence for this associativity, ranging from the richly interconnected structure of the brain itself to the patterns of associative memory and thinking that are frequently observed in human behavior.

Associative structuring is provided to users by *Hypertext*. Hypertext can be simply defined as the creation and representation of links between discrete pieces of information. When this information can be in the medium of graphics or sound, as well as text or numbers, the resulting structure is referred to as *hypermedia*. For intelligent data bases, hypertext serves two roles. First, it is an information-structuring technique that supplements traditional text- and record-oriented data bases. Second, it is a way of structuring the interface that adds associative links to the hierarchical links found in menus.

Links are between *nodes*. Some link types that provide the basic functionality required for intelligent data bases are as follows:

1. *Pan links* simply move to a related node. They allow one to move around or navigate through hypermedia.
2. *Zoom-in links* expand the current node into a more detailed account of the information. *Zoom-out links* return to a higher-level view of the current object (these links are particularly useful in browsing facilities).
3. *Hierarchy links* allow the user to view any hierarchies in which the current object is embedded. If more than one hierarchy is involved, the user chooses which context is appropriate.
4. *Broaden links* display the parents of the current object for the hierarchies that it is embedded within. *Specialize links* display the children of the current object as defined by the hierarchies it is embedded within.
5. *Conditional links.* The availability of activation of these links is conditional upon the stated interests or purposes of the user. Conditional links are hidden unless they are of interest, or a particular user has access or has been cleared to use them.
6. *Index links* move the user from an indexed node to the corresponding index entry for that node. The index can then be used to enter the relational data base, or to find documents that share a particular index term. Indexing hypertext is a good way of controlling the proliferation of links between nodes.

Links are important in hypertext, but nodes are also needed for storing and displaying information. We now consider a basic set of node types:

1. *Text Nodes:* These consist of text fragments. The text itself may (1) be a document, or define the object represented by the node; or (2) represent base-level information, such as that provided in a document.
2. *Picture Nodes:* Pictures may be embedded within text nodes, or they may be nodes in their own right. A picture and text may be mutually documenting, as when one uses a painting to illustrate a biography of Leonardo da Vinci, or when one pursues biographic details of the artist after viewing a picture.
3. *Sound Nodes:* Like pictures, sounds may be embedded within text, or exist as nodes in their own right. Sounds may generally be treated in similar fashion to pictures, although they are less likely to be zoomed. Sounds, too, represent uninterpreted information.
4. *Mixed-Media Nodes:* These nodes contain some combination of text, pictures, and sounds. In many cases, the same information may be represented in a combination of linked nodes or as a single mixed-media node.

5. *Buttoned Text* (or mixed-media): The distinction between buttons and links is that a link connects a pair of nodes, whereas a button executes a procedure.

Links and nodes may be combined with the object-oriented features of an intelligent data base for more power. An object consists of a name and a set of attributes. An object belongs to a class, and has one or more methods associated with it either directly, or through the class that it belongs to. Objects, and classes, may be linked to other objects and classes. The user sees the interface in terms of these objects connected by links into a network.

Let us consider a simple example of how rules, objects, and hypertext are combined in an intelligent data base. Consider an information-retrieval workstation on which the following text is displayed:

```
The main competitor to Ford is General Motors.
```

To seek further information, the user must first "select" an object (or enter information about an object). The user is then provided with a number of choices about how to find more information about the object itself or about related objects. Some of the choices will be static or fixed, and generally available whenever the object is displayed. Other choices will be context-sensitive, or determined at run-time.

Now, to get more information, the user selects the word "General Motors" by highlighting it. The user may at other times type the word "General Motors" and ask about information relating to it. The system displays a set of possible choices for information about the object (i.e., "General Motors"); for example:

Annual report
Subsidiaries
Summary of business activities
Stock analysis
Investment outlook

The user selects *Subsidiaries*, and a list of the company's subsidiaries will appear. The user may then select a subsidiary such as *Hughes Aircraft,* and continue to get an annual report for that company. From within the annual report, the information may be displayed in a variety of forms (e.g., pie chart, table, etc.). The list of the subsidiaries may dynamically change in the underlying data base.

The user may also get the same information by typing "General Motors" and then being shown a set of contexts in which the object "General Motors" may be interpreted (e.g., company, dealerships, etc.). Thus the same information may be retrieved either by unstructured browsing or through structured (query-based) searches.

The browsing capabilities of the interface are supplemented by allowing a series of operations to be performed on an object. These operations are dynamically determined with a set of *rules* based on context, the user, data available, and so on.

These inferentially controlled choices are determined by rules, such as

```
        'X' is a choice for operator, 'Object'
If
        'Object' is-a 'Parent' and
        'X' is a choice for operator, 'Parent';
```

or,

```
        Stock is a choice for operator, 'Object'
If
        'Object' is-a Company and
        (Status of 'Object') = Public;
```

In the second rule, Stock will be a choice if the company is public. Similarly, the information that is actually displayed will change according to the interests of the intelligent data base user.

It is clear that the equation

```
Hypertext = Prepackaged text
```

does not apply in the previous scenario. By combining rule-based reasoning and object orientation, an intelligent data base dynamically constructs links and displays the most relevant information.

22.8 THE COMMERCIAL PERSPECTIVE

Can we estimate how much a crystal ball is really worth? Are there inherent limits to the "value" of information? With the increasing complexity of today's society, can we afford not to build a digital crystal ball? The answer to all three questions is no. In the right hands, information can be exponentially more valuable than any other asset. Almost every aspect of science, technology, and business will benefit from intelligent data bases.

The risk/benefit ratio for automatic delivery is very attractive. A discovery system has to discover just a few unexpected relationships to pay for itself. And since the discoveries may always be rechecked with queries, there is little chance of loss due to errors, but great potential for gain. This is in contrast to expert systems that mimic human experts. If an expert system errs, the consequences may be serious.

With intelligent data bases, data glut should be viewed as an "opportunity" and not as a problem. The payoffs from intelligent data bases are significant. Even if we manage to provide a fraction of data base users with hidden knowledge from their data bases, we will achieve a substantial level of productivity increase within society as a whole.

Our society is becoming more and more dependent upon information. Most large organizations are totally dependent upon their centralized data bases. Today, it is taken for granted that in order to compete in any business one needs a data base management system. As the size of the average data base grows, intelligent data base tools are as much a necessity as the data bases themselves.

BIBLIOGRAPHY

R. Blum, "Discovery, Confirmation and Incorporation of Causal Relationships from a Large Time Oriented Clinical Database: The RX Project," Computers in Biomedical Research, 1982.

W. Gale, Ed., *Artificial Intelligence and Statistics,* Addison-Wesley, Reading, Mass., 1986.

B. Gains, and M. Shaw, Induction of inference rules for expert systems, *J. Fuzzy Sets and Systems,* 1986

J. Holland et al., *Induction,* MIT Press, Cambridge, Mass., 1986.

E. Hunt et al., *Experiments in Induction,* Academic Press, New York, 1966.

J. Mahnke, IXL tool discovers database patterns, *MIS Week* (December 1988).

R. Michalski et al., *Machine Learning,* Vol. 1, Tioga Books, Palo Alto, Calif., 1983; Vol. 2, Morgan Kaufman Publishers, Los Altos, Calif., 1986.

D. Michie, Automating the synthesis of expert knowledge, *ASLIB Proc.* London, 1984.

K. Parsaye, M. Chignell, S. Khoshafian, and H. Wong, *Intelligent Databases,* Wiley, New York, 1989.

K. Parsaye, Acquiring and Verifying Knowledge Automatically, *AI/Expert* (May 1988).

K. Parsaye, Machine Learning: the next step, *Computer-World* (October 1987).

R. Quinlan, "Discovering Rules from Large Collections of Examples," in D. Michie, Ed., *Expert Systems in the Micro Electronic Age,* Edinburgh University Press, Edinburgh, 1979.

J. Tukey, "Statistician's Expert Systems" in W. Gale, Ed., *Artificial Intelligence and Statistics,* Addison-Wesley, Reading, Mass., 1986.

CHAPTER 23

Artificial Intelligence: Between Symbol Manipulation, Commonsense Knowledge, and Cognitive Processes

MARKUS F. PESCHL

23.1 INTRODUCTION

The problem of *knowledge representation* has turned out to be one of the most urgent questions in computer science, because its solution has an important influence on the quality of the solution of the whole task; we do not only think of knowledge representation in artificial intelligence (AI), but also in "traditional" computer programs—in most technical applications, for instance, a mathematical representation of the world is chosen (e.g., differential equations are mapped to computer programs or a continuum is mapped to a discrete grid in numerical mathematics). In each case, however, there has to be found a mapping from the "real world" to the structures that are provided and determined by the computer's software and hardware. The topic of this chapter is to discuss (and make explicit) these implicit assumptions of what we normally assume to be real world, of the mapping process from real world to computer structures, of knowledge representation in symbolic structures, and so forth.

The basic assumption is that there is a kind of *homomorphic relation* between objects, phenomena, features, and so on, of real world and the items of the knowledge representation structure. We will see that this homomorphic relation is the critical point in the whole discussion (i.e., of what kind this relation is). Before reflecting on this problem, I want to make

---
Neural and Intelligent Systems Integration, By Branko Souček and the IRIS Group.
ISBN 0-471-53676-8 ©1991 John Wiley & Sons, Inc.

another critical remark, which is discussed in more detail in [1,2]*: if we are taking of real world, what do we understand by this term and how is it related to the problem of knowledge representation? As we will see, a *constructivist* perspective is assumed in our argumentation—this means that *our own cognitive reality* is the result of a construction process being determined by the nervous system's structure and by the perturbations being received from the environment (real world). We never have direct access to this environment; in other words this means that our perception of this environment is only a *representation* of certain aspects of it.[†] Hence, our own cognitive reality is only *one possible representation of the world* itself.

Most of us would agree to this argumentation. If we are developing knowledge based systems, we assume, however, that our own cognitive reality is *the* reality and in no way take into consideration that each of us is dealing with the *representation* of real world and *not* with real world per se; that is, representing real world in the computer means a sort of second order representation (a representation of representation). The system designer thus implements his or her own interpretation (representation) of the real world—the semantics is determined by the designer and by the user; the system itself does not have any semantic knowledge although modern knowledge representation techniques pretend to have semantic knowledge. They have only semantics on a syntactical level, which does not imply any improvement. One could argue, that the system designer is not implementing his/her own knowledge, but "scientific knowledge"; as we will see in the section "Scientific versus Commonsense Knowledge," it is very problematic to make a strict distinction between "scientific" and commonsense knowledge, because it turns out that both have very similar structures [3–6]. The problems arising from these wrong assumptions will be discussed among others in this chapter. As we have seen from the last two paragraphs the problems we have to deal with in computer science and AI are not only of technical and implementational interest, but also have epistemological and philosophical perspectives and implications. Both—epistemology as well as computer since (AI)—can learn a lot from each other. It is, thus, one aim of this chapter to show, how these disciplines are interrelated and what could be learned (by both of them) from a consequent (interdisciplinary) cooperation.

23.1.1 Knowledge Representation—Commercial Application vs. Interdisciplinary Basic Research

If we are looking at the development in computer science since about 1960, we can see that computers are not only used for number crunching any

---

* It is important to shortly have a look at this problem in order to understand the background and the context of this chapter.
† These aspects are *determined* by the structure of the sensors and nervous system; that is, we perceive, for instance, only a certain spectrum of electromagnetic waves—it is determined by the sensitivity of the photosensors in the retina.

23.1 INTRODUCTION

more, but also for processing, storing, representing, and so on, huge amounts of *knowledge*. Computer science, especially AI, became aware of the fact that computers provide the means for handling knowledge and thus focused its research interest on the investigation of how to represent and process knowledge in order to obtain "intelligent behavior."

One of the theories which have become very famous, imprinting and important for AI is the *physical symbol systems hypothesis* (PSSH) having been propagated by Newell and Simon [7,8], which runs as follows:

> The *Physical Symbol Systems Hypothesis*. A physical symbol system has the necessary and sufficient means for *general intelligent action*.

They even go further when claiming that we humans use the same symbols in our everyday lives as computers use (see quotation in Section 23.4.1). We will have to critically review this hypothesis from an epistemological point of view. Computer science, however, *unreflectedly* uses this paradigm in knowledge based systems, such as in expert system.* One aim of this chapter is to investigate this approach out of a perspective which is not only restricted to computer science or AI aspects, but also to involve philosophical as well as more general considerations to point out the shortcomings and problems of the PSSH.

What are the assumptions being made in this approach?

1. *Cognition is computation on a representation*. Cognition is information processing. Information processing is always based on a representation of what has to be processed; that is, there is a mapping (done by a human designer) from reality† to a representation structure on which an algorithm (e.g., rule application, theorem prover, etc.) executes its calculations (as depicted in Fig. 23.1). This process structure is determined by the von Neumann architecture of most computers. The results of these calculations (i.e., application of an algorithm) is another item in the representational space. These new items have to be "proven" in real world; that is, there is a mapping back from the representational space to real world—this is done (by humans) in most cases by *active acting* in the environment. So far the idea of representations does not represent a problem, because, as we have seen above, each of us is acting in his/her representational space (i.e., his or her "private" cognitive domain).

2. *Symbolic representation*. Problems are arising when we are introducing the ideas of the PSSH (Fig. 23.1): the representational space is realized in the form of *symbols*. What do we understand by a symbolic representation from a more general point of view? We might be seduced by thinking that our

---

* AI even is *not* aware of working in this *paradigm* (in the sense of T. S. Kuhn [9]) and in most cases does not reflect on the assumptions having been made *implicitly*.
† Do not forget what we have said about "reality"; "reality" and "real world" always mean the *representation of reality* in the human's cognitive domain.

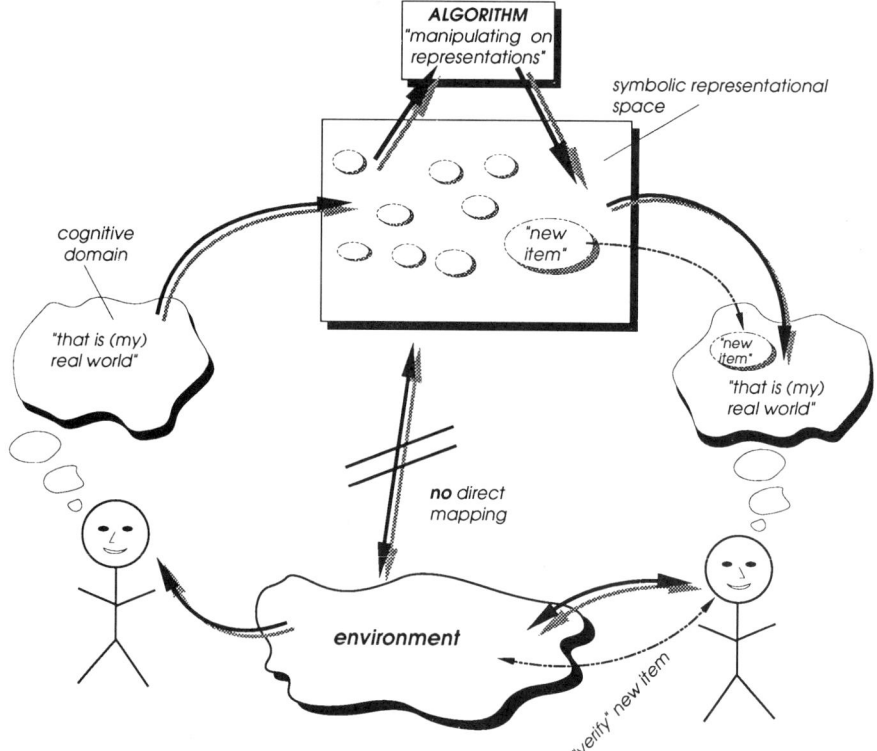

Figure 23.1 *The information processing feedback loop. The human's cognitive domain is mapped to the representational space; the environment is not mapped directly to the representational space. The results of the computations are "tested" in the environment.*

cognitive domain consists of symbols, words, that our thinking is determined by our language, and other factors. Language and linguistic items are only the tip of the iceberg; there is another basis than symbols in our cognitive domain, which we call *implicit knowledge* (we discuss it later). This means that our natural language is based on a more basic level (of neural activities) and can be seen as the result of a mapping process from the subconscious to the consciousness. This language consists of words, sentences, and so on, that get their meaning from experience with the environment. *Symbols,* as they are used in knowledge-based systems, are the result of another mapping process: natural language items (i.e., semantic features) are mapped to a (n arbitrary) pattern, a list of letters, or a bit string; it is a sort of coding done by a human designer who associates with each pattern (symbol) a certain "feeling" (i.e., a certain semantic feature)*—hence, symbols are the result of a mapping process from the linguistic space within the human's cognitive

---

* This pattern *triggers* a certain meaning in the human's cognitive domain.

domain to the representational space which is filled with arbitrary and meaningless strings. What is called real world in Figure 23.1 thus is reduced to the domain of natural language* and what we have described as the representational space is a formal structure of symbols (in the form of rules, semantic networks, etc.). The algorithm does nothing but formal manipulations on these meaningless formal structures.

So, what does the verb *represent* mean? It means to *stand for* something. A word, for instance, stands for an object or a phenomenon in the environment, a certain state in one's nervous system stands for a feeling or impression. What is symbol in a symbol manipulating machine standing for? It is standing for a natural language word—*not* for an object in the environment (Fig. 23.1); this is only a second order relationship. As an implication symbol manipulation does not operate on direct representations of the environment, it even does not operate on the environment's natural language representation; it operates on a meaningless result of a mapping of the domain of the human's natural language. The problems arising from this concept will be discussed in the following sections.

3. *Semantics and syntax*. Symbols and symbolic representation have to be seen under three aspects:

1. Symbols are *physical* entities and can be manipulated as physical items, by applying a rule to a set of symbols and by adding another or changing an existing symbol.
2. The *syntactical* aspect determines the symbol's role within an abstract construct of *mathematical logic,* a formal grammar, or a set of rules; describing the syntactical aspect very critically we could say it determines only the *organization* of how to handle, how to manipulate the symbols within the symbols' representation space.
3. Symbols have a *semantic* "value"; however, this value is determined by the system's syntax itself being determined by the designer/user of the system. Thus each semantic feature is encoded in a syntactical structure; *syntax reflects the (correlated) semantics.* We will see that there is *no* semantics in the machine unless we assume another approach, which is sketched in Section 23.5.3 and in Peschl [2,10]. As has been shown, semantic features are only the result of an active (constructive) interpretation process of the human interacting with the symbolic representation.

The implications of these assumptions will be discussed in this chapter; to anticipate the most important conclusion: the symbolic representation paradigm lacks the possibility of representing non verbal (i.e. implicit or subcon-

---

* This is a *restriction* because it reduces the cognitive domain, which consists of the whole structure of implicit knowledge, to the "poor" domain of consciousness and natural language.

scious) knowledge which is, as we have seen, the *basis* for language, semantics, etc.

One could argue that the problems and questions having been sketched and discussed in the last paragraphs do not have anything to do with artificial intelligence, because it has turned out that formal logic and its application in orthodox (i.e., symbol manipulating) AI has done very well in many domains (e.g., expert systems). However, this is true only, if one exposes oneself to the dictatorship of mathematical (formal) logic—if everything in "real life" has to be categorized and formalized you will never come to "real life." There is, for instance, the old question of how to handle *commonsense* knowledge (this question is very closely related to the question of how to represent implicit knowledge; we will discuss it in Section 23.2.1). So it seems justified to differentiate between two aspects of looking at AI:

   a. Either we are interested in *commercial* systems—the only question being of interest is that the expert system or knowledge based system should work properly; it has to fulfill a certain function (if a certain input is presented, the system has to produce a certain output) following a given specification. The problems and relevance of cognitive processes are of *no* interest in this approach; or
   b. We are interested in a *model for cognition*—a model that adequately describes the phenomenon of cognition, language, and similar entities. This is a problem of basic research, which is, in most cases, done without commercial orientation.

We can uphold this differentiation, however, only as long as (a) gets along with its own strategies and technologies without carrying out (or having to carry out) metareflections. Orthodox AI (i.e., the symbolic approach in AI), however, has reached its *limits* in our days, which seems to be due to the following points:

1. *Requirements:* Requirements are continuously increasing; the system has to represent more and more knowledge, semantic information, commonsense knowledge, and so on. Commercial users are not content with trivial toy systems any more and AI people are promising always more complex and more intelligent systems, because they enthusiastically extrapolate their successes and results from very small and quite primitive systems to large scale and complex systems. In our days it turns out, however, that this approach does not really work out very well and basic problems are arising when applying methods from blocks' world to real world.

2. *Mathematical (formal) logic and symbols:* Mathematical logic provides very interesting and tempting means for the manipulation of knowledge. As we will see in Section 23.2 the unreflected and exclusive use of *symbols* for representing knowledge (in orthodox AI) and the application of

methods from mathematical logic causes many problems for representing commonsense and nonverbal knowledge.

3. *Basic research:* Orthodox AI is not really interested in basic research; this means that what is called "basic research" in orthodox AI are only investigations for improving and optimizing the performance of symbol manipulating systems, of knowledge representation techniques, and so on—it never carries out reflections upon its (symbolic) paradigm or upon principal problems. This is very much due to point (1) as these reflections are considered expensive and unnecessary, because they cannot *directly** contribute anything to the final product. The aim of this chapter is, however, to show that it is, in the long run, necessary and fruitful to take into consideration such ideas, if one is interested in a real improvement in (commercial) AI.

4. *Interdisciplinary cooperation:* The lacking interest in *interdisciplinary cooperation* with other disciplines is one of the main reasons for the problems arising in orthodox AI. The only interest in a commercial product does not allow the costs of an interdisciplinary research group. As has been shown in Peschl [11] real† interdisciplinary cooperation can even cause a paradigmatic shift (in the sense of T. S. Kuhn [9]) as it is happening in the field of parallel distributed processing (connectionism) and AI/cognitive science in our days.

As easily can be seen, the problems arising, when representing commonsense knowledge, are not solvable by exclusively applying methods from computer science. One could argue that AI not only uses computer science methods and knowledge, such as mathematical logic, formal grammars, etc. but also makes use of scientific findings such as of cognitive psychology. If we look, however, a bit closer at cognitive psychology (e.g., Anderson [12], Mandl and Spada [13]), we will find out that there are close relations between the methods of artificial intelligence and cognitive psychology (e.g., propositional networks, problem solving and searching, schema theory, etc.)

So, we can conclude that almost *no interdisciplinary cooperation* can be found. This leads to the problems and limits of symbol manipulation in orthodox AI. If we are interested in representing and processing commonsense knowledge or very fine grained (implicit) knowledge, we have to take into consideration epistemological, psychological, linguistic as well as computer science specific aspects, because the problem of representing knowledge has been investigated for a long time by philosophers, psychologists, and others; there is a long tradition in philosophy and philosophy of science and it is the chance of our century to make use of this unique knowledge and

---

* *Directly* means that these considerations and reflections do not produce any commercial success.

† *Real* interdisciplinary means that the cooperating disciplines *consequently* include the results from their discussions with the other disciplines in their work. In most cases, however, interdisciplinary cooperation is only a question of prestige and is reduced to having a look at the other disciplines' results.

combine it with modern technologies for applying it in, for instance, knowledge-based systems—therefore this knowledge could help us not only to *prematurely* realize and avoid dead ends in scientific research, but also give us hints and show us possible solutions to our problems arising in artificial intelligence and/or cognitive science. In the following sections I want to make evident some problems that arise in "pure" (i.e., unreflected application of) symbolic knowledge representation and in the process of knowledge engineering from a more interdisciplinary point of view. The aim is to show the importance of the consequent integration of computer science (AI), epistemological as well as (cognitive) psychological aspects. This is not only valid for (real) basic research, but also for the development of commercial systems with very high complexity, because at this level of complexity problems cannot be solved by a single discipline with a restricted methodology any more.

23.2 KNOWLEDGE—CONCEPTS AND SUGGESTIONS

23.2.1 Human Knowledge and Expert Knowledge

Before discussing the problem of how to represent knowledge in the computer I am going to try to explain how I see the problem of what is knowledge; it is difficult, however, to define it with any precision as there are so many facets of knowledge, intelligent behavior, and language. Please remember that we are considering knowledge as the result of a construction process and that the knowledge in human brains must not be treated as equivalent to the environment! Be aware that, if we are talking of (human) knowledge, this means already a sort of representation of real world; as will be shown in the following sections, there are different forms of knowledge—it will turn out that these different forms depend only on a different quantity of structuring: scientific knowledge, for instance, is very well structured (being due to the applied methods and the rationalistic tradition), whereas the knowledge of a little child seems to have—from a rationalistic point of view—almost no structure.

At first we will discuss the relation between scientific and commonsense knowledge in order to show that the gap between these two forms of knowledge is not as wide as commonly assumed. Having this in mind I am trying to show how linguistic (explicit) knowledge is embedded in commonsense structures of knowledge.

Scientific vs. Commonsense Knowledge   Normally there is made a clear distinction between expert knowledge and commonsense knowledge. My claim is, however, that we have to give up or at least rethink this unreflected assumption in order to see the problems (being discussed in this treatise) arising (for instance in AI) from this artificial distinction. This distinction is closely related to the problem of the discrimination of scientific and nonscientific (or commonsense) knowledge. As I will show, we can learn a lot from this discussion in philosophy of science and science theory for our problems (this is only one possibility of how interdisciplinary cooperation could be realized in this field). To sum up the most important point beforehand: It turns out that we cannot uphold this strict distinction between commonsense and scientific knowledge in such an extent any more; one has to see that scientific knowledge is a very specialized and well structured form arising from commonsense knowledge which is the result of a process of constructing being determined by our nervous system and our interaction with the environment. The *interweaving* of *knowledge* and *acting* is of great importance for scientific knowledge as well as for commonsense knowledge (Fig. 23.1). Both are using language and linguistic representations (however in different extends of complexity) for communicating their contents. We will have to look at these problems in more detail as their discussion can help us in

- Analyzing, seeing more clearly and understanding better the problems and difficulties orthodox AI is confronted with, and
- In finding alternative and more adequate ways, methods and approaches for commercial systems as well as (basic research) artificial cognitive systems. My claim is that we have to take into consideration results from cognitive science basic research if we want to construct really complex systems and that an interdisciplinary cooperation has become quite inevitable in order to meet the claims for high complexity from industry and users.

As has been stated, we are assuming a *constructivist* point of view. This means that all our scientific and commonsense knowledge is the result of a process structuring our environment—stated another way, by perceiving and interacting with our environment we are throwing a certain structure being determined by our nervous system, by our sensory system, by artificial sensory instruments, and so forth, over it. This process of giving structure to "reality"* (i.e., the process of creating and constructing our own (cognitive) reality) is embedded in the social and cultural context being itself the result of evolution and our construction processes (phylogenesis and ontogenesis). Scientists as well as nonscientists stand within this context and are (in most cases) unconsciously influenced by it.

This implies that there is not really a qualitative distinction between scientific and commonsense knowledge any more; the definitions of what is

---

* The problem of reality and real world has been discussed already—remember that we are differentiating two forms of reality: (1) physical reality or our environment in which we are living and (2) cognitive reality being the result of our (cognitive) construction processes; I will refer to them as "reality" or "environment" and "cognitive reality," respectively.

knowledge being supported in Section 23.3 are valid for both forms. As P. S. Churchland shows, this distinction is not only artificial and not very useful but also quite relative (in regard of time):

> ". . . Primitive theories give way to sophisticated theories, and as the latter become the common coin of everyday life, they may then acquire the status of commonsense. A remarkable thing about the human brain is that it can use those primitive theories to bootstrap its way to ever more comprehensive and powerful theories—to find the reality behind the appearance."
>
> P. S. Churchland, [14, p. 265]

So-called scientific knowledge arises, as Y. Elkana [6] or T. S. Kuhn [9] states, from *critical discourse* between competing theories, or better, between competing groups of researchers each being of a certain theory. This is, however, not only true for scientific knowledge, but also for most forms of commonsense knowledge. That already can be seen when observing, how little children are learning tasks by acquiring knowledge by trial and error; this method of trial and error can be interpreted as critical discourse (not being restricted to verbal categories but rather to acting) with the child's environment. The structure of this dialectrical discourse being maintained by the continuous interplay of thesis and antithesis (and the continuous cycle of analysis and synthesis, of induction and deduction [15, p. 165] can be found in the process of imparting knowledge to somebody as well as in the process of the discourse in a scientific community (i.e., in the discussion between experts). It seems that this gap between commonsense knowledge and scientific knowledge is an artifact and is not as wide as normally assumed; there is a homomorphic relation between the structures of those two (artificial) types of knowledge. We have to investigate, however, to what extent this distinction is justified and valid.

It seems that scientific knowledge is in most cases very formal* knowledge. This view of scientific knowledge is suggested to the reader and/or listener because of its strict and deductive argumentation. It also suggests that the presented knowledge has a definitive and an absolute character without bearing in mind that this knowledge has its validity only within the (scientific) paradigm the scientist works in; this paradigm [9] is determined by the cultural, social, economical and political context. The reason for this fictitious absolute character of scientific knowledge can be found in the fact that most talks, papers, and books lay stress on the context of *justification* and *legitimation* (e.g., deductive description and argumentation of research results and hypotheses). Everybody working in science and in research projects knows, however, that, in most cases, this is not how scientific knowl-

---

* The meaning of "formal" is not restricted to formal in the sense of (mathematical) logic; it rather includes all explicit knowledge being forced into theories, hypotheses, and rules by applying certain (scientific) methods.

edge (e.g., medical) or decisions are produced. We rather have to take into account the context of *discovery;* it is less formal and much more intuitive. Investigating scientific knowledge from this perspective implies that the formal structure (which is generated for presenting the results, discoveries, decisions, etc.) turns out to be quite artificial and seems to be justified only for two reasons:

1. For proofing the theories and hypotheses (using the methods of formal systems), and
2. For providing better means of communicating this knowledge with other scientists (i.e., a sort of special "code" (in the sense of semiotics, e.g., Eco [16,17]) or standard for international scientific communication).

Thus, we have to differentiate strictly between the context of justification and the context of discovery, where context of justification stands for the scientific knowledge as it is presented to the public or scientific community; it also represents, for instance, the formal description of a decision having been made by a scientist or doctor. Context of discovery means the more *intuitive* view of how the so called scientific knowledge has been produced or "invented"; it also includes the questions and problems arising from a more cognitive and not formal perspective when an expert is deciding or solving a problem. To conclude this discussion: What is called scientific knowledge is the result of a process of formalizing and structuring intuitive knowledge. The crucial difference between scientific and commonsense knowledge can be identified as follows:

1. *Formal apparatus:* It is the formal apparatus (e.g., mathematics, logic, argumentation techniques, etc.) making up scientific knowledge. It completely hides and cuts out the intuitive component of knowledge.

2. *Paradigmatic language:* Thus, scientific knowledge represents nothing but a very *sophisticated* kind of *language**—it is accepted (as scientific knowledge) only if the scientist makes use of this language; to put it another way, its use (i.e., following the rules of the actual scientific paradigm) ensures that the knowledge, which has been produced, is compatible with this paradigm and its results are taken for valid, reliable and "objective" (but only within the chosen paradigm).†

3. *Predictability and reproducibility:* At least two minimal *criteria* can be found for making a distinction between scientific and commonsense knowl-

---

* It is a special form of a language-game (in the sense of L. Wittgenstein, e.g., [18]).
† As T. S. Kuhn describes [9], problems are arising if one scientist or a group of scientists are presenting results or theories being inconsistent with the theories of the actual paradigm or which are trying to analyze and reflect its (implicit) assumptions. Kuhn shows that such results can cause a paradigmatic shift or even a scientific revolution.

edge, or better for bing called "scientific"*: scientific knowledge has to be at least

Predictable: that is, theoretical and/or empirical knowledge has to represent a means for predicting phenomena that will take place in the future;

Reproducible: this is also a criterion for predictability; it means that the scientific description of the empirical, theoretical, hypothetical knowledge has to be deterministic within the context of the paradigm.

If these criteria are not met we could state, following P. Feyerabend, "anything goes" [4] (this means that whatever anyone says is scientific knowledge, implying a kind of "scientific anarchism"). We can look at these criteria as criteria of success; that is (scientific) knowledge is said to be successful if it can predict certain phenomena. This idea is influenced by the theory of evolution and is, as we will see in Section 23.3, not only restricted to scientific knowledge (it is, however a minimal requirement of it).

4. *Sophisticated methodology:* One has to apply very sophisticated methods (e.g., complicated gauges, formal systems, etc.) for obtaining scientific knowledge where as commonsense knowledge is obtained by "applying" our own (human) sensory system as instruments for the construction of our cognitive reality. To put it another way: Scientific knowledge is the result of very complex and subtle methods, instruments, and other factors. This implies that the difference to commonsense knowledge is also a question of different complexities of methods. Scientific methods represent a sort of extension of the human sensory apparatus; they are constructs and results of theories being developed in a certain paradigm and they are supporting the scientist in constructing new hypotheses theories, experiments, and methods.

The two most important points seem to lie in the differences of the *structure* of knowledge (i.e., scientific knowledge is very well structured knowledge) and in the differences of the applied methods for the construction of this knowledge. What are the implications for symbol manipulating knowledge based systems from this discussion?

I wanted to make clear that scientific or expert knowledge is, in most cases, as artificial as the commonly assumed wide gap between commonsense and scientific knowledge. It is impossible to make a clear distinction between the two forms of knowledge and thus it is a quite relative discrimination depending on the culture, on the age, and on social conditions. Its basis is always commonsense knowledge. That is why it seems to be a bad idea to investigate and simulate only the formal aspect of knowledge, deci-

---

* These criteria are, of course, also valid for commonsense knowledge.

sion making, or problem solving, as it is done by orthodox AI (what is due to the claims of the PSSH). As we have seen this aspect does not take into consideration the intuitive or commonsense part at all, although it seems to be the more important component of knowledge. Looking at the problem of how to represent these processes in the more or less scientific knowledge space out of this perspective, it turns out that the symbol manipulating and/or deductive way of processing this sort of knowledge is *in*adequate to a great extent, because it passes over the fact of the commonsense component of knowledge in silence and ignorance.* We will discuss this problem in Section 23.3.1. As has been mentioned above (Section 23.1) (orthodox) AI should force real basic research and interdisciplinary cooperation as a consequence of this discourse.†

In discussing the differences between scientific and commonsense knowledge we have worked out that we have to rethink this (strict) distinction—the most important result of these considerations is that we have to be aware of the fact that scientific or expert knowledge (as it is presented in a paper, book, talk, etc.) is always embedded in commonsense knowledge and represents in most cases the results of artificial formalizations of commonsense insights, ideas, or findings. Orthodox AI did not take into consideration this fact for at least two reasons:

1. *Lack of a complex theory of commonsense knowledge:* The solution for this problem could be a very complex form of a theory of action considering not only syntactic, but also semantic‡ as well as pragmatical aspects.

2. *Lack of a comprehensive theory of knowledge representation:* This point is closely related to the first as it is an implication of it. It is the result of our scientific paradigm and of our way of viewing at science (and scientific knowledge); its structure is shaped by the early work of philosophy of science and history of science (clarifying the gap between commonsense and scientific knowledge). One can find, however, in philosophy of science and epistemology a new trend describing the situation more realistic (e.g., [4,6,9,19]). They show the scientist being strongly influenced by social, cultural, and economical factors and describe the embedding of "scientific knowledge" in commonsense knowledge. The problem is that orthodox AI and computer science did not realize this new trend yet, and go on making use of methods having been developed a long time ago (syllogism, etc.).

---

* Or orthodox AI community is not capable of admitting to having failed in the attempt and claim to simulate and model cognitive processes (cognitive processes do not only consist in playing chess or proofing mathematical formulas).

† The parallel distributed processing approach shows the first signs of such an attempt in some projects.

‡ Semantics, as it is used in orthodox AI, is nothing but syntactic semantics; in other words, syntactic structures represent semantics—this is not really a progress, because AI did not leave its syntactic and symbolic level at all and remains in formal logic and manipulation of meaningless (mathematical) symbols.

Nevertheless, some people are called experts; they have used their experience and intelligence to develop highly specialized knowledge in one domain. In many cases, however, it is *not* possible for them to explicitly utter, how they came to a certain decision or how they solved a problem; the reasons for this have been discussed above (e.g., implicit or *tacit* knowledge). This means, the expert is not capable of explaining his or her own process of problem solving in verbal terms in many cases. This implies that he or she cannot tell the full "truth" and uses (or has to use) sometimes farfetched and artificial rules, methods or strategies to explain the process of his or her reasoning. This is what causes this pseudointellectual and pseudoscientific character of systems representing scientific knowledge*—they are abusing this scientific character for going on in doing research in this manner and for going on in processing knowledge in this manner. Although the expert works in a very well defined (scientific) domain, we are dealing with a sort of implicit knowledge, which cannot be explicitly formulated in terms of natural language and even less in terms of formal logic.

23.2.2 (Linguistic) Implicitness and Explicitness of Knowledge

We will see that this unability of explicitly uttering (scientific) knowledge is very closely related to the problems already discussed. There is a homomorphic relation between the questions concerning "scientific vs. commonsense knowledge" and "explicit (verbal) versus implicit (nonverbal) knowledge." Most of our commonsense knowledge can*not* be uttered explicitly—it is only possible to vaguely describe or to paraphrase it (e.g., as a vague feeling); we cannot talk about it as, for instance, about a mathematical proof. This causes the problems having been mentioned above (lack of a complex theory of commonsense knowledge, etc.). That is why we have to look out for alternative methods of

a. Describing and investigating,
b. Representing and processing

commonsense knowledge. I am convinced that we have to take into account aspects of pragmatism and "real" semantics (in contrast to orthodox AI) for adequately answering the questions being asked in (a); that is, we have to be aware of the fact that our knowledge is the result of a process of structuring our environment and constructing it in our cognitive domain—we are "directly" (i.e., there is a physical, nonsymbolic connection between our organism and the environment (being realized by sensors and effectors); our language is also embedded into this "physical" interaction structure [23]) interacting with our environment and (natural) language is nothing but a very

---

* The term *scientific knowledge* is used for justification and for demonstrating that the knowledge these systems are dealing with has an absolute or "objective" character.

complex form of behavior. The parallel distributed processing (PDP) approach in computer science seems to offer at least more adequate possibilities [for (b)] than the orthodox AI's symbol manipulation approach, because it has a neuroepistemological foundation (e.g., [14,20], [21]); it has an empirical as well as philosophical basis whereas orthodox AI is pure speculation (being based on and influenced by concepts of formal logic, linguistics (e.g., [22] and computer science).

As it is well known, the problem of representing and processing commonsense knowledge is by far not solved satisfactorily in orthodox AI (being based on the PSSH). If the expert's domain is not as formal as the domain of, for instance, playing chess or solving mathematical problems, etc. the problems of reasoning in this domain can be reduced, as we have seen, to reasoning in the domain of commonsense knowledge. It seems to be clear that, in most cases, this sort of knowledge depends on the human's experience emerging from the ability of learning, adaptation to the environment, and from being able to interact actively and without symbolic instance* with the environment in a verbal as well as in a nonverbal way:

Verbal Knowledge This is what we normally (and naively) understand by knowledge (for instance in orthodox AI having its basis on Newell's and Simon's PSSH); a (linguistically) explicit description of phenomena, objects, theories, processes, in our environment using words or symbols of a natural or formal language. It is implicitly assumed that there is a direct relation between environment, knowledge and language (Fig. 23.2). I want to make these assumptions more transparent and explicit in the following paragraphs:

1. there is a process of mapping from the domain of the environment (DOE, "reality") to the domain of knowledge (DOK); it is assumed that knowledge is the result of a *passive* process of mapping objects, phenomena, in DOE to knowledge structures, ideas, in DOK, mistakenly implying a (almost isomorphic) correspondence between real world structures and knowledge structures.

2. another assumption is the *isomorphic relation* between the structure of DOK and the domain of (natural) language (DOL), as it is shown in Figure 23.2. This leads to the very popular—however unreflected—ideas of treating thinking equivalently to manipulating (linguistic) symbols (and vice versa). As has been shown above (for instance in section 1) these assump-

---

* "Without symbolic instance" means that interaction between the human (the machine) and the environment does not take place by exchanging symbols (with linguistic meaning), but rather, as suggested in Peschl [2] takes place by physical (nonverbal) interactions (sensory input and motor output). Another point is the *active* interaction which means that the (natural or artificial) cognitive system is (or should be) not only capable of passively perceiving (symbolic) input but also of actively changing its environment.

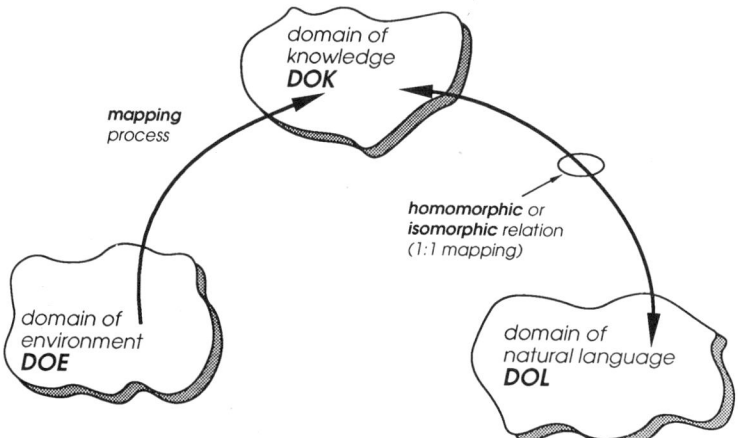

Figure 23.2 *The (naive) assumptions on the relative between the domains of environment, knowledge and (natural) language (being made implicitly by, for instance, orthodox AI).*

tions represent the basis of orthodox AI in its (research) program (or better in its research *ideology*) of the physical symbol systems hypothesis.

One can see that a structural equivalence between Figure 23.2 and the semiotic triangle (as presented by U. Eco [16,17]) can be found. We will, however, have to rethink and perhaps to reformulate this concept in a later discussion in order to get away from the approach having been presented above. It seems to me problematic for the following reason: It is quite superficial and follows only commonsense considerations and commonsense speculations on the questions and problems of what is language, what is knowledge. If we take into account the results of modern neuroscience (and their epistemological implications),* we cannot be satisfied with our commonsense considerations, because they are—from a neuroscientist's and cognitive scientist's more interdisciplinary point of view (as presented in this treatise)—full of contradictions and because they are not capable of answering very serious questions concerning language acquisition, the emergence of meaning, the relation between so called conscious and unconscious aspects of acting, the influence (the existence) of nonverbal aspects of language.

We have to *overcome* the (wrong and misleading) idea of unreflectedly assuming that language is the ultimate structure and substratum of our thinking; it is rather a very specialized form of behavior with high complexity being due to the structure of our brains. It is only the favored use of (natural) language in all our communications which makes us think that all our cogni-

---

* There can be observed an increasing mutual interest in the cooperation between empirical and more speculative disciplines (i.e., neuroscience and epistemology or philosophy of mind).

tive processes take place in language. The neurocybernectician H. Maturana shows that language rests on physical orienting interactions within a consensual domain [23,24] from a natural scientific as well as form an epistemological point of view. We will rather follow this path in our considerations on AI, cognitive science and cognitive modeling, because it promises not to get into the same troubles and dead ends as orthodox AI has got since it completely relies on this linguistically influenced symbolic paradigm. This approach of mapping "reality" to linguistic or knowledge structures is not adequate from a constructivist point of view; we are asked to rethink and to reflect our implicit assumptions of the correspondence of "reality," knowledge and language. The discussion on nonverbal knowledge will be a first step toward this reflection and perhaps toward a new understanding and new concepts of cognition, knowledge, and language.

Nonverbal (or "Tacit" or Implicit) Knowledge "We are knowing more than we capable of expressing explicitly (in natural language)"; this is Michael Polanyi's central thesis in his book *The Tacit Dimension* [25]. It will turn out that this form of knowledge, as will be described below, is very important for our considerations concerning knowledge, knowledge representation, and especially for our investigations of parallel distributed processing and cognitive modeling. Each of us has made the experience that he or she could not express a feeling, a phenomenon, an object, in natural language words—he or she could not find the words, however, had a certain "feeling" for that unexpressable thing.

Another example is the process of translating a text to another language (Fig. 23.3): we are reading a sentence or paragraph in language L_ψ; it is the *context* of the word or phrase w_ψ we want to translate to language L_ξ triggering a certain state. This state is independent from any natural language; we could call it the "meaning" of the "concept" of or the "representative feeling" for a word, sentence, paragraph. If we want to translate word w_ψ to language L_ξ, this (acoustic or visual stimulus of the) word will trigger a certain state (in the context of the sentence's or paragraph's state) being independent from this (acoustic or visual) symbol in our brain. Very simply spoken this state is the representation of this symbol (but not only the symbol's representation) in the receiver's brain. From this "interlingual phase" the process of translation is finished by searching for and finding the correct word (symbol) w_ξ in the "target-language" L_ξ. If we are comparing this concept of translation with the orthodox methods of syntactic symbol manipulation, the concept presented here seems to be more plausible, as it is more adequate in respect to results from neuroscience and it does not offer such a superficial and naive view of a complex problem as orthodox linguistics or orthodox AI does; that is, the complex problem of translating or understanding a language is not reduced to a problem of *syntactical* manipulation (with pseudosemantics), it is rather seen as a cognitive process on the basis of implicit knowledge; as it goes beyond the boundary of natural lan-

646 SYMBOL MANIPULATION, COMMONSENSE KNOWLEDGE, AND COGNITIVE PROCESSES

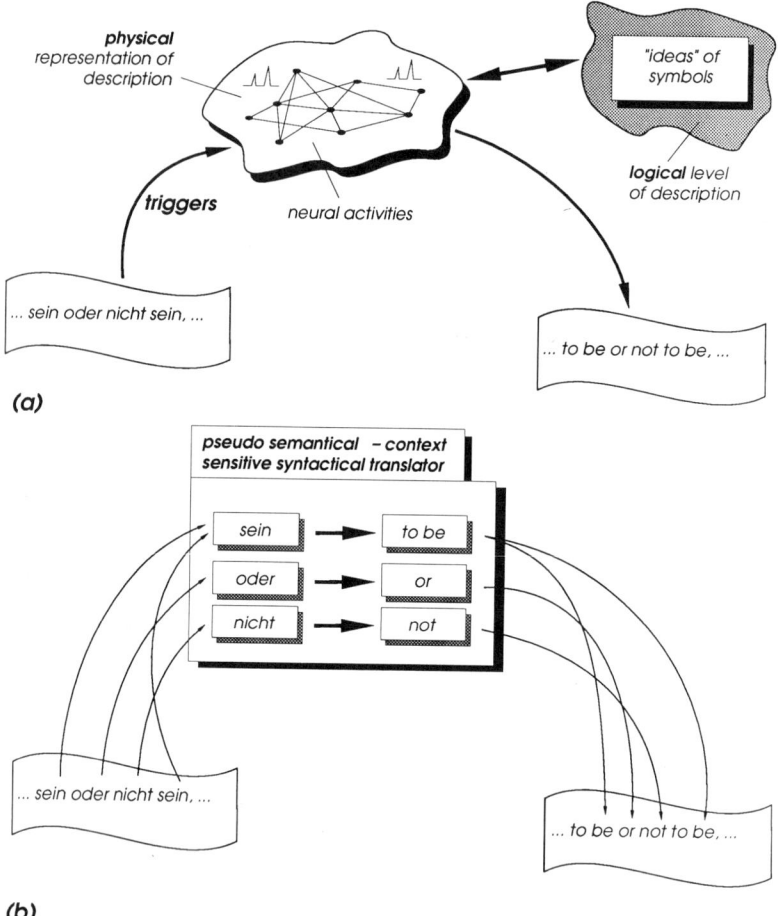

Figure 23.3 The "cognitive" and "neuro" perspective (a) versus the "orthodox" (syntactical) concept (b) of the phenomenon of translation.

guage and symbols and tries to integrate an empirical (and not only speculative) view of neural processes being the basis of all of our cognitive phenomena.

Implicit (tacit) knowledge arises from the direct interaction with the environment (i.e., sensory information and motor action—the cognitive system is part of a feedback loop over the environment); it interacts with it by just exchanging physical actions. It is clear that *language is embedded* in *these physical interactions,* too. So we can see that nonverbal knowledge is the *substratum* of all verbal actions—as Maturana writes, language gets its meaning in the observer's cognitive domain, as he (the observer) associates his own connotations, which have been learned by experience and which are triggered by the sensory input and the inner state of the observer's nervous system [26,23]. It is important to see that this form of implicit (or tacit) knowledge (as Polanyi calls it [25]) is essential to language, verbal knowledge, etc. Thus we should not investigate so much linguistic processes, representing phenomena on the surface (i.e., on the surface of syntactical structure), but rather try to understand, as stated above, in an interdisciplinary discourse verbal communication or language at a *lower level* (of physical interactions)—from *bottom up*.

23.3 CONCLUDING REMARKS ON KNOWLEDGE AND ITS REPRESENTATION

What I wanted to make evident is the close relation between the different kinds of knowledge and of language (communicative behavior). We have seen that scientific knowledge is embedded in common sense knowledge as well as verbal knowledge is embedded in nonverbal knowledge. The conclusion I want to draw is that, if we are interested in developing a real intelligent and "smarter" (expert or knowledge based) system, we first of all have to realize that we can *reduce* the problem of finding a theory of scientific knowledge to finding a theory of nonverbal knowledge (Table 23.1). Mathematical or formal logic, as it is applied in most orthodox symbol manipulating AI systems, in most cases is not the adequate way of describing scientific as well as common sense or nonverbal knowledge.

Hence, we have to look out for or develop a theory of commonsense and/or nonverbal knowledge. As has been mentioned in Section 23.1 this can not be done without the integration of other disciplines such as epistemology, neuroscience and (cognitive) psychology. Once again we have to realize that the traditional methods and techniques being provided by computer science or orthodox AI [i.e., the (unreflected) application of methods from mathematical (formal) logic] are by far *not sufficient* for coming up to the demands and claims of such a theory for several reasons. I do not want to discuss them in detail here, I just want to sketch three of them:

1. *Restriction to (Natural) language:* Variables of mathematical logic always represent a linguistic concept, object or category. They are interpreted in terms of (natural and explicit) language. This causes an *exclusion* of nonverbal knowledge implying, according to our considerations, that it is an unsuitable (or at least only very limited) method for describing common-

TABLE 23.1 Relations between Different Forms of Knowledge

Scientific knowledge	⇒	Commonsense knowledge
⇓		⇓
Verbal knowledge	⇒	Nonverbal knowledge

sense (or nonverbal) knowledge (being the basis for verbal or scientific knowledge). The meaning of the symbols being used by the system is completely dependent on the human interpretation and there is no possibility to "learn" the meaning of a symbol, because of the lacking direct access to reality; in other words, the *pragmatical aspect* is left out and replaced by the human designer/user who is not capable to interact with the system in any other way but by exchanging (for the system) meaningless symbols (i.e., by defining the system's semantics in terms of syntactic interaction).

2. *Restricted context:* As one consequence of this restriction to natural language emerges a very limited possibility of describing the (implicit) background and/or the context which is, as we have seen in Section 23.2.1, very important and has a strong influence on the decisions being made in the process of reasoning.

3. *Restricted learnability:* As we have seen in (1), learning from experience is very important for learning the meaning of a symbol word, of a language. However, the restrictions 1 and 2 do not allow an adequate process of learning, because, for instance, symbols being used in formal logic cannot change their meaning; they can change only their meaning in the interpreter's cognitive domain (as shown in Section 23.4, this can become a problem).

We can conclude that we also have to take into consideration results from cognitive science and cognitive modeling, because we can learn a lot from their basic insights concerning (human) cognitive processes (e.g., reasoning, memorizing, knowledge representation, learning, etc.). If we aim at a very complex and (really) intelligent machine we will have to integrate ideas and results from various disciplines helping us to come up to the required standard.

So, how can we describe the phenomenon of knowledge in such a way that it fits to the claims having been made in the sections above? We can (or better, we have to) reduce the notion of knowledge to, as Maturana writes, "to know is to be able to operate adequately in an individual or cooperative situation" [26, p. 53]. It is very similar to Y. Elkana's approach; he writes that the aim of knowledge (being obtained by experience) lies in maximizing security [6, p. 261].* These are very wide definitions, they include, however, all kinds of knowledge and allow a more realistic view of the problem of how to represent knowledge in respect to modeling cognition as well as how to design successful complex commercial applications.

23.1.1 Representing Knowledge (in a Computer)

If we are talking about knowledge based systems, we usually think of symbols and symbol manipulation. I want to call up, however, to rethink and

---

* Y. Elkana writes: "Der Mensch ist von Unsicherheit umgeben, und sein Alltagsleben ist öde, rauh und trostlos. Das Ziel des Wissens liegt darin, Sicherheit zu finden, . . ." [6], p. 261).

23.3 CONCLUDING REMARKS ON KNOWLEDGE AND ITS REPRESENTATION

reflect this method whether it is adequate for representing knowledge which is not well suited for formalization. I think this is justified, as demands are made on providing, say, expert systems or "intelligent" human-computer interfaces with (more) common sense facilities in order to make them "smarter." It is difficult, however, to represent common sense knowledge with explicit symbols, which can be manipulated by a computer, for two reasons:

1. As has been discussed, commonsense knowledge to a great extent contains implicit knowledge—knowledge that cannot be verbally uttered.
2. There is a mapping, however, from natural language symbols to "computer symbols"—thus it is not possible to map implicit knowledge to symbol structures in a computer, although this kind of knowledge seems to be a very important and influencing factor for making decisions, reasoning, etc. in a domain being not as formal as, for instance, the domain of mathematical problems.

The problem we are confronted with is that machine knowledge is—in the best case—restricted to knowledge which can be expressed by natural language. Symbols, such as *house* or *table* do not mean anything to the computer, unless they have explicit (symbolic) relations or rules adding connotations to the symbols. Even if this is the case, they do not have any meaning in the sense as we use it. So, what is the symbol manipulating machine lacking?

To answer this question shortly: It is the *experience* with the real world (i.e., the pragmatical approach to its environment) which implies that language is *used* (in the sense of Section 23.2.1). Thus language has a direct relation to the objects used in the environment. This means that the human who uses for instance the symbol *house* has his or her own connotations of this symbol, which he or she has learned by *experience* with the object house in interacting with his or her environment.* Computers using the symbolic approach do not have this commonsense knowledge, as they do not have active (i.e., sensory and motor action) access to their environment. The implications for knowledge engineering will be discussed in Section 23.4.

What solution could be found to this problem? I can only offer an approach which seems to be very difficult to realize at the present state of research in computer science. I would suggest to make use of the PDP paradigm offering very interesting possibilities, such as the subsymbolic representation of knowledge [27,28], the capability of learning from experience [29]. I am aware of the fact, however, that PDP networks are still in

---

* Humans knowing a language need not learn every meaning by experience with the environment; linguistic learning (i.e., learning by hearing or reading about a phenomenon, object, etc.) even is predominant over learning by experience.

their infancy and do not offer efficient mechanisms and means for *formal reasoning* or formal manipulations being due to the completely different approach of processing. *Hybrid representation* of knowledge could be a compromise; this way of representing knowledge makes use of the advantages of symbol manipulation as well as of the PDP approach.

As we have seen, intelligent behavior does not only rest on symbol manipulation, but also is influenced by "intuitive," nonlinguistic processes. If we combine the symbol manipulation's advantage of being good in formal deriving and handling symbolic knowledge with the advantage of the PDP approach of processing nonformalized knowledge, which is learned by examples, a way is offered for finding heuristics and applying them to symbolic knowledge without having to explicitly define them and applying them to symbolic knowledge (i.e., to integrate also a kind of implicit knowledge in the system.*

23.4 KNOWLEDGE ENGINEERING, LANGUAGE, AND MAKING USE OF SYMBOL MANIPULATING SYSTEMS

The main problem the process of knowledge engineering has to face is that it has to fill a kind of what I call a *cognitive distance*—the gap between the expert and the (expert) system, on the one hand, and the gap between the computer system and the user, on the other hand. Figure 23.4 shows this problem of linguistic mapping: as the expert, the knowledge engineer and the user have different experiences and a different cognitive background, a certain symbol (such as *house*) triggers different associations in their cognitive domains. This can lead to difficulties and misinterpretations when using knowledge-based systems in an unreflected manner.

23.4.1 Knowledge Engineering—Organized Loss and Distortion of Knowledge?

As we can see, there is not only a *loss* of information but also a *change* of information in the process of knowledge engineering and of using a knowledge based system. It has become clear (from the argumentation in the last sections) that, if we are talking about information, we do not think of the invariant information of the shape (pattern) of symbols that are exchanged between the designer/user and the computer system; we rather think of it in terms of *system relative* meanings; that is, each system (designer, user, etc.) "associates" its own connotations which are triggered by the symbol. Why does this loss and change of information take place? We can explain this phenomenon only by reflecting on our own use of language as means for

---
* This seems to be a viable way *only* for commercial applications; it does not, however, have any value—as orthodox AI—for adequately describing cognition.

Figure 23.4 *The problem of the cognitive distance between the expert and the user.*

describing our environment in the view of Sections 23.1 and 23.3.1. We will see that it is *not enough* to use some common sense arguments, such as "language consists of symbols being mapped to the computer," "the human mind makes use of symbols for thinking," or

> ". . . Thus it is a hypothesis that these symbols are in fact the same symbols that we humans have and use everyday of our lives. Stated another way, the hypothesis is that humans are instances of physical symbol systems, . . ."
>
> Newell & Simon, [7, p. 116]

etc., in order to get an adequate view of this problem.

H. Maturana offers an epistemological concept of language which could help us in better understanding the problem of the process of knowledge engineering, as depicted in Figure 23.4. It already has been sketched in Section 23.3.1 and is a well known problem in semiotics and philosophy of language. Language takes place in a *consensual domain* which means that each participant taking part in the communication is capable of using certain symbols (acoustic, visual, etc.) with a "public" meaning overlapping to such an extent with the "private" meaning that communication is successful; that is, the participant has reached his or her aim (i.e., has triggered the aspired action in the partner).

"Public" meaning is the result of structural coupling of the contributing organisms; "private" meaning arises as a result of *individual experiences* with the environment [as stated in Section 23.2.1; the *active use* of symbols brings life (i.e., meaning) into them]. We can only speak of meaning, however, if we are aware of the fact that meaning gets its "meaning" only in communicative structures (i.e., in an consensual domain). It is clear that these two forms of meaning are interrelated with each other and that there exists a strong interaction between them. "Private" meaning represents the basis for the "public" meaning—both of them find themselves in permanent flow—the *use* in *interacting* with the *environment* and with the communication partner determines the meaning of words, symbols, sentences, etc. So, what we have called the "cognitive distance" is nothing but the difference between the "public" and the "private" meaning or between two different "private" meanings of a symbol, of a sentence, and so on.

It is clear that the cognitive distance between two communicating persons is due to the facts that they do not possess the same experiences with one symbol, as their lives were different to that point when they meet and talk with each other. Each symbol they are using refers to a set of experiences (giving rise to a set of connotations) in each of the participant's cognitive domain. Hence, it is the degree of overlapping of experiences being responsible for how they do understand the other. The advantage of the dialogue with a human is that one may ask how a certain word was meant or one may ask to describe a symbol (or better the use of a symbol) in other words in order to get a better "feeling" of how the other uses this symbol (in a given context); that is, in order to reduce the cognitive distance and obtain a better overlapping of connotations.

As natural language is capable of expressing one phenomenon, object, etc. in different words (i.e., using different symbols referring to the object), it is possible to reduce this gap—in most cases we will *not* obtain, however, *total* coherence. It is the knowledge engineer's task to

1. Reduce this gap as far as possible (i.e., to intensively ask the expert for his or her use of language);
2. To find a mapping (to symbol structures being capable of being processed by a computer) which contains all the relevant* information having been extracted in (1).

Case 1 is easier than 2 as the possibility of asking the human expert (in order to get a better understanding of what the expert really meant; i.e., to find a better matching of the mental models) is given, whereas in case 2 it is the knowledge engineer's responsibility to

3. Find an adequate form of representation in the computer—an efficient and sufficient (with respect to 4) way of representing the knowledge obtained in (1);
4. and, what is even more difficult, to find a representation being capable of triggering the suitable associations in the user's cognitive domain† when he or she is using this knowledge-based system. It is clear that this is not only a problem of AI, but also represents one of the main problems in human-computer interaction.

---

* "Relevant" does not mean that it is the knowledge engineer's intention to represent only this part of the expert's knowledge which is "representable" by the means of the given (computer) system.
† "Suitable" means that the user's and the expert's cognitive domains (experiences) are as corresponding as possible.

23.4.2 Knowledge Transfer to the Computer

As long as there is a dialogue between humans (e.g., expert-user, expert-knowledge engineer), no serious problems can arise, because there is always the possibility of asking if something is not understandable. If an artificial symbolic instance (such as a knowledge-based system) is in between the expert and the end user practical as well as epistemological problems are arising. Remember that the "richness" of the knowledge engineer's model is less than that of the expert; it is even worse in the computer's model of knowledge due to the following points:

- The expert's inability (in most cases) to structure his or her knowledge in such a way that
    - It is possible for the knowledge engineer to get an overview* in order to prestructure knowledge.
    - At least the knowledge engineer is provided with the kind of *tacit (implicit)* knowledge being of great importance for understanding the expert's domain and even of greater importance for the user who does not know, in most cases, very much about the expert's domain.
    - It is suitable, after some formalizations, for being mapped to symbolic structures that can be processed by a computer.
- The knowledge engineer's inability (in most cases) to listen in such a way to the expert that he or she is not always thinking of how and which means of representations are suitable for the domain. In other words, the knowledge engineer does not really get engaged in all facets of the domain in many cases, which is, however, very important for its deep understanding. "Deep understanding" means to get involved in the domain's "tacit dimension" [25], which in many cases can be achieved only by getting in "physical" touch with it. This is due to what we have said about language and its use in Section 23.2. This seems to be impossible for several reasons:
    - Financial: neither the expert nor the knowledge engineer's company can afford, in most cases, a practical training in the expert's domain.
    - Time restrictions: most projects are developed under pressure of time (and money) and do not allow this time-consuming approach.
    - Knowledge engineer: in most cases the knowledge engineer him/herself is not interested in getting involved in the expert's domain to such a "physical extent."
- Knowledge-based systems offer—compared to the complexity of most domains—only very poor means for representing (expert) knowledge.

---
* "Overview" means that the knowledge engineer should get a *feeling* for the domain.

As already mentioned, the most important disadvantage seems to lie, in the best case, in the fact that only explicit knowledge (i.e., knowledge which can be verbally uttered) can be mapped to and processed by a symbol-manipulating system. This is the case if we think, for instance, of rule-based systems or of systems representing knowledge by using semantic networks or frames. None of those representing mechanisms takes into account implicit knowledge, which is hardly surprising as this kind of knowledge does not fit in a common sense view of language, knowledge, representation, and is not well formalizable (as it is difficult to linguistically describe it). The last argument, however, is due to the approach having been suggested by the PSSH—of course, it is very nice and tempting to think that our knowledge only consists of explicit symbols, but it is not the full truth, if we think of domains being not so well formalizable (i.e., domains being closely related to commonsense knowledge).

Another reason for the lacking possibility of representing common sense (or implicit) knowledge lies in the restrictions of the von Neumann (hardware) architecture (F. J. Varela [30]), which specifies and (pre)determines certain data and processing structures.*

- If we take up Whitaker's [31] idea of seeing an expert system as a facsimile of the expert, we can say that it is a facsimile of the "second order," as (in most cases) the knowledge engineer himself or herself is a facsimile (of first order)—the knowledge-based system thus is a "copier" offering, however, the possibilities of representing only the most important fragments of knowledge ("the tips of the icebergs," a skeleton structure).

- The process of *mapping* natural language knowledge to formal language symbols itself is problematical in many ways, as

    The *context* gets lost; in other words, single symbols are picked out of a fine grained network of explicit and implicit knowledge. These symbols are connected to each other (by rules or in semantic networks by edges) in a (in most cases) quite artificial manner.† Hence, there is built up a pseudocontext (a very reduced context) in the representational space (Fig. 23.5; this figure was inspired by E. Oeser's talk on knowledge technology in January 1990, Vienna).

    The knowledge engineer is *influenced* by the (restricted) possibilities and forms of representing knowledge (on the computer system) in his thoughts. As he/she knows very well that it is his or her task to

---

* The PDP approach seems to try to get rid of these boundaries in order to introduce a new processing paradigm.
† As already mentioned in Section 23.2.1 it is not possible for the experts to explicitly utter their knowledge which they implicitly know, so they have to "invent" artificial rules, reasoning strategies, statements, etc. in order to "satisfy" the knowledge engineer.

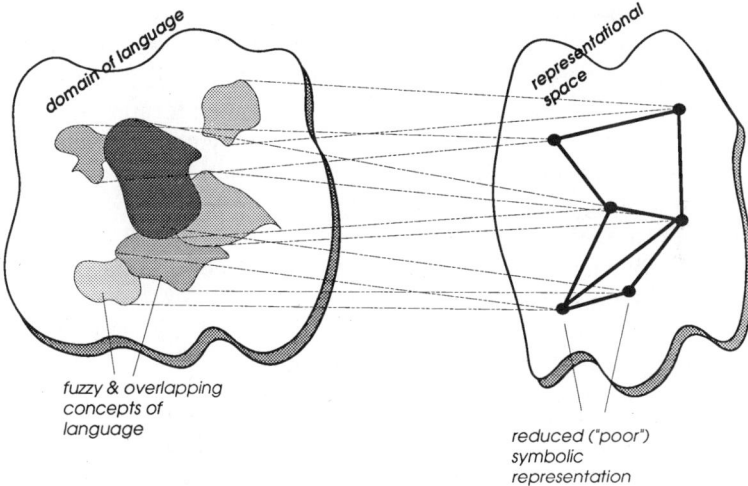

Figure 23.5 *The reduction of fuzzy natural and overlapping language concepts in the domain of a natural language to a very restricted and "poor" symbolic structure.*

fill the gap between the expert's unformalized utterances and the very formal representing mechanisms, he or she decides (or has to decide) to choose the way of thinking in terms of these formal representing mechanisms as it is necessary and the only viable way of implementing the knowledge into the system. Therewith the knowledge engineer bars himself or herself the way to deeply understand the expert's problems, knowledge, intentions, etc.

There is, of course, the loss of the direct relation of the symbol to the physical object. As mentioned, the *use* of a symbol determines its meaning. This is, however, not an AI specific problem, as we face it whenever we are writing down (a part of) our knowledge. There is the loss of all of our "private" connotations—only the structure having been mapped to the knowledge based system (i.e., the pseudocontext) is left and has to be interpreted by the user, which leads us to the last station on our way of describing the process of knowledge transfer in (symbolic) knowledge based systems being discussed in Section 23.4.4.

23.4.3 Knowledge Processing in Computers

We do not have to discuss this topic in detail, because it is well known and it is mainly of technical interest. It should be emphasized, however, that there exists an alternative to symbolic knowledge processing: the parallel distributed processing approach and its combination with symbol manipulation (i.e., "hybrid systems") seems to be a promising approach out of the dilemma of orthodox AI.

Just let us make aware what is happening when symbolic knowledge is processed by a (symbol manipulating) computer:* the second-order representation (symbolic representation of the human linguistic representation of the "real world") is manipulated by an algorithm—a formal apparatus following strict rules operates on the (meaningless) symbols which represent parts of the "real world." The processes of operation as well as the symbols have their meaning only in the designer's/user's intentions and cognitive domain; they are, from the computer's "point of view," only syntactic structures producing new syntactic structures which are interpreted in the last step of using a knowledge-based system.

23.4.4 Using Knowledge-based Systems

To sum it up in short: it is the user's task to *repair* and *compensate* the errors having been made in the process of knowledge engineering. Expressing it drastically, the user is confronted with the "bare" symbols on the screen; i.e. they possess (on the screen; i.e., uninterpreted) only these "pseudoconnotations" having been programmed to the system by the knowledge engineer. The user in many cases has not even access to these connotations. Stated another way, it is the user who has to fill the skeleton of knowledge being presented by the system with his or her own connotations. The disadvantage is that the user does not know whether his or her connotations may be used and whether they are used in the expert's sense.

Thus, the knowledge having been produced by some symbol manipulations leaves its pseudocontext and passes over to the user's context where it triggers associations being based on the user's experiences. In domains being not well suited for formalization,† hence, the possibility of *misinterpretation* of the computer's results becomes evident, as used symbols are not as denotative as, for instance, in a strict mathematical domain. It is the user's task to *reconstruct* the meaning, the intentions and the context of the expert's cognitive domain.

There is not only the loss of information, but what is even worse a *garbling* of information. It is clear that this could have disastrous consequences, because, in the case of the user, language, or better its symbols, are not only abstract descriptions of knowledge any more, but they are applied to reality and they have "real-world consequences" (especially if the user *blindly* and *unreflectedly* follows the knowledge-based system's instructions). As it is the aim of an expert system to replace the expert and his or her knowledge by a computer in such a way that everybody being capable of using a computer can work with this knowledge, we have to expect that the user, who is (by definition) not an expert of this domain, blindly relies on the system (and uses it for legitimizing his or her decisions)!

So, we have followed the process of "lifting off" knowledge from real world—taking it out of the expert's cognitive context and mapping it to

---

* We can not consider the case of parallel distributed processing here.
† This means that there is no definite and unmistakable terminology in this domain.

symbolic structures. After some (more or less complex) manipulations on those symbols (in the representational space)* the result is interpreted by the user, who himself or herself more or less reflectedly applies it to the real world. It is the user's *responsibility* to judge whether the results are useful and correct or not, but how should he or she do this, if he or she is not an expert?

23.5 CONCLUSION

23.5.1 Interdisciplinary Cooperation and Rethinking Knowledge Engineering

As has been shown the development of really complex and intelligent systems not only requires technical and computer science knowledge, but also needs an epistemological foundation in order to avoid, for instance, the basic (paradigmatic) problems orthodox AI has to face in these days. *Consequent interdisciplinary cooperation* between various disciplines enables success in basic research as well as in commercial applications. Problems of commercial programs often could be solved or even avoided by a more interdisciplinary research strategy or by taking seriously and considering results from other disciplines. This chapter represents one approach to facing these problems from an epistemological as well as from a computer science point of view.

What conclusions can be drawn from this critical review of symbol manipulating knowledge based systems?

- As has been done in this chapter, we have to *critically* apply, develop and use the symbol manipulation (i.e., the PSSH) approach. It has been shown that an epistemological investigation of the problems of what is language, what is knowledge, what is cognition, can help us in better understanding the problems of orthodox AI and in getting developing alternative forms of knowledge representation.
- We have further to focus our attention to the process of knowledge engineering, because it is the *knowledge engineer's responsibility* to find an adequate method of mapping the expert's knowledge to the computer in such a way that he or she *reflectingly* and effectively makes use of available representation mechanisms on the one hand and is aware of the problems of interpreting the "produced knowledge" in the user's cognitive context on the other hand.†

---

* There is no relation to reality in the computer.
† It is the knowledge engineer's responsibility to confess in certain cases that it is not possible to formalize a certain domain to such an extent that he or she is capable of mapping it to a computer system and of taking responsibility for "real world consequences" arising from the expert system's "conclusions."

- We have to look out for ways providing the means of integrating implicit knowledge. One approach has been mentioned when I suggested a kind of hybrid knowledge representation (combining the advantages of symbol manipulation techniques with the advantages of parallel distributed processing).
- We have to be aware of the fact that, when using the symbol manipulation approach, an *accumulation, amplification* and *propagation of errors* (similar to numerical errors in numerical mathematics) takes place in the process of repetitive mapping. To specify this point more precisely:

  As we have seen, there is an error (loss of information) arising when mapping the knowledge at first to the knowledge engineer and then to the computer (Fig. 23.4).

  This clipped information represents the basis on which manipulations are executed. Another factor amplifying the error is the process of reasoning itself.*

  The result being not at all perfect has to be (is mis-) interpreted by the user.

  In Maturana's terms we could conclude in saying that there is a loss and change of connotations when applying a knowledge-based system.
- We have to look out for means of how to *narrow the cognitive distance*—the gap between expert, knowledge engineer and user. This is closely related to the claim of integrating implicit knowledge, because if we could handle implicit knowledge, we could solve the problem of narrowing the gap in a more adequate way.

23.5.2 Suggestions

Finally I want to make some suggestions for how we could solve some of the problems having been addressed in this chapter.

- We could introduce a *very well defined terminology* for a certain domain. This means that the expert, the knowledge engineer as well as the user has a *denotative* description of the symbols, words, being used by the system—it is compulsory upon each of them.
- Another possibility is the *restriction of the domain* to a very formal and unambiguous part of the whole domain—this would be, however, contradictory to the claim for "smarter" systems that make (or should make) more use of commonsense knowledge.
- Representing knowledge by making use of icons: this means that knowledge is represented pictorially; this is not possible in some cases, however. Nevertheless it would reduce the possible connotations which are

---

* This is also a question of cognitive psychology; its discussion would go beyond the scope of this treatise.

triggered by plain symbols. It does not solve the problem of how knowledge is mapped to the system and how it could be processed, as it is not possible to process pictorial information only. So one could think of another form of hybrid representation in such a way that symbols are associated with iconic representations.
- Private user interface: the interface between the user and the system is adaptive to each user's use of language. This would require a mechanism being capable of learning fine nuances in language (perhaps in a "subsymbolic" way [27]).
- The system should provide the capability of giving more precise information, if the user asks (like in a help system). "More precise information" means that the result is described in another way (i.e., for instance by using other words).

23.5.3 Alternatives and Prospects

The suggestions having been made above represent possible improvements, however, they do not solve the principal problems which have been addressed in this chapter. Therefore I would like to sketch the following alternative approach, which seems to be quite unrealistic from today's point of view, but looking at the development of cognitive science, we can observe the beginning of a process of rethinking.

This (in the field of AI) completely new paradigm rests on the ideas of self-organization and on basic research results from neuro science, epistemology and parallel distributed processing. We have to give up the common idea of representation (as it has been discussed and criticized in this chapter); the alternative is a bottom up understanding of what is knowledge and knowledge representation being based on (neuro)cybernetical and epistemological concepts. We are thinking of an autonomous organism being directly coupled (over sensors and effectors) to its environment and being equipped with a PDP network for information processing. As has been shown in this chapter we are assuming a constructivist point of view; this artificial organism is learning from its environment by *constructing* its own reality (i.e., knowledge resp. knowledge representation). There has been done some basic work (M. Peschl [2,10] in this approach, however we are standing at the very beginning. Of course we can not offer such spectacular results as orthodox AI did (for instance, a chess-playing computer or a medical expert system, etc.)—the results we are offering are located at the other end of *basic processes*—physical structural coupling, development of a consensual domain (as the basis for developing a common language), development of very simple forms of language, symbols, and so on.

We are following a way which is based on the *acting* in the environment (the neurocybernetician and epistemologist F. Varela suggests this approach [30] in another context). In this approach we are applying a learning by doing strategy and building up (constructing) the representation by experiencing the environment—PDP network learning strategies offer the possibility of learning by making use of self-organization techniques. They are capable of fulfilling the claims being made by epistemology (constructivist ideas, etc.) and neuroscience (on an abstract level, P. S. and P. M. Churchland [14,32]. We can *integrate* these disciplines in an interdisciplinary discourse that promises to be very fruitful. In the long run we think of exposing such an artificial organism to an expert's environment and thus could "breed" an artificial expert having learned from experience and having the implicit knowledge which is so important for solving most of our (expert as well as commonsense) problems.

REFERENCES

[1] M. F. Peschl, An Alternative Approach to Modelling Cognition, in *Proc. of Man and Machine Conf.*, John von Neumann Society for Computing Sciences, Hungary, 1989, pp. 49–56.

[2] M. F. Peschl, "Cognition and Neural Computing—An Interdisciplinary Approach," in *Proc. of Internat. Joint Conference on Neural Networks*, Lawrence Erlbaum, Hillsdale, N.J., 1990, pp. 110–113.

[3] P. Feyerabend, *Die Aufklärung hat noch nicht begonnen*, in P. Good (ed.), Von der Verantwortung des Wissens, Suhrkamp, es 1122, Frankfurt/M., 1983, pp. 24–40.

[4] P. Feyerabend, *Wider den Methodenzwang (Against Method)*, Suhrkamp, stw 597, Frankfurt/M., 1983.

[5] P. Feyerabend, *Wissenschaft als Kunst*, Suhrkamp, NF 231, Frankfurt/M., 1984.

[6] Y. Elkana, *Anthropologie der Erkenntnis*, Suhrkamp, Frankfurt, 1986.

[7] A. Newell and H. A. Simon, Computer science as empirical inquiry: Symbols and search, *Communications of the ACM*, Vol. 19 (3), 113–126, (March 1976).

[8] A. Newell, Physical Symbol Systems, *Cognitive Science* 4, 135–183, (1980).

[9] T. S. Kuhn, *Die Struktur wissenschaftlicher Revolutionen*, Suhrkamp Taschenbuch, Frankfurt, 1967.

[10] M. F. Peschl, A Cognitive Model Coming up to Epistemological Claims—Constructivist Aspects to Modeling Cognition (1990), in *Proc. of Internat. Joint Conf. on Neural Networks* Vol. 3, pp. 3-657–3-662.

[11] M. F. Peschl, Auf dem Weg zu einem neuen Verständnis der Cognitive Science (On a new Understanding of Cognitive Science), *INFORMATIK FORUM*, 4.Jg., Heft 2, pp. 92–104, (June 1990).

[12] J. R. Anderson, *Kognitive Psychologie*, Heidelberg, Spektrum der Wissenschaft Verlag, 1988.

[13] H. Mandl and H. Spada, Eds., *Wissenspsychologie*, Psychologie Verlagsunion, München-Weinheim, 1988.

[14] P. S. Churchland, *Neurophilosophy. Toward a Unified Science of the Brain*, MIT Press, Cambridge, Mass., 1986.

[15] E. Oeser, Evolution und Selbstorganisation als Integrationskonzepte der Wissenschaftsforschung, *Zeitschrift f. Wissenschaftsforschung,* 2(4), pp. 163–173 (1988).

[16] U. Eco, *Einführung in die Semiotik,* Uni-Taschenbücher UTB 105, Wilhelm Fink Verlag München, 1972.

[17] U. Eco, *Zeichen. Eine Einführung in einen Begriff und seine Geschichte,* Suhrkamp, es 895, Frankfurt, 1977.

[18] L. Wittgenstein, *Philosophische Untersuchungen:* Teil I and II, Suhrkamp–Taschenbuch Wissenschaft: stw 501, Frankfurt/Main, 1984.

[19] J. Kriz, H. E. Lück, and H. Heidbrink, *Wissenschafts- und Erkenntnistheorie (Philosophy of Science and Epistemology),* Leske und Bundrich, Opladen, 1987.

[20] P. M. Churchland and P. S. Churchland, Could a machine think?, *Scientific American,* 26–31, (January 1990).

[21] E. Oeser and F. Seitelberger, *Gehirn, Bewußtsein und Erkenntnis,* Wissenschaftliche Buchgesellschaft Darmstadt, 1988.

[22] N. Chomsky, *Regeln und Repräsentationen (Rules and Representations),* Suhrkamp, stw 351, Frankfurt/M., 1981, (Columbia University Press, 1980).

[23] H. R. Maturana, Biologie der Sprache: die Epistemologie der Realität (Biology of Language: The Epistemology of Reality); in H. R. Maturana, *Erkennen: Die organization und Verkörperung von Wirklichkeit,* Vieweg-Verlag, 1982, pp. 236–271.

[24] H. R. Maturana, Repräsentation und Kommunikation (Representation and communication functions), in H. R. Maturana, *Erkennen: Die Organization und Verkörperung von Wirklichkeit,* Vieweg-Verlag, 1982, pp. 272–296.

[25] M. Polanyi, *The Tacit Dimension (Implizites Wissen),* Doubleday, Garden City, N.Y. (1966); Suhrkamp–Taschenbuch Wissenschaft, stw 543, Frankfurt/M, 1985.

[26] H. R. Maturana, "Biology of Cognition," in H. R. Maturana and F. J. Varela, *Autopoiesis and Cognition,* D. Reidel Publishing Company, Dordrecht and Boston, 1980, pp. 2–60.

[27] P. Smolensky, On the proper treatment of connectionism, *Behavioral and Brain Sciences* 11, 1–74, (1988).

[28] D. E. Rumelhart and J. L. McClelland, *Parallel Distributed Processing, Explorations in the Microstructure of Cognition, Vol. 1: Foundations,* MIT Press, Cambridge, Mass., 1986.

[29] G. E. Hinton, Connectionist learning procedures, Technical Report CMU-CS-87-115, Carnegie–Mellon University, Pittsburgh, Pa., 1987.

[30] F. J. Varela, *Kognitionswissenschaft—Kognitionstechnik. Eine Skizze aktueller Perspektiven (Cognitive Science),* Suhrkamp, stw 882, Frankfurt/Main, 1990.

[31] R. Whitaker and O. Östberg, *Channeling Knowledge: Expert Systems as Communications Media,* AI & Society, Vol. 2, Springer, 1988, pp. 197–208.

[32] P. M. Churchland, *A Neurocomputational Perspective—The Nature of Mind and the Structure of Science,* MIT Press, Cambridge, Mass., 1989.

INDEX

Accommodation, 39
Action potential, 39
Adaptive classifier algorithms, 420
Adaptive control algorithms, 573
Advanced software, 503
Animal behavior, 54
Artificial lizard, LIZZY, 224
Associator, 73
Automatic discovery, 615
Automatic error detection, 621

Backpropagation, 82, 112, 239, 263
Bit slice, 311

Cell recruitment learning algorithm, 389
Character recognition, 154
Cognition, 631
Competitive learning, 88
Component-based software, 506
Concept learning, 71
Conceptual map, 99
Conjugate gradient algorithm, 115
Coprocessor neurocomputer, 248
Correlation, 328
Crossover, 213

Data base, intelligent, 615
Discrete neuronal model, 161, 343
Distributed computer systems, 528
Distributed computer systems programming, 531

Distributed knowledge processing, 437
Distributed memory, 262

Energy histograms, 146
Entity model, 513, 517
Entropy, 324
Expert knowledge, 636

Fifth generation systems, 457
  kernel language, 476
  knowledge base, 467
Flexible automation, 563, 606
Fourier transform, 326
Fraser representation, 441, 444
Fuzzy and relational comparators, 396
Fuzzy evidential logic, 368
Fuzzy set comparator, 333

Gabor transform, 327
Generation of natural languages, 193
Genetic algorithm, 212
Genetic programming, 207, 216
Global memory, 552
Granularity of parallelism, 445

ICOT, 457, 501
Image recognition, 150, 330, 415
Intelligent tools, 618
Intelligent tracking control, 419
Intelligent user interface, 624

Integration:
  in business, 20, 503
  in data delivery, 22, 615
  levels, 32
  in process control, 25, 529
  of Reasoning, Informing and Serving, 28, 435
  in robotics, 25, 563
  of software, 10, 503, 564

Japan fifth generation, 457

Kalman algorithm, 116
Knowledge compiler, 474
Knowledge representation, 629, 647

Learning connections, 360
Loosely coupled systems, 440

Minkowski-r power metrics, 179
MIMD processing, 251
Models in robotics, 576, 595
Multi-level processing control, 407

Neural hardware, 281, 311, 333
Neural-knowledge-fuzzy hybrids, 320
Neural model, 40
Neural network architecture, 314
Neural network model, 42, 236, 284
Neural network simulations, 49
Neurocomputers, 235, 247

Object detection, 412
Object-oriented processing, 443, 449
Object recognition, 135
Object splitting/grouping, 418
Object tracking, 411

Paracell language, 528, 536, 542, 558
Parallel inference machine, 462, 528, 546
Parallel neural networks, 238
Parallel software development machine, 494
Pattern matching, 399

Pattern recognition, 416
Postprocessing, 333
Predicate theory, 354
Preprocessing, 322
Probabilistic discrete neuronal model, 161
Propositional theory, 350

Query processing, 622

Reasoning, 341
Refractory period, 39
Resting voltage, 39
Robot controllers, 593
Robotics, 563, 576
Robotic simulator-integrator, 563, 595
Robot-neural network hybrids, 589
Rule-based reasoning, 341, 350, 355, 394

SIMD processing, 148, 250
Simulated annealing, 136
Software development systems, 572, 596
Software integration, 10, 503, 564
Software protection, 11
Software superdistribution, 12
Stock control software, 519
Supervised learning, 73, 179
Symbol manipulating systems, 340, 382, 407, 436, 457, 650
Symbol processing, 383, 629
Synopses, 40

Temporal learning, 182
Threshold, 39
Tightly coupled systems, 439
Transformations, 326
Transputers, 281, 286

Unsupervised learning, 87, 289

Variable slopes, 124

Walker, 220

5

97